Cryptic Female Choice in Arthropods

A mating pair of syrphid flies. *Photo* Carlos Toscano-Gadea

Alfredo V. Peretti · Anita Aisenberg
Editors

Cryptic Female Choice in Arthropods

Patterns, Mechanisms and Prospects

Springer

Editors
Alfredo V. Peretti
Instituto de Diversidad y Ecología Animal
CONICET—Universidad Nacional de Córdoba
Cordoba
Argentina

Anita Aisenberg
Laboratorio de Etología, Ecología y Evolución
Instituto de Investigaciones Biológicas Clemente Estable
Montevideo
Uruguay

ISBN 978-3-319-17893-6 ISBN 978-3-319-17894-3 (eBook)
DOI 10.1007/978-3-319-17894-3

Library of Congress Control Number: 2015937214

Springer Cham Heidelberg New York Dordrecht London

Printed on acid-free paper

Springer International Publishing AG Switzerland is part of Springer Science+Business Media
(www.springer.com)

Foreword

Cryptic Female Choice: A Tale About a Boy Who Loved Flies

More than 30 years ago, I first presented the term "cryptic female choice" as a label for certain behaviors I discovered in female *Harpobittacus nigriceps*, a species of hangingfly (Mecoptera) (Thornhill 1983). Since then, cryptic female choice (CFC) theory has diversified and expanded, and in the process, so too have the hypotheses, concepts, and topics that comprise the broader field of sexual selection research.

Traditionally, the process of sexual selection was restricted to variation among conspecific males in their mating success and associated reproductive success resulting from the males' trait differences affecting competition for mates or being chosen as a mate by the opposite sex. This is the Darwinian or classical view of sexual selection in evolutionary biology. Parker's (1970) ideas about sperm competition expanded this classical perspective to include sexual selection favoring male traits that solve the adaptive problem of a mate's insemination by a sexual competitor and the resultant competition between ejaculates of different males for the egg(s) of a single female. The boundaries of sexual selection research were expanded again when Hrdy (1977) recognized that male infanticidal behavior toward offspring still under maternal care and sired by another male can increase the mating success of infanticidal males and hence cause sexual selection to act on males.

CFC theory expanded Darwinian intersexual selection theory by recognizing that female choice could take place after mating started, too. CFC theory gives female choice a larger role in sexual selection and the evolution of male traits and mating systems. Cryptic choice by females extends female control in decisions affecting male reproductive success beyond the premating context and through all stages of the reproductive process. CFC traits cause variation in male reproductive success by their expression after mating starts, and CFC can continue to cause sexual selection on males during egg or offspring production by a female, and even after eggs are laid or offspring are born, if females differentially invest based on the traits of the sires of their offspring. At the beginning of the discussion of CFC in insects in the literature, CFC was based on evolved adaptations of females that function to choose high-quality sires for offspring and/or defend against sexual coercion by males that circumvent female choice. Sire quality was discussed as male traits that increase mating success of sires' sons and/or the fitness of both sexes of sires' offspring (Thornhill 1983; Thornhill and Alcock 1983).

The most compelling argument for the process of CFC (as well as Darwinian female choice) arises from Trivers (1972) insights about parental investment and sexual selection. As he emphasized, parental investment is any investment by a parent in an offspring that improves offspring survival and hence offspring reproductive success at the expense of the parent's ability to invest in other offspring. Parental investment in a given generation then fundamentally causes the number and survival of offspring in the next generation and therefore becomes the focus of sexual/mating competition by the opposite sex. Males compete with other males to capture the parental investment of females that can be obtained through siring eggs. Females, by contrast, as the sex that typically has the largest obligate parental investment, have adaptations that function to adaptively allocate their limited parental investment through tight control of that investment and its efficient expenditure. Females are hence designed to assess ecological circumstances that affect efficient expenditure of their parental investment and to differentially allocate parental investment depending on how circumstances affect its adaptive use. These decisions are designed to be sensitive to variation in conditions that affect optimal allocation of parental investment and hence the reproductive profits females receive from their investments. Sire and mate quality is one such condition that is widely important. Hence, females are selected to assess the quality of their mates and the sires of their offspring and bias investment toward high quality ones. It is often maladaptive for females that this assessment and related investment allocation process end when mating begins. Maximum female reproductive success is typically dependent upon females continuing their control of the relative success of different mates subsequent to the premating context and for females to extend their influence over which mate and sire receives their limited parental investment. In particular, females gain in reproductive success by extending this control of the paternity of offspring when (a) females engage in what John Alcock and I called convenience polyandry in which females adaptively mate with multiple males to reduce the costs of rejecting them (Thornhill and Alcock 1983); (b) circumstances prevent the female's full assessment of male quality before mating; (c) males circumvent female choice by sexual coercion; or (d) females can obtain non-genetic material benefits from males (e.g., nuptial gifts) by mating with multiple males.

About a decade after publication of my paper on CFC, Eberhard (1996) made a strong case for the importance of CFC in *Female Control: Sexual Selection by Cryptic Female Choice* by reviewing much of the evidence at the time in favor of CFC and detailing many mechanisms by which it may occur. Arnqvist (2014) recently analyzed the historical trend in scientific papers that contain "cryptic female choice" since the term first appeared in 1983. The published research referencing the term was almost nil before the mid-1990s and then began a steady climb that is still seen; this trend is due fundamentally to Bill Eberhard's book which has promoted many new avenues of CFC research and had significant scientific impact. Researchers in a broad range of circumstances and studying diverse animal taxa have published evidence for CFC, and the current collection of book chapters treats that research across arthropods and offers many new findings and research directions.

Darwin did not live to see the heuristic success of his theory that female choice played an important role in the evolution of male traits. Female choice was perhaps Darwin's most controversial theory in his entire treatment of life's history as evolutionary history (Bajema 1984). Darwin held fast to his ideas about the important role of female choice in evolution up to his death. His last defense of the evolutionary perspective was a defense of female choice and was read to the public at a meeting of the Zoological Society of London just hours before he died (Bajema 1984, p. 150). No doubt Darwin would be thrilled to know of the many hundreds of empirical documentations of female premating choice and male adaptations that function to impress females in many taxa that followed the explosion of sexual selection research beginning in the early 1970s and continuing today. And no doubt he would be thrilled to know his nascent insights about female choice were even more powerful and far-reaching than he recognized, which is seen in the evidence for CFC. I suspect this collection of chapters on CFC in arthropods would greatly impress Darwin and be among his favorite books.

To be asked to write the forward for the book at hand brought forth a plethora of memories about my scientific career and even my childhood. Like many biologists, I have wondered how my particular scientific interests came about and were maintained and magnified. Although I have researched numerous topics and taxa over 45 years of scientific research and publishing, my primary interest has been in sexual selection processes, including sexual selection by female choice, which generated my research on CFC. Here is a tale about my life with flies that accounts for my research turning to CFC.

As a boy, I observed and collected a range of plants and animals, but my favorites were insects and of those I found robber flies, horse and deer flies, scorpionflies, and hangingflies particularly amazing and beautiful. When I was 12 years old, my mother told me about an Englishman named Charles Darwin who had some ideas (she called them "big ideas") about the things I was interested in, plants and animals. She took me to the public library in our Alabama town where we found a copy of Darwin's *The Descent of Man and Selection in Relation to Sex*. I believe that my childhood experiences with Darwin's ideas, evidence, and methods had an enduring influence on my life. As I grew older and advanced in my studies of biology, I continued to be fascinated with Darwin's theory of sexual selection. In 1972, after my first year of a doctoral programme in biology at the University of Michigan, Richard Alexander, my Ph.D. mentor at the time, suggested I read some published papers on sperm competition by Geoff Parker and an in the press manuscript on parental investment and sexual selection by Robert Trivers. These papers supercharged my interest in sexual selection, and Trivers paper in particular sparked my interest in female mate choice. I remembered at that time my observations in Alabama several years earlier of nuptial feeding in a species of hangingfly (*Hylobittacus apicalis*). I had observed that courting males offer a prey arthropod to a female and mating females feed upon the gift throughout mating. I found this same species locally around Ann Arbor, Michigan in large populations. Scorpionflies (*Panorpa* species; Mecoptera) were abundant in the same woodland habitats, and they engaged in nuptial feeding too. I was quickly hooked on these insects and the opportunity they provided to study female mate choice. My research that followed included documenting adaptive premating female mate choice under both field conditions and in laboratory experimental conditions. The field study was the first demonstration of female mate choice under field conditions in an insect (Thornhill 1976).

My research continued through the 1970s and early 1980s to reveal that females of *H. apicalis* prefer males as mates that offer large prey as gifts and reject males with small gifts as mates. This was the pattern when males carrying large prey were abundant (under high population density of *H. apicalis*). Under high density, females rely primarily on male-provided nuptial gifts and thereby avoid hunting prey themselves, which has the survival cost of falling prey to a spider after flying into its web during hunting. In contrast, under low density, females are less choosy in premating choice, often not rejecting as mates any males with prey, and switching to behaviors during and after mating with a male with small prey that I thought should bring about sexual selection on males. In comparison with the high-density females, the low-density females mated briefly, which reduced male insemination success. As well, the low-density females remained sexually receptive and re-mated with males until a male with a large gift was encountered. The female would mate with this male offering the large gift and then became sexually unreceptive and began laying eggs, which I assumed to be primarily fertilized by the last male mated. My research showed that the large prey females preferred as nuptial gifts were quite rare in the habitat, whereas small prey were common. I interpreted the female behaviors during and after mating as female avoidance of inept hunters as sires of offspring and simultaneously a preference for adept hunters and hence sires of high genetic quality (Thornhill 1980, 1984a).

My research in the 1970s and early 1980s on several species of *Panorpa* scorpionflies was also providing evidence of both premating and cryptic female choice (Thornhill 1981, 1984b). *Panorpa* are scavengers of dead arthropods and not predators. Males compete for arthropod carcasses to use as nuptial gifts or to feed upon for fueling their salivary glands, allowing them to provide a female with a hardened saliva mass as a nuptial gift. My research showed premating female choice occurred by female preference for mates with large or high-quality gifts. Cryptic choice occurred in females' re-mating and egg-laying patterns that would bias egg-siring success in favor of males that provided large or high quality gifts and disfavor males that used sexual coercion without nuptial feeding.

About the same time, Alcock (1979) and Gwynne (1984) were studying mate choice in Australian hangingflies (*Harpobittacus*) and publishing findings on premating female choice based on nuptial gift size similar to my findings discussed above on premating choice in *H. apicalis*. I spent 1982 in Australia conducting field and laboratory studies of *H. nigriceps*. It was in my 1983 paper on *H. nigriceps* from that research that I coined the term cryptic female choice for the behaviors of female *H. nigriceps* similar to those during and after mating in *H. apicalis* and *Panorpa* I have mentioned as well as CFC for large-bodied males. In this period, John Alcock and I were collaborating on our book, *The Evolution of Insect Mating Systems* (1983). Our book featured discussion of premating female choice as well as female choice during and after mating. We included too evidence of CFC in hangingflies. An emphasis in our book was that female choice was likely an important and widespread cause of evolution in insects despite the limited direct evidence for its occurrence in nature at the time and the controversy about female choice that first arose in Darwin's time and that emerged again with the revival of interest in the topic sparked by Trivers' (1972) paper.

The thinking that led to original CFC theory arose independently from a number of researchers. Both in my 1983 paper and in *The Evolution of Insect Mating Systems* other researchers were mentioned and cited who published observations from animal mating systems that fit the concept and mechanisms of what is now cryptic female choice. These included observations on insects by Jim Lloyd and John Sivinski. In the 1983 paper, I emphasized animal systems but suggested that CFC was likely in plants as well. Cryptic female choice is the only kind of sexual-selection-based female choice in plants. In the same year (1983), Nancy Burley and Mary Willson's impressive research monograph, *Mate Choice in Plants*, was published. They discussed the same basic concept of CFC as John Alcock and I did, but without using the term CFC. Bill Eberhard's foundational contributions, in particular his book on female control I have mentioned, came later but served to extend and clarify CFC theory in important ways.

This book on CFC in arthropods is the first monograph to present research findings on CFC across a taxonomic group for the last few decades. Because of the book's breadth and its many empirical findings and hypotheses, I predict it will play an important role in the history of research on CFC as well as the history of research in sexual selection in general.

Randy Thornhill
Department of Biology, University of New Mexico
Albuquerque, NM, USA

References

Alcock J (1979) Selective mate choice by females of *Harpobittacus australis* (Mecoptera: Bittacidae). Psyche 86:213–217

Arnqvist G (2014) Cryptic female choice. In: Shuker D, Simmons LW (eds) The evolution of insect mating systems. Oxford University Press, New York, pp 204–220

Bajema CJ (ed) (1984) Evolution by sexual selection theory: prior to 1900. Benchmark papers in systematic and evolutionary biology, vol 6. Van Nostrand Reinhold, New York

Eberhard WG (1996) Female control: sexual selection by cryptic female choice. Princeton University Press, Princeton

Gwynne DT (1984) Nuptial feeding behavior and female choice of mates in *Harpobittacus similis* (Mecoptrea: Bittacidae). J Aust Entomol Soc 23:271–276

Hrdy SB (1977) The Langurs of Abu: female and male strategies of reproduction. Harvard University Press, Cambridge

Parker GA (1970) Sperm competition and its evolutionary consequences in the insects. Biol Rev 45:525–567

Thornhill R (1976) Sexual selection and nuptial feeding behavior in *Bittacus apicalis* (Insecta: Mecoptera). Am Nat 110:529–548

Thornhill R (1980) Mate choice in *Hylobitticus apicalis* (Insecta: Mecoptera) and its relation to some models of female choice. Evolution 34:519–538

Thornhill R (1981) *Panorpa* (Mecoptera: Panorpidae) scorpionflies: systems for understanding resource-defense polygyny and alternative male reproductive efforts. Ann Rev Ecol Syst 12:355–386

Thornhill R (1983) Cryptic female choice in the scorpionfly *Harpobittacus nigriceps* and its implications. Am Nat 122:765–788

Thornhill R (1984a) Alternative female choice tactics in the scorpionfly *Hylobittacus apicalis* (Mecoptera) and its implications. Am Zool 24:367–383

Thornhill R (1984b) Alternative hypotheses for traits believed to have evolved by sperm competition. In: Smith RL (ed) Sperm competition and the evolution of animal mating systems. Academic Press, New York, pp 151–178

Thornhill R, Alcock J (1983) The evolution of insect mating systems. Harvard University Press, Cambridge

Trivers R (1972) Parental investment and sexual selection. In: Campbell B (ed) Sexual selection and the descent of man 1871–1971. Aldine, Chicago, pp 136–179

Preface

Sexual selection is currently the target of multiple and controversial theoretical and experimental studies. Selection on mating and post-mating patterns can result from several mechanisms, including sperm competition, extreme sexual conflict (i.e., sexual coercion and/or sexually antagonistic coevolution), cryptic female choice, or a combination of them. More than 30 years have passed since Randy Thornhill proposed that processes occurring after copulation has began and that are under direct female control can influence male chances of paternity. The cornerstone book by Eberhard (1996) compiled impressive evidence regarding the many possible mechanisms of female control of paternity. During the last decades, much subsequent research focused on female roles during and after mating, documenting the importance of female decisions for male reproductive success. Discrimination among males during or after copulation is called cryptic female choice because it occurs after intromission, the event that was formerly used as the definitive criterion of male reproductive success. As in most cases mechanisms of cryptic female choice occur inside the female, this phenomenon is usually difficult to detect and confirm. Because it sequentially follows intra- and intersexual interactions that occur before copulation, cryptic female choice has the power to alter or negate precopulatory sexual selection.

However, though female roles in biasing male paternity after copulation have been proposed for a number of species distributed in many animal groups, cryptic female choice continues to be sometimes underestimated. Furthermore, during recent years, the concept of sexual conflict has been frequently misused linking it irrevocably with sexual coercion and sexually antagonistic coevolution in opposition to sexual selection by cryptic female choice, without exploring other alternatives.

The present book revisits cryptic female choice in arthropods through detailed contributions from across the world to answer key behavioral, ecological, and evolutionary questions. The reader will find a critical summary of major breakthroughs in taxon-oriented chapters, offering many new perspectives and cases to explore, sometimes sharing unpublished data. The choice of focusing this book in arthropods was not deliberate, but we based our decision in the demonstrated value of this group for sexual selection studies. The possibility of cryptic female choice is explored in many groups of arthropods such as spiders, harvestmen, flies, butterflies, crickets, earwigs, beetles, eusocial insects, and crustaceans.

The book includes 18 chapters written by researchers from areas related to animal behavior, behavioral ecology, and evolution. We start with a prologue written by Randy Thornhill, which is followed by a first chapter that provides a baseline introduction to cryptic female choice concepts by William G. Eberhard. The following chapters provide a survey of the research done on cryptic female choice during the last decades on different model organisms, always within arthropods. The results of each chapter are discussed giving final remarks and suggesting directions for future research.

This book would not have been possible without the enthusiastic support provided by the authors of each chapter and reviewers. Aditi Pai, Adolfo Cordero Rivera, Alex Córdoba-Aguilar, Bernard Sainte-Marie, Bruno A. Buzatto, Carlos Cordero, Darryl Gwynne, Diana Pérez-Stamples, Fernando G. Costa, Flavia Barbosa, Gilbert Barrantes, Gustavo Requena, Ignacio Escalante, Jennifer Hamel, Juergen Heinze, Karim Vahed, Laura Sirot, Luc F. Bussière, Margaret Bloch Qazi, Maria José Albo, Mary Jane West-Eberhard, Michael Jennions, Oliver Yves Martin, Rafael L. Rodríguez, Rhainer Guillermo, Rodrigo H. Willemart, Solana Abraham, and Yoshitaka Kamimura provided helpful suggestions that improved the final versions of the chapters included in the book. Finally, we would like to acknowledge our families for their patience, love and unconditional support.

We hope that the book provides a source for the discussion of ideas and avenues for future research on sexual selection, transmitting our passion for this astonishing animal group.

Alfredo V. Peretti
Anita Aisenberg

Contents

Editors and Contributors

About the Editors

Alfredo Vicente Peretti is Principal Researcher of the Consejo Nacional de Investigaciones Científicas y Técnicas de la Argentina, Instituto de Diversidad y Ecología Animal (CONICET-IDEA) and associate professor of the Departamento de Diversidad Biológica y Ecología, Facultad de Ciencias Exactas Físicas y Naturales, Universidad Nacional de Córdoba, Argentina. His research on arachnids started in 1991 and initially focused on reproductive behavior in scorpions. His main areas of research are sexual selection, communication, and evolution of genitalia in arthropods.

Anita Aisenberg is an Assistant Professor of the Laboratorio of Etología, Ecología y Evolución, of the Instituto de Investigaciones Biológicas Clemente Estable, Montevideo, Uruguay, and she is member of the Sistema Nacional de Investigadores (ANII) and the Programa de Desarrollo de Ciencias Básicas (PEDECIBA), Uruguay. Since 2001, she has been working on spider behavior. Her main interests are related to sexual selection, reproductive behavior, and evolution of sex roles.

Contributors

Anita Aisenberg Laboratorio de Etología, Ecología y Evolución, Instituto de Investigaciones Biológicas Clemente Estable, Montevideo, Uruguay

Maydianne C. B. Andrade Department of Biological Sciences and Department of Ecology & Evolutionary Biology, University of Toronto Scarborough, Toronto, ON, Canada

Boris Baer Centre for Integrative Bee Research (CIBER), The University of Western Australia, Crawley, Australia

Joaquín Baixeras Instituto Cavanilles de Biodiversidad y Biología Evolutiva, Universidad de Valencia, Paterna, Spain

Flavia Barbosa Institut de Recherche Sur La Biologie de L'Insecte (IRBI), Université François Rabelais, Tours, France

Gilbert Barrantes Escuela de Biología, Universidad de Costa Rica, Ciudad Universitaria, Costa Rica

R. D. Briceño Escuela de Biología, Universidad de Costa Rica, Ciudad Universitaria, Costa Rica

Lucía Calbacho-Rosa Instituto de Diversidad y Ecología Animal, CONICET—Universidad Nacional de Córdoba, Córdoba, Argentina

Carlos Cordero Departamento de Ecología Evolutiva, Instituto de Ecología, Universidad Nacional Autónoma de México, D.F., Mexico City, Mexico

Alex Córdoba-Aguilar Departamento de Ecología Evolutiva, Instituto de Ecología, Universidad Nacional Autónoma de México, México D.F., México

Luiz Ernesto Costa-Schmidt Instituto de Diversidad y Ecología Animal, CONICET—Universidad Nacional de Córdoba, Córdoba, Argentina

Karina Cuevas-Yáñez Departamento de Ecología Evolutiva, Instituto de Ecología, Universidad Nacional Autónoma de México, México D.F., México

Stefan Dennenmoser Max-Planck Institute for Evolutionary Biology, Plön, Germany

William G. Eberhard Escuela de Biología, Universidad de Costa Rica, Ciudad Universitaria, Costa Rica

Tatyana Y. Fedina Department of Molecular and Integrative Physiology, University of Michigan, Ann Arbor, USA

Daniel González-Tokman Departamento de Ecología Evolutiva, Instituto de Ecología, Universidad Nacional Autónoma de México, México D.F., México

Marie E. Herberstein Department of Biological Sciences, Macquarie University, Sydney, Australia

Yoshitaka Kamimura Department of Biology, Keio University, Hiyoshi, Yokohama, Japan

Sara M. Lewis Department of Biology, Tufts University, Medford, USA

Glauco Machado Departamento de Ecologia, Instituto de Biociências, Universidade de São Paulo, São Paulo, SP, Brazil

Rogelio Macías-Ordóñez Red de Biología Evolutiva, Instituto de Ecología, Xalapa, Veracruz, México

Emily C. MacLeod Department of Biological Sciences and Department of Ecology & Evolutionary Biology, University of Toronto Scarborough, Toronto, ON, Canada

Ángela Nava-Bolaños Departamento de Ecología Evolutiva, Instituto de Ecología, Universidad Nacional Autónoma de México, México D.F., México

Adriana Nava-Sánchez Departamento de Ecología Evolutiva, Instituto de Ecología, Universidad Nacional Autónoma de México, México D.F., México

Alfredo V. Peretti Instituto de Diversidad y Ecología Animal, CONICET—Universidad Nacional de Córdoba, Córdoba, Argentina

Gustavo S. Requena Department of Ecology and Evolutionary Biology, Yale University, New Haven, CT, USA

Miguel Rivas Departamento de Ecología Evolutiva, Instituto de Ecología, Universidad Nacional Autónoma de México, México D.F., México

Rafael L. Rodríguez Behavioral and Molecular Ecology Group, Department of Biological Sciences, University of Wisconsin–Milwaukee, Milwaukee, WI, USA

Jutta Schneider Zoological Institute, Biozentrum Grindel, University of Hamburg, Hamburg, Germany

Laura K. Sirot Department of Biology, College of Wooster, Wooster, OH, USA

Estefanía Stanley Laboratorio de Etología, Ecología y Evolución, Instituto de Investigaciones Biológicas Clemente Estable, Montevideo, Uruguay

Martin Thiel Facultad Ciencias del Mar, Universidad Católica del Norte, Coquimbo, Chile

Carlos Toscano-Gadea Laboratorio de Etología, Ecología y Evolución, Instituto de Investigaciones Biológicas Clemente Estable, Montevideo, Uruguay

Gabriele Uhl Zoological Institute and Museum, General and Systematic Zoology, University of Greifswald, Greifswald, Germany

Karim Vahed University of Derby, Derby, UK

Mariana F. Wolfner Department of Molecular Biology and Genetics, Cornell University, Ithaca, NY, USA

Chapter 1
Cryptic Female Choice and Other Types of Post-copulatory Sexual Selection

William G. Eberhard

Abstract This chapter discusses sexual selection by cryptic female choice (CFC) and other possible types of selection on traits involved in male–female interactions during and following copulation. Morphological, behavioral, and probably also physiological traits all show the typical earmarks of sexual selection: puzzlingly extravagant, apparently non-utilitarian design; and rapid divergent evolution. I discuss ways to attempt to distinguish CFC from other possibilities, and their potential overlap. Differentiating narrow-sense from broad-sense male–female conflict may help clear up some current confusion. The most central differences between the leading hypotheses concern the expected effects of selectively granting paternity on a female's fitness. Unfortunately, convincing tests of these effects have not been feasible due to technical limitations; published claims regarding such measurements must be treated with caution. Several types of data that provide less direct tests, including defensive designs of females, the presence of female sense organs specialized to sense courtship stimuli from males, physical damage inflicted on the female by the male during copulation, and physical male–female struggles, are discussed. Different types of selection may operate simultaneously in some species, and all may be applicable in particular cases; the major questions concern the relative frequencies of species in which each type of selection occurs. The hypotheses nevertheless provide useful theoretical contexts for understanding multiple aspects of reproductive biology. One promising area for future studies, in which arthropods can provide both experimental and comparative data, is the role of non-genitalic male copulatory courtship structures in stimulating or physically coercing females during sexual interactions.

W.G. Eberhard (✉)
Smithsonian Tropical Research Institute, Panama, Republic of Panama
e-mail: william.eberhard@gmail.com

W.G. Eberhard
Escuela de Biología, Universidad de Costa Rica, San Pedro, Costa Rica

W.G. Eberhard
Biological Sciences and Museum of Natural History,
Louisiana State University, Baton Rouge, USA

A.V. Peretti and A. Aisenberg (eds.), *Cryptic Female Choice in Arthropods*,
DOI 10.1007/978-3-319-17894-3_1

1.1 Introduction

Studies of sexual selection broke into new territory with Parker's realization (1970) that the competition among males for chances to fertilize eggs often continues even after copulation has begun. This extension brought Darwin's ideas on sexual selection to bear on a new suite of male–female interactions and offered explanations for many otherwise puzzling male and female traits, such as "post-copulatory" male traits (those brought into play only after copulation has already begun). These often have the typical earmarks of pre-copulatory sexual selection (Darwin 1871): puzzlingly extravagant and apparently non-utilitarian designs; and rapid divergent evolution that often results in species-specific differences, even among closely related species. They include the male's genitalia as well as a multitude of other "contact courtship" structures on various parts of the male body that are also sexually dimorphic and that are specialized to contact females during sexual interactions (Eberhard 1985), and the paradoxical male courtship behavior that often occurs after copulation has already begun ("copulatory courtship" of Eberhard 1994). Still other traits that had not previously been thought to have sexually selected functions include male seminal components that affect female reproductive physiology or that form copulatory plugs, surface molecules on the gametes that are involved in the interactions between sperm and eggs (in animals with both internal and external fertilization); and substances and structures that affect the interactions between pollen and female tissues in plants (Swanson and Vacquier 2002; Swanson et al. 2001, 2003; Karn et al. 2008; Clark et al. 2006; Zigler 2008; Zigler and Lessios 2003; Zinkl et al. 1999; see also Sirot and Wolfner 2015, Chap. 13).

These analyses of post-copulatory sexual selection have spawned a substantial literature that includes several book-length reviews (Smith 1984; Birkhead and Møller 1992, 1998; Eberhard 1985, 1996; Baker and Bellis 1995; Simmons 2001; Arnqvist and Rowe 2005; Leonard and Córdoba-Aguilar 2010), including the present volume. There are several hypotheses for how selection can play a role in the evolution of the morphological, physiological, and behavioral traits involved in post-copulatory events. I will describe these hypotheses and some techniques for distinguishing among them, and then give a brief update of the field and a look into the future.

1.2 Cryptic Female Choice and Alternative Theories Regarding Post-copulatory Selection

1.2.1 Sexual Selection

Three classes of post-copulatory sexual selection have been proposed. One of these, cryptic female choice ("CFC") (e.g., Eberhard 1985, 1996), is the theme of this book. To evaluate the possible importance of CFC, it is necessary to take into account the other two, direct male–male competition (sperm competition or "SC") (e.g., Simmons 2001) and sexually antagonistic coevolution between males and females ("SAC") (e.g., Holland and Rice 1998; Arnqvist and Rowe 2005). Ideally, a discussion of these hypotheses would begin with clear, generally accepted definitions of each. Unfortunately, the hypotheses have changed over time in how they are used, and the oldest criteria for some hypotheses are not necessarily the best. There have also been (and continue to be) widespread applications of terms like "conflict" that employ different meanings and that sometimes stretch ideas beyond their conceptual boundaries (see discussions by Simmons 2001; Jennions 2005; Brennan and Prum 2012; Kokko and Jennions 2014); Kokko and Jennions (2014) have recently advocated combining CFC and SC under SAC sensu lat.

I will not try to resolve this apparently intractable tangle (I strongly doubt that there is any set of definitions that would enjoy general acceptance). But there are, however, ways to distinguish CFC from alternative hypotheses which I believe can help guide productive thinking. I will use some examples from other chapters in this book to illustrate this approach.

1.2.1.1 Sperm Competition (SC)

The earliest general discussion of post-copulatory sexual selection was that of Parker (1970), who used the term "sperm competition" to cover all facets. Later authors made the useful distinction (since adopted by Parker) between sperm competition (SC) and CFC, echoing the distinction that Darwin made when he divided pre-copulatory sexual selection into direct male–male combat, and female choice (Table 1.1). SC is now used mostly in this narrower, direct male–male interaction sense. Strictly speaking, SC in this sense does not occur unless temporal and spatial overlap of and interactions among ejaculates from competing males occurs (or potentially occurs) in the female (Simmons 2001). In practice, however, SC is usually expanded to include paternity biases that result from male activities that directly influence the interactions between sperm in the female. SC can occur when a male directly manipulates the sperm of rival males by removing them from the female, packing them into corners in the female, diluting them, killing them, leaving barriers in the female that prevent future inseminations, or removing such barriers. Examples of SC in this book include soldier flies (Barbosa 2015, Chap. 14), depositing and removing mating plugs in Lepidoptera and spiders (Aisenberg et al. 2015, Chap. 4; Andrade and MacLeod's Chap. 2; Cordero and Baixeras's Chap. 12; Schneider et al.'s Chap. 3), and sperm removal in spiders and Apachyidaeearwigs (Calbacho-Rosa and Peretti 2015, Chap. 5; Kamimura 2015, Chap. 10, Chap. 10).

Table 1.1 Correspondence between Darwin's categories of pre-copulatory sexual selection and post-copulatory selection via sperm competition (SC) and cryptic female choice (CFC)

Prior to copulation	Male–Male battles	Female choice
During and following copulation	Sperm competition	Cryptic female choice

Selection of the other two types, SI and SAC, can occur both before and following copulation

1.2.1.2 Cryptic Female Choice

CFC, the post-copulatory equivalent of Darwin's female choice, includes paternity biases which result from differences in the expression of female activities (including morphogenesis) that come into play during and following copulation with different males and that favor males which have particular traits over others which lack these traits. The female is thought to benefit from exercising choice among males by obtaining superior sires for her offspring. Historically, CFC was first discussed as such in relation to biases in oviposition following matings with different males in a scorpion fly (Thornhill 1983) (see Birkhead 1998 for brief earlier mentions). Thornhill defined CFC as female-influenced processes occurring during and/or after copulation that bias offspring production more toward one male than other mates (or potential mates) (Thornhill 1983). More than 20 such female mechanisms have been described (Table 1.2; Eberhard 1996, 2010). Chapters in this book add the possibility of storing sperm from different males in different storage sites and then biasing their use, which may occur in a spider (Andrade and MacLeod's Chap. 2), and several interesting possible mechanisms at the molecular level in Diptera (Sirot and Wolfner 2015, Chap. 13). CFC mechanisms include such basic female reproductive process as sperm transport and dumping, sperm maintenance in storage organs, ovulation, oviposition, and acceptance of mating attempts by future males. Increased performance by the female of any one of these processes (e.g., increased sperm transport, more rapid oviposition) as a result of copulating with some males as compared to others can bias paternity. From the male's perspective, any improvement in his ability to induce the female to alter one of these processes in a way that favors his chances of paternity can result in an advantage in competition with other males who mate with the same female. Though not often emphasized, CFC can be affected by male traits sensed by the female prior to the initiation of copulation (during classic, pre-copulatory courtship) as well as by copulatory courtship traits.

A pair of finer distinctions need to be mentioned. Because the sperm competition interactions listed above are generally played out inside the female's body, the effects of the female on the outcomes of many SC interactions could be considered to constitute CFC in a more inclusive sense than I will use here (Arnqvist 2014). Just as in pre-copulatory sexual selection (Andersson 1994; Wiley and Posten 1996), it can be difficult to draw a line between direct male–male competition and female choice. The importance of making distinctions comes not from application of the names per se, but in understanding the reasons why traits evolved (their functions) (Arnqvist 2014).

A second point concerns the evolutionary diversification of males and females expected under CFC. One of the major lessons for sexual selection of studies of the morphology and physiology of different portions of animal nervous system is their extensive interconnectedness. This interconnectedness opens up diverse possibilities to males that are under sexual selection to stimulate females. Not only are there many female processes (>20), but there are likely many different types of stimuli that influence each of these processes. Arnqvist (2014) missed this point when he supposed that CFC on male genitalia would be unable to explain the evolutionary diversification in male genitalia because male "... genital morphology

Table 1.2 A probably incomplete list of possible mechanisms known to occur in species of arthropods with internal insemination by which a female could exercise CFC by varying the intensity of her response to the male, thus imposing paternity biases favoring the current male copulating with (largely after Eberhard 2010)

1. Permit penetration by the male's genitalia deep enough to allow sperm deposition at the optimum site for storage or fertilization
2. Refrain from terminating copulation prematurely
3. Transport sperm to storage or fertilization sites
4. Modify internal conditions (e.g., pH) inside reproductive tract to reduce defenses against microbial invasion that kill sperm
5. Nourish or otherwise maintain sperm alive in storage sites
6. Nourish or otherwise maintain sperm alive on their way to storage sites
7. Refrain from discarding sperm from current male
8. Discard sperm from previous males
9. Move sperm from previous male to sites where the current male can remove them
10. Accede to male manipulations that result in discharge of his spermatophore
11. Grow more immature eggs to maturity (vitellogenesis)
12. Ovulate
13. Produce eggs with more nutrients
14. Oviposit all available mature eggs
15. Prepare uterus for implantation
16. Refrain from removing copulatory plug produced by male
17. Aid male in the formation of a copulatory plug that impedes future intromissions
18. Produce a copulatory plug that impedes further intromissions
19. Modify morphology following first copulation to make subsequent insemination more difficult
20. Refrain from removing spermatophore before all sperm are transferred
21. Abort previously formed zygotes
22. Refrain from aborting zygotes from current sperm
23. Refrain from mating with other males in the future
24. Invest more in caring for offspring

would be fine-tuned to an 'optimal' genital configuration and there would be little selection for morphological innovation ...". Selection on male morphology and behavior could change when new male variants arose that were able to take advantage of the interconnectedness of female nervous systems to cause further positive stimulation of decision centers for potential CFC processes.

1.2.1.3 Sexually Antagonistic Coevolution (SAC)

SAC can result from coevolutionary races between males and females to control one or more of the many processes in the female that affect the male's chances of paternity. SAC occurs when sexual selection resulting from competition among males favors the male's ability to manipulate or influence the female to respond to him in ways that are advantageous for the male but that are disadvantageous for the female. The disadvantages to the female are related to traits that are under natural rather than sexual selection. For instance, a female might be induced by the male to lay her eggs sooner after copulation ended (before another male might be able to mate with her) and thereby failed to wait until she had found optimum oviposition sites. Natural selection on the female would favor development of traits that reduced these disadvantages by reducing her susceptibility to the male traits. An example of such female changes would be, for instance, waiting longer before laying eggs, or being more selective while searching for oviposition sites immediately following copulation. Such a female change could then reduce the male's ability to influence females and result in a further round of sexual selection on the males to exert additional influence on the female with respect to oviposition. This could result in an endless coevolutionary race between males and females. Such coevolution would explain the typically rapid and divergent pattern of evolution of the traits involved; it could occur in both pre- and post-copulatory contexts.

Male manipulations are likely to be particularly effective in SAC when the male uses stimuli that function for the female in another, non-sexual context, because female defenses against such stimuli are less likely to be as effective. An example of such a "sensory trap" occurs in the fiddler crab *Uca pugilator* (Christy 1995). Females use objects that project above the flat horizon of the sandy beaches as visual markers for possible shelters where they can hide from predators if attacked; they tend to approach such objects while wandering on the beach while sampling courting males. Males build pillars beside their burrows that utilize this female defensive response to draw them near enough for the male to court.

Unfortunately, the more general phenomenon of male–female conflict of interests has often been mistakenly taken to be equivalent to or an indicator of SAC. In fact, conflict of reproductive interests between the sexes is not limited to cases in which SAC occurs, but will occur any time that a female encounters and is courted by a conspecific male but does not permit him to fertilize all of her eggs; male–female conflict is inherent, for instance, in all classic Darwinian female choice situations (e.g., Kokko and Jennions 2014). Thus, male–female conflict over post-copulatory events includes SC, CFC, and SAC. In practice, conflation of male–female conflict with SAC has blurred the distinction between SAC and other hypotheses (below) and has made SAC seem more important than it would have if the different hypotheses were carefully distinguished. Brennan and Prum (2012) make the useful distinction between "narrow-sense" male–female conflict (corresponding to SAC sensu stricto, as I have defined it above, and as I will use it throughout this chapter), and "broad-sense" conflict (corresponding to SAC sensu lato) in this wider sense that includes sexual selection in general.

1.2.2 Natural Selection Favoring Species Isolation (SI)

One further, naturally rather than sexually selected function that has been proposed for many species-specific post-copulatory traits is species isolation (SI).

Isolation could come about due to mechanical incompatibility between heterospecific pairings (mechanical lock and key), or failure of heterospecific males to trigger essential female reproductive processes (sensory lock and key) (Eberhard 1985; Shapiro and Porter 1989; Simmons 2014). Natural selection on both males and females (especially on females, because of their generally greater investment of time and materials in their gametes and offspring) will favor the ability to avoid wasting these resources as a result of mating with members of other species, because hybrids between species are often selectively inferior in terms of survival and reproduction. SI selection on a female is thought to favor her ability to distinguish conspecific from heterospecific males before her eggs are fertilized (and thus lost) by heterospecific crossing. As in CFC and SAC, multiple pre- and post-copulatory events (accept a copulation or not, transport sperm or not, dump sperm or not, etc.) could be influenced by SI. There are reasons to expect, however, that SI should be most important before copulation begins: it is advantageous to both sexes to distinguish and avoid heterospecific pairings as early as possible in the mating sequence; often this will be before male and female come into physical contact and begin to copulate (Alexander et al. 1997; Peretti 2010). It is nevertheless possible that even when pre-copulatory isolation mechanisms have evolved, post-copulatory selection could still favor additional species-specific, fail-safe devices that further reduce the chances of selectively disastrous cross-specific fertilizations that might result from occasional failures to avoid heterospecific pairings earlier in the mating process. Or it could be supposed (less convincingly) that post-copulatory female discriminations are vestiges of former selection favoring SI, but that have not yet disappeared, even though pre-copulatory isolation mechanisms are now in place ("ghosts of species isolation past"). SI arguments can be confidently rejected, however, as explanations for traits of species that were never in their evolutionary history in contact with closely related species (e.g., cave and island endemics, parasites that mate in isolation from all heterospecific relatives inside their hosts) (Eberhard 1985).

1.3 How to Distinguish Between CFC and the Other Hypotheses

1.3.1 CFC and SC

In some respects, SC is the easiest of the hypotheses to evaluate, because it can be confidently eliminated in many cases in which neither the male nor his semen has direct access to the sperm of rival males. In many of the groups discussed in this book, the traits that are discussed do not involve access by the male or his seminal products to rival sperm. Two additional points need to be made. Failure to eliminate the SC hypothesis is not, of course, conclusive evidence that SC occurs. As already noted, it is sometimes difficult to disentangle SC from CFC (see also Barbosa 2015, Chap. 14; Kamimura 2015, Chap. 10; Sirot and Wolfner 2015, Chap. 13).

For example, SC via dilution of a rival male's sperm by producing larger ejaculates (e.g., Simmons 2001) depends on the design of the sperm storage site in the female, and this can evolve under CFC. By having a voluminous, extensible receptacle (a trait which can obviously vary over evolutionary time, and which could also possibly be altered facultatively by the female in some species by contracting or not contracting particular muscles—see, e.g., Miller 1984, 1990 on odonates), a female can bias paternity via dilution, favoring males with larger volumes of sperm, thus exercising CFC in their favor. The female's body is the playing field on which sperm competition occurs, so changes in a female's morphological, physiological, or behavioral traits can bias the outcomes of SC. In other words, just as there is overlap between direct male–male aggression and female choice in pre-copulatory interactions (e.g., the "passive female choice" of Wiley and Posten 1996), the post-copulatory distinctions between SC and CFC are not always clean. This "female playing field" aspect of SC was recognized early on (Parker 1970), but has been accorded little importance in many (most) discussions of SC.

A second, less well-recognized possibility is that some aspects of SC may produce stimuli that trigger CFC processes. Possible examples include movements during sperm removal, such as the "pumping movements" in some odonates, other movements of sperm-removing structures, and the presence of a mating plug in a female's genitalia. Thus, the seemingly paradoxical presence of "useless" sperm removal penile spines, and the "horns" on aedeagal heads in some odonate species in which the male genitalia do not reach sites in the female where sperm are present (Andrés and Cordero-Rivera 2000; Córdoba-Aguilar 2003; Cordero-Rivera and Córdoba-Aguilar 2010) may be explained as relics of sperm removal movements in which the incidental stimulation of the female eventually came to be the function.

SC and CFC can occur simultaneously, as illustrated by the fact that a male's access to the sperm of rival males is sometimes indirect. For instance, the male of the damselfly *Calopteryx hemorrhoidalis* cannot reach the sperm of rival males stored in the female's spermathecae; but he is able, by rubbing his genitalia against a sensory plate in the oviduct, to induce her to move them to her oviduct, where he is then able to trap and remove them. The male himself actively removes rival sperm, so SC selection has played a role. But the female's responses are biased, as she moves more sperm to the oviduct when the male's penis is thicker, so this aspect of male genitalic morphology is subject to CFC (see Córdoba-Aguilar et al. 2015, Chap. 9). The number of sensory structures in the oviduct plate that the male rubs with his penis is reduced, when compared with the sensory structures of a relative in which the male does not induce females to discard sperm with his penis (Córdoba-Aguilar 2005). This reduction could be due to SAC or to CFC (by reducing her sensitivity, a female under CFC could discriminate more strongly in favor of particularly stimulating males; alternatively, under SAC, she could reduce her reproductively damaging responses to male stimulation).

Proper attribution to the different types of selection in cases like these can be difficult. For instance, in order to adduce a causal role for CFC in addition to that for SC for evolutionary changes in the design of the female's sperm storage organ in a species in which sperm dilution occurs, one would need to demonstrate that traits of the female storage organ evolved due to the advantage that the female derived from having sons which were particularly effective at diluting rival ejaculates—not an easy task! In general, thorough, rigorous proofs of hypotheses regarding function can be very difficult (Williams 1966, 1997).

1.3.2 CFC and SI

SI has been ruled out pretty convincingly (in my opinion) as a general explanation, at least for the evolution of genitalia. Large accumulations of data in various taxonomic groups do not conform to several general predictions regarding the expected consequences of greater or lesser physical isolation from forming cross-specific pairs (e.g., strict allopatry, host differences in parasitic species) (Eberhard 1985, 2004b). This does not mean, however, that selection for SI can confidently be ruled out for all possible cases. Support for the SI rather than the CFC hypothesis for a particular male trait relies on the supposition that formation of cross-specific pairs is not rare (or, at least, was not rare in the moderately recent past) and that such pairs arrive at the stage in mating when that particular trait comes into play. In some cases, this possibility can be ruled out with reasonable confidence. If species do not overlap both geographically and temporally, heterospecific pairing is not a problem, and selection favoring SI will not occur. Even when there is geographic overlap, SI can sometimes be ruled out. For instance, observation of artificially combined cross-specific pairs in species of the spider *Leucauge* showed that SI selection on the sexually dimorphic male chelicerae, which only come into play after the spiders come into contact, is unlikely: males never even initiate long-distance pre-copulatory courtship and thus cannot be expected to make intimate physical contact with cross-specific females (see Aisenberg et al. 2015, Chap. 4). But in most species strong evidence on this point is lacking. The SI via sensory lock-and-key hypothesis survives some of the types of evidence that weigh against mechanical, lock-and-key SI, but it is also contradicted strongly by some of the others. Thus, it also fails as a general explanation, but must also be considered on a case-by-case basis.

On the other side of the coin, evidence that females reject cross-specific males on the basis of male stimulation, as in the male claspers of *Enallagma* damselflies (Robertson and Paterson 1982; McPeek et al. 2008, 2009), is not sufficient to demonstrate SI. Female ability to make such a discrimination could have evolved under sexual selection, and only incidentally also impede heterospecific pairing. There is evidence favoring a sexually selected CFC function for similar stimulation in another damselfly, *Hetaerina americana*; females grasped (and thus stimulated) with the claspers of a conspecific male lay more eggs when the male releases her (A. Córdoba-Aguilar, personal communication). Similarly, morphological incompatibility of female and male genitalia in carabid beetles (Sota and Kubota 1998; Kamimura and Mitsumoto 2012) is also not sufficient to demonstrate SI by lock-and-key arguments. In these beetles, the expected evolutionary sequence of genital changes under SI, beginning with females which have erected barriers to heterospecific males being favored, seems not to have occurred, because females suffer mechanical damage when they couple with heterospecific males. The resulting barrier to gene flow appears likely instead to be an incidental consequence of some other types of selection. Evidence of intra-specific uniformity and interspecific differences in genital morphology (McPeek et al. 2008, 2009 on odonate clasping organs) is also unconvincing, because such patterns are not incompatible (as is sometimes claimed) with other hypotheses. Sexual selection does not necessarily imply intra-specific divergence (though this could occur). In addition, there could be divergence in copulation behavior resulting from sexual selection, even when there is no intra-specific divergence in morphology (Eberhard 1985).

1.3.3 CFC and SAC

1.3.3.1 Direct Measurements of Fitness Effects

The crucial difference between CFC and SAC concerns the sizes of the payoffs that a female is presumed to obtain from failing to cooperate with some males. CFC focuses on the possibility that the females gain from favoring some males over others because she obtains sons that will have better abilities to induce positive CFC responses by females in the following generation. In contrast, SAC focuses on the possibility that the female gains from avoiding losses in her naturally selected abilities (in the current generation) to produce offspring. Each hypothesis supposes that the payoffs that accrue from the type of benefit that it emphasizes are greater under natural conditions (where the animals evolved) than those from the alternative type of benefit. In other words, a sexual conflict explanation for a male trait will prevail when the benefit to the female from reducing costs imposed by the male trait is greater than the genetic benefit that the female obtains from acquiring superior genes for that trait in her sons when she accepts some males but not others as sires. Vice versa, the CFC hypothesis supposes that the indirect genetic benefits are larger than the direct, naturally selected costs imposed by the male. The difference between the two hypotheses depends on the relative magnitudes of these quantities.

Unfortunately, it is generally impossible, because of practical limitations, to make quantitative measurements in nature of these two types of variable that are precise enough to compare the relative sizes of these two alternative payoffs convincingly. It seems to me undeniable, for instance, that some kinds of sexual interactions must lower the female's fitness, as in the wounds in the walls of the female's reproductive tract produced by the male genitalia of *Drosophila* spp. (Kamimura 2007, 2010; Yassin and Orgogozo 2013; Masly and Kamimura 2014) and *Callosobruchus* spp. beetles (Crudgington and Siva-Jothy 2000; Hotzy et al. 2012). But the crucial question in these cases of whether the size of this reduction in the fitness of the female is greater or less than the reproductive payoff she obtains in terms of her sons' abilities to sire offspring under natural conditions remains unknown. It cannot be simply assumed (as is sometimes the case) that just because the female is physically damaged by the male, the costs of this damage are greater than the benefits to the female from his siring her offspring (see B.II below).

Similarly, the elegant studies of Tallamy and colleagues (Tallamy et al. 2002, 2003) demonstrated that a female of the beetle *Diabrotica undecimpunctata howardi* gains from exercising CFC in favor of males that perform superior copulatory courtship (more rapid antennal vibrations) when she relaxes her oviduct muscles and allows the male to inflate his genitalia inside her and transfer sperm only when the male has vibrated his antennae more rapidly. She obtains superior sons, which execute more effective copulatory courtship. Nevertheless, the magnitude of this gain, taking into account population densities and male–female encounter rates in the wild, is not known. Nor is the relation of this magnitude known with respect to the potential costs to the female's reproduction of her rejection of some males (e.g., delay in acquiring sperm due to rejection of some males, metabolic costs of rejection behavior, possible increased susceptibility to predation during copulation). It is possible that the same trait can result in both types of effects on the female's fitness, and selection from SAC and CFC can even sometimes reinforce each other (Cordero and Eberhard 2005). Again, the absolute magnitudes of the costs and benefits are crucial; knowledge of whether or not a cost or benefit exits is not enough. Demonstrating that one type of selection has occurred does not justify discarding the other. Similarly, Arnqvist (2014) points out the difficulty of disentangling selection on putative CFC traits.

An additional, important problem is that even though direct measurements of these variables can be made in captivity, the data are not adequate for testing CFC versus SAC because the captive conditions (e.g., *Drosophila* culture bottles) undoubtedly have large effects on the magnitudes of both types of payoffs. For instance, demonstration that a male seminal product increases the female's oviposition rate and reduces her lifespan in captivity does not solve the question of whether the earlier oviposition does or does not compensate for a reduced lifespan in nature, or even whether females ever live long enough in nature for the longevity effects to manifest themselves (and of course, it also fails to give any indication of whether the magnitude of a possible net loss to the female in ovipositions is or is not large enough to compensate the gains she may derive in nature from genetically superior sons). Or, to balance this comment with a study that arrived at different conclusions, the substantial indirect benefit to females of the cricket *Acheta domesticus*, may balance or overshadow the female survival costs of mating with more attractive males under captive conditions. But despite the authors' care in employing captive conditions at least somewhat similar to those used for immediately preceding generations, doubt remains regarding the ecological realism of these experiments. The upshot of all this uncertainty is that the relative importance of SAC and CFC is very difficult or perhaps impossible to judge by direct measures (see Eberhard 2009, 2010 for further discussion).

Can the controversies concerning CFC be resolved? For the reasons just given, I believe that the jury is out regarding direct measurements of the crucial variables relating to the central point of the CFC–SAC conflict, the question of whether females gain or lose in total reproductive output (in terms of both quantity and quality of offspring) from particular aspects of their responses to interactions with males in nature; and I believe it is likely to remain out for the indefinite future. The necessary data on female fitness under ecologically realistic conditions are simply too difficult to obtain. I am thus pessimistic regarding measuring the crucial variables that would be required to resolve the theoretical conflicts via direct measurements. Claims that one or the other hypothesis has been definitely confirmed by direct observations of reproductive payoffs should be critically examined in light of the potential problems discussed here.

1.3.3.2 Indirect Tests

One contrast between CFC and SAC concerns the degree of active female cooperation with the male. Under CFC (and classic Darwinian female choice in general), the female will often promote and facilitate pairing and paternity for certain preferred males. Under SAC, females (except possibly virgin females) are expected to resist males until further resistance is more costly than giving in. Thus, CFC predicts, in contrast to the "antagonistic coevolution" of SAC, that there will be "synergistic male–female coevolution" (West-Eberhard 2014). This active selective female cooperation that is predicted under CFC could be manifested in both morphological and behavioral traits.

Female Sense Organs and Responses to Their Stimulation

One contrast concerns the evolution of the female sense organs that receive male stimuli. Under the SAC hypothesis, females are predicted not to have sense organs and responses to stimulation of these organs whose only function is to receive and respond to manipulative stimuli from the male (she would produce more surviving offspring if she received and/or responded less, not more). Rather the male is expected to stimulate preexisting female sensory receptors and to exploit the responses to stimulation of these receptors that evolved to perform other functions for the female; the male takes advantage of this combination of female sensitivity and responses to manipulate her for his own ends (e.g., via "sensory traps").

Under CFC (and SI via sensory lock and key), in contrast, a female gains from distinguishing among males by modulating her responses to the stimuli that she receives from the male. One mechanism by which such female discrimination could evolve is through acquiring specialized organs to sense particular male signals. Contrary to some discussions (e.g., Arnqvist 2014), CFC is not necessarily expected to be associated with female sensitivities and responses that under natural selection. To be sure, the early stages of the evolution of a male's use of particular stimuli in CFC would depend on the existence of female sense organs that can sense the male's signal, and on female responses to such stimulation. But subsequent evolution under CFC could favor both changes in her responses (e.g., changes in thresholds) and modifications of the sense organs themselves that function to improve her perception of differences in the morphology and behavior of males.

Arthropod species with specialized male contact courtship organs are especially interesting in this context, because it is possible to distinguish specialized female tactile sense organs predicted by CFC and sensory lock and key, such as modified setae (Aisenberg et al. 2015, Chap. 4), campaniform organs (Eberhard 2001), or peg-like basiconica mechanoreceptors (Robertson and Paterson 1982), and to check whether they are restricted to the area contacted by the male's specialized organ. In some cases, these female sense organs might coevolve with the form or behavior of the male contact courtship organs, differing between species in concert with differences in the design or behavior of the male's contact courtship organs. Morphological coevolution of female receptors of this sort is not the only possible female mechanism by which a female could carry out CFC, however; for instance, she could alter her CNS to change the process of analyzing stimuli or criteria for executing responses. Thus, the CFC (and SI) prediction is that sometimes but not always there will be female tactile sense organs specially modified to sense male contact courtship devices.

The existence of specialized female sense organs whose only apparent function is to sense the species-specific stimuli of the male constitutes strong evidence against SAC, which predicts that they will not occur. One possible female receptor organ of this type is the sexually dimorphic, elongate setae on sternum of a female spider (Aisenberg et al. 2015, Chap. 4), but experimental tests of the role of these setae in male–female interactions are still lacking. A more completely studied example is the species-specific arrays of female sensillae on the pronotum, whose stimulation is known to be crucial for copulation to occur in *Enallagma* damselflies (Robertson and Paterson 1982). There are, on the other hand, at least two cases in which species-specific female sense organs of this sort were searched for but do not occur (on the chelicerae of female *Leucauge* spiders, the wings of female *Archisepsis* flies) (Aisenberg et al. 2015, Chap. 4; Eberhard 2001). These data are too sparse to reveal whether there is a general pattern; an extensive SEM survey of male–female dimorphism in tactile receptors in the areas of the female body that are contacted by male courtship devices (e.g., the species surveyed in Eberhard 2004a) would be very interesting.

SAC also predicts that the areas near female sense organs stimulated by males will often show signs of previous coevolutionary conflict; one form would be potentially defensive structures whose designs are appropriate to fend off or protect against male contact courtship devices. Such female structures could be, for instance, erectable spines or retractable covers that would hold the male's structure away from her body (Eberhard 2004a). A concrete example of a defensive structure of this sort (though it functions in a male–male rather than a male–female context) is the set of large flanges on the midline edges of the prothoracic mesostigmal grooves in male (but not female) *Enallagma* damselflies, which probably defend the male from being grasped by other males (Robertson and Paterson 1982).

In addition, SAC predicts that male contact courtship devices should often have designs that are appropriate to increase the male's ability to overcome (push aside, pry under, etc.) such female defenses. In contrast, as explained above, under CFC the female could gain by sensing the male's contact courtship structures, so neither female defensive designs nor male offensive designs to overcome female defenses are expected (though female choice by mechanical fit might sometimes result in somewhat similar designs). Again, female discrimination could also occur via changes in her CNS rather than her morphology, so the predictions of both SAC and CFC are for trends rather than consistent presence or absence.

These predictions were tested in a broad survey of insects and spiders. Contrary to SAC predictions, lack of obvious female defensive designs was the rule (Eberhard 2004a, b). This same lack of apparent female morphological defenses also occurs in a spider and in tsetse flies discussed in this book (Aisenberg et al. 2015, Chap. 4; Briceño and Eberhard 2015, Chap. 15). The recent discovery of subtle female modifications in *Drosophila* associated with potentially damaging male genital structures (e.g., Kamimura and Mitsomoto 2012; Yassim and Orgogozo 2013) (see next section) introduces a note of caution, however, regarding lack of morphological female defenses. In this genus previous reliance on male rather than female genitalia to distinguish species in taxonomic studies (Lachaise et al. 2000) gave a misleading impression of a lack coevolution of female structures that are adjusted to the elaborate male genitalia.

Even when there are female defenses against male-inflicted damage, interpretation of SAC versus CFC is not simple. A female defense does not rule out CFC: females could gain indirectly, through improved offspring quality, by imposing paternity biases with defensive structures. The crucial question regarding function hinges again on the relatively sizes of the costs and benefits. It must also be kept in mind that the predictions concern relative frequencies, not presence/absence. Thus, the SAC prediction is for female defensive structures to be present in some but not all groups, this is because there are other, non-mechanical ways such as changes in the female's CNS for females to overcome male manipulations. In contrast, CFC predicts that female fending-off structures that would prevent males from making contact with the female's sense organs may be rare. The female's best interests would seem to usually but not always be served by evaluating the stimuli from males, not by excluding the male non-selectively from her sense organs (unless the female's choice criterion is the male's ability to circumvent such a barrier).

Physical Damage to the Female

Several types of data have been used in discussions of SAC to argue that male genitalia and seminal products damage the reproductive output of females. Some types of evidence are weak because the male effects on female reproduction have only been measured in captivity and do not deal with the possibility that these effects may be selectively irrelevant in nature; even such basic effects as reductions in life expectancy or total egg production in captivity could be irrelevant under natural conditions if predation or difficulty in finding suitable oviposition sites impose important limitations (see discussion in Eberhard 2010). Another type of evidence, physical damage to female structures, such as hole in the lining of her reproductive tract due to abrasive or perforating male genitalia or harmful seminal products, is more convincing. Damage of this sort ("traumatic" copulation or insemination) has been documented in various animal groups (summarized in Arnqvist and Rowe 2005; Lange et al. 2013; Masly and Kamimura 2014; Arnqvist 2014; for an additional case in a different taxonomic group, see Eberhard 1992). If such holes in her reproductive tract were advantageous to the female, the expectation is that she would not seal them up again following copulation, as is often the case (Crudgington and Siva-Jothy 2000; Flowers and Eberhard 2006; Kamimura 2010; Yassin and Orgozo 2013). Measuring the magnitude of the damage, in terms of lost offspring, has the same practical problems as do other such measurements in captivity (e.g., the rate of female infections that result from wounds under the typically highly crowded but also otherwise relatively aseptic conditions of captivity may not be representative of those in nature). Nevertheless, it seems reasonable to deduce that the effect of these traumas on the female is negative rather than positive or neutral.

What can be learned regarding SAC versus CFC in species in which females are damaged by males? A recent burst of elegant studies, with especially well-documented details of the morphological interactions between male and female genitalia in several species of *Drosophila* (Kamimura 2007, 2010; Kamimura and Mitsumoto 2012; Yassin and Orgogozo 2013; Masly and Kamimura 2014), offers interesting illustrations of possible evolutionary responses of females to damage. In three of nine species that were checked, there were strong female sclerites (hypothesized to be "vaginal shields") in areas contacted by the male's cerci during copulation (Yasssin and Orgogozo 2013), as would be predicted by SAC. One other "shield" also occurs on the vulva and another in the uterus (one species each); both are contacted by other potentially damaging male genital structures.

Apparently the male's cercal teeth do not normally produce wounds, however, in either these or other *Drosophila* species which lack "shields" at these sites (Kamimura 2010). Should one take the presence of possible female defensive structures in some species as confirmation of SAC? Or should their absence in other species in which males have genital structures with spines and pointed processes be taken as refutation of SAC? And what about the likelihood that the female is stimulated by the process of producing this damage (below)? Or the possibility that areas of thickened cuticle have other functions, such a providing as substrates on which the male can tap or vibrate, or as a filter that makes the female less sensitive to stimuli from the male and thus better able to bias her responses in favor of those males which are especially good at producing such stimulation? Surely labeling a female structure with a suggestive term like "shield," which implies that the function has been established, is not helpful. I am not suggesting that I understand the functions of these fascinating male and female structures. Rather I wish to emphasize that such understanding will require keeping the different possible explanations clearly in mind, and avoiding the temptation to rush to judgment before there is good reason to choose between them.

A similarly open mind is needed to understand the most common type of female "defensive" structures in *Drosophila* spp.—the membranous furrows and pouches of the female tract, where spiny or abrasive portions of the male genitalia rest during copulation (Kamimura 2010; Kamimura and Mitsumoto 2012; Yassin and Orgogozo 2013; Masly and Kamimura 2014). Crosses of female *D. sechellia* with males that had posterior genitalic lobes of different sizes and shapes suggest that these pouches can have defensive effects (Masly and Kamimura 2014), probably because they reduce the force with which the male structure presses against the lining of the female tract and thus reduce the damage that it can inflict. But there are complications with this interpretation. In female *D. sechellia*, smaller rather than larger male posterior lobes were more likely to cause wounds (Masly and Kamimura 2014). In addition, paradoxically, the walls of the female pouches in *Drosophila* are generally very thin and membranous (Kamimura 2007, 2010; Yassin and Orgogozo 2013; Masly and Kamimura 2014); in this respect, the pouches seem designed to *allow* rather than prevent wounding. An effective female defense could be accomplished by a simple thickening or hardening of the portion of wall that is contacted by the male's genital structure. Perhaps such hardening would make the lining stiffer and thus impede oviposition, a second function of the female reproductive tract. But a small female compensation, such as extending the nearby oviduct wall slightly, would be so simple and cheap that it seems reasonable to expect such a defense. In any case, I do not see how to eliminate the possibility that the pouches represent filters; they could function to allow the female to favor the males that have mechanically compatible lobes (Yassin and Orogogozo 2013), or those that have the greatest abilities to stimulate or penetrate the walls of these pouches.

Finally, it seems that stimulation of the female occurs nearly inevitably, while the male is in the process of wounding her. It is remarkable that a stimulating function has not been considered for these male genital structures in *Drosophila,* nor in those of other insects with abrasive male genitalia such as seed beetles (Hotzy et al. 2012). In the context of possible stimulation, it is also striking that some other *Drosophila* male genital structures are spinous (the aedeagus) or sharply peaked (the ventral branches), but do not make wounds in the female (Kamimura 2010). These structures could serve as stimulators, as hold-fast devices (Kamimura 2010), or both. The lack of transfer of male material through the wounds made by the male genitalia in one species, (*D. melanogaster*) (Kamimura 2010), also favors these possibilities over a "wounding to introduce seminal products into the female body cavity" interpretation, which is more appropriate in other species in which such transfer actually occurs (Hotzy et al. 2012). This transfer function also seems improbable with respect to the perforations produced by the posterior lobes *D. sechillia* (Masly and Kamimura 2014), because these male structures are basal with respect to his aedeagus, which is introduced much deeper into the female and near whose tip the ejaculate emerges. The function of producing perforations in the female is thus not clear in these species, and stimulation has not been ruled out. In sum, the exciting discovery of various types of genital wounds in *Drosophila* spp. (and other animals) presents a complex panorama in which various interpretations in combination with or instead of SAC are feasible.

Male–Female Behavioral Struggles

Another easily misunderstood trait related to the relative importance of SAC and CFC involves using direct observations of physical struggles between males and females. Struggles preceding and during copulation are often taken as indicators that sexual conflict hypotheses should be applied. But this kind of evidence must be considered carefully, for at least two reasons. In the first place, an overt male–female struggle is not a reliable indicator of SAC. Take, for example, the sepsid fly *Archisepsis diversiformis*. The male jumps onto the female at feeding and oviposition sites, usually with little or no preliminary interactions, and clamps the bases of the female's wings firmly with elaborate species-specific modifications of his front femur and tibia. Almost invariably, the female immediately shakes her body violently, as if to throw the male off. Most mountings (some of which last for 5 min or more) end with the male dismounting after having failed to copulate (Baena and Eberhard 2007). At first glance, this looks like a classic case of a male forcing the female to mate by wearing down her resistance: Stronger, more vigorous males would be able to resist the female's attempts to dislodge them, and by imposing the cost of staying mounted, eventually force females to mate. But the details of the interaction speak clearly against this interpretation. In those pairs in which mating occurred, copulation generally began within about 30 s of the mounting; in none of the cases in which the male stayed mounted for a long time while the female shook, did he succeed in mating. In addition, video recordings showed that the male was almost never thrown off physically; rather he stepped off of the female during a break in her shaking (Baena and Eberhard 2007). And finally, studies of the genital mechanics of this species (and other sepsid flies, where female shaking behavior is also the rule) showed that the male is physically incapable of forcing his genitalia into the female's body. At nearly all times during a female's life, the distal-most segments of her abdomen (her "proctiger") rest on the dorsal surface of her vulva, covering the opening to her reproductive tract. In order for a male to achieve intromission, the female must first lift her proctiger, allowing him physical access to her vulva. The male has no genital structure capable of prying up this female "door," so he must rely entirely on active female cooperation to gain intromission (Eberhard 2002). In sum, copulations in this species are not reasonably attributed to the male overcoming female resistance, despite the forceful male–female struggles that consistently precede copulation, and the timing of her cooperation shows that it is not appropriately attributed to physical coercion by the male. Rather copulation results only in those cases in which the female cooperates actively with the male.

The inverse problem also exists: lack of overt female resistance is not a reliable signal of lack of male–female conflict (Kokko and Jennions 2014). For instance, as noted by Arnqvist and Rowe (2005), sexual conflict can be "hidden" when male and female interactions are evenly balanced. One illustrative image is the static pose of evenly matched arm wrestlers, who are fighting intensely but hardly moving (Jennions 2005). I would argue, however, that this limitation is not always severe, because of the potential power of resolution of well-focused, detailed behavioral observations to resolve questions of this sort (Peretti and Córdoba-Aguilar 2007). It would easy, for instance, to discriminate a photograph of a true, evenly matched arm wrestling battle from photographs of "battles" in which the participants' arms were linked but their torsos were not positioned appropriately to apply force to the opponent, or their legs were positioned inappropriately (e.g., legs were crossed) and did not give support and power to the torso and arm. In the same way, additional details of male–female interactions can sometimes be used to evaluate the SAC prediction of a history of male–female coevolution. One type of indirect evidence of SAC would come from the mechanical designs of females. Females should show signs of having evolved previous defenses against males, while the species-specific traits in modern males should be designed to enable them to overcome these female defenses (see Eberhard 2004a).

This technique for evaluating these SAC predictions has at least two important limitations. Firstly, recognizing the "defensive" nature of a design in the female is not always easy, because the effectiveness of a defense depends on the design of the males (see the discussion of female pouches and male genital spines in *Drosophila* above). Secondly, a female's defense may sometimes occur in her nervous system rather than in her external morphology, as noted above. This type of defense leaves no easily read vestige in the behavior or external morphology of the female. Nevertheless, there are general classes of relatively inexpensive and effective defensive structures (e.g., derived, moveable devices that can cover the portion of the female that needs to be defended when she is accosted by a male), and they should be common under SAC. The data on insects and spiders clearly do not fit this prediction. Female devices with designs appropriate to facultatively protect areas of the female's body that are contacted by species-specific male "copulatory courtship structures" seem to be rare (Eberhard 2004a); the most common female characteristic in a survey of 106 structures in 84 taxonomic groups was a complete lack of any possibly defensive female structure (seen in 54 % of the groups). This argues against the generality of SAC.

1.4 The Current State of Affairs

It is now clear that post-copulatory interactions between a female and the male and products from the male have important impacts on the relative reproductive success of males, and result in sexual selection. No single-species study will be enough to resolve questions regarding the importance of CFC relative to alternative hypotheses for explaining post-copulation events (e.g., Arnqvist and Rowe 2005; Jennions 2005). SC surely exists in some cases, though the possible additional role of CFC in these cases has generally not been explored adequately (see Kamimura 2015, Chap. 10; Sirot and Wolfner 2015, Chap. 13). SI via mechanical lock and key is surely not a generally applicable explanation of genital diversity, but nevertheless, it cannot be ruled out a priori, and cases must be examined one by one. SI via sensory lock and key survives some of these types of evidence, but not others, so is also an unlikely general explanation but must be tested case by case. The greatest uncertainty surrounds the controversy between CFC and SAC interpretations. To judge by perusing the literature and scanning titles in recent congresses, conflict-based interpretations appear to have prevailed in the minds of many. At a very superficial level of analysis, this emphasis is reasonable: "broad-sense" male–female conflict (Brennan and Prum 2012) is almost trivially true for male–female interactions in nature. Male and female reproductive interests are seldom identical.

At the more interesting narrow-sense level of conflict, however, I believe that the controversy between SAC and CFC has not been resolved. It is possible that the selection on males to win out over other males has produced profound conflict between the sexes, favoring traits that inflict damage to the female's naturally selected reproductive interests, and that the most important payoff to females from rejecting some males is from avoiding this damage. It is also possible that the most important payoff to the female is obtaining superior survivorship or signaling genes for her offspring and that many cases of apparent male–female conflict are better understood as selective female cooperation with some males but not others. I thus believe that the current popularity of conflict-based hypotheses is largely due to a lack of appropriate analyses. The fact that both kinds of selection could act simultaneously on the same trait and that different traits of the same animal could evolve due to different types of selection (Cordero and Eberhard 2005; Kokko et al. 2003; Kokko and Jennions 2014; Barbosa 2015, Chap. 14) emphasize the likelihood that different hypotheses are correct in different cases.

1.5 Peering into the Future

Perhaps a little simplified history can help illuminate the future. I believe that this book marks a typical "middle age" in the ontogeny of a possibly fruitful scientific idea. CFC was born as an extension or refinement of the transformative realization of Parker (1970) that sexual selection can occur even after copulation has begun. The basic addition of CFC to Parker's original insight is that female choice can occur in many different forms after copulation has begun, in addition to direct male–male battles via sperm competition.

The first general discussions of CFC and its derivatives (Thornhill 1983; Eberhard 1985, 1996) were based mostly on suggestive evidence that had been collected in other contexts (as also occurred in early discussions of SC and SAC—Arnqvist and Rowe 2005; Jennions 2005). Some early doubts regarding CFC (e.g., Møller 1997), emphasized this inevitably incomplete and "cobbled together" nature of much of the evidence. The obvious way to evaluate such doubts was (and is) to gather further observations, using techniques and experimental designs specifically designed to test both the criticisms and the hypothesis itself.

The first general presentations of CFC were convincing enough to earn the hypothesis a place in the standard list of possible explanations that are currently proposed by specialists in the field with respect to post-copulatory selection (it is sometimes less thoroughly explored than some other hypotheses, however—see Eberhard 2012). Nevertheless, the acceptance of CFC at a more general level among evolutionary biologists has lagged behind. For example, CFC is still a very poor sister to SC and SAC in one large recent evolution textbook (Zimmer and Emlen 2013), where it is not even mentioned in the index.

Some of this neglect is probably related to the recurrent "amnesia" among biologists in thinking about the importance of social rather than environmental factors in sexual selection, and the reasons given by Darwin to distinguish between sexual and natural selection (West-Eberhard 2014). CFC, in contrast with SAC and SI, has a relatively weak logical connection with the ability of the animal to deal with its external environment. For instance, the size of a male's genitalia, a likely candidate for the action of CFC (Eberhard 1985), shows relatively little response in size to changes in nutrition; male genitalia are likely to be poor signals of a male's size or health (Eberhard et al. 1998; Emlen et al. 2012). The male copulatory courtship behavior that is favored by CFC by females in the beetle *Diabrotica undecimpunctata howardi* has no perceptible association with hatching rate, survivorship, developmental rate, size, age at first maturity, or fecundity of the offspring, but is nevertheless correlated with the copulatory courtship behavior of her male offspring (Tallamy et al. 2003). In general, undue emphasis on natural rather than sexual selection can lead one to miss the importance of sexually selected payoffs.

Another possible problem for general acceptance of CFC may be its emphasis on the female's sometimes potentially invincible powers to control post-copulatory sexual events, in contrast to the perhaps more intuitively appealing views of male control (SC) or male–female battles over control (SAC). In his book on sperm competition, for instance, Simmons (2001, p. 279) doubted even the possibility of sorting out male and female effects: "Interpretations of male versus female control can rarely be more than a point of view, neither of which can be said to be right or true." This seems to me to fly in the face of common sense. A male may, in the end, have little or no ability to impose his own reproductive interests after he has done all he can in terms of courtship, seminal products, etc. If, for instance, the female nevertheless fails to open up her reproductive tract for him and he has no morphological structure with which he can force his way in (Tallamy et al. 2002; Baena and Eberhard 2007), or if she unceremoniously dumps his ejaculate from her body as soon as he withdraws his genitalia (for an easily intuited example, see Baker and Bellis 1995 on humans), she will have the decisive last word (see also Kokko and Jennions 2014).

Conclusions drawn from incomplete tests of CFC are still another problem. Take as an example a case that is deservedly well known, in which exquisitely fine experimental modifications were made of the spines on one male genital structure (the ventral cercal lobe) of *Drosophila* flies (Polak and Rashed 2010) that grasps the female oviscapts (Eberhard and Ramirez 2004; Kamimura and Polak 2010). The conclusion that these structures do not function to influence CFC, on the basis of the lack of any effect of removing this genital structure on whether sperm is transferred and is or is not used preferentially with respect to the sperm of other males in the same female, failed to take into account many other possible mechanisms of CFC (increased resistance to further mating, greater numbers of eggs matured and laid after copulation before another copulation, reduced delay in laying eggs already mature, etc.). CFC, because of the multitude of possible mechanisms by which it can occur, is an especially difficult hypothesis to test experimentally.

Whatever the reason for lingering doubts, future resolution of the question of whether CFC is of wide general importance will be determined by the accumulation of relevant data that test the predictions that CFC and rival hypotheses make. Perhaps the major role of the present book is to show that a surprising amount of scattered observations have accumulated in which CFC was specifically tested and supported. The current book complements another recent, taxonomically more wide-ranging compilation (Leonard and Córdoba-Aguilar 2010). It is especially interesting to see the applicability of these ideas at the level of molecules and genes (Sirot and Wolfner 2015, Chap. 13). The difficulties in resolving the relative importance of different theories that were discussed above do not mean that there are not exciting new directions for empirical study of the phenomena that these theories aspire to interpret. In fact, as noted by Jennions (2005), perhaps the greatest payoff from these controversies will be to provide motives for studying exciting topics in reproductive biology in the context of testing theories, rather than simply describing esoteric phenomena.

I will close by emphasizing one topic that I find particularly promising, that of "non-genital contact courtship devices" in arthropods (Fig. 1.1) (Eberhard 1985). These are mentioned in several chapters in this book. They are male structures that are specialized to contact the female on her outer surface, and they often bear the typical signatures of sexual selection: puzzlingly ornate designs that are not obviously functional, and rapid divergence between closely related species (Eberhard 1985, 2004a). I believe that these structures hold special promise for improved understanding of sexual selection in general for several reasons. The sense organs that females use to perceive male tactile stimuli are located at specific, spatially localized sites on her body, and they probably often function more or less exclusively in the single task of sensing the male. The sites of these female sense organs are easily determined in many arthropods and can thus provide insights into female choice criteria in different species (e.g., Robertson and Paterson 1982). The numbers and locations of female tactile sense organs can be compared to see whether they mirror differences in the sizes, shapes, and locations of male contact courtship organs in these species. And in each species, the female's sense organs can also be compared with the homologous tactile sense organs of conspecific males. Do females evolve special receptors to sense the species-specific aspects of male contact courtship structures (as is expected to be common under CFC, but not under SAC)?

Furthermore, it is often possible to experimentally mask or cripple the particular female tactile receptors that are sensitive to male contact courtship stimuli, impeding the female's ability to sense the male but leaving her otherwise intact and able to respond normally to most other stimuli. Such selective blinding of the female to the male's tactile charms has powerful effects in tsetse flies (Briceño and Eberhard 2015, Chap. 15), an orb-weaving spider (Aisenberg et al. 2015, Chap. 4), and a sepsid fly (Eberhard 2002). When combined with experimental alterations of the corresponding male morphology, strong conclusions can be obtained with respect to function.

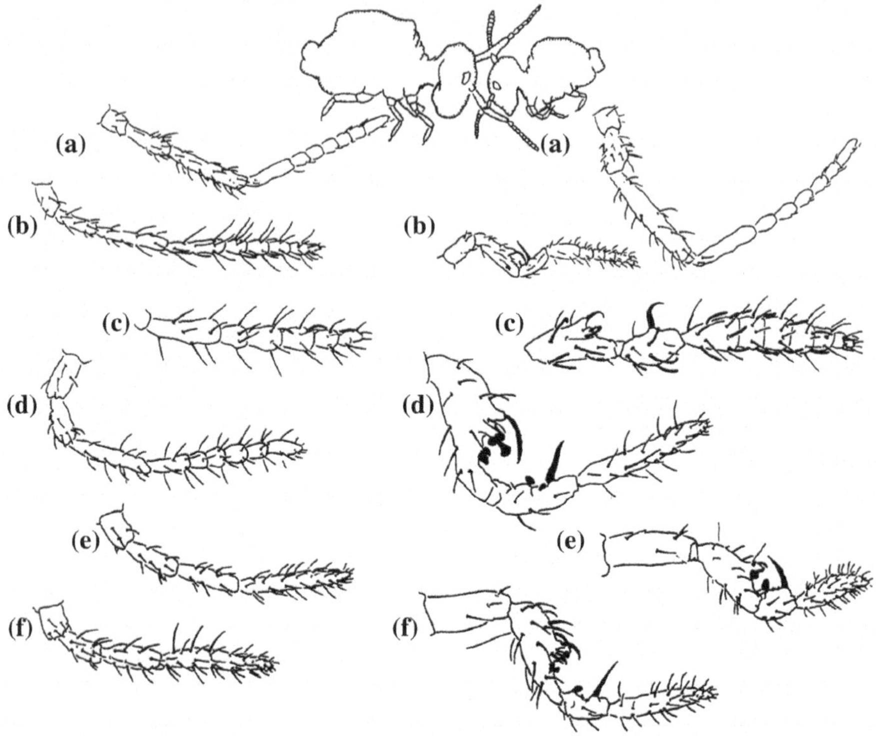

Fig. 1.1 An illustration, in the male contact courtship organs in symphypleona collembolans, of the common trend for females to lack morphological adjustments to the species-specific modifications in male structures. The male uses the basal segments of his antenna to grasp the female's antennae (*upper drawing*). In each pair of drawings below, the female antenna is on the *left*, and the male on the *right*. **a** *Bovicornia greensladei*, **b** *Yosiides himachal*, **c** *Smithuridia sphaeridioides*, **d** *Debouttevillea marina*, **e** *Denisiella* sp., **f** *Jeannenotia stachi*. The lack of differences in the female antennae that correspond to the differences in the male antennae is not in accord with predictions of the mechanical lock and key species isolation or sexually antagonistic coevolution hypotheses concerning coevolution in female morphology. *Upper drawing* after Mayer (1957); *lower drawings*, to different scales, after Massoiud and Betsch (1972)

The study of non-genital contact courtship devices is especially feasible in arthropods, because many of their surface tactile sense organs—setae and cuticular stress sensors such as campaniform organs and lyriform organs—are easily seen and can be easily masked or inactivated experimentally in highly specific ways. A survey of such female-specific sense organs among closely related species (as in Robertson and Paterson 1982), combined with a known phylogeny and experiments to determine the effects of experimental "blinding" females (e.g., Krieger and Krieger-Loibl 1958 on damselflies), could yield exciting insights into the evolution of post-copulatory sexual selection, and sexual selection in general.

Acknowledgments I thank Anita Aisenberg and Alfredo Peretti for the invitation to write this chapter; Y. Kamimura, M.J. West-Eberhard, M. Jennions, R.L. Rodriguez, and A. Aisenberg for useful comments on a preliminary draft; and the Smithsonian Tropical Research Institute and the Universidad de Costa Rica for financial support.

References

Aisenberg A, Barrantes G, Eberhard WG (2015) Post-copulatory sexual selection in two tropical orb-weaving *leucauge* spiders. In: Peretti AV, Aisenberg A (eds) Cryptic female choice in arthropods. Springer, Cham

Alexander RD, Marshall DC, Cooley JR (1997) Evolutionary perspectives on insect mating. In: Choe JC, Crespie BJ (eds) The evolution of mating systems in insects and arachnids. Cambridge University Press, Cambridge, pp 4–31

Andersson M (1994) Sexual selection. Princeton University Press, Princeton, NJ

Andrés JA, Cordero-Rivera A (2000) Copulation duration and fertilization success in a damselfly: an example of cryptic female choice? Anim Behav 59:695–703

Arnqvist G (2014) Cryptic female choice. In: Shuker D, Simmons LW (eds) The evolution of insect mating systems. Oxford University Press, New York, pp 204–220

Arnqvist G, Rowe L (2005) Sexual conflict. Princeton University Press, Princeton

Baena M, Eberhard WG (2007) Appearances deceive: female "resistance" behaviour in a sepsid fly is not a test of male ability to hold on. Ethol Ecol Evol 19:27–50

Barbosa F (2015) An integrative view of postcopulatory sexual selection in a soldier fly: interplay between cryptic mate choice and sperm competition. In: Peretti AV, Aisenberg A (eds) Cryptic female choice in arthropods. Springer, Cham

Baker RR, Bellis MA (1995) Human sperm competition, copulation masturbation and infidelity. Chapman and Hall, New York

Birkhead TR (1998) Cryptic female choice: criteria for establishing female sperm choice. Evolution 52:212–1218

Birkhead TR, Møller AP (1992) Sperm competition in birds—evolutionary causes and consequences. Academic Press, New York

Birkhead TR, Møller AP (1998) Sperm competition and sexual selection. Academic Press, NY

Brennan PLR, Prum RO (2012) The limits of sexual conflict in the narrow sense: new insights from waterfowl biology. Phil. Trans. R. Soc. B 367:2324–2338

Briceño RD, Eberhard WG (2015) Species-specific behavioral differences in male tsetse fly genital morphology and probable cryptic female choice. In: Peretti AV, Aisenberg A (eds) Cryptic female choice in arthropods. Springer, Cham

Calbacho-Rosa L, Peretti AV (2015) Copulatory and post-copulatory sexual selection in haplogyne spiders, with emphasis on Pholcidae and Oonopidae. In: Peretti AV, Aisenberg A (eds) Cryptic female choice in arthropods. Springer, Cham

Christy J (1995) Mimicry mate choice and the sensory trap hypothesis. Am Nat 146:171–181

Clark NL, Aagaard JE, Swanson WJ (2006) Evolution of reproductive proteins from animals and plants. Reproduction 131:11–22

Cordero C, Eberhard WG (2005) Interaction between sexually antagonistic selection and mate choice in the evolution of female responses to male traits. Evol Ecol 19:111–122

Cordero-Rivera A, Córdoba-Aguilar A (2010) Selective forces propelling genitalia evolution in Odonata. In: Leonard J, Córdoba-Aguilar A (eds) The evolution of primary sexual characters in animals. Oxford University Press, New York, pp 332–352

Córdoba-Aguilar A (2003) Predicting mechanisms of sperm displacement based on genital morphometrics in the Calopterygidae (Odonata). J Ins Behav 16:153–167

Córdoba-Aguilar A (2005) Possible coevolution of male and female genital form and function in a calopterygid damselfly. J Evol Biol 18:132–137

Córdoba-Aguilar A, González-Tokman D, Nava-Bolaños A, Cuevas-Yáñez K, Rivas M (2015) Female choice in damselflies and dragonflies. In: Peretti AV, Aisenberg A (eds) Cryptic female choice in arthropods. Springer, Cham

Crudgington H, Siva-Jothy M (2000) Genital damage, kicking and early death. Nature 407:855–856

Darwin C (1871) The descent of man and selection in relation to sex, 6th edn. Reprinted. Modern Library, New York

Eberhard WG (1985) Sexual selection and animal genitalia. Harvard University Press, Cambridge

Eberhard WG (1992) Courtship before and during copulation by three species of *Macrodactylus* (Coleoptera, Scarabeidae, Melolonthinae). Ethol Ecol Evol 5:19–63

Eberhard WG (1994) Evidence for widespread courtship during copulation in 131 species of insects and spiders, and implications for cryptic female choice. Evolution 48:711–733

Eberhard WG (1996) Female control: sexual selection by cryptic female choice. Princeton University Press, Princeton

Eberhard WG (2001) The functional morphology of species-specific clasping structures on the front legs of male *Archisepsis* and *Palaeosepsis* flies (Diptera, Sepsidae). Zool J Linn Soc 133:335–368

Eberhard WG (2002) The function of female resistance behavior: intromission by male coercion vs. female cooperation in sepsid flies (Diptera: Sepsidae). Rev Biol Trop 50:485–505

Eberhard WG (2004a) Rapid divergent evolution of sexual morphology: comparative tests of antagonistic coevolution and traditional female choice. Evolution 58:1947–1970

Eberhard WG (2004b) Male–female conflict and genitalia: failure to confirm predictions in insects and spiders. Biol Rev 79:121–186

Eberhard WG (2009) Postcopulatory sexual selection: Darwin's omission and its consequences. Proc Nat Acad Sci USA 106(Suppl 1):10025–10032

Eberhard WG (2010) Evolution of genitalia: theories, evidence, and new directions. Genetica 138:5–18

Eberhard WG (2012) Experiments with genitalia: a commentary. Trends Ecol Evol 26:7–21

Eberhard WG, Ramirez N (2004) Functional morphology of the male genitalia of four species of *Drosophila*: failure to confirm both lock and key and male-female conflict predictions. Ann Ent Soc Am 97:1007–1017

Eberhard WG, Huber BA, Rodriguez RL, Briceño RD, Solis I, Rodriguez V (1998) One size fits all? Relationships between the size and degree of variation in genitalia and other body parts in twenty species of insects and spiders. Evolution 52:415–431

Emlen M, Warren IA, Johns A, Dworkin I, Lavine LC (2012) A mechanism of extreme growth and reliable signaling in sexually selected ornaments and weapons. Science 337:860–864

Flowers W, Eberhard WG (2006) Fitting together: copulatory linking in some *Neotropical Chrysomeloidae*. Rev Biol Trop 54:829–842

Holland B, Rice W (1998) Chase-away sexual selection: antagonistic seduction vs resistance. Evolution 52:1–7

Hotzy C, Polak M, Rönn JL, Arnqvist G (2012) Phenotype engineering unveils the function of genital morphology. Curr Biol 22:2258–2261

Jennions MD (2005) Sexual tension: does conflict lead to costly mate choice? Evolution 60:415–417

Kamimura Y (2007) Twin intromittent organs of *Drosophila* for traumatic insemination. Biol Lett 3:401–404

Kamimura Y (2010) Copulation anatomy of *Drosophila melanogaster* (Diptera: Drosophilidae): wound-making organs and their possible roles. Zoomorphology 1129:163–174

Kamimura Y (2015) What is indirect cryptic female choice? Theoretical considerations and an example from a promiscuous earwig. In: Peretti AV, Aisenberg A (eds) Cryptic female choice in arthropods. Springer, Cham

Kamimura Y, Mitsumoto H (2012) Lock-and-key structural isolation between sibling *Drosophila* species. Ent Sci 15:197–201

Kamimura Y, Polak M (2010) Does surgical manipulation of *Drosophila* intromittent organs affect insemination success? Proc R Soc B Biol Sci 278:815–816

Karn RC, Clark NL, Nguyen ED, Swanson WJ (2008) Adaptive evolution in rodent seminal vesicle secretion proteins. Molec Biol Evol 25:2301–2310

Kokko H, Jennions MD (2014) The relationship between sexual selection and sexual conflict. In: Rice W, Gavrilets S (eds) Additional perspectives on the genetics and biology of sexual conflict. Cold Spring Harb Perspect Biol, Woodbury 6(9):a017517

Kokko H, Brooks R, Jennions MD, Morly J (2003) The evolution of mate choice and mating biases. Proc R Soc B Biol Sci 270:653–664

Krieger F, Krieger-Loibl E (1958) Beitrage zum Verhalten von *Ischnura elegans* und *Ischnura pumilio* (Odonata). Z Tierpsychol 15:82–93

Lachaise D, Harry M, Solignac M, Lemeunier F, Bénassi, V Cariou ML (2000) Evolutionary novelties in islands: *Drosophila santomea*, a new *melanogaster* sister species from São Tomó. Proc R Soc B Biol Sci 267:1487–1495

Lange R, Reinhardt K, Michaels NK, Anthes N (2013) Functions, diversity, and evolution of traumatic mating. Biol Rev 88:585–601

Leonard J, Córdoba-Aguilar A (2010) The evolution of primary sexual characters in animals. Oxford University Press, New York

Masly JP, Kamimura Y (2014) Asymmetric mismatch in strain-specific genital morphology causes increased harm to *Drosophila* females. Evolution. doi:10.1111/evo.12436

Massoiud Z, Betsch J-M (1972) Étude sur les insectes collemboles, II: les caractéres sexuels secondaires des antennes des Symphypléones. Rev. Ecol. Biol. Sol 9:55–97

Mayer H (1957) Zur Biologie und Ethologie einheimischer Collembolen. Zool. Jahrb. Abt. System. 85:501–570

McPeek M, Shen AL, Torrey JZ, Farid H (2008) The tempo and mode of 3-dimensional morphological evolution in male reproductive structures. Am Nat 171:E158–E178

McPeek MA, Shen L, Farid H (2009) The correlated evolution of three-dimensional reproductive structures between male and female damselflies. Evolution 63:73–83

Miller PL (1984) The structure of the genitalia and th volumes of sperm stored in male and female *Nesciothemis farinosa* (Foerster) and *Orthetrum chrysostigma* (Burmeister) (Anisoptera: Libellulidae). Odonatologica 13:415–428

Miller PL (1990) Mechanisms of sperm removal and sperm transfer in *Orthetrum coerulescens* (Fabricius) (Odonata: Libellulidae). Physiol Ent 15:199–209

Møller AP (1997) Female control; sexual selection by cryptic female choice. uart Rev Biol 72:202–203 By William G. Eberhard

Parker GA (1970) Sperm competition and its evolutionary consequences in the insects. Biol Rev 45:525–567

Peretti AV (2010) An ancient indirect sex model. Single and mixed patterns in the evolution of scorpion genitalia. In: Leonard J, Córdoba-Aguilar A (eds) The evolution of primary sexual characters in animals. Oxford University Press. New York, pp 218–248

Peretti A, Córdoba-Aguilar A (2007) On the value of fine-scaled behavioural observations for studies of sexual coercion. Ethol Ecol Evol 19:77–86

Polak M, Rashad A (2010) Microscale laser surgery reveals adaptive function of male intromittent genitalia. Proc R Soc B Biol Sci 277:1371–1376

Robertson RR, Paterson HEH (1982) Mate recognition and mechanical isolation in *Enallagma* damselflies. Evolution 36:243–250

Schneider J, Uhl G, Herberstein ME Cryptic female choice within the genus *Argiope*: a comparative approach. Chapter 3 this volume

Shapiro AM, Porter AH (1989) The lock and key hypothesis: evolutionary and biosystematics interpretations of insect genitalia. Ann Rev Ent 34:231–245

Simmons LW (2001) Sperm competition and its evolutionary consequences in the insects. Princeton University Press, Princeton, NJ

Simmons LW (2014) Sexual selection and genital evolution. Austral Entom 53:1–17

Sirot L, Wolfner M (2015) Who's zooming who? Seminal fluids and cryptic female choice in diptera. In: Peretti AV, Aisenberg A (eds) Cryptic female choice in arthropods. Springer, Cham

Smith R (1984) Sperm competition and the evolution of animal mating systems. Academic Press, New York

Sota T, Kubota K (1998) Genital lock-and-key as a selective agent against hybridization. Evolution 52:507–1513

Swanson WJ, Vacquier VD (2002) The rapid evolution of reproductive proteins. Nat Rev Genet 3:137–144

Swanson WJ, Clark AG, Waldrip-Dail HM, Wolfner MF, Aquando CF (2001) Evolutionary EST analysis identifies rapidly evolving male reproductive products in *Drosophila*. Proc Nat Acad Sci USA 98:7375–7379

Swanson WJ, Nielsen R, Yang Q (2003) Pervasive adaptive evolution in mammalian fertilization proteins. Mol Biol Evol 20:18–20

Tallamy DW, Powell BE, McClafferty (2002) Male traits under cryptic female choice in the spotted cucumber beetle (Coleoptera: Chrysomelidae). Behav Ecol 13:511–518

Tallamy DW, Darlington MB, Pesek JD, Powell BE (2003) Copulatory courtship signals male genetic quality in cucumber beetles. Proc R Soc B Biol Sci 270:77–82

Thornhill R (1983) Cryptic female choice and its implications for the scorpionfly *Harpobittacus nigriceps*. Am Nat 122:765–788

West-Eberhard MJ (2014) Darwin's forgotten idea: the social essence of sexual selection. Neurosci Biobehav Rev 46:501–508

Wiley RH, Poston J (1996) Indirect mate choice: competition for mates and coevolution of the sexes. Evolution 50:1371–1381

Williams GC (1966) Adaptation and natural selection: a critique of some current evolutionary thought. Princeton University Press, Princeton

Williams GC (1997) The pony fish's glow: and other clues to plan and purpose in nature. Basic Books, New York

Yassin A, Orgogozo V (2013) Coevolution between male and female genitalia in the *Drosophila melanogaster* species subgroup. PLoS ONE 8(2):e57158

Zigler KS (2008) The evolution of sea urchin sperm binding. Int J Develop Biol 52:791–796

Zigler KS, Lesios HA (2003) Evolution of binding in the pantropical sea urchin *Tripneustes*: comparisons to binding of other genera. Mol Biol Evol 20:220–231

Zimmer C, Emlen D (2013) Evolution: making sense of life. Roberts, Greenwood Village

Zinkl GM, Zwiebel BI, Grier DG, Preuss D (1999) Pollen-stigma adhesion in *Arabidopsis*: a species-specific interaction mediated by lipophilic molecules in the pollen exine. Development 126:5431–5440

Chapter 2
Potential for CFC in Black Widows (Genus *Latrodectus*): Mechanisms and Social Context

Maydianne C.B. Andrade and Emily C. MacLeod

Abstract Understanding the causes of variation in paternity is an important goal of research in sexual selection. While much progress has been made in understanding the dynamics of pre-copulatory mate choice and competition, post-copulatory processes that affect paternity are much less well understood. Distinguishing cryptic female choice (CFC, biases in paternity due to female post-copulatory preferences) from sperm competition (SC, competition among ejaculates for fertilization) and particularly from sexual antagonistic coevolution (SAC, intersexual conflict over mating frequency or fertilization) is at best challenging, and at worst, may be impossible. One way forward is to identify taxa in which comparative approaches, field studies, and detailed mechanistic study are all tractable, as these may allow the design of critical tests of distinguishing predictions of these intertwined hypotheses at different levels of analysis. Here, we focus on the widow spiders, the approximately 30 species in the genus *Latrodectus*, as a promising group for future work on CFC. *Latrodectus* spiders are distributed worldwide and show a fascinating range of mating systems set in variable ecological and social contexts. We outline how some of the aspects of *Latrodectus* natural history and biology may favor CFC rather than (or in addition to) pre-copulatory choice. Moreover, inter- and intraspecific variation in the benefits of CFC may provide the opportunity for comparative and intraspecific studies. We then identify likely mechanisms of CFC in this genus and end by providing suggestive evidence for CFC in a few of the species that have been studied in detail. Although there is no evidence for sperm selection in the *Latrodectus* to date, female behavior can lead to biases in paternity via changes in copulation frequency, duration, and remating as a function of male courtship effort. In addition, although first-male sperm precedence mediated by the deposit of sperm plugs is common in this genus, females may be able to manipulate the success of this internal plug and

M.C.B. Andrade (✉) · E.C. MacLeod
Department of Biological Sciences and Department of Ecology & Evolutionary Biology, University of Toronto Scarborough, Toronto, ON, Canada
e-mail: mandrade@utsc.utoronto.ca

A.V. Peretti and A. Aisenberg (eds.), *Cryptic Female Choice in Arthropods*,
DOI 10.1007/978-3-319-17894-3_2

retain the ability to cede paternity to later mates. Since relatively few *Latrodectus* species have been studied in detail, this discussion is necessarily suggestive, rather than conclusive. Nevertheless, the available data suggest increased focus on this genus may prove productive for examining causes and consequences of CFC.

2.1 Introduction

Cryptic female choice (CFC) may be defined as the occurrence of paternity biases among males that arise as a result of behavioral, physiological, or morphological traits of females that favor certain mates over others (Eberhard 1996). The question of how to identify CFC and distinguish it from other post-copulatory processes that may cause biased paternity has led to considerable debate and disagreement (see Chap. 1 for an overview; Eberhard 1993; Arnqvist and Rowe 1995, 2005; Birkhead 1998, 2000; Eberhard 2000; Kokko et al. 2003, 2006). For example, direct competition between the ejaculates of males within females (sperm competition, SC, Parker 1970; Birkhead 2000; Simmons 2001) and CFC could both shape sperm use patterns in similar ways since sperm competitiveness may be linked to other aspects of male phenotype (Birkhead 2000; Evans et al. 2013). Similarly, when there is sexual conflict over mating frequency and fertilization, sexually antagonistic coevolution (SAC) may lead females to resist male mating attempts and male adaptations for controlling fertilization (Arnqvist and Rowe 2005). Since males with certain phenotypes may be more likely to overcome this resistance, biases in fertilization linked to male phenotypes may result, and the outcomes may resemble those expected under CFC (Birkhead 1998, 2000). Although these processes may be intertwined in reality, assessing how mating outcomes affect the fitness of females and their offspring in nature may be informative since only CFC explicitly predicts net genetic benefits to offspring of choosy females (Chap. 1, Kokko et al. 2003, 2006). Similarly, comparative analyses of morphology and behavior may allow identification of intersexual evolutionary arms races (SAC, e.g., Arnqvist and Rowe 2002), the evolution of sensory adaptations of females that enhance assessment of or stimulation by males (CFC, Eberhard 1994, 2000), or the evolution of male traits that increase competitive fertilization success (SC, Simmons 2001). A review of these debates and possible resolutions is beyond the scope of this chapter. While acknowledging the challenges inherent in identifying CFC, the primary goal of this chapter is to discuss existing evidence for the importance of CFC in one taxon: the widow spiders (genus *Latrodectus*), and to argue that several features of this group may make it a good subject for tests of theory regarding the evolution and effects of post-copulatory processes affecting paternity. There are currently few empirical studies that directly examine paternity biases in *Latrodectus*, so this review does not include distinguishing tests, but is rather a review of aspects of the available literature that suggests the potential importance of CFC. We start by suggesting three general features of taxa in which studies of CFC are feasible and likely to be productive for testing distinguishing predictions of CFC, SAC, and SC.

First, it is desirable to identify species in which paternity biases arising from SC can be distinguished from biases induced by female-directed processes such as

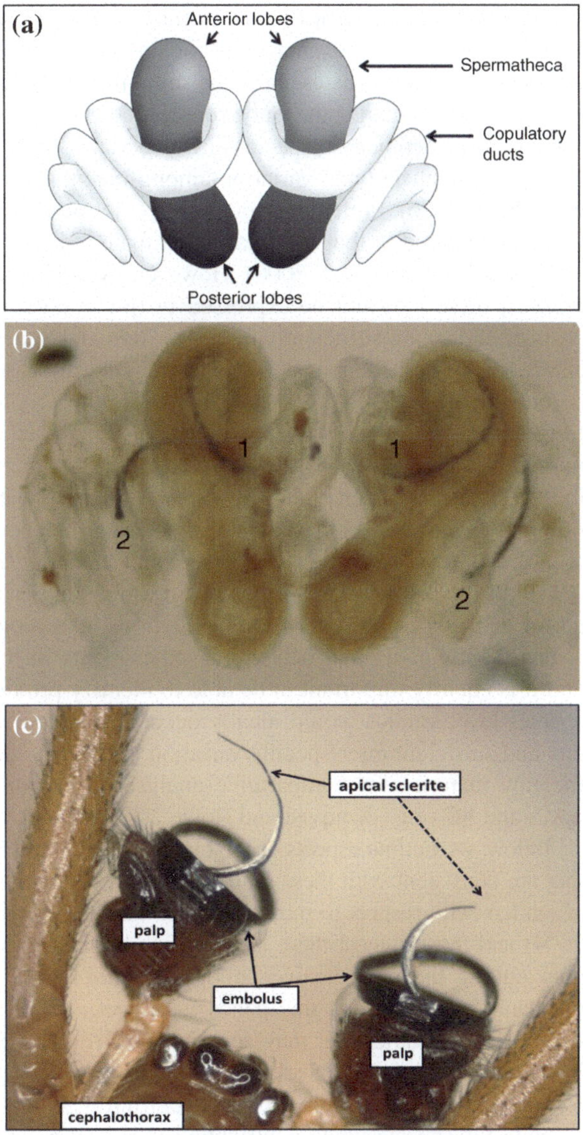

Fig. 2.1 Reproductive organs of *Latrodectus hesperus* female (**a, b**) and male (**c**), which are representative of genital morphology in this genus. A line drawing of the ventral view of the internal genitalia shows the anterior and posterior lobes of the two spermathecae and the coiled copulatory ducts (**a**). The dissected, cleared reproductive organs of a field-caught female (**b**) contains four sclerites (bases labeled with numbers). Two sclerites are placed at the entrance to the spermathecae and are likely to be successful sperm plugs (*1*), and two sclerites are deposited within the copulatory ducts and will not function as plugs (*2*). Sclerites are the broken apical portions of the embolus of the male's pedipalp (*palp*, **c**), which is inserted in the copulatory ducts. *Arrows* indicate the base of the apical sclerite (which is missing on the *right palp*, *dotted arrow*). The membranous *pars pendula* is visible in the terminal coil of each embolus. Line drawing (**a**) created by Simon Ford

variation in sperm storage or differential sperm use at fertilization. In many spiders, sperm are transferred in a non-motile (encapsulated) state, (Michalik and Ramírez 2014), ejaculates from different copulations are stored in separate storage organs (spermathecae, e.g., Fig. 2.1, Levi 1959; Bhatnagar and Rempel 1962; Snow and Andrade 2004; Berendonck and Greven 2005; Neumann and Schneider 2011), and direct SC is thus minimized, making spiders attractive subjects for studying (other) post-copulatory processes (Eberhard 2004). *Latrodectus* have independent spermathecae, each of which is filled in independent copulations (see Sect. 2.2.3.1), so the influence of SC on sperm use patterns is likely to be low (Berendonck and Greven 2005).

Second, one can identify species (or contexts) in which post-copulatory choice is likely to be more advantageous for females than pre-copulatory (mate) choice. Although there are debates about the strength of selection on female choice for indirect (genetic) benefits (the only source of benefits for CFC), females commonly show mate preferences in species in which males do not provide resources, but there is significant phenotypic variation among potential sires (Andersson 1994; Kokko et al. 2003, 2006). Preferences in these species arise if the fitness of choosy females is high compared to that of females that mate indiscriminately (Jennions and Petrie 1997; Kokko et al. 2003, 2006). However, even if females show strong mate preferences, these may not be expressed in (pre-copulatory) mate choice in nature if costs of choice are sufficiently high (De Jong and Sabelis 1991; Jennions and Petrie 1997; Kokko and Mappes 2005). This may be the case if, for example, rejecting a potential mate leads to costly delays in offspring production or a high risk of infertility for virgin females (De Jong and Sabelis 1991), such as when the encounter rate with males is low (Kokko et al. 2003, 2005; Heubel et al. 2008). In cases like these, females may mate relatively indiscriminately to ensure fertilization, but selection may favor females that can adjust paternity independent of male mating success (e.g., De Jong and Sabelis 1991; Kokko and Mappes 2005; Bleu et al. 2012).

Third, it is useful to examine taxa in which studies are tractable at different levels of analysis to allow tests of predictions at ecological and evolutionary timescales. Mechanistic and behavioral experiments in the laboratory are invaluable, and it is equally important to be able to identify proxies for fitness in nature. Such species-level studies would ideally occur in a taxon with a well-resolved phylogeny and sufficient inter-specific variation in ecology and social behavior to provide the raw material for comparative analyses of evolutionary predictions of CFC, SAC, and SC (e.g., Arnqvist and Rowe 2002).

Below, we outline aspects of the biology and natural history of *Latrodectus* spiders that are consistent with these latter two features (since the advantages of these types of spiders with respect to minimizing SC have been established elsewhere, Eberhard 2004) and thus suggest these spiders may be good models for studies of CFC. We then outline the reproductive biology of *Latrodectus* spiders to identify potential mechanisms of CFC in this genus (see Sect. 2.2.3). Finally, we summarize data from existing studies that may suggest females are employing CFC in a few of the well-studied species (see Sect. 2.3). One goal of this chapter was to highlight avenues for research that is likely to produce fruitful studies of sexual selection within *Latrodectus* species while contributing to a larger comparative data set. Currently, empirical investigations of mating behavior, ecology, and reproductive success are rare in *Latrodectus*, with experiments reported in just 7 of the 30 species in the genus (*L. geometricus*, L. *hasselti*, *L. hesperus*, *L. mactans*, *L. pallidus*, *L. revivensis*, and *L. tredecimguttatus*). These studies include investigations of intersexual and intra-sexual interactions as well as their outcomes (correlates of mating success, sperm use patterns, sperm plugs, courtship behaviors, intersexual aggression, and cannibalism), studies of female distribution and success in the field, as well as analyses of how changes in ecological or social context that affect female fitness are linked to mating outcomes. It is our hope that more such work on additional species will eventually allow leveraging of the variation found in this genus to test distinguishing predictions of CFC compared to other post-copulatory mechanisms (particularly SAC).

2.2 Natural History and Biology of *Latrodectus* Spiders in the Context of CFC

Here, we review features of *Latrodectus* natural history and biology that appear to be common to most species that have been studied to date, and highlight areas of known variability as well features that may be relevant to understanding post-copulatory processes. What follows is a piecing together of likely patterns from studies or descriptions from different species, with leading superscripts on citations indicating the *Latrodectus* species from which each description is derived (Table 2.1).

2.2.1 Comparative Context and Variation

Spiders in the genus *Latrodectus* (informally, the 'widow' spiders, family Theridiidae) comprise approximately 30 recognized species with worldwide distribution (Levi 1959; Garb et al. 2004), living under a wide range of ecological conditions (Lubin et al. 1993; Carrel 2001; Salomon et al. 2010; Trubl et al. 2012; Simo et al. 2013). Widows are well known in popular culture because of their neurotoxic venom (Jelinek 1997; Ushkaryov et al. 2004), the tendency of several species to thrive in anthropogenically disturbed habitats (Trubl et al. 2012; Vetter et al. 2012), and

Table 2.1 *Latrodectus* species for which information is reported in the chapter with their associated in-text reference numbers (noted as leading superscripts on the relevant citations[a])

Species	Superscript reference number
L. mirabilis	1
L. antheratus	2
L. corallinus	3
L. diaguita	4
L. mactans	5
L. hesperus	6
L. variolus	7
L. revivensis	8
L. bishopi	9
L. hasselti	10
L. curacaviensis	11
L. pallidus	12
L. tredecimguttatus	13
L. geometricus	14
L. indistinctus	15

[a]Species notes are not provided for review papers comparing more than 4 species

because of their lurid reputation for females that kill and consume courting males (D'Amour et al. 1936). In reality, there is a wide range of mating systems in this genus (e.g., Breene and Sweet 1985; Forster 1992; Andrade 1996; Knoflach and van Harten 2002; Segev et al. 2003; Segoli et al. 2006), ranging from polygyny and polyandry to the most extreme examples of male monogyny enforced by polyandrous female cannibals (observed in two species to date: Forster 1992; Segoli et al. 2008a). For example, in the Australian *L. hasselti*, males are monogynous and facilitate cannibalism by females (Forster 1992; Andrade 1996), but neither of these behaviors is found in the sister taxon *L. katipo* (Kavale 1986; Forster 1992). Similarly, there is variation in the frequency of polyandry, with females of some species rarely remating (Herms et al. 1935) but others mating with multiple males (Andrade 1996; Knoflach and van Harten 2002; Segoli et al. 2006; MacLeod 2013). In the absence of polyandry, CFC and SC do not exist, but SAC may be intense (Arnqvist and Rowe 2005). Thus, this type of variation may provide a strong basis for using phylogenetically independent contrasts to infer causes and consequences of CFC, SC, and SAC (e.g., Harvey and Pagel 1991 and see suggestions below).

Recent analyses suggest *Latrodectus* spiders may have undergone relatively recent, rapid speciation. The latter is consistent with the broadly similar morphological (e.g., Levi 1959, 1983), behavioral (Kaston 1970; Ross and Smith 1979a; Breene and Sweet 1985; Forster 1992; Neumann and Schneider 2011), and venom (Garb and Hayashi 2013) characteristics across most species in the genus. While genital morphology is unusually (for spiders) similar across species (Levi 1959, 1966; Knoflach and van Harten 2002), there is subtle, but potentially important interspecific variation in the relative diameter of male compared to female genital structures (Bhatnagar and Rempel 1962; Knoflach and van Harten 2002). Since these variations are likely to affect the success of cuticular sperm plugs often left inside females by mating males in all *Latrodectus* species (see Fig. 2.1 and Sect. 2.2.3.1 below, Levi 1959, 1961; Bhatnagar and Rempel 1962; Knoflach and van Harten 2002; Berendonck and Greven 2005), these could reflect a history of CFC, SC, and/or SAC (e.g., Eberhard 1985, 1996; Arnqvist and Rowe 2005). This genus could thus allow comparative study of hypotheses regarding links between the evolution of post-copulatory sexual selection, genitalic structures, mating systems, and ecology (Eberhard 1985, 1996; Simmons 2001; Arnqvist and Rowe 2005). Although the phylogenetic topology of the genus is not yet completely clear (Garb et al. 2004), recent work (Garb and Hayashi 2013) including sequencing of the genome of *L. hesperus* (BCM-HGSC i5 k Pilot Project, Accession: PRJNA168123 ID: 168123) suggests a more well-resolved phylogenetic analysis based on robust molecular data will soon be available.

2.2.2 Ecology and Phenology: CFC May Be More Likely that Mate Choice

Latrodectus are multivoltine with an annual (males, some females) or biennial (some females) life cycles and overlapping generations ([5,6,7]Kaston 1970; [1,2,3,4]

Estévez and González 1991; [6]Salomon et al. 2010). Males develop more rapidly (in fewer instars) and have much shorter life spans than females (typically, weeks compared to a year or more, [14,15]Smithers 1944; [5,6,7]Kaston 1970; [10]Andrade and Banta 2002). Females are sedentary, but males are nomadic and search for potential mates as adults. Most reproductive activity occurs over the course of a few months, the duration and timing of which varies with geographic region ([14,15]Smithers 1944; [5,6,7]Kaston 1970; [5,6,13,14]Levy and Amitai 1983; [8]Anava and Lubin 1993; [9]Carrel 2001; [8]Segev et al. 2003; [6]Salomon et al. 2010; [6]Salomon 2011), but which encompass 'spring through fall' in temperate environments. All developmental stages of both sexes may be present in populations throughout their active season ([12]Segoli et al. 2006; [6]Salomon 2011). Nevertheless, population structure often changes in predictable ways over time (Elias et al. 2011), such that the number of mate-searching males ([8]Anava and Lubin 1993; [8]Segev et al. 2003) and thus the potential for and cost of pre-copulatory choice would vary over time for virgin females (De Jong and Sabelis 1991; Kokko et al. 2003; Kokko and Mappes 2005; Heubel et al. 2008; Bleu et al. 2012). For example, challenging climatic conditions combined with minimal prey availability (e.g., hot dry weather, winter, [6]Salomon 2011) characterize 'low' activity periods where spider development may be retarded ([5]D'Amour et al. 1936; [10]Downes 1987), and males or juveniles may be more likely to die than adult or subadult females. In the same habitats, seasonal changes associated with increased prey abundance (i.e., spring, wet season) correspond with increased spider activity and density and thus the availability of potential mates ([6]Salomon et al. 2010; [6]Salomon 2011). Thus, there may be cyclical variation in the net fitness effects of pre-copulatory compared to post-copulatory choice for females, mediated by the risk of remaining unmated (De Jong and Sabelis 1991). Mate rejection (pre-copulatory choice) at times of year when there are relatively few males ([8]Anava and Lubin 1993; [6]Salomon et al. 2010) could lead to significant delays in mating, or the risk of not mating at all (De Jong and Sabelis 1991; Kokko and Mappes 2005; Elias et al. 2011; Bleu et al. 2012). In contrast, under the same conditions, females that permit the first-arriving male to mate but bias paternity toward better males that are encountered later will have secured fertility insurance while also 'trading up' to preferred sires for their offspring (e.g., De Jong and Sabelis 1991; Pitcher et al. 2003; Bleu et al. 2012). In addition, if there are seasonal shifts in the frequency of preferred males, this may also make CFC beneficial. For example, in at least two *Latrodectus* species, male body size, mass, or condition show significant differences between the early and late season ([8]Segev et al. 2003; S. Fry, personnel communication), and wide variation in male body size has been reported in many species ([5,6,7]Kaston 1970; [8]Segev et al. 2003; [6]Brandt and Andrade 2007). Some of this variation arises from developmental plasticity ([10]Kasumovic and Andrade 2006; [10]Stoltz et al. 2012). However, the maintenance of considerable phenotypic variation under standardized laboratory rearing (e.g., males range from 6.1 to 27.1 mg in *L hesperus*) suggests additive genetic variance underlies some of these differences, and this may represent a source of indirect benefits for choosy females ([10]Brandt and Andrade 2007). Thus, mechanisms of CFC may arise because there are benefits to manipulating paternity for females that become sexually mature at a time of year when males with less-preferred phenotypes are common.

There is evidence that the risk of remaining unmated is significant in nature for females of several species of *Latrodectus* ([12]Segoli et al. 2006). Even when males are numerous, a patchy distribution of males on females' webs is common ([10]Kasumovic and Andrade 2009). Whereas some females have more than five (and up to 8) males present simultaneously in the field ([8]Anava and Lubin 1993; [10]Andrade 1996), the median in most species appears to be 1 or 2 males per female, and observations include many females with no males on their webs ([5,7,9,11]McCrone and Levi 1964; [8]Anava and Lubin 1993; [10]Andrade 1996; [8]Segev et al. 2003; [12]Segoli et al. 2006). Even in enclosures with experimental release of males, some females do not attract any mates (e.g., [10]Kasumovic and Andrade 2009) and this risk is higher for females on poor diets ([6]MacLeod and Andrade 2014). Consistent with this, among field-collected females in two species, 12 % ([12]Segoli et al. 2006) to 17 % (Andrade and Kasumovic 2005) were unmated. Even if these data represent delayed mating of some females (rather than no mating), this also entails a significant risk in nature, where mortality from extrinsic sources is unpredictable and even relatively small mating disadvantages can affect female fitness (Kokko and Mappes 2005). Moreover, virgin females suffer from decreased longevity relative to mated females in at least one species ([10]Stoltz et al. 2010), which adds to the risk of forgoing mating. We predict that these risks will lead to the evolution of mechanisms for CFC rather than precopulatory mate choice in some *Latrodectus* species.

Since the risk of remaining unmated is likely to vary with the local or seasonal availability of males, the occurrence of CFC could vary in predictable ways within or across species. This variability provides the opportunity for interesting comparative or intraspecific tests of links between variation in male availability and CFC (Kokko and Mappes 2005) that we have suggested here. For example, in species where male density is relatively invariable across the mating season ([12]Segoli et al. 2006), or where the risk of not mating is negligible, the evolution of CFC may be less likely than in species with variation in the density of males (or of those males with preferred phenotypes; [8]Anava and Lubin 1993) and a higher risk of not mating. Moreover, within species with variable male density, we would predict females would employ CFC more commonly at times of year or in microhabitats where males are scarce (e.g., Elias et al. 2011), but may more frequently reject their first suitor when the local density of males is high. *Latrodectus* females have access to cues of mate availability since males frequently cohabit with immature females in nature ([8]Segev et al. 2003; [12]Segoli et al. 2006), multiple males may be present on an adult female's web simultaneously ([5]D'Amour et al. 1936; [8]Anava and Lubin 1993; [10]Andrade 1996), and females may detect chemical cues released by nearby males (Elias et al. 2011). In other spiders, females adjust their mating behavior as a function of their juvenile experience with males via direct social interactions or cohabitation (Hebets 2003; Johnson 2005). In *Latrodectus* however, first-male sperm precedence is common (see Sect. 2.2.3.3) and so 'trading up' would require cryptic mechanisms (CFC) for shifting paternity to favor better males that are encountered later (see Sect. 2.3).

2.2.3 Reproductive Biology and Possible Mechanisms of CFC

Here, we describe *Latrodectus* genitalia and focus on those features that may be relevant to post-copulatory processes. *Latrodectus* genitalia show relatively small differences among species, so our description of common structures is likely to apply broadly across the genus ([8,12,13]Levi 1966; Knoflach and van Harten 2002). We then describe mating behavior (Sect. 2.2.3.2), which seems to be quite similar across the genus in terms of sequence and elements. Finally, we discuss paternity (Sect. 2.2.3.3) and mechanisms that females might employ to effect paternity biases if they mate initially with a non-preferred male (Sect. 2.3).

2.2.3.1 Genital Morphology and Possible Mechanisms of CFC

Latrodectus females have paired, independent spermathecae (sperm storage organs, Fig. 2.1), each of which is entered through a coiled copulatory duct that is accessed via one of two copulatory pores ([6]Bhatnagar and Rempel 1962). Sperm exits each spermatheca via a separate fertilization duct ([6]Bhatnagar and Rempel 1962; Austad 1984). Males have paired intromittent organs (pedipalps, Fig. 2.1) with a heavily sclerotized, coiled structure (the embolus) that completely penetrates one of the female's copulatory ducts during separate (ipsilateral) copulatory insertions (Knoflach and van Harten 2002). The copulatory pores are located on a ventral abdominal sclerotized genital plate (epigynum) which has numerous mechanoreceptive hairs overhanging the genital opening ([5,6,7]Kaston 1970; [8]Berendonck and Greven 2002, [8]2005). Spermathecae are 'dumbbell' shaped with a relatively narrow medial channel joining the anterior and posterior lobes (Fig. 2.1). During copulation, the embolus is usually threaded through the copulatory duct until the tip extends into the anterior spermathecal lumen ([6]Bhatnagar and Rempel 1962). SC and morphology studies suggest that fertilization success is maximized if males ejaculate in this location, whereas sperm storage is less likely if males fail to fully insert the embolus and instead ejaculate in the copulatory ducts ([5]Abalos and Baez 1963; [10]Snow and Andrade 2004).

The embolus consists of two parts: the heavily sclerotized but flexible truncus ([6]Bhatnagar and Rempel 1962) with a median channel along its length into which is fused the membranous pars pendula (or ejaculatory duct, [6]Bhatnagar and Rempel 1962), through which sperm flows. The pars pendula terminates at a rounded hump or recurved tooth which marks the base of the apical sclerite, the distal-most part of the truncus through which sperm is released. The hump or tooth at the base of the sclerite marks a point at which the embolus will often break during copulation (reviewed in: Uhl et al. 2010), leaving the apical sclerite lodged inside the female's genitalia (Fig. 1.1, [6]Bhatnagar and Rempel 1962; [8]Berendonck and Greven 2002). The sclerite could function as a sperm plug, blocking insemination (but not mating) by subsequent males, if its base is lodged in the end of the copulatory duct ([6]Bhatnagar and Rempel 1962; [10]Snow and Andrade 2005; [10]Snow et al. 2006; [13]Neumann and Schneider 2011). However, male success at plugging females is variable. Although in some species males always lose one sclerite with each copulation ([14]Segoli et al. 2008a), in others sclerite loss is variable ([12]Segoli et al. 2008b; [13]Neumann and Schneider 2011; [6]MacLeod 2013). Moreover, even when sclerites are lost, males sometimes place them in the copulatory ducts or entirely inside the spermatheca, locations where they fail to block rivals ([12]Segoli et al. 2006, 2008b; [10]Snow et al. 2006; [6]MacLeod 2013). The rate of plug placement failure apparently varies among species, with failure rates approaching 50 % of copulations in some species (Knoflach and van Harten 2002; [10]Snow and Andrade 2004; [13]Neumann and Schneider 2011; [6]MacLeod and Andrade 2014). Thus, although plug-mediated first-male sperm precedence does occur in *Latrodectus*, paternity outcomes are variable and male or female traits, or their interaction could affect this variability (see Sect. 2.2.3.3, [10]Andrade 1996; [10]Snow and Andrade 2005; [10]Snow et al. 2006; [13]Neumann and Schneider 2011).

Some of the variation in plugging success in *Latrodectus* is likely related to interspecific variation in the size and structure of the copulatory ducts at their point of entrance into the spermathecal lumen (Knoflach and van Harten 2002). In some species, the point of entry of the copulatory ducts to the spermatheca is a narrow, well-defined collar that is sufficiently small relative to the size of the sclerite to make blockage of the entrance likely ([6]Bhatnagar and Rempel 1962; [8]Berendonck and Greven 2005). In other species, the point of entry is relatively broad and much less likely to be plugged by a single sclerite (Uhl 2002; Knoflach and van Harten 2002). This is consistent with the observation that mated females of some species are typically found with only one sclerite lodged at the entrance to each spermatheca (e.g., [5,6,13,14]Levy and Amitai 1983; [8]Berendonck and Greven 2002; Knoflach and van Harten 2002; [6]MacLeod 2013) and any other sclerites present are located in the copulatory ducts ([6]Bhatnagar and Rempel 1962; [10]Snow et al. 2006; [6]MacLeod 2013). In contrast, in other species, the rate at which multiple sclerites are found in the same spermathecal opening ('failed plugs') is high; at least 50 % of field-collected *L. pallidus* females had at least one failed plug ([12]Segoli et al. 2008b). Although not systematically documented, illustrations of the genitalia of field-caught females show failed plugs in several other species which also appear to have relatively wide points of entry to the spermathecae (*L. renivulvatus*, L. *geometricus*, L. *dahli*, Knoflach and van Harten 2002). Variation in this aspect of female morphology across species is likely to have a significant effect on whether plugs can affect sperm use patterns, but this has never been tested. A comparative analysis of the rate of plug failure as well as the relative size of male and female genital structures across species may reveal past selection via sexual conflict over the control of fertilization (SAC) or via sexual selection (CFC or SC). Such an analysis would also identify those species in which plugging failure rates are relatively high and in which CFC may thus be more likely. This type of study would be facilitated by the fact that sclerites are not disturbed by fertilization and remain lodged in females even after their death (e.g., sclerites are visible in the dissected spermathecae of females preserved for decades in alcohol).

There has long been speculation that having independent sperm storage organs that are inseminated in separate copulations could make sperm selection particularly feasible in some invertebrates (Siva-Jothy and Hooper 1996; Simmons 2001). Berendonck and Greven (2005) looked in *L. revivensis* for internal structures that might permit differential sperm usage from the two spermathecae and thus suggest a mechanistic basis for CFC. At fertilization, sperm empties from the posterior lobe of the spermatheca through a common fertilization duct into the uterus externus, the site of fertilization ([6]Bhatnagar and Rempel 1962; [1,3,11]De La Serna De Esteban et al. 1987; [8]Berendonck and Greven 2005), via a process that is not well understood in spiders (Foelix 1982). In *L. revivensis*, the tube through which sperm enters the common fertilization duct is heavily sclerotized on one side and membranous on the other side and may function as a valve that controls the movement of sperm at fertilization ([8]Berendonck and Greven 2005). In its neutral state, the valve is closed, but muscle contractions could pull it open, theoretically creating negative pressure in the fertilization duct, and drawing sperm out of the spermathecae ([8]Berendonck and Greven 2005). This may be achieved by a band of muscle that connects the internal surface of the epigynum to the terminal portion of the copulatory duct and the ventral side of the common fertilization duct, and/or by a second, paired muscle that extends from the epigynal plate to the common fertilization duct ([2,3]Lauriade Cidre 1988; [8]Berendonck and Greven 2005). However, this valve would draw sperm from both spermathecae simultaneously, and in the absence of any recognizable mechanisms for differential sperm release, Berendonck and Greven (2005) argued that CFC by sperm selection is unlikely in *L. revivensis* or other species with comparable morphology.

However, we suggest there are other ways that the fertilization valve could affect paternity. For example, since the spermathecae often contain proteins and fluids even before mating ([8]Berendonck and Greven 2002 [8]2005; Useta et al. 2007; Michalik and Ramírez 2014), females that opened this valve during copulation (rather than during fertilization) might draw fluids out of the spermatheca, thus affecting pressure in the spermathecae and copulatory ducts. Such a change could alter the likelihood of effective insemination or even plug placement by the female's current mate. Since males insert in one spermatheca at a time, such a mechanism could allow females differential control over sperm storage and plugging, and thus over first-male sperm precedence (Sect. 2.2.3.3). The possibility that these contractions could be triggered by male stimulation of the epigynal hairs during courtship and mating is particularly intriguing. Since males typically move the embolus across the epigynum multiple times before insertion is achieved, repeated contact between the pedipalps and these hairs is common ([5,6,7]Kaston 1970) and may serve a stimulatory function ('copulatory courtship,' e.g., Huber 2005; Eberhard and Huber 2010). Whether the fertilization valve is activated during copulation, is linked to stimulation of the epigynal hairs, and whether paternity is affected could be studied with a combination of mating trials, sensory-hair ablation (e.g., Aisenberg et al. 2015), muscle physiology, and freeze-fixation (Huber 1993).

2.2.3.2 Mating Behavior and Possible Mechanisms of CFC

At sexual maturity, males abandon the webs on which they developed and localize females using airborne pheromones released from the female's silk ([8]Anava and Lubin 1993; [6]Kasumovic and Andrade 2004; [10]Andrade and Kasumovic 2005; [6]MacLeod and Andrade 2014). Female production of sex pheromones varies with mating status, age or developmental stage ([10]Stoltz et al. 2007; [10]Perampaladas et al. 2008; [10]Jerhot et al. 2010), and diet ([6]MacLeod and Andrade 2014). This likely explains why females attract fewer males when they are immature than when they are mated ([10]Andrade and Kasumovic 2005; Kasumovic et al. 2009b; [6]MacLeod 2013), and (in some species) when they have had food withheld ([6]Johnson et al. 2011; [6]MacLeod and Andrade 2014). Nevertheless, in nature, males also mate with previously mated females ([8]Segev et al. 2003; [12]Segoli et al. 2008b; [6]MacLeod 2013) and are often found cohabiting with subadult females ([5]D'Amour et al. 1936; [8]Anava and Lubin 1993; [10]Andrade and Kasumovic 2005). Males may encounter juveniles by chance while searching for adult females (subadults do not produce sex pheromones in some species, [8]Anava and Lubin 1993; [10]Jerhot et al. 2010), or they may respond to chemical cues produced by juveniles ([6]Ross and Smith 1979b).

Despite guidance from chemical signals, male mortality is very high during mate searching (>80 %) in some species ([10]Andrade 2003; [8]Segev et al. 2003; [12]Segoli et al. 2006), and it seems likely to be costly in others where spiders that change web location suffer approximately 40 % mortality ([8]Lubin et al. 1993). The challenges of seeking females may thus impose strong selection on males (e.g., Foellmer and Fairbairn 2005). If so, mate searching may act as a 'first filter' that narrows the types of males to which virgin females are exposed, decreasing the benefits of CFC. Laboratory studies show that larger *Latrodectus* males have a mobility advantage over smaller males ([10]Brandt and Andrade 2007; but see Moya-Larano 2002), but field studies show no clear effect of male phenotype (size, mass, or condition) on mate-searching success ([10]Andrade 2003; [8]Segev et al. 2003; size: [12]Segoli et al. 2006). This may be an issue of statistical power since few males are successful (e.g., [10]Andrade 2003; [12]Segoli et al. 2006) or it may be that the most relevant phenotypic traits for searching success were not assessed (e.g., intermediate-sized males have the highest metabolic efficiency, [10]De Luca et al. 2015). However, regardless of any possible phenotypic advantages of intermediate or large-sized males, the more rapid development of relatively small males may allow them to reach newly matured females more rapidly ([10]Kasumovic and Andrade 2006, 2009; Elias et al. 2011). Thus, it is not yet clear whether females are courted by a restricted subset of males in nature.

After reaching a female, males may cohabit on or near the female's web for a variable period, particularly if the female is not yet sexually mature or receptive ([5,7,9,11]McCrone and Levi 1964; [8]Segev et al. 2003; [12]Segoli et al. 2006). Most descriptions of mating are based on laboratory pairings of single males and females, which is common in the field (but see [10]Stoltz et al. 2008; [12]Harari et al. 2009; [10]Stoltz et al. 2009; [10]Stoltz and Andrade 2010). In these cases, males often begin courtship shortly after contacting the web (triggered by contact pheromones on the silk, [6]Ross and Smith 1979b; [10]Stoltz et al. 2007; [10]Jerhot et al. 2010; [6]Scott et al. 2012). Courtship provides ample opportunity for female assessment of males and for male stimulation of females in more than one modality (e.g., tactile, vibratory, and chemical). Courtship has a number of vibratory elements common to all species thus far described (for details see [5,6,7]Kaston 1970; [6]Ross and Smith 1979a; [10]Forster 1992). Males may also cut and bind portions of the female's web (web reduction), which may reduce pheromone release or alter the female's response to the courting male (Watson 1986; [6]Scott et al. 2012). In some species, males loosely string silk across the dorsal surface of the female's abdomen ('the bridal veil'), a behavior whose function is unclear (see speculations in [5,6,7]Kaston 1970; [6]Ross and Smith 1979a).

During courtship, females are typically quiescent, hanging dorsal-side-down from the underside of the horizontal snare of their webs with their legs flexed. Females may eventually straighten their legs and allow their body to sag away from the web, a receptive posture that facilitates the male mounting their ventral surface. Males may then begin a period of courtship that includes vibratory and tactile movements while on the female's abdomen ([10]Forster 1995). Eventually, males scrape one pedipalp against the female's epigynum, eventually loosening the coil of the embolus and attempting to insert it into one of the female's copulatory pores (Fig. 2.1). There may be multiple copulation attempts, with intervening periods of courtship on the web and abdomen before the male is successful. Courtship duration is impressively long in some species (e.g., mean of 5 h in *L. hasselti*), but on the order of 10 min or less in other species ([10]Forster 1995; Knoflach and van Harten 2002). Intersexual interactions during extended courtships could reflect selection via CFC, or could be due to SAC if male stimulation is met with female resistance. However, under SAC, we might expect consistent, overt signs of aggression rather than quiescence in *Latrodectus* since extreme female-biased size dimorphism (females at least 100× heavier than males: [5,6,7]Kaston 1970) minimizes the cost of aggression to females. Non-receptive females can (and do) easily knock males out of their webs with no risk of injury ([10]Andrade 1996; [10]Stoltz et al. 2008).

After copulating with one palp, males of many species return to the web and court again before attempting to mount the female a second time and insert their second palp ([5,6,7]Kaston 1970; [6]Ross and Smith 1979a; [10]Forster 1992; [10]Andrade 1995; [14]Segoli et al. 2008a). The duration and occurrence of the inter-copulatory dismount varies. In some species (and perhaps under some circumstances), males insert both palps in sequence without leaving the female's abdomen ([5]D'Amour et al. 1936; [5,7,9,11]McCrone and Levi 1964). The presence of an inter-copulatory interval in some species is intriguing as it is likely to be costly to males given the risk of the appearance of a competitor at any time ([10]Stoltz and Andrade 2010). From the female's perspective, however, this period provides an additional opportunity for assessment and discrimination at a point when her mate has inseminated only one of the female's two spermathecae (Snow and Andrade 2005). As we argue below (Sect. 2.3.1), this provides one opportunity for females to bias paternity.

Females receive ample sperm at a single copulation to fertilize their lifetime supply of eggs ([10]Andrade and Banta 2002; [8]Berendonck and Greven 2005) and repeatedly mating with the same male does not increase their fertility ([10]Andrade and Banta 2002; [12]Segoli et al. 2008b). This means virgin females have the option of mating once and leaving one spermatheca empty to accept future mates without suffering any fertility costs. Perhaps this explains why, once mated (even after just one copulation), females' sexual receptivity decreases ([10]Andrade 1996; Segoli et al. 2008b; [13]Neumann and Schneider 2011; [6]MacLeod 2013). Nevertheless, females clearly copulate more than once in many species, which includes repeated mating with individual males (most often, once with each palp) and multiple mating with more than one male ([10]Andrade 1996; [10]Andrade and Banta 2002). Conservative estimates of multiple mating can be derived from assessing the number and position of sclerites inside females' genitalia given that each male can deposit a maximum of two, with one within each tract ([10]Andrade 1996; [12]Segoli et al. 2008b; [13]Neumann and Schneider 2011; [6]MacLeod 2013). Studies with systematic sampling of females suggest at least 12–47 % of females are polyandrous ([12]Segoli et al. 2006; [6]MacLeod 2013), and this is consistent with descriptive studies in which a subset of females collected across species show patterns of sclerite deposition consistent with polyandry ([6]Bhatnagar and Rempel 1962; [5]Abalos and Baez 1963; Mueller 1985; Knoflach and van Harten 2002; [12]Segoli et al. 2008b). Molecular studies of paternity would be invaluable to estimate polyandry more precisely, but no such studies are available for any *Latrodectus* species (see Sect. 2.2.3.3).

2.2.3.3 Sperm Use Patterns

The few studies of paternity in *Latrodectus* use sterile male techniques in the laboratory, in which females are mated to irradiated males as well as unmanipulated males, and paternity pattern is inferred from the hatch rate of eggs (Boorman and Parker 1976; Sillen-Tullberg 1981). These studies show that, when two males inseminate the same spermatheca, first-male sperm precedence is the dominant pattern (100–80 % first-male paternity) as expected for species with sperm plugs ([10]Snow and Andrade 2005; [10]Snow et al. 2006; [13]Neumann and Schneider 2011; [6]MacLeod 2013). However, reversals to second-male precedence (0–20 % paternity of first male) are not uncommon (e.g., found in 18.8 % of cases, [10]Snow and Andrade 2005; 33.6 %, [6]MacLeod 2013), although sperm mixing (20 to 80 % paternity) is rare ([10]Snow and Andrade 2005; [6]MacLeod 2013). There is some evidence that reversals of first-male priority occur when the first male is unsuccessful at leaving his sclerite in a plug position ([10]Snow et al. 2006), but in other studies, there is no straightforward relationship between sclerite loss and paternity ([13]Neumann and Schneider 2011; [6]MacLeod 2013). This suggests other mechanisms may also determine paternity. One possible explanation for first-male precedence is the relative position of the ejaculates inside the female. The female's spermatheca contains proteins that may form a matrix that holds the first ejaculate in each spermatheca closer to the fertilization duct ([8]Berendonck and Greven

2005). Some of the variation in SC results may arise because the sterile male technique leads to erroneous inferences when baseline levels of male sterility are high and/or sample sizes are small (Rugman-Jones and Eady 2001; Garcia-Gonzalez 2004). This makes even clearer the necessity of paternity studies using molecular methods to corroborate or challenge these apparent patterns.

Although first-male precedence is common when competing males inseminate the same spermathecae, when matings proceed naturally, sperm use patterns may vary considerably (e.g., ranging from 0 to 100 % paternity of the first male to mate, [10]Andrade 1996). In *L. hasselti*, paternity was positively correlated with copulation duration (Andrade 1996), even though sperm transfer required only a fraction of the time spent in longer copulations (Snow and Andrade 2004). Such a result suggests paternity could be mediated by the activity of ejaculatory proteins transferred during longer copulations (Michalik and Ramírez 2014), as these can affect fertilization in other taxa (SC, e.g., Chapman et al. 1995). Another possibility is that females permit longer copulations with preferred males (e.g., [13]Neumann and Schneider 2011) and these males are then more likely to place plugs successfully, or females later bias paternity in their favor ([10]Snow and Andrade 2004). However, the wide range of variation in paternity may also be attributable to the range of different mating outcomes that are possible when two males attempt to mate with the same female. This includes variation in copulation frequency of each male (inseminating both or only one spermatheca), which male inseminates each spermatheca first, and whether first-mating males in each spermatheca deposit an effective plug. Attempts to disentangle these sources of variation are often hampered by relatively small sample sizes. However, experimental matings with manipulated insemination patterns (males forced to inseminate the same or opposite organs, [10]Snow and Andrade 2005) corroborate inferences from morphological studies ([8]Berendonck and Greven 2005) that suggest sperm/ejaculates are released from both spermathecae equally and thus enter the uterus internus without bias at fertilization (also see Sect. 2.2.3.1). Thus, if each male inseminates a different spermatheca, the expectation is a fair raffle paternity pattern (Parker 1990) with an average near 50 % (regardless of plug deposition). Experiments of this type do show paternity is centered on 50 %, but variation is still common ([10]Snow and Andrade 2005). Deviations from shared paternity under a fair raffle likely arise from differences in the number of sperm inseminated by each male (Parker 1990) since sperm loads are variable among individuals and within individuals over time ([10]Snow and Andrade 2004; [6,10]Modanu et al. 2013). Examining links between patterns of deposition and positioning of sclerites within females (Sect. 2.2.3.1) and paternity also reveals sources of variation. In *L. tredecimguttatus* and *L. hesperus*, relatively large males more frequently place successful plugs ([13]Neumann and Schneider 2011; [6]MacLeod 2013), perhaps because they have longer emboli. Alternatively, females may prefer larger males and affect plug placement success using either a physiological mechanism like the one suggested above (Sect. 2.2.3.1), or behavioral mechanisms. For example, if female quiescence during mating is necessary for successful threading of the embolus into the spermatheca (Sect. 2.2.3.2), then agitation may interfere with plug placement ([10]Snow et al. 2006; but see [13]Neumann and Schneider 2011).

2.3 CFC in *Latrodectus*: Potential Mechanisms and Evidence

Despite the gaps in our knowledge, the reproductive biology of *Latrodectus* spiders suggests at least three possible mechanisms of CFC. These may be employed most commonly if virgin females mate indiscriminately with the first male they encounter (for reasons outlined in Sect. 2.2.2) and then utilize one or more of these mechanisms to allow later-mating males to have higher paternity than would be expected given the common occurrence of first-male sperm precedence (Sect. 2.2.3.3).

First, females may adjust the frequency with which they copulate with particular males. When insemination of the two spermathecae occurs in two discrete copulations separated by an intervening period of additional courtship, females mating with non-preferred males can terminate mating after one, rather than two copulations with relatively little cost or risk. With one spermatheca filled, females will have fertility insurance ([10]Andrade and Banta 2002). If the first mate places a successful plug, a later mate can achieve at least 50 % paternity if they inseminate the opposite organ ([10]Snow and Andrade 2005; [10]Snow et al. 2006). Moreover, if the first male fails to place a plug (Sect. 2.2.3.1), and the female's second mate inseminates both spermathecae, paternity will then favor the second male overall (e.g., see calculations in [10]Kasumovic and Andrade 2009). Second, females may adjust the duration of copulation with a particular male and affect paternity ([10]Snow and Andrade 2005). While the mechanism by which this functions is not yet clear, paternity is correlated with copulation duration in at least one species in the genus ([10]Andrade 1996), and a positive correlation between copulation and paternity is common in spiders and other invertebrates (Simmons 2001; Schneider and Andrade 2011). Third, females may be able to affect the positioning of the embolus during ejaculation and sclerite loss. If the first mate is not of a preferred phenotype, females could block plug placement to ensure later males have higher paternity.

2.3.1 CFC and Control of Copulation Frequency

If females restrict first mates to one copulation and subsequently remate only with preferred males, they not only avoid the risk of remaining unmated and retain the option of 'trading up' to more preferred males (Bleu et al. 2012), but they also employ this mechanism after acquiring significant information derived from male courtship behavior, copulatory behavior, and potentially sperm or ejaculate characteristics. This is particularly true because the ability of males to sustain prolonged, vigorous vibratory courtship prior to the first mating attempt may provide significant information about male traits such as metabolic competence, body condition, and juvenile provisioning (which reflects both development history and diet [10]Kasumovic et al. 2009a; [10]Stoltz et al. 2012; Kasumovic and Seebacher 2013;

[10]De Luca et al. 2015). Thus, CFC in *Latrodectus* may occur when female-mediated variation in male copulation frequency (equivalent to repeated mating success) is linked to some aspect of male phenotype, and those same traits are linked to the likelihood or frequency of the females copulating with a subsequent suitor. Below, we consider two ways this may manifest: (1) decreased female receptivity to second copulation attempts from some, but not all males; (2) female cannibalism of some males after a single mating, preventing subsequent copulations.

2.3.1.1 Remating and Receptivity to Second Males

While this has not been studied specifically from the perspective of CFC in *Latrodectus*, it is clear that the 2 copulations with each male is not necessarily the norm ([15]Smithers 1944; [5,6,7]Kaston 1970; [8]Anava and Lubin 1993; [14]Segoli et al. 2008a; [12]Harari et al. 2009; [10]Stoltz et al. 2009; [6]MacLeod 2013). For example Kaston (1970) reports that females become aggressive toward males after one copulation in 2/3 of matings, causing males to retreat and fail to accomplish a second copulation (based on observations in *L. variolus* and *L. hesperus*). Experimental studies of *L. hesperus* and *L. hasselti* suggest that in both species, courtship duration predicts the copulation frequency permitted by females. In recent work, *L. hesperus* females were more likely to copulate twice with males that courted for longer (compared to shorter) durations ([6]MacLeod and Andrade in preparation). As the first step in a SC study (detailed methods in [10]Snow and Andrade 2005; [6]MacLeod 2013), naive virgin males and females were paired in the laboratory, and mating sequences recorded. Among the 182 males that mated in these trials, 19 % ($n = 38$) copulated only once despite their repeated attempts at second copulations. There was no relationship between copulation frequency and male mass or duration of the first copulation (all $p > 0.05$), but the males with which females copulated only once spent significantly less time courting prior to their first copulation attempt (7.7 ± 0.72 min, $n = 34$) than did males with which females copulated twice (15.0 ± 0.97 min, $n = 146$; generalized linear model with logit link function: Wald $\chi^2 = 19.177$, $p < 0.001$). Moreover, when females from this study were paired with a second male on the next day, females that allowed only one copulation with their first mate were more likely to copulate with a second male (62 % remated, $n = 34$) than were females that allowed two copulations with their first mate (26 % remated, $n = 146$; Fisher's exact test, $p = 0.0002$). This is the opposite of what would be expected if some females simply had low receptivity to male mating attempts in general (e.g., as may be expected from SAC).

2.3.1.2 Premature Cannibalism and Receptivity to Second Males

In *L. hasselti*, a similar effect of courtship duration on female behavior was observed, but the mechanism was more extreme; males were wrapped in silk and killed by females during or after the first copulation ('premature cannibalism,'

[10]Snow and Andrade 2005; [10]Stoltz et al. 2008, 2009; [10]Stoltz and Andrade 2010). This could occur in any *Latrodectus* species where cannibalism occurs after the commencement of copulation ([14,15]Smithers 1944). *L. hasselti* females were more likely to mate twice with her first mate if he courted for longer, or was relatively large ([10]Snow and Andrade 2005; [10]Stoltz et al. 2008, [10]2009; [10]Stoltz and Andrade 2010). In these experiments, females were presented with two males simultaneously, so an alternative mate was available for evaluation while females made decisions about premature cannibalism of their first mate. However, as was seen in *L. hesperus*, cryptic preferences in *L. hasselti* also operate in the absence of direct comparisons or competition. In pairings of individual males with naive virgin females, males that were prematurely cannibalized were those with significantly shorter courtships ([10]Stoltz et al. 2008) or smaller body size ([10]Snow and Andrade 2005) than males that survived their first mating. Thus, CFC may be triggered by some absolute rather than relative criterion with respect to male size and courtship duration. As was the case with *L. hesperus*, remating behavior of *L. hasselti* females was also affected by interactions with their first mate. In *L. hasselti*, this manifested as a strong inverse relationship between the courtship duration of the first male and the female's copulation frequency with a second male ([10]Stoltz et al. 2008). Thus, smaller males who invest less in courtship are more likely to be cannibalized after just one copulation, and their mates are more likely to copulate multiple times with a rival.

In *L. hasselti*, decreased receptivity to new males has also been linked to female cannibalism of first males that succeed in mating twice. In a laboratory study, females that consumed their first mate during and after their second copulation were less likely to copulate with a subsequent male compared to females that did not cannibalize their first mate ([10]Andrade 1996). This result has not been replicated, however, and it is unclear whether such an effect might arise as a female decision or via chemical manipulation by males (e.g., Andersson et al. 2004; Aisenberg and Costa 2005), particularly since cannibalistic matings are longer and thus may include the transfer of more ejaculatory substances. Chemical manipulation is not likely to explain premature cannibalism, however, as males are often wrapped in silk and disabled, but not consumed by females when attacked during the inter-copulatory interval. Additional study of these aspects of mating and reproduction in this species and in *L. geometricus* (the only other *Latrodectus* in which this type of cannibalism has been reported: [14]Segoli et al. 2008a) would be valuable.

2.3.2 CFC and Control of Copulation Duration

There have been no direct demonstrations of female control over copulation duration in *Latrodectus*, and it seems likely that copulation duration is affected by both male and female behaviors. Nevertheless, observations suggest female behavior may be primarily responsible for limiting the duration of relatively brief copulations ([13]Neumann and Schneider 2011). While females are generally quiescent following embolus insertion in successful matings, females may also become agitated during copulation and strike at the male, move their legs across the male's body ([5,6,7]Kaston 1970; [13]Neumann and Schneider 2011; MCB pers. obs.; ECM pers. obs.), or attempt to bite the male's legs (*L. mirabilis*, L. Baruffaldi, pers. comm.). In copulations involving agitated females, males typically pull the embolus free and retreat shortly after female activity commences ([5,6,7]Kaston 1970; [14]Segoli et al. 2008a). In *L. tredecimguttatus*, the occurrence of female attacks was linked to significantly shorter copulations (14 vs. 23 min, [13]Neumann and Schneider 2011). In *L. hasselti*, there is also evidence for female control over copulation duration; males are able to survive partial cannibalism ([10]Andrade et al. 2005) during their first copulation, continue to copulate as they are consumed, and in fact copulate for longer as they are being cannibalized (11 vs. 25 min: [10]Andrade 1996; 13 vs. 19 min: [10]Snow and Andrade 2004; 13 vs. 17 min: [10]Snow and Andrade 2005). Copulation typically ends when the female wraps the male in silk. Thus, in this case, it appears that females permit longer copulations when they choose to cannibalize a male.

The mechanism underlying links between paternity benefits and longer copulation durations may be more complex than a direct relationship with sperm transfer. In *L. hasselti*, complete sperm transfer occurs well before the termination of most copulations ([10]Snow and Andrade 2004), an effect also shown in another spider where copulation duration is positively related to paternity (*A. bruennichi*: Schneider et al. 2006). Thus, increased paternity shares are not simply due to having more sperm in the 'raffle' (Parker 1990) after longer copulations. This leaves several possible explanations for the link to paternity. First, the additional time may allow males to transfer more ejaculatory fluids or proteins (e.g., Vocking et al. 2013), and these may mediate fertilization success (e.g., via SC or SAC: Simmons and Siva-Jothy 1998). Second, females may preferentially store more sperm (Useta et al. 2007) or preferentially use sperm from males that copulate for longer (CFC, e.g., Argiopidae: Welke and Schneider 2009) and perhaps are better at stimulating the female through copulatory courtship (Eberhard 1994). Third, males that copulate for longer may be higher 'quality' spiders with a range of traits that reflect superior resource acquisition and allocation, including more sperm ([10]Snow and Andrade 2005). The mechanisms underlying the relationship between copulation duration, sperm transfer, and paternity is thus ambiguous and would benefit from additional study.

Even if rapid sperm transfer is the rule for *Latrodectus* males (which has not yet been established), there may be species in which there is sufficient female-mediated limitations on sperm transfer to make this an important mechanism affecting paternity. For example, Kaston (1970) reported that copulation durations for *L. hesperus*, L. *mactans*, and *L. variolus* range from 1 to 32 min, with a median between 4 and 8 min. In Neumann and Schneider's (2011) study of *L. tredecimguttatus*, some copulations were also extremely brief (<1 min). Even in *L. hasselti*, a copulation of 2 min predicts transfer of only 67 % of the male's sperm load (calculated from equation provided in Fig. 2 legend in [10]Snow and Andrade 2004). Given the wide variation in the number of sperm carried by males (<500 to >120,000 per palp in *L. hasselti* and *L. hesperus*: [10]Snow and Andrade 2004; [6,10]Modanu et al. 2013), this could lead to a substantial disadvantage in paternity for some males. Studies that focus on the details of these potential mechanisms could provide significant insight into whether female truncation of some copulations might be a mechanism of CFC (see Herberstein et al. 2011). Such studies would elucidate how variation in sperm storage and paternity is related to male traits in natural copulations, and when copulation duration is manipulated to mimic the range of natural variation. This is particularly true if such studies take advantage of the 'double-barreled' nature of the genital tracts and apparent correlation between the number of sperm in the male's two palps ([10]Snow and Andrade 2004; [6,10]Modanu et al. 2013) to assess and manipulate factors affecting sperm transfer, storage, and copulation duration (see [10]Snow and Andrade 2004 for possible methods).

2.3.3 CFC and the Mechanics of Ejaculation and Plug Placement

In order to ejaculate into female spermatheca and place apical sclerites at the spermathecal lumen, males must fully insert their coiled embolus (2–4 coils depending on the species: Levi 1959) into the female's insemination ducts ([5]Abalos and Baez 1963; [8]Berendonck and Greven 2002, [8]2005; [10]Snow et al. 2006). This apparent structural challenge has led to the hypothesis that brief copulations, or other female-based disruptions to the process, might significantly decrease male insemination or plugging success ([8]Berendonck and Greven 2002; [13]Neumann and Schneider 2011). This idea has not been well studied, but insights can be gained from two sources. First, Neumann and Schneider (2011) found a significant negative relationship between copulation duration and successful sclerite placement in *L. tredecimguttatus*, where copulation duration was shortened by female attacks. Interpreting the effect of this result on paternity is complicated because SC data from this study were equivocal, and so it is not yet clear whether correctly placed sclerites function as plugs in *L. tredecimguttatus* ([13]Neumann and Schneider 2011). Second, in an experimental study, when Snow and Andrade (2004) artificially terminated copulations of *L. hasselti* after 5 min, the female's spermatheca already contained about 90 % of the male's total sperm load. Males apparently had no difficulty reaching into the spermatheca in that time despite having three coils in their embolus (the maximum recorded in the genus is four in *L. geometricus*, the minimum is one in L. dahli: [5,6,13,14]Levy and Amitai 1983; Knoflach and van Harten 2002). It would be valuable to have additional research that combines observational study with manipulated copulation duration with SC studies and post hoc dissection of genitalia. A focus on the location of sperm masses and sclerites would also be useful. Comparative study could capitalize on such studies since males vary across species in the number of coils (complexity) of their embolus and thus presumably in the mechanical challenge of mating.

We speculate above that females may have internal mechanisms that affect male success at placing plugs, or perhaps the location of ejaculation

(Sect. 2.2.3.1). We close by describing one study that, while not examining polyandry, suggests that virgin females are able to manipulate plug placement by males and may do so as a function of the perceived availability of potential mates (see Sect. 2.2.2, [10]Biaggio 2007). In a laboratory study of the effects of cohabitation on female behavior, subadult females were held alone or with another spider (a male or an early-instar juvenile) for the duration of their penultimate instar to mimic cohabitation ([10]Biaggio 2007). Cohabitation cages allowed vibrational and chemical cues to be transmitted between cohabitants, but no direct contact. After their adult molt, focal females were placed in a mating trial with a naïve adult male. Regardless of cohabitation history, females mated readily across all treatments and there was no difference in the number of copulations achieved by males (General Linear Model, $F_{2,55} = 1.371$; $p = 0.262$). However, the number of sclerites in the 'plug' position was significantly lower in females that had cohabited with males compared to the other two treatments ($F_{2,55} = 4.02$; $p = 0.023$), although there was no relationship between plug position and any of the measured traits of males or intersexual interactions ($p > 0.5$ for male size, male condition, courtship duration, and copulation duration). This study suggests that females may respond to cues indicating the availability of males (cohabitation) and adjust the likelihood of monopolization of paternity by their first mate (see predictions in Sect. 2.2.2).

2.4 Conclusions

In this chapter, we sought to highlight the potential of spiders in the genus *Latrodectus* for the study of CFC. Our current understanding of the biology and evolutionary ecology of these spiders as it relates to paternity is rudimentary. Nevertheless, the available information suggests future studies in this genus may yield interesting insights into the contexts in which CFC evolves, and how female-mediated effects on sperm use compare to those arising from SC or SAC. We argue that *Latrodectus* females may benefit from choice, but employ cryptic methods to bias paternity toward favored males, rather than rejecting potential males outright because of a significant risk of remaining unmated in nature. The few studies currently available suggest larger, more vigorous males are favored, but even less-preferred males often mate successfully. We predict that cryptic mechanisms of adjusting paternity may be more likely, or be employed more frequently, in species where spatiotemporal variation in the availability of males generates groups of females likely to encounter few males, or in contexts where the first male encountered is frequently of a non-preferred phenotype. We suggest cryptic choice may operate via female physiological responses to mating males that adjust the likelihood of sperm storage or effective sperm plug placement by males. This predicts evolutionary links between the risk of remaining unmated and the frequency with which 'successful' versus 'failed' plugs are found in field-caught females. More commonly, however, paternity biases may be introduced by female-determined variation in the number of copulations ceded to first compared to subsequent mates and variation in the duration of those copulations. Although these ideas are intriguing and *Latrodectus* could be an important model for understanding CFC, our ability to make general inferences is hampered since only 7 species have been the subject of formal study of reproductive behaviors or ecology. This chapter suggests some targeted behavioral, physiological, and comparative studies that, if conducted on a range of species, may leverage the fascinating variation in this group to enrich our understanding of the evolution of CFC and interactive effects of CFC and other post-copulatory processes.

Acknowledgments We are grateful to the efforts, patience, and comments of A. Aisenberg, an anonymous reviewer, and Simon Ford, who created the illustration in Fig. 2.1. The Canada Research Chairs program (award to MCBA) provided financial support for ECM and MCBA during preparation of this manuscript.

References

Abalos JW, Baez EC (1963) On spermatic transmission in spiders. Psyche 70:197–207

Aisenberg A, Barrantes G, Eberhard W (2015) Hairy kisses: tactile cheliceral courtship affects female mating decisions in *Leucauge mariana* (Araneae, Tetragnathidae). Behav Ecol Sociobiol 69:313–325

Aisenberg A, Costa FG (2005) Females mated without sperm transfer maintain high sexual receptivity in the wolf spider *Schizocosa malitiosa*. Ethology 111(6):545–558

Anava A, Lubin Y (1993) Presence of gender cues in the web of a widow spider, *Latrodectus revivensis*, and a description of courtship behaviour. Bull Br Arachnol Soc 9:119–122

Andersson J, Borg-Karlson AK, Wiklund C (2004) Sexual conflict and anti-aphrodisiac titre in a polyandrous butterfly: male ejaculate tailoring and absence of female control. Proc R Soc Lond B Biol Sci 271(1550):1765–1770

Andersson M (1994) Sexual selection. Princeton University Press, Princeton, NJ

Andrade MC (1995) Sexual cannibalism in the redback spider (*Latrodectus hasselti* Thorell): Natural selection for female feeding and sexual selection for male sacrifice. MSc thesis, University of Toronto, Toronto

Andrade MC (1996) Sexual selection for male sacrifice in the Australian redback spider. Science 271:70–72

Andrade MC (2003) Risky mate search and male self-sacrifice in redback spiders. Behav Ecol 14(4):531–538

Andrade MC, Banta EM (2002) Value of male remating and functional sterility in redback spiders. Anim Behav 63(5):857–870

Andrade MC, Gu L, Stoltz JA (2005) Novel male trait prolongs survival in suicidal mating. Biol Lett 1(3):276–279

Andrade MC, Kasumovic MM (2005) Terminal investment strategies and male mate choice: extreme tests of Bateman. Integr Comp Biol 45(5):838–847

Arnqvist G, Rowe L (1995) Sexual conflict and arms races between the sexes: a morphological adaptation for control of mating in a female insect. Proc R Soc Lond 261:123–127

Arnqvist G, Rowe L (2002) Antagonistic coevolution between the sexes in a group of insects. Nature 415:787–789

Arnqvist G, Rowe L (2005) Sexual Conflict. Princeton University Press, Princeton

Austad S (1984) Evolution of sperm priority patterns in spiders. In: Smith R (ed) Sperm Competition and the evolution of animal mating systems. Academic Press, Toronto, pp 221–249

Berendonck B, Greven H (2002) Morphology of female and male genitalia of Latrodectus revivensis Shulov, 1948 (Araneae, Theridiidae) with regard to sperm priority patterns. In: Toft S, Scharff N (eds) European Arachnology 2000. Aarhus Univ Press, Aarhus C, pp 157–167

Berendonck B, Greven H (2005) Genital structures in the entelegyne widow spider *Latrodectus revivensis* (Arachnida; Araneae; Theridiidae) indicate a low ability for cryptic female choice by sperm manipulation. J Morphol 263(1):118–132

Bhatnagar RDS, Rempel JG (1962) The structure, function, and postembryonic development of the male and female copulatory organs of the black widow spider *Latrodectus curacaviensis* (Muller). Can J Zool 40:465–510

Biaggio MD (2007) Implications of subadult cohabitation for the mating system of redback spiders (*Latrodectus hasselti*). MSc thesis, University of Toronto Toronto

Birkhead TR (1998) Cryptic female choice: criteria for establishing female sperm choice. Evolution 52(4):1212–1218

Birkhead TR (2000) Defining and demonstrating postcopulatory female choice - again. Evolution 54(3):1057–1060

Bleu J, Bessa-Gomes C, Laloi D (2012) Evolution of female choosiness and mating frequency: effects of mating cost, density and sex ratio. Anim Behav 83(1):131–136

Boorman E, Parker GA (1976) Sperm (ejaculate) competition in *Drosophila melanogaster*, and the reproductive value of females to males in relation to female age and mating status. Ecol Entomol 1(3):145–155

Brandt Y, Andrade MC (2007) Testing the gravity hypothesis of sexual size dimorphism: are small males faster climbers? Functional Ecol 21(2):379–385

Breene RG, Sweet MH (1985) Evidence of insemination of multiple females by the male black widow spider, *Latrodectus mactans* (Araneae, Theridiidae). J Arachnol 13:331–335

Carrel JE (2001) Population dynamics of the red widow spider (Araneae : Theridiidae). Fla Ent 84(3):385–390

Chapman T, Liddle L, Kai J, Wolfner M, Patridge L (1995) Cost of mating in *Drosophila melanogaster* females is mediated by male accessory gland products. Nature 373:241–244

D'Amour FE, Becker FE, Riper WV (1936) The Black Widow Spider. Quarter Rev Biol 11(2):123–160

De Jong MCM, Sabelis MW (1991) Limits to runaway sexual selection: the wallflower paradox. J Evol Biol 4(4):637–656

De La Serna De Esteban CJ, Lauria L, Spinelli M (1987) Female genitalia of 3 species of the genus *Latrodectus* Araneae Theridiidae, *Latrodectus antheratus, Latrodectus mirabilis* and *Latrodectus corallinus*. Physis Secciόn C los Continentes y los Organismos Terrestres 45(108):31–37

De Luca PA, Stoltz JA, Andrade MCB, Mason AC (2015) Metabolic efficiency in courtship favors males with intermediate mass in the Australian redback spider, *Latrodectus hasselti*. J Ins Physiol 72(0):35–42

Downes MF (1987) Postembryonic development of *Latrodectus hasselti* Thorell (Araneae, Theridiidae). J Arachnol 14:293–301

Eberhard WG (1985) Sexual selection and animal genitalia. Harvard University Press, Cambridge

Eberhard WG (1996) Female Control: Sexual selection by cryptic female choice. Princeton University Press, Princeton, New Jersey

Eberhard WG (1993) Evaluating models of sexual selection: genitalia as a test case. Am Nat 142:564–571

Eberhard WG (1994) Evidence for widespread courtship during copulation in 131 species of insects and spiders, and implications for cryptic female choice. Evolution 48:711–733

Eberhard WG (2000) Criteria for demonstrating postcopulatory female choice. Evolution 54(3):1047–1050

Eberhard WG (2004) Why study spider sex: special traits of spiders facilitate studies of sperm competition and cryptic female choice. J Arachnol 32(3):545–556

Eberhard WG, Huber BA (2010) Spider genitalia: precise maneuvers with a numb structure in a complex lock. In: Leonard J, Cordoba-Aguilar A (eds) The evolution of primary sexual characters in animals. Oxford, New York, pp 249–284

Elias DO, Andrade MC, Kasumovic MM (2011) Dynamic population structure and the evolution of spider mating systems. Adv Ins Physiol 41:65

Estévez AL, González A (1991) Vital statistics of *Steatoda retorta* Gonzalez 1987, and comparative analysis with those of *Latrodectus mirabilis* (Holmberg), *Latrodectus antheratus* (Badcock), *Latrodectus corallinus* Abalos, *Latrodectus diaguita* Carcavallo and *Tidarren sisyphoides* (Walckenaer) (Araneae, Theridiidae). Stud Neotrop Fauna Environ 26(2):75–81

Evans JP, Rosengrave P, Gasparini C, Gemmell NJ (2013) Delineating the roles of males and females in sperm competition. Proc R Soc Lond B Biol Sci 280:20132047

Foelix R (1982) Biology of Spiders. Harvard University Press, Cambridge

Foellmer MW, Fairbairn DJ (2005) Selection on male size, leg length and condition during mate search in a sexually highly dimorphic orb-weaving spider. Oecologia 142(4):653–662

Forster L (1992) The stereotyped behaviour of sexual cannibalism in *Latrodectus hasselti* Thorell (Araneae: Theridiidae), the Australian Redback spider. Aust J Zool 40:1–11

Forster L (1995) The behavioural ecology of *Latrodectus hasselti* (Thorell), the Australian redback spider (Araneae: Theridiidae): a review. Records of the Western Australia Museum Supplement No. 52:13–24

Garb JE, Gonzalez A, Gillespie RG (2004) The black widow spider genus *Latrodectus* (Araneae: Theridiidae): phylogeny, biogeography, and invasion history. Molec Phylogen Evol 31(3):1127–1142

Garb JE, Hayashi CY (2013) Molecular Evolution of alpha-Latrotoxin, the Exceptionally Potent Vertebrate Neurotoxin in Black Widow Spider Venom. Molec Biol Evol 30(5):999–1014

Garcia-Gonzalez F (2004) Infertile matings and sperm competition: the effect of "nonsperm representation" on intraspecific variation in sperm precedence patterns. Am Nat 164(4):457–472

Harari AR, Ziv M, Lubin Y (2009) Conflict or cooperation in the courtship display of the white widow spider, *Latrodectus pallidus*. J Arachnol 37(3):254–260

Harvey PH, Pagel MD (1991) The comparative method in evolutionary biology. Oxford University Press, Oxford

Hebets EA (2003) Subadult experience influences adult mate choice in an arthropod: exposed female wolf spiders prefer males of a familiar phenotype. Proc Natl Acad Sci USA 100(23):13390–13395

Herberstein ME, Schneider JM, Uhl G, Michalik P (2011) Sperm dynamics in spiders. Behav Ecol 22(4):692–695

Heubel KU, Lindstrom K, Kokko H (2008) Females increase current reproductive effort when future access to males is uncertain. Biol Letters 4(2):224–227

Huber BA (1993) Genital mechanics and sexual selection in the spider *Nesticus cellulanus* (Araneae: Nesticidae). Can J Zool 71:2437–2447

Huber BA (2005) Sexual selection research on spiders: progress and biases. Biol Rev 80(3):363–385

Jelinek GA (1997) Widow spider envenomation (latrodectism): a worldwide problem. Wildern Environ Med 8 (4):226–231

Jennions MD, Petrie M (1997) Variation in mate choice and mating preferences: a review of causes and consequences. Biol Rev Cam Phil Soc 72(2):283–327

Jerhot E, Stoltz JA, Andrade MC, Schulz S (2010) Acylated serine derivatives: a unique class of arthropod pheromones of the Australian redback spider, *Latrodectus hasselti*. Angewandte Chemie International Edition 49(11):2037–2040

Johnson JC (2005) Cohabitation of juvenile females with mature males promotes sexual cannibalism in fishing spiders. Behav Ecol 16(1):269–273

Johnson JC, Trubl P, Blackmore V, Miles L (2011) Male black widows court well-fed females more than starved females: silken cues indicate sexual cannibalism risk. Anim Behav 82(2):383–390

Kaston BJ (1970) Comparative biology of American black widow spiders. San Diego Soc Nat Hist Trans 16:33–82

Kasumovic MM, Andrade MC (2004) Discrimination of airborne pheromones by mate-searching male western black widow spiders (*Latrodectus hesperus*): species-and population-specific responses. Can J Zool 82(7):1027–1034

Kasumovic MM, Andrade MC (2006) Male development tracks rapidly shifting sexual versus natural selection pressures. Curr Biol 16(7):R242–R243

Kasumovic MM, Andrade MC (2009) A change in competitive context reverses sexual selection on male size. J Evol Biol 22(2):324–333

Kasumovic MM, Brooks RC, Andrade MC (2009a) Body condition but not dietary restriction prolongs lifespan in a semelparous capital breeder. Biol Letters 5(5):636–638

Kasumovic MM, Bruce MJ, Herberstein ME, Andrade MC (2009b) Evidence for developmental plasticity in response to demographic variation in nature. Ecology 90(8):2287–2296

Kasumovic MM, Seebacher F (2013) The active metabolic rate predicts a male spider's proximity to females and expected fitness. Biol Letters 9(2):20121164–20121164

Kavale J (1986) The comparative biology of two *Latrodectus* species. University of Otago, Zoology Dunedin

Knoflach B, van Harten A (2002) The genus *Latrodectus* (Araneae: Theridiidae) from mainland Yemen, the Socotra Archipelago and adjacent countries. Fauna of Arabia 19:321–362

Kokko H, Brooks R, Jennions MD, Morley J (2003) The evolution of mate choice and mating biases. Proc R Soc B-Biol Sci 270(1515):653–664

Kokko H, Jennions MD, Brooks R (2006) Unifying and testing models of sexual selection. Ann Rev Ecol Evol Syst 37(1):43–66

Kokko H, Mappes J (2005) Sexual selection when fertilization is not guaranteed. Evolution 59(9):1876–1885

Lauria de Cidre L (1988) El sistema genito feminina en dos especies del genero *Latrodectus* (Araneae, Theridiidae). Rev Brasil Biol 48:771–776

Levi H (1983) On the value of genitalic structures and coloration in separating species of widow spiders (*Latrodectus* sp.)(Araneae: Theridiidae). Verh Naturw Ver Hamburg (NF) 26:195–200

Levi HW (1959) The spider genus *Latrodectus* (Araneae: Theridiidae). Trans Am Microscop Soc 78:7–43

Levi HW (1961) Evolutionary trends in the development of palpal sclerites in the spider family Theridiidae. J Morph 108:1–10

Levi HW (1966) The three species of *Latrodectus* (Araneae), found in Israel. J Zool, Proc Zool Soc London 150:427–432

Levy G, Amitai P (1983) Revision of the widow-spider genus *Latrodectus* (Araneae, Theridiidae) in Israel. Zool J Linn Soc 77(1):39–63

Lubin Y, Ellner S, Kotzman M (1993) Web relocation and habitat selection in a desert widow spider. Ecology 74(7):1915–1928

MacLeod EC (2013) New insights in the evolutionary maintenance of male mate choice behaviour using the western black widow, *Latrodectus hesperus* PhD thesis, University of Toronto, Toronto

MacLeod EC, Andrade MCB (2014) Strong, convergent male mate choice along two preference axes in field populations of black widow spiders. Anim Behav 89:163–169

McCrone JD, Levi HW (1964) North American widow spiders of the *Latrodectus curacaviensis* group (Araneae: Theridiidae). Psyche 71:12–27

Michalik P, Ramírez MJ (2014) Evolutionary morphology of the male reproductive system, spermatozoa and seminal fluid of spiders (Araneae, Arachnida)—current knowledge and future directions. Arthropod Struct Dev 43(4):291–322

Modanu M, Michalik P, Andrade MC (2013) Mating system does not predict permanent sperm depletion in black widow spiders. Evol Develop 15(3):205–212

Moya-Laraño J, Halaj J, Wise DH (2002) Climbing to reach females: romeo should be small. Evolution 56(2):420–425

Mueller GH (1985) Abgebrochene Emboli in der Vulva der 'Schwarzen Witwe' *Latrodectus geometricus* C.L. Koch 1841 (Arachnida: Araneae: Theridiidae). Entomol Z 95:27–30

Neumann R, Schneider JM (2011) Frequent failure of male monopolization strategies as a cost of female choice in the black widow spider *Latrodectus tredecimguttatus*. Ethology 117(12):1057–1066

Parker GA (1990) Sperm competition games: raffles and roles. Proc R Soc Lond B 242:120–126

Parker GA (1970) Sperm Competition and its evolutionary consequences in the insects. Biol Rev 45:525–567

Perampaladas K, Stoltz JA, Andrade MCB (2008) Mated redback spider females re-advertise receptivity months after mating. Ethology 114(6):589–598

Pitcher TE, Neff BD, Rodd FH, Rowe L (2003) Multiple mating and sequential mate choice in guppies: females trade up. Proc R Soc B Biol Scien 270(1524):1623–1629

Ross K, Smith RL (1979a) Aspects of the courtship behavior of the black-widow spider, *Latrodectus hesperus* (Araneae, Theridiidae), with evidence for the existence of a contact sex-pheromone. J Arachnol 7(1):69–77

Ross K, Smith RL (1979b) Aspects of the courtship behaviour of the black widow spider, *Latrodectus hesperus* (Araneae: Theridiidae), with evidence for the existence of a contact sex pheromone. J Arachnol 7:69–77

Rugman-Jones PF, Eady PE (2001) The sterile male technique: a source of error in estimating levels of sperm precedence? Entomol Exp Appl 98(2):241–244

Salomon M (2011) The natural diet of a polyphagous predator, *Latrodectus hesperus* (Araneae: Theridiidae), over one year. J Arachnol 39(1):154–160

Salomon M, Vibert S, Bennett RG (2010) Habitat use by western black widow spiders (*Latrodectus hesperus*) in coastal British Columbia: evidence of facultative group living. Can J Zool 88(3):334–346

Schneider J, Andrade MC (2011) Mating behaviour and sexual selection. In: Herberstein ME (ed) Spider behaviour: flexibility and versatility. Cambridge University Press, Cambridge, pp 215–274

Schneider JM, Gilberg S, Fromhage L, Uhl G (2006) Sexual conflict over copulation duration in a cannibalistic spider. Anim Behav 71:781–788

Scott C, Vibert S, Gries G (2012) Evidence that web reduction by western black widow males functions in sexual communication. Can Ent 144(5):672–678

Segev O, Ziv M, Lubin Y (2003) The male mating system in a desert widow spider. J Arachnol 31(3):379–393

Segoli M, Arieli R, Sierwald P, Harari AR, Lubin Y (2008a) Sexual cannibalism in the brown widow spider (*Latrodectus geometricus*). Ethology 114(3):279–286

Segoli M, Harari AR, Lubin Y (2006) Limited mating opportunities and male monogamy: a field study of white widow spiders, *Latrodectus pallidus* (Theridiidae). Anim Behav 72:635–642

Segoli M, Lubin Y, Harari AR (2008b) Frequency and consequences of damage to male copulatory organs in a widow spider. J Arachnol 36(3):533–537

Sillen-Tullberg B (1981) Prolonged copulation: a male 'postcopulatory' strategy in a promiscuous species, *Lygaeus equestris* (Heteroptera: Lygaeidae). Behav Ecol Sociobiol 9:283–289

Simmons L, Siva-Jothy M (1998) Sperm competition in insects: mechanisms and the potential for selection. In: Birkhead T, Moller A (eds) Sperm competition and sexual selection. Academic Press, London, pp 341–434

Simmons LW (2001) Sperm competition and its evolutionary consequences in the insects. Princeton University Press, Princeton, N.J

Simó M, Dias MDD, Jorge C, Castro M, Dias MA, Laborda A (2013) Habitat, redescription and distribution of *Latrodectus geometricus* in Uruguay (Araneae: Theridiidae). Biota Neotropica 13(1):371–375

Siva-Jothy MT, Hooper RE (1996) Differential use of stored sperm during oviposition in the damselfly *Calopteryx splendens xanthostoma* (Charpentier). Behav Ecol Sociobiol 39(6):389–393

Smithers RHN (1944) Contribution to our knowledge of the genus *Latrodectus* (Araneae) in South Africa. Ann S African Mus 36:263–312

Snow LS, Abdel-Mesih A, Andrade MC (2006) Broken copulatory organs are low-cost adaptations to sperm competition in redback spiders. Ethology 112(4):379–389

Snow LS, Andrade MC (2004) Pattern of sperm transfer in redback spiders: implications for sperm competition and male sacrifice. Behav Ecol 15(5):785–792

Snow LSE, Andrade MCB (2005) Multiple sperm storage organs facilitate female control of paternity. Proc R Soc B Biol Scien 272(1568):1139–1144

Stoltz J, Elias D, Andrade MC (2009) Male courtship effort determines female response to competing rivals in redback spiders. Anim Behav 77(1):79–85

Stoltz JA, Andrade MCB (2010) Female's courtship threshold allows intruding males to mate with reduced effort. Proc R Soc Lond B Biol Sci 277(1681):585–592

Stoltz JA, Andrade MCB, Kasumovic MM (2012) Developmental plasticity in metabolic rates reinforces morphological plasticity in response to social cues of sexual selection. J Insect Physiol 58(7):985–990

Stoltz JA, Elias DO, Andrade MC (2008) Females reward courtship by competing males in a cannibalistic spider. Behav Ecol Sociobiol 62(5):689–697

Stoltz JA, Hanna R, Andrade MCB (2010) Longevity cost of remaining unmated under dietary restriction. Function Ecol 24(6):1270–1280

Stoltz JA, McNeil JN, Andrade MC (2007) Males assess chemical signals to discriminate just-mated females from virgins in redback spiders. Anim Behav 74(6):1669–1674

Trubl P, Gburek T, Miles L, Johnson JC (2012) Black widow spiders in an urban desert: population variation in an arthropod pest across metropolitan Phoenix. AZ. Urban Ecosyst 15(3):599–609

Uhl G (2002) Female genital morphology andsperm priority patterns in spiders (Araneae). In: Toft S, Scharff N (eds) 19th european colloquium of arachnology aarhus, Denmark, 2000. Aarhus University Press, Aarhus, Denmark, pp 145–156

Uhl G, Nessler S, Schneider J (2010) Securing paternity in spiders? A review on occurrence and effects of mating plugs and male genital mutilation. Genetica 138(1):75–104

Useta G, Huber BA, Costa FG (2007) Spermathecal morphology and sperm dynamics in the female *Schizocosa malitiosa* (Araneae: Lycosidae). Euro J Entomol 104(4):777–785

Ushkaryov YA, Volynski KE, Ashton AC (2004) The multiple actions of black widow spider toxins and their selective use in neurosecretion studies. Toxicon 43(5):527–542

Vetter RS, Vincent LS, Danielsen DWR, Reinker KI, Clarke DE, Itnyre AA, Kabashima JN, Rust MK (2012) The prevalence of brown widow and black widow spiders (Araneae: Theridiidae) in Urban Southern California. J Med Entomol 49(4):947–951

Vocking O, Uhl G, Michalik P (2013) Sperm dynamics in spiders (Araneae): Ultrastructural analysis of the sperm activation process in the garden spider *Argiope bruennichi* (Scopoli, 1772). PLoS ONE 8(9). doi:10.1371/journal.pone.0072660

Watson PJ (1986) Transmission of female sex pheromone thwarted by males in the spider *Linyphia litigiosa*. Science 233:219–221

Welke K, Schneider JM (2009) Inbreeding avoidance through cryptic female choice in the cannibalistic *Argiope lobata*. Behav Ecol 20(5):1056–1062

Chapter 3
Cryptic Female Choice Within the Genus *Argiope*: A Comparative Approach

Jutta Schneider, Gabriele Uhl and Marie E. Herberstein

Abstract The orb-web spider genus *Argiope* (Araneae) offers an excellent opportunity to detect cryptic female choice and to identify the species-specific traits that might lead to its evolution and maintenance. All studied species of the genus *Argiope* are characterized by low male mating rates. While males of some species are strictly mono- or bigynous, others plastically switch between these two strategies. All studied species show sexual cannibalism during copulation. Generally, males die after their second copulation, but the probability of surviving the first copulation differs considerably between species and so does copulation duration. Males of most species break off pieces of their genitalia during copulation that act as mating plugs, but how often this occurs and how effective these plugs are is highly variable. Females that mate multiply can influence the relative paternity success of males through their partial control of copulation duration and their likely control of sperm storage. There is evidence that females cryptically favor small males over large ones, unrelated males over siblings, and males that courted over those that do not. We will sketch variation within and between species in mating systems and related traits, and we will discuss how this relates to cryptic female choice. We will review the existing evidence for cryptic female choice and suggest future avenues in elucidating possible mechanisms that facilitate cryptic female choice and the cues that females may base their choices on.

J. Schneider (✉)
Zoological Institute, Biozentrum Grindel, University of Hamburg,
Martin-Luther-King Platz 3, 20146 Hamburg, Germany
e-mail: jutta.schneider@uni-hamburg.de

G. Uhl
Zoological Institute and Museum, General and Systematic Zoology,
University of Greifswald, J.-S. Bachstrasse 11, 17487 Greifswald, Germany
e-mail: gabriele.uhl@uni-greifswald.de

M.E. Herberstein
Department of Biological Sciences, Macquarie University, Sydney NSW 2109, Australia
e-mail: marie.herberstein@mq.edu.au

A.V. Peretti and A. Aisenberg (eds.), *Cryptic Female Choice in Arthropods*,
DOI 10.1007/978-3-319-17894-3_3

3.1 The Genus *Argiope*—an Excellent Model for Cryptic Female Choice

Evidence for cryptic female choice is notoriously difficult to obtain due to its hidden nature and interactions with other processes such as sperm competition (see introduction chapter). Spiders have been at the forefront of research on cryptic female choice, offering one of the early publications that demonstrated paternity bias by the female based on the relative size of her mates (Elgar et al. 2000). This study was conducted on the orb-web spider *Argiope keyserlingi*. Subsequent studies on this species and congeners have revealed a rich diversity in mating behaviors and intriguing details of the complex nature of male and female mating strategies. The genus *Argiope* has a world-wide distribution and consists of over 80 species (Levi 1983). We already have data on the mating biology for 15 species (Table 3.1), some with direct evidence of cryptic female choice and others providing indirect evidence. All studied species are characterized by a female-biased sexual size dimorphism, low male mating rates, and the regular occurrence of cannibalism during or after copulation. In this book chapter, we relate these features of the *Argiope* mating system to the potential for and evidence of cryptic female choice.

Table 3.1 Data on *Argiope* species with origins from around the globe

Species: Argiope	Origin	Mating	Scape present	% cannibalized	% genital damage	References
aemula	Asia	Hub	No	38		Sasaki and Iwahashi (1995), Levi (1983) and Jäger (2012)
aemula	Taiwan	Thread	No	56	85	Cory and Schneider (unpublished)
aetherea	Australia/Asia	Thread	No	21	0	Nessler (2009) and Levi (1983)
aetheroides	Taiwan	Thread	No	15	87	Cranyon, Welke and Schneider (unpublished)
argentata	Costa Rica	Hub/thread	No	33	95	Uhl (unpublished)
argentata	Uruguay	Hub/thread	No	70	100	Ghione and Uhl (unpublished) and Ghione and Costa (2011)
aurantia	America	Hub	Yes	23	94	Foellmer (2008) and Nessler (2009)

(continued)

Table 3.1 (continued)

Species: Argiope	Origin	Mating	Scape present	% cannibalized	% genital damage	References
australis	Africa	Hub	No	13	68	Welke and Schneider, Strauß, Nessler and Schneider, Uhl and Wegner (unpublished)
blanda	Costa Rica	Thread	No	58	84	Uhl (unpublished)
bruennichi	Europe	Hub	Yes	72	91	Nessler et al. (2009) and Nessler (2009)
keyserlingi	Australia	Thread	No	40	96	Herberstein et al. (2005b) and Nessler (2009)
lobata	Israel	Hub	No	8	51	Hirt and Schneider (unpublished)
lobata	Spain	Hub	No	46	45	Nessler et al. (2009), Welke and Schneider (2009) and Nessler (2009)
mascordi	Australia	Thread	No	20	80	Levi (1983) and Nessler (2009)
perforata	Asia (Taiwan)	Thread	No	0	100	Levi (1983), Cory, Mueller and Schneider (unpublished)
submaronica[a]	Costa Rica	Thread	No	14	100	Uhl (unpublished)
sector	Africa	Hub	No	36	92	Nessler (2009)
trifasciata	Spain	Hub	No	25	100	Cory, Magdziak and Schneider (unpublished)
trifasciata	USA	Hub	No	44	69	Levi (1983) and Nessler (2009)
trifasciata	Costa rica	Hub/thread	No	48	50	Uhl (unpublished)

[a]Previously *savignyi*

Species can be divided into those that mate in the hub and on a thread and a single species to date was found to do both. Females of those species that possess a protrusion that covers the genital openings (scapus) could theoretically use the scape to control genital coupling. Species and populations within species show intriguing variation in the average proportion of males that are cannibalized after their first copulation. The probability that males damage their genitalia during their first copulation also ranges from 0 (*A. aetherea*) to 100 %

3.2 Mating System and the Potential for Cryptic Female Choice

We follow the definition of cryptic female choice, as set out in the introductory chapter to this book: paternity biases resulting from female activities during and following copulation. In the context of the biology of *Argiope*, such female actions include the following: preventing monopolization by a single male, preventing the formation of a mating plug, advertising for additional males, controlling copulation duration and hence the ability of males to transfer sperm, and finally controlling sperm storage and sperm activation.

Male and female mating rates are likely to have a strong influence on the evolution and maintenance of cryptic female choice. Obviously, polyandry is a requirement for cryptic female choice, but in *Argiope*, male mating rates are very low, selecting for male strategies that protect male reproductive investment, possibly at the cost of female mating rates, and hence her opportunity to bias paternity. Below, we describe male mating rates in spiders generally and *Argiope* specifically, and how they might relate to females exerting choice over fertilization. We also sketch male mating strategies and male mate choice to explain the specific setting for the potential of cryptic female choice in the genus *Argiope*.

3.3 Male Mating Rates: Monogyny and Bigyny

Monogynous mating systems are relatively rare overall, although they are taxonomically widespread (Schneider and Fromhage 2010). Monogyny is different from monogamy because males mate only once while the females mate multiply. For unknown reasons, monogyny is relatively common in spiders and has evolved several times independently (Miller 2007). Monogynous mating systems have been documented in *Tidarren*, *Nephilengys*, *Nephila*, *Latrodectus*, *Herennia*, *Dolomedes*, and *Argiope* (Schwartz et al. 2013, 2014). Monogyny is associated with remarkable adaptations in males, such as self-sacrifice and genital mutilation, that appear to promote the male's ability to monopolize an individual female (Schneider and Fromhage 2010; Schneider and Andrade 2011). Males in monogynous mating systems invest maximally in a single female even though they do not provide paternal care. A male-biased sex ratio will promote the evolution of monogyny even in the absence of paternal investment (Fromhage et al. 2005, 2008). The logic is that if there are more males than females, and if both sexes mate multiply, each male faces sperm competition. The stronger the male bias, the lower the average paternity share of a male. Under such conditions, selection can favor a monogynous strategy if it provides a male with a larger than average paternity share. Males can achieve this by monopolizing fertilization of all the eggs in a female or at least by increasing their relative paternity share with a single female above the expected average. As a consequence, monogynous males are under strong selection to win male–male competition with their one and only female. While it is conceivable that females benefit via indirect mate choice from this strong male–male competition if the ability to monopolize a female is heritable, monopolization by a male is likely to result in selection on the female to counteract monopolization and thus asserts cryptic female choice in its broadest definition.

The genus *Argiope* is characterized by very low male mating rates where males mate with only one or sometimes two females in their entire lifetime. This is due to a paternity protection mechanism that most *Argiope* species share: During copulation, males place a mating plug into the insemination duct of the female. The plug consists of a genital fragment, and breakage of this fragment renders the pedipalp dysfunctional (Uhl et al. 2007; see below for more details). Hence, males have the option to use both pedipalps with the same female in a monogynous mating system or use each one with another female in a bigynous mating system (Fromhage and Schneider 2012). As the female has paired genital openings each leading to a separate spermatheca (Vöcking et al. 2013; Herberstein et al. 2011a), a male can only plug both by mating twice with the same female (monogyny) and thus monopolize the female.

The bigynous mating tactic promises the highest reproductive success provided that a male is able to mate with two virgins who will not re-mate. Under such condition, the male will gain full paternity of the eggs of two females. The probability that a male can achieve the maximal reproductive success depends on the degree of male competition and the re-mating probabilities of females. A male that is cannibalized during the first copulation will be at the other end of the spectrum. His reproductive success will be particularly low, if he was unable to transfer the whole sperm load of a single pedipalp and the female re-mates with another male who may transfer more sperm. In the center of the fitness scale are those males that succeed in fertilizing all eggs of a single female. Whether or not the latter reproductive success is above the population average depends on the male bias in the effective sex ratio.

Many species of *Argiope* are exclusively mono- or bigynous. For example, *A. keyserlingi* males appear to never mate twice with the same female but guard the female after copulation and then leave to search for a second mating option (Herberstein et al. 2005a; Zimmer et al. 2014). Other species apparently flip between both strategies. In *A. bruennichi*, males that survive their first copulation may or may not search for a second female (Welke et al. 2012). In a comprehensive study, *A. bruennichi* females were observed continuously over an entire mating season in a natural population (Zimmer et al. 2012). Visiting males were marked if they survived copulation and left the female. Each male that arrived at a female web was closely inspected for pedipalp damage as evidence for a previous mating. Half of the observed males mated with two females, hence opted for a bigynous tactic. The other half was monogynous and was cannibalized by their only mate. They achieved this in two ways: either through a single long copulation or by a brief first copulation followed by a second long copulation with the same female. A model revealed that such a conditional mating strategy can evolve under conditions that apply to *A. bruennichi* (Fromhage and Schneider 2012). The most important variables are (i) a high variation in female quality, and (ii) seasonal as well as (iii) spatial differences in selection regimes. The model further revealed that bigyny alone can also be stable under a restricted set of conditions, and it remains to be seen whether the theoretical assumptions capture the biology of the Australian *A. keyserlingi* with a bigynous mating system (Herberstein et al. 2005a).

3.4 It Takes Two to Tango: Male Mate Choice

Conditional bigyny clearly is a form of male mate choice: If the first virgin female that a male encounters is of low quality (i.e., small), she only receives a short copulation and the male moves on to locate another female. But if the first female is of high quality and promises high fecundity returns, the male will perform both copulations with her (Welke et al. 2012). Hence, males flexibly adjust their mating rate and their mating investment to female quality without rejecting a low-quality female as their first mate. The observations suggest that these males of *A. bruennichi* and perhaps of other *Argiope* species use a trading-up strategy to perform sequential mate choice. Their first priority is to secure a mating with any female first and become choosy only for their second and last copulation. A laboratory study using the same species supports the findings from the field. Virgin males that were assigned to mate with a sister did not resist copulation but mated for shorter, presumably to escape cannibalism and to follow a bigynous strategy (Welke and Schneider 2010). Yet another study of *A. bruennichi* suggests that males base their mating decisions on information that they extract from their environment. Males that were exposed to pheromones of virgin females were more likely to die during their first copulation than males that did not receive such cues (Nessler et al. 2009b).

However, male *Argiope* can also reject females. During two field studies of *A. bruennichi*, males were regularly observed to enter and apparently inspect the female and her web. However, in some cases, males left the web after inspection without attempting copulation even though these females were virgin and not different in size or age from average of females in the population. The presence of a larger female nearby could partly explain such male rejections although laboratory choice trials did not support male preferences for more fecund females (Schulte et al. 2010). Furthermore, genetic relatedness has been excluded recently as an alternative explanation (Zimmer and Schneider, unpublished). Hence, it remains unclear why and when males reject females. It is worth noting that in laboratory settings, males vary tremendously in how soon they commence courtship after they were placed onto a virgin female's web. Variation occurs between species but also within species. While some individuals will resume courtship within seconds or minutes, others will not move for hours, which may also indicate a level of mate rejection.

The potential for simultaneous male mate choice has also been investigated in *Argiope* using dichotomous mate choice paradigms. Such setups are ideal experimental tools to address male perception of differences between females based on volatile or silk-bound chemical cues or signals. *Argiope keyserlingi* males prefer virgin over mated females both in the field and in the laboratory (Gaskett et al. 2004). Males are also more attracted to females with narrow abdomens that are likely recently molted and still virgin (Herberstein et al. 2002). More recent experiments further showed that males prefer once-mated over twice-mated females (Zimmer et al. 2014). *Argiope bruennichi* males objected to the same paradigm discriminated against mated females, but female size and weight are not strong predictors of mating preferences (Schulte et al. 2010).

The above evidence is suggestive of male mate choice, but we cannot exclude that females produce or withhold signals that prevent males from or entice them to attempt mating. A female that appears motionless to the observer may in fact signal her unreceptivity via chemical or other communication channels obscured from direct observation. During the mate approach and courtship, information is exchanged between the two sexes and as we know very little about chemical, tactile, and vibratory courtship in orb-web spiders (Uhl and Elias 2011; Chinta et al. 2010), we can often only speculate about signal presence let alone signal content. We argue the need to keep this in mind whenever discussing sexual selection and attributing decisions to one sex only. We need many more experiments, ideally in laboratory and field settings, to unravel sexual communication and mating decisions in both sexes. First, promising attempts were made in *A. keyserlingi*, in which males generate characteristic shudders in the web with the possible function of pacifying the female (Wignall and Herberstein 2013a, b).

It is likely that other species of *Argiope* show further variation in their mating strategies, and it is even possible that populations of the same species show variation. This within and between species variation makes *Argiope* a particular interesting taxon for studying the causes and consequences of male mating strategies in the context of monogynous/bigynous mating systems. Below, we further explore the male adaptations for monopolizing females.

3.5 Limits to Female Mating Rate: One-Shot Genitalia and Mating Plugs

The potential for cryptic female choice and a female's capacity to influence fertilization success will depend on the male's ability to monopolize the female. As we discussed above, males of *Argiope* benefit from adaptations that protect a male's ejaculate and monopolize the female. Indeed, male *Argiope* lodge a piece of the sperm transfer organ (pedipalp) in the female genital opening, which protects the ejaculate against rivals but renders the pedipalp unusable for further copulations (Uhl et al. 2010; see Figs. 3.1 and 3.2). These one-shot genitalia of *Argiope* are morphologically complex with sclerites that attach and connect to corresponding female genital structures (Uhl et al. 2007). The sperm is transferred through the embolus that connects to the female copulatory opening. The coupling mechanisms are so specific and complex that only one pedipalp can be used at a time. As the female has paired copulatory openings (each leading to a separate spermatheca), only one

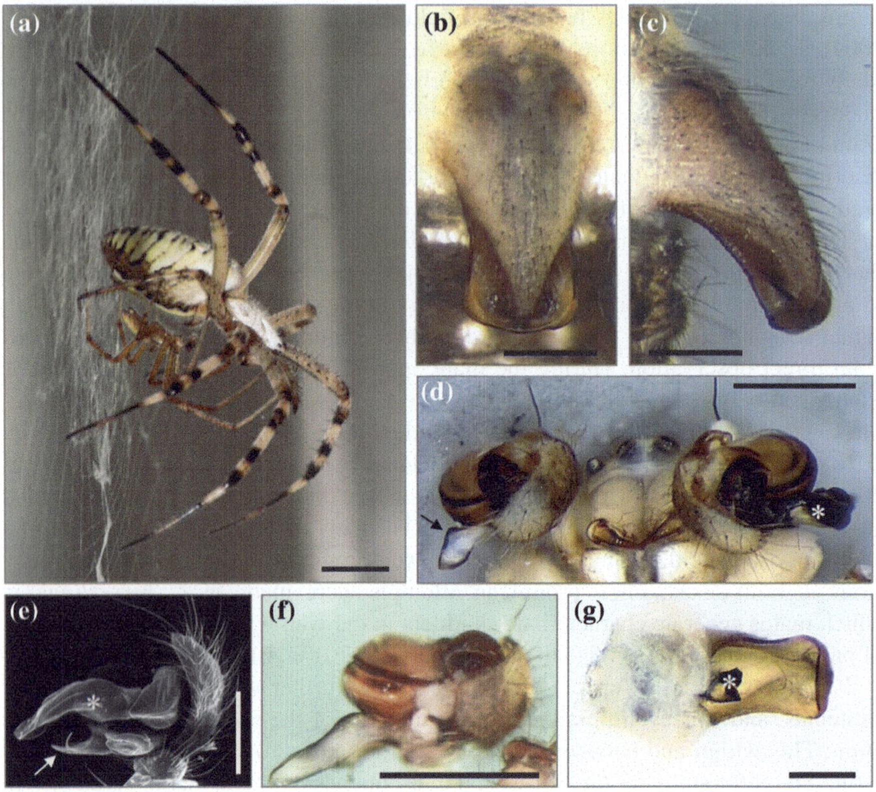

Fig. 3.1 *Argiope bruennichi*. **a** Mating in the hub requires female cooperation. Receptive females assume a stilted posture and lift the scape (size bar: 1 cm). **b** Scapus that covers the genital opening, ventral side. **c** Scapus from lateral in lifted position. **d** Male pedipalps, *right* damaged (*arrow*), *left* intact (embolus marked by *asterisk*), *ventral view*. **e** SEM image of male right intact pedipalp. Embolus marked by *asterisk*. *Arrow* points to median apophysis with spur. **f** Damaged right pedipalp, embolus missing. **g** Scapus removed from female. Embolus (*asterisk*) found in atrium plugging a copulatory opening. Scale bars **a–g** 0.5 mm (*Photographs* S. Nessler and G. Uhl)

of these can be inseminated during a single copulation. Furthermore, the complex coupling co-occurs with breakage of the embolus or its tip, and the placement of this mating plug inside the female copulatory duct (Nessler et al. 2007b).

The breakage of the sperm transferring embolus is the likely functional explanation for rendering the pedipalp dysfunctional (Fig. 3.1d). However, there is intriguing variation in the size of the broken sclerite between and even within species. *Argiope bruennichi* and *A. australis* can break off a large piece or just the small tip of the embolus. Other species break off pieces of medium size. The sclerites can be found inside the female genital openings and act as mating plugs in several species (Nessler et al. 2007a; Herberstein et al. 2012; Foellmer 2008; Uhl et al. 2010). Perhaps, there is a trade-off between the greater ease of breaking off larger pieces and the risk that females or rivals remove or bypass the plug, which may be

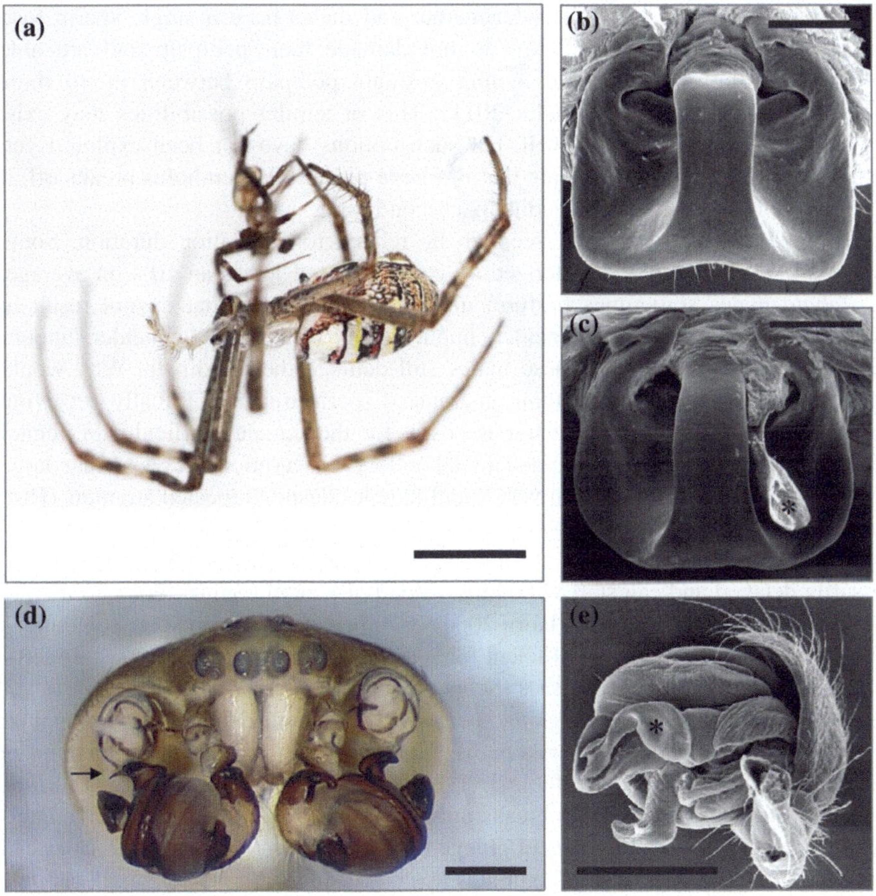

Fig. 3.2 *Argiope keyserlingi*. **a** Mating on the mating thread requires female cooperation. Receptive females move from the hub to the mating thread where they cling to the thread mainly with legs 3 and 4 (size bar: 1 cm). **b** External female genital structure, *ventral side* (size bar: 0.2 mm). **c** Left genital opening plugged by massive embolus (*asterisk*) (size bar: 0.2 mm). **d** Intact male pedipalps, *ventral view*. *Arrow* points to the median apophysis with spur (size bar: 0.5 mm). **e** SEM image of male left pedipalp, intact. Embolus marked by *asterisk* (size bar: 0.5 mm). *Photographs* A. Wignall (**a**), G. Uhl (**b–d**)

more difficult for smaller pieces that are lodged deep inside the female genitalia. However, there is no tested explanation for the variation in sclerite size, yet.

Even though most studied *Argiope* species damage their genitalia, generalizations about the consequences for pedipalp reuse should be drawn with caution. We know that males of several (though not all) monogynous spider species will charge their pedipalps only once: The testes actually degenerate after maturation in *Argiope* and *Nephila* (Herberstein et al. 2005b; Michalik and Rittschof 2011). It remains to be seen whether the degeneration of testes and one-shot genitalia are indeed a general and a correlated trait of the genus *Argiope*. For example, in

N. senegalensis, the testes also degenerate and males have a single sperm load in each pedipalp. However, they do not damage their pedipalp and are able to partition the sperm portion within a single pedipalp between up to three females (Schneider and Michalik 2011). This or similar possibilities may exist in some *Argiope* species as well, but such options have not been explored yet. Nevertheless, it is safe to assume that if a large piece of the embolus breaks off, it is impossible to use the same pedipalp a second time.

Pedipalp damage does not seem to be related to copulation duration. Some species of *Argiope* are characterized by copulations of less than 10 s on average. *A. lobata* males sometimes perform ultrashort copulations that do not result in sperm transfer (Hirt and Schneider, unpublished; Welke and Schneider, unpublished). Curiously, some of these males still damage their pedipalp. Why would males risk genital damage without guaranteed sperm transfer? Equally, receiving a plug without any sperm transfer is costly for the female. Difficulty in genital coupling may be an explanation and laboratory observations suggest that large males are not able to mate with very small females despite repeated attempts (Hirt, unpublished observations). However, this phenomenon has not been investigated systematically in any *Argiope* species yet. Nevertheless, sperm transfer is not necessarily delayed and can start within the second of genital contact. For example, in *A. aurantia* (Foellmer and Fairbairn 2004) or *A. bruennichi*, even short copulations (3 and 5 s, respectively) are sufficient for sperm transfer and may even secure fertilization of two successive egg sacs (Schneider et al. 2005).

While most or even all *Argiope* species exhibit genital damage and plugging, frequency and functionality varies between species. Frequencies of genital damage vary between 40 and 100 % (Table 3.1). In *A. keyserlingi*, *A. aurantia*, and *A. bruennichi,* almost every pedipalp breaks during copulation. The rates of genital damage are correlated with the rates of plugged genital openings although the latter can be lower if not every broken off embolus tip ends inside the female genital opening.

Below, we sketch possible causes and consequences of plug variation. The functionality of the plugs can differ due to their tenure within the female copulatory tract, which is likely to depend on how well a male was able to securely place his plug. While plugs remain lodged inside the female beyond oviposition in some species, plugs in *A. keyserlingi* do not last for more than few days (Herberstein et al. 2012). Similarly, in *A. australis*, 70 % of plugs (consisting of the whole embolus) remained inside the genital opening for less than one week (Hirt and Schneider, unpublished). It is possible that females remove plugs but they may also fall out by themselves. Anecdotal observations suggest that males can remove plugs of other males in *A. australis* (Strauss, Nessler and Schneider, unpublished data).

The rate of successful plugging will determine the maximal female mating rate and consequently the opportunity for sperm competition and cryptic female choice. At the same time, genital damage is a costly adaptation for males predicting strong selection on successful application. Nevertheless, we observe a disparity between genital damage and plugging that is intriguing and suggestive of sexual conflict. An explanation is needed when a large proportion of males damage their genitalia but fail to collect the potential benefit of plugging. For example,

A. lobata has a very low plugging success; even though 40 % of the males damage their genitalia during copulation, less than half of these males manage to place a successful mating plug (Nessler et al. 2009a). In other species, such as *A. bruennichi*, plugging success is very high and the plug is very effective in preventing sperm transfer by another male. When the plugging rate is high and effective, female mating rates are limited to two copulations (although more copulation are possible if a male did not succeed in plugging). Natural mating rates of females are only known for *A. bruennichi* and the majority of females mate with a single male. Average mating rate in a high-density population was estimated as 1.2 (Zimmer et al. 2012). Males can monopolize paternity with a single female if they succeed to copulate twice and apply two effective mating plugs. Failure to monopolize both spermatheca may result in significant paternity losses of 50 % or more if two males inseminate two different spermathecae of the same female (Schneider and Lesmono 2009; Nessler et al. 2009a). How much relative paternity each male gains depends on the number of sperm transferred, which can be predicted by the duration of copulation (*A. bruennichi*: Schneider et al. 2006; *A. keyserlingi*: Herberstein et al. 2011a) but not always (*A. lobata*: Welke and Schneider 2009).

It is tempting to speculate why plugging rates differ between species. If male pedipalps are one shot despite lack of damage, it seems difficult to imagine an adaptive value for the male to refrain from plugging. Males may benefit from not plugging if it increases their survival after the first insertion by escaping cannibalism. This is maybe particularly advantageous if the female is of low value or if attractive alternatives are present. Failure of plugging could also be female induced and constitute cryptic female choice (see below for elaboration on this issue). Female resistance and the timing of her attack may interfere with plug positioning (see below how sclerites on the male pedipalp evolved to facilitate plug positioning). We know little about if or how females contribute to plugging. In *A. argentata*, female active cooperation is not required for copulation and plugging because males were found to plug dead females (Ghione and Costa 2011). Clearly, more studies are required to draw conclusions about possible trade-offs with genital damage and limitations of plugging success.

3.6 Programmed Death of Males

Sometimes during their first copulation but more generally during their second and final copulation, males of all *Argiope* species die. Death may be inflicted by the female, but it may also occur without any active contribution of the female. The programmed death has been termed "spontaneous death" (Foellmer and Fairbairn 2003) and was first observed in *A. aurantia*. The males died in copula within minutes of the onset of genital contact. The heartbeat of the male stopped, while the genitals remained attached to the female. The exact mechanism behind the programmed death is unknown. What can males gain from spontaneous death? Undoubtedly, males achieve a maximal copulation duration and the locking mechanism may ensure that the entire sperm load of the pedipalps is transferred. In addition, the plugging success may be secured. However, experimental investigation of the function of spontaneous death is challenging and requires within species variation in the propensity of spontaneous death which has not been reported yet. The ubiquity of male suicidal mating investment during their second copulation regardless of its cause leaves no variation to be explored in the context of female choice. Therefore, we will below focus our discussion of female influences on the copulations of virgin males.

3.7 Copulation Mechanism

Copulation in spiders entails coupling of the male pedipalps to the female genital region. The paired pedipalps are highly complex in most spider species, consisting of several sclerites and inflatable membranous parts that change their relative position when active (Grasshoff 1973; see Figs. 3.1 and 3.2). Although external female counterparts are generally less complex, they are nevertheless species specific and often exhibit rims, hooks, grooves, and long peculiar protrusions. Internally, female genitalia can be very complicated with twisted or spiral ducts or folds that lead to the multiple spermathecae. Cryptic female choice is a likely cause of such pronounced male and female genital complexity (Eberhard 1996; Hosken and Stockley 2004). Since spider females are predatory and generally larger than their males, male physical sexual coercion is unlikely to occur, despite the fact that courtship in some species may appear fierce (Sentenská, Uhl, Lipke, Michalik, Pekár unpublished) although coercion may occur by means not related to physical strength.

In *Argiope*, female external genitalia can be allocated to two groups, those that possess a so-called scapus, a shoehorn-shaped structure that is directed posteriorly, thereby covering the genital openings (Fig. 3.1). In the other group, the scapus is lacking and genital openings are thus more readily accessible (Fig. 3.2). In *Argiope* species with a scapus, such as *A. bruennichi*, females clearly have to cooperate in mating by abducting the scapus from the ventral body wall, thereby allowing access to the copulatory openings that are hidden deep below (Uhl 1994). Under laboratory conditions, females showed this acceptance behavior almost invariably suggesting that male courtship behavior elicits an immediate reflex action at least in females that were individually laboratory-reared and thus deprived of potentially important information on population density and the availability of mating partners. The scape-possessing species *A. bruennichi* and *A. aurantia* are sister species as well as derived from non-scape-possessing ancestors according to a recent phylogenetic hypothesis by (Cheng and Kuntner 2014) and both species mate in the hub of the female web.

The genus *Argiope* comprises species that mate directly in the hub and species that mate on a mating thread and some (*A. argentata*) are flexible as to the mating location (Elgar 1991; Table 3.1). Hub-mating entails an extensive courtship approach by the male eventually down the ventral surface of the female's abdomen. Hub-mating females that are receptive generally show a specific stilted posture by which they increase the distance between their venter and the web plane, seemingly for generating easier access to her genital region (Robinson and Robinson 1980). Receptive hub-mating females of the scape-possessing species simply need to adduct the scape to avoid mating with a male that does not meet her mate choice criteria. In hub-mating *Argiope* species without scape such as *A. argentata* and *A. blanda*, females were observed to fend the males off with rapid leg movements (Uhl, unpublished). In *Argiope* species that mate on a mating thread (common, e.g., in Australian and Asian *Argiope* species), males remove a segment of the female web and bridge the gap by a multistrand mating thread on which the male performs vibratory courtship (Wignall and Herberstein 2013a). Females interested in mating leave the hub and enter the mating thread where they drop into an acceptance posture while clinging to the thread with at least four legs (Robinson and Robinson 1980). Males then start insertion approaches. If these are initially unsuccessful, females may move back to the hub and males resume courtship. Receptivity of the female is more obvious in thread mating species since she has to move actively onto the mating thread to copulate. Sometimes males produce several mating threads within a single web. However, this does not necessarily guarantee successful mating. Overall, it is clear that *Argiope* females have ample opportunity to reject their mates, however, detailed studies, including controlled experiments that explain which male traits lead to mate rejection are still lacking. Further, whether or not a female rejects a given male may depend on population density, i.e., on information on the presence of other receptive females. It is very likely that not only males but also females can perceive sexual pheromones emitted by conspecific females (Chinta et al. 2010).

When copulation occurs, it can consist of one or two insertions, each performed with a different pedipalp and a second insertion always requires that the male courts the female again. Immediately upon insertion, the male presses his body tightly against the female, which may help to resist female attacks. In *A. bruennichi*, females almost invariably crouch upon insertion and attack the males by scraping movements of the 3rd legs. With these movements, the female seems to push the male toward her mouth region. Relatively smaller males may have a survival advantage by slipping away between the female's legs, while relatively large males may survive copulation because the female cannot easily wrap them. Both hypotheses are not mutually exclusive and remain to be tested.

Insertions are performed in the ipsilateral mode, meaning that the right pedipalp inserts into the right female copulatory opening and the left pedipalp into the left opening. In *A. bruennichi*, two appendages of the genital bulb are responsible for external coupling to the scapus (Uhl et al. 2007). The so-called conductor does not conduct the apical sperm transferring sclerite, the embolus, during mating as the term suggests. Instead, the conductor couples to the scapus from below, whereas the median apophysis is pressed into a lateral groove on the outer surface of the scapus. Consequently, conductor and median apophysis work together to clamp the scapus. Most of the *Argiope* species possess a spur on the median apophysis that appears in most cases as a slightly bent picker of variable length (Figs. 3.1e and 3.2d, e). Experimental removal of the spur has surprising species-specific consequences. In *A. australis*, spur-less males are no longer able to copulate at all (Strauss, Nessler and Schneider unpublished data). In *A. bruennichi*, on the other hand, absence of the spur reduces male paternity success not because coupling is impeded or copulation duration reduced. Interestingly, males without the spur are no longer able to lodge the mating plug into the female genital opening (Nessler et al. 2007b). In *A. bruennichi,* it seems that a missing spur negatively impacts the lever function of the sperm transferring sclerites that may need to be bent in a specific way in order to place the mating plug exactly into the entrance of the copulatory duct. In *A. lobata* and in *A. australis* on the other hand, ablation of the spur seems to prevent successful coupling of the genitalia (Strauss, Nessler and Schneider, unpublished data). It is likely that the coupling mechanics in species that mate on a mating thread are different to those that mate at the hub; however, to date, we do not have comparative data.

Since male mating and paternity success is tightly linked to effective coupling and concerted interaction between male sclerites and female structures, female anatomy strongly determines male success. In addition to species-specific differences, female external genitalia can vary substantially intraspecifically, particularly in those species that are distributed over a wide range (Levi 1968, 1983). For example in *A. bruennichi*, the shape of the scapus varies significantly over the Palearctic distribution and some populations are considerably more variable in genital morphology than others and the differences in scapus shape reflect the genetic distances between populations (Krehenwinkel et al. submitted). Similar differences are expected in the world-wide distributed *A. trifasciata*, but here, shape differences are more difficult to measure due to the lack of a scapus. Widely distributed species are interesting targets for sexual selection research. By staging heteropopulation matings, predictions derived from cryptic female choice theory may be tested against those derived from sexually antagonistic coevolution (Hosken et al. 2002). Males from different populations may be better at manipulating females during courtship, mating, and genital interactions due to a lack of counter adaptations or males from same populations may be preferred due to better coupling abilities and cryptic female preferences.

3.8 Female Sexual Selection Strategies

3.8.1 Female Choice

It is well established that pre- and post-copulatory sexual selection can enhance each other but can also work in different directions (Hosken et al. 2008; Danielsson 2001). Mate choice in *Argiope* can occur pre- or post-copulatorily. Even though the focus of this chapter is cryptic female choice, we think that pre-copulatory choice is critical to complete the picture and better evaluate the importance of cryptic female choice. Therefore, we will briefly review what we know about pre-copulatory female choice in *Argiope* before we will review the evidences for cryptic female choice.

Generally, female spiders attract males to their webs by releasing a sex pheromone (Gaskett 2007). The volatile pheromone of *A. bruennichi* is produced by virgin females and constitutes a citric acid which is released from body and silk (Chinta et al. 2010). Males leave their webs after maturation and search for females. Once they encounter female silk, they quickly locate the edge of the capture web, and enter while performing characteristic vibrations (shudders) that likely signal their presence (Wignall and Herberstein 2013a, b). Females may be able to modulate the timing and the amount of sex pheromone. Behavioral observations suggest that females start producing the pheromone two days after the final molt (Schneider, Sauerland, Gatz and Zimmer, unpublished data), avoiding male attention while still soft and unable to fend off males. Opportunistic matings with molting females in the field do occur but do not seem to be promoted by female signaling (Uhl, Zimmer, Renner and Schneider, unpublished data), at least not in *A. bruennichi*. Nevertheless, males may find immature females haphazardly and stay nearby until the female matures. It is unclear how males gain information about the reproductive state of immature females. After mating, females stop producing the pheromone (Chinta et al. 2010). It remains to be investigated which sex is responsible for this change in signaling behavior.

A recent experimental study of *A. lobata* compared the probability of copulation within two hours of the initial encounter between virgin and previously mated females (Zimmer and Schneider, unpublished data). The results support the notion that females are not selective as virgins (although the males sometimes take long to engage in courtship). In double mating trials, virgin females and virgin males readily copulated with siblings. However, once-mated individuals of both sexes were reluctant to accept a sibling, independently of the genetic relatedness to their first mate. Non-sibling partners were generally accepted as second mating partners, and this was independent of whether a different or the same male had been their first mating partner (Zimmer & Schneider, unpublished manuscript). Much like *A. bruennichi* males (see above), *A. lobata* females seem to use a trading-up strategy by securing a sperm load as first priority regardless of the quality of the sperm and then become choosy during the second copulation (Zimmer and Schneider, unpublished data). These results correspond well with the results of another study on the same species in which females discriminate against sperm of related males but only if they were the second mate (see Welke and Schneider 2009). Reduced receptivity of mated females has been reported in other *Argiope* species such as *A. keyserlingi* (Herberstein et al. 2002).

To summarize our current knowledge on precopulatory female choice in *Argiope*, we can state that mated females are more reluctant to mate than virgin females and that male availability may be relevant in how likely females and males accept a randomly allocated mating partner. Future experiments should vary context by exposing females to cues that signal high or low mate availability and variation in mate quality to test how flexible female choice decision can be and whether the plasticity is adaptive.

3.8.2 Cryptic Female Choice

Cryptic female choice requires polyandry and as we have stressed above, not all females will mate multiply. Several scenarios are responsible for monandry: (i) males plug both genital openings and monopolize the female or (ii) females receive only a single mating from a male (due to male bigyny or male death) and do no re-mate. The latter scenario depends on male availability, on male willingness to mate with a mated female but also on female investment in mate attraction and mate acceptance (see above).

Given that females have the option to mate multiply and the possibility to choose between males, cryptic female choice can occur. Females of some *Argiope* species can likely control whether a male will survive his first copulation and will thereby manage the potential of another male gaining paternity. Thereby, sexual cannibalism could be directly selective if females target males that are of low quality. Alternatively, choice could be indirect in that females attack every male but only high-quality males survive the attack (Prenter et al. 2006). Double mating experiments with *A. lobata* do not support the notion that females discriminate against previous mates to actively increase the probability of polyandry. While once-mated females are generally less receptive than virgins, re-mating females were similarly likely to mate with her previous mate as with a novel male (Zimmer and Schneider, unpublished data). Importantly, this study prevented cannibalism experimentally reducing female selection against certain male phenotypes during her first copulation. This study supports the indirect mate choice scenario where females may test the ability of males to escape an attack and succeed in mating twice. Many more such experiments that also consider male influence (see above) are required to truly understand the selection pressures behind these mating interactions.

In addition to selective sexual cannibalism, females can modulate the timing of an attack and if copulation duration determines sperm transfer, females may control the number of spermatozoa that a male can transfer into a given spermatheca. A third mechanism that may be under female influence is the occurrence and success of plugging. By preventing male plugging, females can receive sperm from another male into an already used spermatheca. Once sperm from different males co-occur within the female, either in the same or in different spermathecae, females may be able to influence sperm storage and sperm activation. In the following section, we discuss the above options and review the evidences. We will focus exclusively on the female perspective and ignore the many potential interactions with male strategies and decisions for the sake of simplicity. We will restrict our assessment of sexual cannibalism, plugging, and copulation duration to matings between virgins. This is because males generally die during their second copulation (Gaskett et al. 2004).

3.8.3 Female Aggression and Sexual Cannibalism

In addition to the intention to test male survival ability, females may use various types of aggressive behavior to terminate copulation. How quickly genitalia disengage depends on the genital coupling mechanism and the degree of male control.

Female attacks during copulation and sexual cannibalism seem to be general features of the genus *Argiope* with only few (if any) exceptions. Table 3.1 shows that proportions of males that survive their first copulation vary between 100 and

28 % across species (note that the data on *A. perforata* are based on very small sample sizes). Potential explanations for the species-specific variation are speculative. The sexual conflict hypothesis states that female aggressiveness and resulting male mortality during first copulations should increase with increasing plug effectiveness (Schneider 2014). Alternatively, population density may modulate the costs and benefits of polyandry; under low density, a virgin female that kills her first male may risk an insufficient sperm supply if she will not find another mate. More explanations are conceivable, but the data do not yet reveal reliable patterns.

The most aggressive species known to date is *A. bruennichi* with rates of male mortality between 40 and 72 % reported in the literature (Schneider et al. 2006; Nessler et al. 2009b). The variation in male survival is likely due to male strategies (see above) since the female attack behavior is highly stereotyped and the attack occurs immediately upon genital coupling. While virgins are very receptive and were never observed to attack a courting male (at least under laboratory conditions), females immediately extrude large amounts of silk from the spinnerets and attempt to wrap the male after genital contact. Males that do not jump off within the first 10 s have no chance of survival. Males that copulate for less than 10 s increase their chance of survival to 60 % (Schneider et al. 2006; Nessler et al. 2009b; Welke and Schneider 2010). The fixed pattern of female behavior leaves no room for adjustments and may have evolved to prevent monopolization by the male. As it happens, *A. bruennichi* males are very good at placing a highly effective mating plug (Nessler et al. 2007a). Even though rates of polyandry are relatively low in *A. bruennichi* (Zimmer et al. 2012), a single sperm load is sufficient to fertilize all eggs of that females (Schneider et al. 2005), reducing the potential costs of sexual cannibalism.

In *A. lobata*, which also mates in the hub, female attacks are much more flexible and male mortality is about 46 % in spiders collected in Spain (Table 3.1). Interestingly, cannibalized males are more likely to leave a plug inside the female genital opening (Nessler et al. 2009a). Thus, cannibalism prevents males that were able to monopolize the contents of one spermatheca from returning and plugging the second one as well. Surviving males can return to inseminate the second spermatheca of the female, but they would not prevent a rival from reusing the first spermatheca. In order to disentangle the various influences on male reproductive success and the potential influence of females, a similar study with a much larger sample sizes is required. The timing of female attacks may also influence plugging success. For example, in *A. australis* males, the occurrence of genital damage increases with increasing copulation duration. Females in this species are very aggressive and end first copulations after less than 10 s (Strauss, Nessler and Schneider, unpublished data). As mentioned above, females may also actively remove plugs. It is unknown whether they can do this selectively.

A double mating study of *A. keyserlingi* has shown that smaller males mate for longer, transfer more sperm, and gain a higher relative paternity than larger males (Elgar et al. 2000). The proposed mechanism behind this pattern was the timing of female attacks that ended copulation. Later studies on the same species suggested that occurrence of sexual cannibalism changed the relationship between copulation duration and sperm transfer (Herberstein et al. 2011a). Experimental prevention of sexual cannibalism in *A. keyserlingi* did not affect duration of copulation nor average sperm transfer. However, the slope of the regression of copulation duration and sperm transfer was only positive if the male was cannibalized (Herberstein et al. 2011a). These data are intriguing and suggest complex dynamics and interacting female and male influences on relative paternity of males.

3.8.4 Controlling Copulation Duration

In other species, copulation duration is a good approximation of sperm transfer and ultimate paternity success—half of the sperm load within a pedipalp is emptied after 10 s of copulation and the whole load is transferred after 30 s in *A. bruennichi* (Fromhage et al. 2003; Schneider et al. 2006). Hence, it is not useful to generalize patterns of sperm transfer from one species to another. In species with relatively longer copulations, the first seconds of copulation may, for example, not be used for sperm transfer as new data on *A. lobata* suggest (Hirt and Schneider unpublished data). Furthermore, copulation can have additional functions such as transfer of accessory substances or copulatory courtship that may play an important role in post-copulatory sexual selection (Chapman 2006; Chapman et al. 2003). Different to insects, the functions of seminal fluids of spiders are still waiting to be investigated (Herberstein et al. 2011b). Moreover, in spiders, there are secretions present within the spermatheca that are produced by the female (e.g., Uhl 1994, 2000; Vöcking et al. 2013). Their function for sperm storage and activation is likely but also unknown. Hence, to date, there are no data to suggest additional functions of copulation duration for any *Argiope* species. Nevertheless, we can probably safely assume that the duration of copulation is relevant and that whichever sex is in control of this trait has a benefit.

3.8.5 Control of the Transfer and Storage of Sperm

The transfer of sperm will likely be a function of time, of male investment and perhaps condition and of female cooperation, but we lack studies that explore these interactions in greater detail. However, sperm storage likely is controlled by females alone. Spermatozoa are stored inside spermathecae until oviposition, which in *Argiope* can amount to several months. Males transfer encapsulated sperm and a large variety of secretions (Vöcking et al. 2013). Decapsulation and activation of sperm occurs non-synchronously suggesting that some sperm is not used during a fertilization event perhaps to be saved for the next clutch of eggs (Vöcking et al. 2013). Unfortunately, this is all we currently know about possible processes inside the female genital tract that influence sperm storage and activation and whatever follows. So far, we only have data on the number of spermatozoa inside the spermathecae and on paternity patterns.

The problem is that the number of sperm in a female's spermatheca could be a result of the number of sperm transferred by the male or the number of sperm stored by the female. However, we can perhaps infer from male and female reproductive strategies which is more likely. Because of the one-shot male genitalia, we have assumed in the past that males will always attempt to empty their pedipalp, because the remaining sperm will be wasted. Assuming this, the number of sperm in a spermatheca is likely to reflect the sperm stored by the female.

Selective sperm storage of polyandrous females has been demonstrated in an experimental study with *A. lobata*. Females received two matings from two different males that were either siblings or unrelated in all possible combinations (Welke and Schneider 2009). The prediction was that females should store less sperm from a related male in order to minimize negative effects on the offspring. Due to the two independent spermathecae connected via independent ducts to separate genital openings, it is easy to assign a male's sperm load to a specific spermatheca. In *Argiope*, the fixed insertion pattern makes it even easier as the side a male copulates into can be determined before copulation. By inducing autotomy of one of the two pedipalps, the two males that mated with the same female were matched to each copulate into a different side. The number of sperm stored in each spermatheca was subsequently counted. We found no difference in the number of sperm females had in storage from their first mates regardless of whether they came from a brother or an unrelated male. However, if the second mate was a brother, significantly fewer sperm were found in the spermatheca than if the male was unrelated (Welke and Schneider 2009). There were no differences in copulation duration and rates of cannibalism between sibling and non-sibling pairs suggesting that males did not modulate their investment and increased their chances of survival as found in *A. bruennichi* (Welke and Schneider 2010).

A study using *A. bruennichi* demonstrated cryptic female in another context, namely in terms male investment in courtship (Schneider and Lesmono 2009). Undisturbed male *A. bruennichi* spend an average of 5 min on courtship during which they move around and touch the female. However, as soon as there is another male in the web, courtship is drastically shortened to less than 60 s, and males will scramble for getting prior access to the female and immediately engage in copulation. This behavior was exploited to experimentally manipulate the courtship of focal males and compare their relative paternity to control males who went through the normal courtship sequence. In the experimental treatment, a male with both pedipalps autotomized functioned as eunuch competitor and induced that the focal male curtailed courtship. Males were allowed a single copulation, and it was documented which pedipalp was used. Second males were introduced and allowed a single copulation that occurred into the same or a different spermatheca. Relative paternity was determined using the sterile male technique (see Schneider and Andrade 2011 for an explanation of the method). Paternity of non-courting males was significantly lower than of courting males. No differences were found between the treatment groups in any phenotypic measurement nor in the duration of copulation. However, the effects only occurred in trials, in which both males mated into different genital openings. Males that did not court were equally successful in placing a mating plug and effectively prevented second males from gaining paternity (Schneider and Lesmono 2009). The results strongly suggest that females select against males that did not court. A non-adaptive alternative explanation that cannot be ruled out could be that males require courtship for increasing sperm transfer. To date, no study investigated possible constraints on male sperm transfer in *Argiope* nor in any other spider.

3.9 Conclusions and Outlook

Spiders of the genus *Argiope* lend themselves in particular for studying cryptic female choice not only because of their specific biology and mating system but also because a solid foundation of research is already laid. Experimentation is particularly easy in *Argiope* due to a fixed copulation pattern and the defined male mating rates of maximally two. In addition, *Argiope* can be studied in the field and in the laboratory. Variation is present between populations within species and between species providing promising opportunities to investigate ecological causes underlying the variation in sexual selection and sexual conflict. The broad distribution of the genus and even some species provides ideal opportunities for relating environmental factors such as climate, seasonality, and phenology to mating strategies.

In mono-/bigynous mating systems, both sexes have similar interests in finding a high-quality partner as maximal mating rates of both sexes are very similar. While this is a fortunate setup in many ways, it can make experiments more difficult to interpret as mate rejection or different mating investment occurs in males and in females and might be hard to tell apart. The often cryptic nature of sexual communication via chemical, vibratory, and tactile channels is difficult to disentangle but offers promising future research avenues—female modulation of signals and cues is just as interesting as male reactions. Only by studying communication between the sexes will we be able to fully comprehend the cues that females may base their choices on. Furthermore, the particular mating system of *Argiope* permits comparisons and analyses of variation in female mating rates without confounding variation in male mating rates since males of most or even all species possess one-shot genitalia and hence the same limitations.

Comparative data are slowly accumulating and can soon be analyzed using a comparative approach based on a phylogenetic hypothesis (Cheng and Kuntner 2014). In this book chapter, we presented evidence for cryptic female choice defined as variation in paternity in several stages of the mating process, starting with the duration of copulation, over the application of a plug and the potential of re-mating as well as selective sperm storage.

Nevertheless, the greatest gaps in our knowledge concern the processes between genital contact and fertilization, which are indeed particularly important in the context of cryptic female choice. We know little about causes of variation in sperm transfer and have no insights in female influences on sperm storage, sperm activation, and transport to the site of fertilization, although the latter aspects must be under female control. First efforts have been made (Vöcking et al. 2013) and are to be continued. In short, the genus *Argiope* offers exciting opportunities and avenues of research that unravel male and female influences of paternity.

References

Chapman T (2006) Evolutionary conflicts of interest between males and females. Curr Biol 16(17):R744–R754

Chapman T, Arnqvist G, Bangham J, Rowe L (2003) Sexual conflict. Trends Ecol Evol 18(1):41–47

Cheng RC, Kuntner M (2014) Phylogeny suggests nondirectional and isometric evolution of sexual size dimorphism in argiopine spiders. Evolution 68–10:2861–2872

Chinta SP, Goller S, Lux J, Funke S, Uhl G, Schidz S (2010) The sex pheromone of the wasp spider *Argiope bruennichi*. Angew Chem Int Edit 49(11):2033–2036

Danielsson I (2001) Antagonistic pre- and post-copulatory sexual selection on male body size in a water strider (*Gerris lacustris*). Proc R Soc B Biol Sci 268(1462):77–81

Eberhard WG (1996) Female control: sexual selection by cryptic female choice. Princeton University Press, Princeton

Elgar MA (1991) Size dimorphism in sexually cannibalistic orb-weaving spiders covaries with courtship behaviour. Evolution 45:444–448

Elgar MA, Schneider JM, Herberstein ME (2000) Female control of paternity in the sexually cannibalistic spider *Argiope keyserlingi*. Proc R Soc B Biol Sci 267(1460):2439–2443

Foellmer MW (2008) Broken genitals function as mating plugs and affect sex ratios in the orb-web spider *Argiope aurantia*. Evol Ecol Res 10(3):449–462

Foellmer MW, Fairbairn DJ (2003) Spontaneous male death during copulation in an orb-weaving spider. Proc R Soc B Biol Sci 270:S183–S185

Foellmer MW, Fairbairn DJ (2004) Males under attack: sexual cannibalism and its consequences for male morphology and behaviour in an orb-weaving spider. Evol Ecol Res 6(2):163–181

Fromhage L, Elgar MA, Schneider JM (2005) Faithful without care: the evolution of monogyny. Evolution 59(7):1400–1405

Fromhage L, McNamara JM, Houston AI (2008) A model for the evolutionary maintenance of monogyny in spiders. J Theor Biol 250(3):524–531

Fromhage L, Schneider JM (2012) A mate to die for? A model of conditional monogyny in cannibalistic spiders. Ecol Evol 2(10):2572–2582

Fromhage L, Uhl G, Schneider JM (2003) Fitness consequences of sexual cannibalism in female *Argiope bruennichi*. Behav Ecol Sociobiol 55(1):60–64

Gaskett AC (2007) Spider sex pheromones: emission, reception, structures, and functions. Biol Rev 82(1):26–48

Gaskett AC, Herberstein ME, Downes BJ, Elgar MA (2004) Changes in male mate choice in a sexually cannibalistic orb-web spider (Araneae: Araneidae). Behaviour 141:1197–1210

Ghione S, Costa FG (2011) Female attack is not necessary for male copulatory organ breakage in the sexually cannibalistic spider *Argiope argentata* (Araneae: Araneidae). J Arachnol 39(1):197–200

Grasshoff M (1973) Bau und Mechanik der Kopulationsorgane der Radnetzspinne *Mangora acalypha* (Arachnids, Araneae). Zeitschrift für Morphologie der Tiere 74(3):241–252

Herberstein ME, Barry KL, Turoczy MA, Wills E, Youssef C, Elgar MA (2005a) Post-copulation mate guarding in the sexually cannibalistic St Andrew's cross spider (Araneae Araneidae). Ethol Ecol Evol 17(1):17–26

Herberstein ME, Gaskett AC, Schneider JM, Vella NGF, Elgar MA (2005b) Limits to male copulation frequency: sexual cannibalism and sterility in St Andrew's cross spiders (Araneae, Araneidae). Ethology 111(11):1050–1061

Herberstein ME, Schneider JM, Elgar MA (2002) Costs of courtship and mating in a sexually cannibalistic orb-web spider: female mating strategies and their consequences for males. Behav Ecol Sociobiol 51(5):440–446

Herberstein ME, Schneider JM, Harmer AMT, Gaskett AC, Robinson K, Shaddick K, Soetkamp D, Wilson PD, Pekar S, Elgar MA (2011a) Sperm storage and copulation duration in a sexually cannibalistic spider. J Ethol 29(1):9–15

Herberstein ME, Schneider JM, Uhl G, Michalik P (2011b) Sperm dynamics in spiders. Behav Ecol 22(4):692–695

Herberstein ME, Wignall AE, Nessler SH, Harmer AMT, Schneider JM (2012) How effective and persistent are fragments of male genitalia as mating plugs? Behav Ecol 23(5):1140–1145

Hosken DJ, Blanckenhorn WU, Garner TWJ (2002) Heteropopulation males have a fertilization advantage during sperm competition in the yellow dung fly (*Scathophaga stercoraria*). Proc R Soc B Biol Sci 269(1501):1701–1707

Hosken DJ, Stockley P (2004) Sexual selection and genital evolution. Trends Ecol Evol 19(2):87–93

Hosken DJ, Taylor ML, Hoyle K, Higgins S, Wedell N (2008) Attractive males have greater success in sperm competition. Curr Biol 18(13):R553–R554

Jäger P (2012) Review on the spider genus *Argiope* Audoin 1826 with special emphasis on broken emboli in female epigynes (Araneae: Araneidae: Argiopinae). Beiträge zur Araneologie 7:272–331

Levi HW (1968) The spider genera *Gea* and *Argiope* in America. Bull Mus Comp Zool 136:319–352

Levi HW (1983) The orb-weaver genera *Argiope*, *Gea* and *Neogea* from the Western Pacific region (Araneae: Araneidae, Argiopinae). Bull Mus Comp Zool 150:247–338

Michalik P, Rittschof CC (2011) A comparative analysis of the morphology and evolution of permanent sperm depletion in spiders. Plos One 6(1). doi:10.1371/journal.pone.0016014

Miller JA (2007) Repeated evolution of male sacrifice behavior in spiders correlated with genital mutilation. Evolution 61(6):1301–1315

Nessler S (2009) Adaptive value of male genital mutilation, sexual cannibalism, and mating strategies in the orb-weaving spider genus *Argiope* (Araneae: Araneidae). University of Hamburg, Hamburg

Nessler S, Uhl G, Schneider J (2009a) Sexual cannibalism facilitates genital damage in *Argiope lobata* (Araneae: Araneidae). Behav Ecol Sociobiol 63(3):355–362

Nessler SH, Uhl G, Schneider JM (2007a) Genital damage in the orb-web spider *Argiope bruennichi* (Araneae: Araneidae) increases paternity success. Behav Ecol 18(1):174–181

Nessler SH, Uhl G, Schneider JM (2007b) A non-sperm transferring genital trait under sexual selection: an experimental approach. Proc R Soc B Biol Sci 274(1623):2337–2341

Nessler SH, Uhl G, Schneider JM (2009b) Scent of a woman—the effect of female presence on sexual cannibalism in an orb-weaving spider (Araneae: Araneidae). Ethology 115(7):633–640

Prenter J, MacNeil C, Elwood RW (2006) Sexual cannibalism and mate choice. Anim Behav 71:481–490

Robinson MH, Robinson B (1980) Comparative studies of the courtship and mating behaviour of tropical araneid spiders. Pacific Insects Monogr 36:1–218

Sasaki T, Iwahashi O (1995) Sexual cannibalism in an orb-weaving spider *Argiope aemula*. Anim Behav 49(4):1119–1121

Schneider J, Fromhage L (2010) Monogynous mating strategies in spiders. In: Kappeler P (ed) Animal behaviour: evolution and mechanisms. Springer, Heidelberg, pp 441–464

Schneider JM (2014) Sexual cannibalism as a manifestation of sexual conflict. In: Rice RW, Gavrilets S (eds) The genetics and biology of sexual conflict. Cold spring harbor perspectives in biology. Cold Spring Harbor Laboratory Press, New York

Schneider JM, Andrade MCB (2011) Mating behaviour and sexual selection. In: Herberstein M (ed) Spider behaviour: flexibility and versatility. Cambridge University Press, Cambridge, pp 215–274

Schneider JM, Fromhage L, Uhl G (2005) Extremely short copulations do not affect hatching success in *Argiope bruennichi* (Araneae, Araneidae). J Arachnol 33(3):663–669

Schneider JM, Gilberg S, Fromhage L, Uhl G (2006) Sexual conflict over copulation duration in a cannibalistic spider. Anim Behav 71:781–788

Schneider JM, Lesmono K (2009) Courtship raises male fertilization success through post-mating sexual selection in a spider. Proc R Soc B Biol Sci 276(1670):3105–3111

Schneider JM, Michalik P (2011) One-shot genitalia are not an evolutionary dead end—regained male polygamy in a sperm limited spider species. BMC Evol Biol 11:197

Schulte KF, Uhl G, Schneider JM (2010) Mate choice in males with one-shot genitalia: limited importance of female fecundity. Anim Behav 80(4):699–706

Schwartz SK, Wagner WE, Hebets EA (2013) Spontaneous male death and monogyny in the dark fishing spider. Biol Lett 9(4). doi:10.1098/rsbl.2013.0113

Schwartz SK, Wagner WE, Hebets EA (2014) Obligate male death and sexual cannibalism in dark fishing spiders. Anim Behav 93:151–156

Uhl G (1994) Ultrastructure of the accessory glands in female genitalia of *Pholcus phalangioides* (Fuesslin, 1775) (Pholcidae, Araneae). Acta zool (Stockholm) 75(1):13–25

Uhl G (2000) Two distinctly different sperm storage organs in female *Dysdera erythrina* (Araneae: Dysderidae). Arthropod Struct Dev 29:163–169

Uhl G, Elias DO (2011) Communication. In: Herberstein M (ed) Spider behaviour: flexibility and versatility. Cambridge University Press, Cambridge, pp 127–189

Uhl G, Nessler S, Schneider J (2010) Securing paternity in spiders? A review on occurrence and effects of mating plugs and male genital mutilation. Genetica 138(1):75–104

Uhl G, Nessler SH, Schneider J (2007) Copulatory mechanism in a sexually cannibalistic spider with genital mutilation (Araneae: Araneidae: *Argiope bruennichi*). Zoology 110(5):398–408

Vöcking O, Uhl G, Michalik P (2013) Sperm dynamics in spiders (Araneae): ultrastructural analysis of the sperm activation process in the garden spider *Argiope bruennichi* (Scopoli, 1772). Plos One 8(9). doi:10.1371/journal.pone.0072660

Welke K, Schneider JM (2009) Inbreeding avoidance through cryptic female choice in the cannibalistic orb-web spider *Argiope lobata*. Behav Ecol 20(5):1056–1062

Welke KW, Schneider JM (2010) Males of the orb-web spider *Argiope bruennichi* sacrifice themselves to unrelated females. Biol Lett 6(5):585–588

Welke KW, Zimmer SM, Schneider JM (2012) Conditional monogyny: female quality predicts male faithfulness. Front Zool 9. doi:10.1186/1742-9994-9-7

Wignall AE, Herberstein ME (2013a) The influence of vibratory courtship on female mating behaviour in orb-web spiders (*Argiope keyserlingi*, Karsch 1878). Plos One 8(1). doi:10.1371/journal.pone.0053057

Wignall AE, Herberstein ME (2013b) Male courtship vibrations delay predatory behaviour in female spiders. Sci Rep 3. doi:10.1038/srep03557

Zimmer SM, Schneider JM, Herberstein ME (2014) Can males detect the strength of sperm competition and presence of genital plugs during mate choice? Behav Ecol 25(4):716–722

Zimmer SM, Welke KW, Schneider JM (2012) Determinants of natural mating success in the cannibalistic orb-web spider *Argiope bruennichi*. Plos One 7(2). doi:10.1371/journal.pone.0031389

Chapter 4
Post-copulatory Sexual Selection in Two Tropical Orb-weaving *Leucauge* Spiders

Anita Aisenberg, Gilbert Barrantes and William G. Eberhard

Abstract This chapter focuses on descriptive and experimental studies of the sexual biology of two spider species, *Leucauge mariana* and *Leucauge argyra*. We examine general questions related to female effects on paternity by taking advantage of several unusual traits: direct female participation in forming copulatory plugs and physical clasping by the female rather than the male during copulation in both species; and derived traits in the genitalia of males and females and occasional female cannibalism of conspecific males by trapping the male's genitalia in adhesive copulatory plugs in *L. argyra*. These derived traits, combined with other aspects of sexual interactions that are typical of spiders in general, such as complete, easy female avoidance of unwanted sexual advances of males, copulatory courtship by males, and imperfect male ability to remove copulatory plugs from the female's genitalia, constitute a rich mixture from which clear lessons can be drawn. Post-copulatory sexual selection probably acts on male genitalia and their behavior in both species. Two derived male genitalic clasping and clamping devices *L. argyra* may be associated with the extremely aggressive female behavior. One genital structure in male *L. mariana* may have evolved under sexual selection by sperm competition to remove copulatory plugs from female genitalia. In contrast to expectations from theory based on male–female conflict, the female genitalia of *L. argyra* has a derived trait that selectively facilitates rather than impedes male genital clamping. In contrast to

---

A. Aisenberg
Laboratorio de Etología, Ecología y Evolución,
Instituto de Investigaciones Biológicas Clemente Estable,
Avenida Italia 3318, CP 11600 Montevideo, Uruguay
e-mail: anita.aisenberg@gmail.com

G. Barrantes · W.G. Eberhard
Escuela de Biología, Universidad de Costa Rica, Ciudad Universitaria, Costa Rica
e-mail: gilbert.barrantes@gmail.com

W.G. Eberhard (✉)
Smithsonian Tropical Research Institute, Life Sciences and Museum of Natural History,
Louisiana State University, Baton Rouge, LA 70808, USA
e-mail: william.eberhard@gmail.com

A.V. Peretti and A. Aisenberg (eds.), *Cryptic Female Choice in Arthropods*,
DOI 10.1007/978-3-319-17894-3_4

lock-and-key expectations, another prominent, derived female genital structure in *L. argyra* does not mesh with any male structure during copulation; it may instead function in sensing male genital movements. Additional non-genital male traits associated with copulation are also probably under post-copulatory sexual selection in both species. Experimental manipulations of male chelicerae and of possible female receptors of stimuli from the male chelicerae demonstrated that stimulation from secondary sexual modifications of the male chelicerae influences several female reproductive processes in *L. mariana*, including receptivity to remating, female interruption of copulation, and copulatory plug formation. Additionally, two male copulatory courtship behavioral patterns correlate with greater plug formation in *L. mariana*, and one pre-copulatory courtship behavioral pattern in males of *L. argyra* correlates with plug formation, suggesting additional possible male effects on post-copulatory female reproductive processes. Other male sexual behavior correlates with the timing of female emission of plug material that may affect a male's chances of surviving the encounter. In sum, female *Leucauge* have powerful effects on whether copulation occurs, when it will end, whether a mating plug will be formed, and whether the male will survive the encounter. Several genital and non-genital male traits likely evolved under a mix of post-copulatory sexual selection (sperm competition and probably cryptic female choice [CFC]) to elicit favorable modulation of these female effects.

4.1 Introduction

Sexual selection by cryptic female choice (CFC), as described in Chap. 1, is feasible in many different animal groups (Eberhard 1996). Whether or not CFC is actually widespread in nature, however, is yet to be determined. The kinds of data needed to test for CFC and to discriminate it from other post-copulatory processes such as sexually antagonistic coevolution (SAC), sperm competition, and species isolation are only available in a limited number of animal groups (see Chap. 1). The present chapter summarizes descriptive and experimental studies that have expanded the taxonomic range of CFC studies to include two species in the large, widespread tetragnathid spider genus *Leucauge*, *L. mariana*, and *Leucauge argyra* (Fig. 4.1). The first section includes background information that provides a context in which to evaluate the significance of the results of the behavioral correlations and experimental manipulations that are described in the second section.

Studies of *Leucauge* sexual biology illustrate how concentrating attention on a particular group can reveal multiple mechanisms of post-copulatory sexual selection acting on multiple male characters. Questions related to female control of paternity are especially susceptible to study in *Leucauge*, due to several unusual traits (Eberhard and Huber 1998a; Aisenberg and Eberhard 2009; Aisenberg and Barrantes 2011): Females participate directly in forming copulatory plugs; the female physically clasps the male during copulation rather than vice versa; there are several major derived genital traits in the males and females of *L. argyra*; and females occasionally cannibalize conspecific males in this species by trapping the male's genitalia in adhesive copulatory plugs. These derived traits are

Fig. 4.1 **a** *Leucauge mariana*; **b** *Leucauge argyra*; **c**, **d** mating plugs on the genitalia of field-collected *Leucauge mariana* (*left*) and *L. argyra* (*right*)

accompanied by some other aspects of sexual interactions that are typical of spiders in general, including easy and effective female avoidance of unwanted sexual advances of males; copulatory courtship by males (Huber 1998, 2005); transfer of immobile, encapsulated sperm; and a male ability to remove some but not all copulatory plugs from female genitalia (e.g., Masumoto 1993). We will give general background information that is important to interpret observations of reproductive behavior, then focus on several different topics that were examined in different studies, and close with a general discussion.

4.2 Background on *Leucauge* Sexual Biology

Members of this large genus of 150 or more species are medium-sized orb weavers (approximately 40–100 mg) which build more or less horizontal webs. To date, published detailed descriptions of the sexual behavior are available for only two species, *L. mariana* and *L. argyra*; there are also brief observations of *Leucauge regnyi* by Alayón (1979), and unpublished observations of *Leucauge venusta* by Castro (1995). These species were chosen not because they were known to be interesting with respect to CFC, but for the more practical reason that they are both very common in easily accessible habitats and copulate readily in captivity where their genitalia can be observed under a dissecting microscope.

Both are very abundant in early second-growth vegetation at some sites in Costa Rica. Population densities of *L. mariana* in the Valle Central are highly seasonal (Méndez 2002); those of *L. argyra*, which occur in a habitat with more highly seasonal rainfall along the Pacific coast of Costa Rica, nevertheless seem to be more stable (careful counts have not been made, however). In the past, both species presumably lived in natural second growth 'vegetation', such as along the edges of creeks and rivers, landslides, and tree-fall gaps. The localized but persistent nature of second-growth sites along water courses in the past may have led to locally dense populations. This is important, because the chances that multiple males would find and attempt to mate with any given female probably increase with greater population density.

The longevity of mature males (henceforth "males") has never been measured in the field, but there are two indications that they may survive for up to several weeks; field-captured males (of undetermined ages) generally survive for a week or more in captivity; and males in the field evict medium-sized nymphs from their orbs for several hours in the field and use it for prey capture (often, when a male is found at the hub of an orb, there is an immobile nymph on a peripheral frame or anchor line). In *L. mariana*, males also sometimes ingest silk from the webs of mature females (WE unpub.). Long-term survival of males would increase the density of males and thus the chances that a female will be courted by multiple males during her lifetime. Incidental observations in the field and in captivity suggest that females also live for several weeks. Females probably lay multiple clutches of eggs in nature, because females which laid eggs in captivity subsequently built orbs.

Several other aspects of the sexual biology of *Leucauge* in the field also provide perspectives for the observations that are described below. Males of both *L. mariana* (Eberhard et al. 1993; Eberhard and Huber 1998; Méndez 2002) and *L. argyra* (A. Aisenberg, G. Barrantes, and W. Eberhard, unpub.) have two different mating strategies. Males often occupy the small tangle web that is built by a penultimate female nymph just prior to molting (males were never found in similar webs built by penultimate male nymphs, which are of similar size). Any female nymph found with an accompanying male invariably molts within 1–2 days, and the male then mates with her in the first hours after she has molted. Males also occur on the orb webs of mature females (henceforth "females"), and court and mate with them there. Males fight with each other on the webs of both adult and penultimate females (Méndez 2002; A. Aisenberg, G. Barrantes, W. Eberhard, unpub.). Aggressive male defense of both penultimate and mature females suggests that both the first male to mate with a female and also later males achieve some paternity.

Insemination (at least in virgin *L. mariana* females) occurs early rather than late during the approximately 15-min copulation (Eberhard and Huber 1998a). The male intromittent structure, the embolus, is a thin, hollow, hairlike sclerite. The sperm in both species are stored in the female in the large, soft-walled "chamber I" of the spermatheca, which is connected with additional, complex, hard-walled chambers deeper in the female's body that differ greatly in form in *L. mariana* and *L. argyra* (Eberhard and Huber 1998a; Triana and Quesada, in prep.). Measurements of the lengths of the portion of the male embolus that can be everted and of the insemination duct of the female in *L. mariana* showed that the embolus probably reaches into the lumen of chamber I during copulation (Eberhard and Huber 1998a).

Two details of the sexual morphology and behavior of these species are especially important in the context of possible post-copulatory sexual selection (we will use "post" in the usual sense in such discussions, of any event occurring after the beginning of a copulation) and merit special attention: mating pairs are held together with a cheliceral clasp, in which the female grasps the male, rather than vice versa; and females in the field often have "plugs" of material adhering to their genitalia on or near the openings to their insemination ducts (Fig. 4.1).

4.3 Pre-copulatory Behavior

Male *L. mariana* court females with diverse behaviors, most of which probably produce vibrations in the web that are perceived at a distance by the female (we will use "courtship" to refer to behavior that was repeated both within and between pairs, that obviously resulted in stimuli being received by the other spider, and that had no obvious mechanical function in bringing and keeping the spiders together; the term "copulation" is used to include the period during a cheliceral clasp, including but not restricted to genitalic contact; the term "insertion" designates the entrance of the embolus and conductor into the mouth of the insemination duct on the epigynum). Males performed at least seven types of courtship behavior (Eberhard and Huber 1998a; Aisenberg 2009; Aisenberg and Eberhard 2009): jerks, rocking, abdomen bobbing, palp rubbing, twanging, line tapping, and tapping the female. None of these types of courtship physically coerces the female in any way.

Sexually receptive females of *L. mariana* gave four behavioral responses: (1) Turn to face males: Sometimes they turned only after the males had performed repeated bouts of courtship behavior; males never responded to females that were facing in some other direction by moving to place themselves in front of females. All orientation movements, when they occurred, were performed by females; (2) Open chelicerae: Females often repeatedly opened and closed their chelicerae (both the basal segments and the fangs) prior to linking with males; presumably, these were intention or exploratory movements associated with cheliceral clasping. Occasionally, when females had not opened their fangs, males made repeated small lunging movements nearby, in an apparent attempt to induce females to open their fangs; (3) Assume mating position: Just prior to copulation, females lowered their body and spread their anterior legs, thus making space for males to approach them close enough for cheliceral clasp and insertion of their palps; and (4) Bend abdomen ventrally: Ventral flexion of females' abdomen was probably critical, at least when males were small with respect to females. In two cases, females returned their abdomen to the more typical orientation, while males were attempting to insert their palps, and in one of these pairs, the male was then unable to reach her epigynum with his palp.

In pre-copulatory courtship in *L. argyra*, the males perform jerks, palpal rubbing, twangs, tap lines, and tap the female, as well as an additional behavior, foreleg rubbing (Aisenberg and Barrantes 2011). All the courtship behavioral patterns were usually performed while the male was on the same line or lines on which the female was resting, and could thus convey vibratory stimuli to her.

The important point, with respect to possible CFC as well as the species isolation and male–female conflict hypotheses (see below), is that a female *Leucauge* has received a rich set of stimuli from the male before he ever contacts her; and she has had ample opportunity to reject his advances by failing to respond in any of several ways. The fact that females must be persuaded to perform four different behaviors for copulation to occur suggests that pre-copulatory male courtship

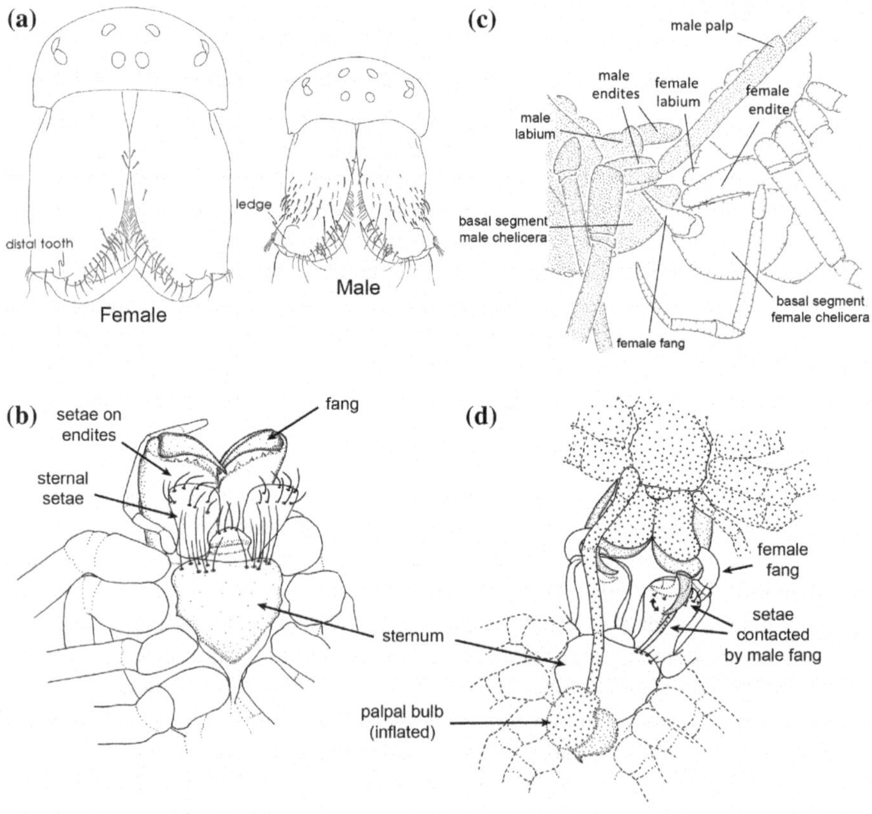

Fig. 4.2 Cheliceral dimorphism in anterior view (**a**), female endite and sternal setae in ventral view (**b**), and cheliceral clasp in lateral (**c**) and ventral (**d**) view in *L. mariana*. The anterior surface of the male chelicerae (**a**) has more setae and a small plate or ledge near the base of the fang. The anterior edge of the sternum of the female (**b**) has longer setae than those of the male, and these are deflected by movements of the male fangs during a clasp (**c**, **d**). The female's fang clamps the distal portion of the basal segment of the male's chelicera (**c**) (**a** from Eberhard and Huber 1998a; **b–d** from Aisenberg et al. 2015)

functions to induce these female acceptance responses. In addition, there is no way that a male can force a female to copulate in either species. For a mating to occur, the female must turn to face the male, spread her anterior legs, open her chelicerae to allow him to insert his basal cheliceral segment between her fang and basal segment, grasp the male's chelicerae with her fangs (Fig. 4.2), and tilt her abdomen ventrally to bring her epigynum within reach of the male's palpal bulb. The male has no way to force any of these female responses, even with a newly molted female. When a female fails to respond to a male's courtship, he eventually desists and leaves (interactions with newly molted females have not been observed often enough, however, to comment on male persistence in this context).

4.4 Copulation

Copulation is a complex process in *Leucauge*, and males perform highly repetitive leg and cheliceral movements in addition to the repeated movements with both their entire palps and with particular sclerites of the palpal bulb. Again, the most complete descriptions are available for *L. mariana* (Eberhard and Huber 1998a), and the descriptions that follow refer to this species (for comparisons, see Aisenberg and Barrantes 2011, Sect. 4.4.4 on *L. argyra*).

4.4.1 Movements of the Female and Male's Body and Legs

During copulation, males performed at least three types of the apparent courtship behavior that also occurred prior to copulation: leg tapping, abdomen bobbing, and rocking. Tapping often occurred during the first moments after the female grasped the male chelicerae and the male attempted to insert his palp, and also during the withdrawal of one palp and insertion of the other. Males also performed an additional behavior, leg pushing, which did not occur prior to cheliceral clasping. Each of the male's four anterior legs contacted the corresponding legs of the female, and the male's legs were repeatedly extended synchronously to push gently against the female's legs. Usually, the male contacted the female with the distal portions of his legs I and II (tarsi, metatarsi). In most cases, the male's legs III and IV were immobile. Bursts of leg pushing began when the basal hematodocha of the palp was inflated.

These behavioral patterns are important in the context of possible CFC in two respects. First (assuming that male behavior is adaptive), they demonstrate that there has likely been selection on males to elicit further female cooperation with their mating attempts, even after the pair has coupled. Secondly, they continue the theme seen in pre-copulatory interactions, in that the male's behavior was not physically coercive. Tapping, abdomen bobbing, rocking, and leg pushing in no way physically force the female to respond in any particular way.

4.4.2 Cheliceral Clasping

Most species in Tetragnathidae have large and conspicuous chelicerae which are larger in males, and the female and male clasp their chelicerae prior and during mating (Bristowe 1929; Eberhard and Huber 1998a; Álvarez-Padilla et al. 2009, 2011). Male–female dimorphism (with male chelicerae larger) has evolved at least five times in this family (Álvarez-Padilla and Hormiga 2011). The morphological designs and use reflect the typical male–female roles in *Tetragnatha* and *Pachygnatha*. The basal segment of the male chelicera has a distal, spurlike processwhich appears to serve the defensive function of arresting the female's fang and preventing it from closing, while the male's own fang clamps the basal segment of the female and appears to hold the pair of spiders together (Bristowe 1929). Cheliceral locking may thus anchor the male more securely to the female's body during mating, and this could facilitate the coupling of their relatively simple genitalia (Levi 1981; Kraus 1984).

The chelicerae of *Leucauge* are used differently. It is the female who seizes the male's chelicerae during copulation, rather than vice versa (Castro 1995; Eberhard and Huber 1998a). The female spreads her basal segments and opens her fangs, and the male then inserts the distal portion of each basal segment (with his fangs closed) between the female's fang and her basal segment; she then closes her fang, clamping the distal portion of the male's basal segment (Fig. 4.2). The chelicerae are not sexually dimorphic in length (Fig. 4.3), but, at least in *L. mariana*, *L. argyra*, and *L. venusta*, the male chelicerae have sexually dimorphic macrosetae, distal lateral plates, and corrugations (Figs. 4.2 and 4.3a). The designs of these modifications do not show any signs of being mechanically useful to defend against the female or to stabilize male–female cheliceral grasps. These sex-specific structures seem to be more appropriate for stimulating the female. There is further diversity in the genus; another, unidentified species of *L.* sp. (perhaps near *moerens*?) did not perform cheliceral clasps and lack sexual dimorphism in male and female chelicerae (Barrantes et al. in press).

Male chelicerae also show other modifications in tetragnathids. A sexually dimorphic, rugose cuticular texture of the male chelicerae has evolved at least five times (Álvarez-Padilla and Hormiga 2011). The basal segments of the chelicera of male *L. argyra* have a rugose area near the distal, lateral corner, but those of *L. mariana* are smooth (Fig. 4.3). Males of both of these species as well as other *Leucauge* and the related genera *Mesida* and *Opadometa* have more abundant macrosetae on the anterior surfaces of their chelicerae than do females (Álvarez-Padilla and Hormiga 2011). This sexually dimorphic cheliceral trait may be species-specific in at least some *Leucauge* species (H.W. Levi unpublished figures; Castro 1995).

In both *L. mariana* and *L. argyra*, the female's chelicerae seize the male's chelicerae before palpal insertion begins, and the cheliceral clasp is usually maintained throughout most if not all the copulation (Eberhard and Huber 1998a; Aisenberg 2009; Aisenberg and Eberhard 2009; Aisenberg and Barrantes 2011). Mating ends when the female releases the male's chelicerae. Male chelicerae are narrower in the region (just short of the tip) that the female clasps with her chelicerae (Fig. 4.2a; Eberhard and Huber 1998a; Aisenberg et al. 2015).

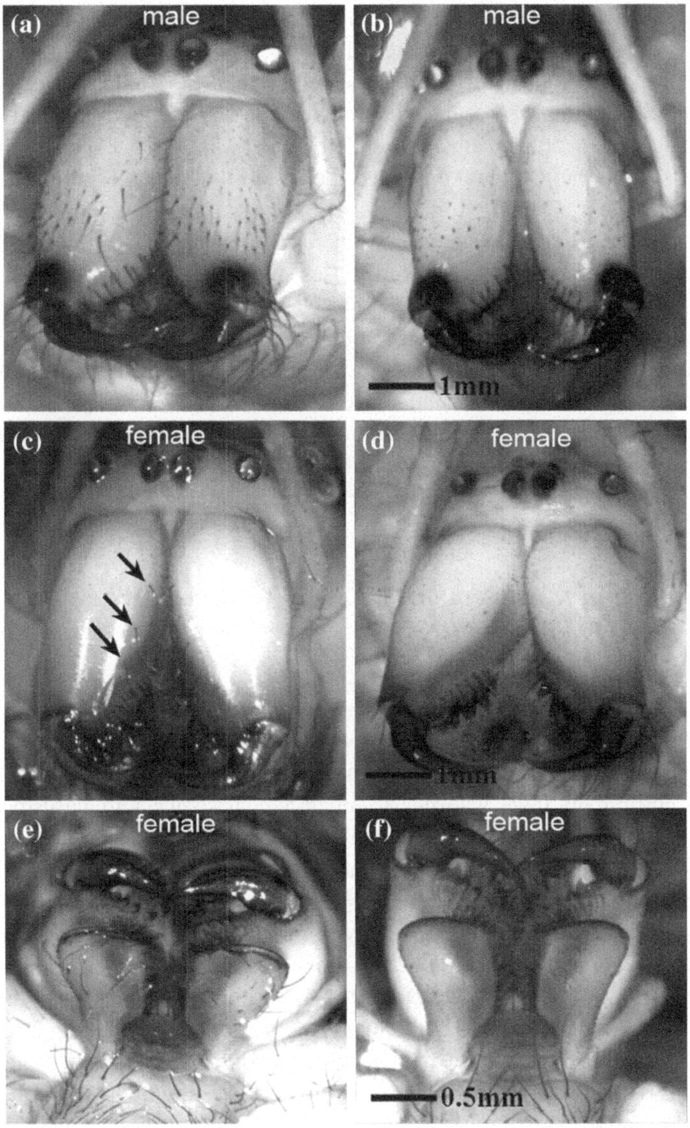

Fig. 4.3 Frontal views of the chelicerae of female (**a**, **b**) and male (**c**, **d**) *Leucauge mariana*; **a**, **c** are unmodified, while **b**, **d** have their setae shaved off. The endites and labium of female, before (**e**) and after (**f**) setae, were shaved off. The setae along the margins of both the chelicerae and the endites, which presumably function when the spider bites and ingests prey, were left intact (figure from Aisenberg et al. 2015)

The morphological details of the cheliceral clasps observed in *L. mariana* were as follows (Eberhard and Huber 1998a; Aisenberg et al. 2015). The female always opened her chelicerae wide as the male approached (usually with his own chelicerae closed) and then grasped the distal portions of the basal segments of the male's chelicerae by closing her fangs. The inner surface of her fang grasped the posterior surface of the male's chelicerae rather than his endites (Fig. 4.2d). The distal portion of the male's basal segment projected beyond the female's chelicerae, and his fangs often opened and closed while he was being clasped (Fig. 4.2d). In some cases, his fangs, when extended, deflected long setae on the anterior margin of the female's sternum (Fig. 4.2b, d). The spiders often separated and then the female clasped the male again several times during a given pairing. Between clasps, the spiders moved apart, in some cases several body lengths. The male often courted again before the next cheliceral clasp. In some cases, the female's behavior just after a pair broke apart appeared to be aggressive, and she made rapid bursts of movement and gave relatively violent jerks on lines running toward the male. The male nevertheless often courted and successfully induced her to approach again (or to allow him to approach) and to assume the acceptance posture. Copulations with virgin females were longer and included more cheliceral clasps than copulations with non-virgins (Aisenberg and Eberhard 2009; Aisenberg and Barrantes 2011).

4.4.3 *L. mariana*: Movements of Male Genitalia When the Female Lacked a Plug

The movements of the male's genitalia in copulations with females that lacked a mating plug form an important baseline for comparisons with the movements of *L. argyra* (Sect. 4.4.4), those involving in plug removal (Sect. 4.4.5) and the effects of cutting off male sclerites (Sect. 4.5.2.2), so we will describe them in detail (based on Eberhard and Huber 1998a). To copulate, the male of *L. mariana* extended a palp one or more times so that the dorsal surface of his cymbium contacted the ventral surface of the female's abdomen just anterior to her epigynum. At least, some of the many setae on the cymbium, especially those on its basal half, were interlaced among the setae on the surface of the female's abdomen near her epigynum and may have helped couple the cymbium mechanically to the female. After the cymbium was seated, the basal hematodocha expanded. This moved the more distal portions of the palp away from the cymbium and rotated them nearly 180°. The conductor and embolus moved toward and usually contacted the female's epigynum, inserting (or apparently attempting to insert) the conductor and the embolus into the opening of an insemination duct (the "atrium"). There were two types of insertions of the male palps—"long" and "short."

4.4.3.1 "Long" Insertions

Long insertions (average duration >1 min) usually occurred early in copulations, while "short" insertions (as short as a second or so) tended to occur later (with exceptions). In a long insertion, the palp usually made only a single long insertion,

Table 4.1 Female reproductive processes that are capable of affecting male paternity success and that were correlated with particular male sexual traits in *Leucauge mariana* and *L. argyra*

Correlations between variation in male traits and female responses		
Male trait	Female response	Species
More bouts of copulatory courtship (leg pushing)	Form copulatory plugs more frequent	*L. mariana*
More duration of copulatory courtship (leg pushing)	Form copulatory plugs more frequent	*L. mariana*
More short insertions of palps	Form copulatory plugs more frequent	*L. mariana*
More twangs during pre-copulatory courtship	Form copulatory plugs more frequent	*L. argyra*
Experimental manipulations		
Remove male setae chelicerae	Remates more	*L. mariana*
Remove female setae to sense male chelicerae setae	Remates more, forms copulatory plug less frequent, and interrupts copulation more	*L. mariana*
Remove tip of conductor male pedipalp[a]	Less sperm transferred to female	*L. mariana*
Additional female effect		
Earlier, more abundant production of plug liquid	Trap and kill male	*L. argyra*

[a]Control treatments were not exactly equivalent, leaving doubts in interpreting results

but there were repeated hematodochal expansions and collapses. The conductor and the embolus, which were driven against the epigynum by the movements produced by the first hematodochal inflation, remained in contact with the epigynum, while the basal and medial hematodochae repeatedly collapsed partially and then reinflated (Table 4.1). The base of the embolus moved toward the tip of the conductor at the beginning of the insertion and then remained there immobile. During each inflation, the conductor twisted slightly around the point where its tip contacted the atrium (the embolus was apparently inside the insemination duct—see below). The twisting movement caused the hook process on the conductor to sweep antero-laterally on the female's epigynum until it was arrested by encountering the hood at the anterior margin of the atrium. The tip of the embolus projected 155–165 μm beyond the tip of the conductor in three males. This distance was nearly the same as the distance travelled by the base of the embolus toward the tip of the conductor, confirming that the movement of the embolus base caused the embolus to be exerted. Thus, the tip of the embolus must have passed through the insemination duct and then entered deep into chamber I of the spermatheca, because the length of the insemination duct of the female was only about 60–80 μm (Eberhard and Huber 1998a). Since the base of the embolus did not move after the first hematodochal inflation, the embolus presumably remained inserted in chamber I throughout each long insertion. A long insertion ended when the conductor and embolus pulled away from the epigynum, and the cymbium and the entire palp were withdrawn from the female's abdomen.

Sometimes, a male appeared to have difficulty freeing the conductor and embolus from the epigynum following a long insertion, so that only after he had pushed the female with his legs (and sometimes, the female had released her cheliceral grip) did his palp come free with a snap. It is possible that the triangular tip of the embolus (Fig. 4.4; Méndez and Eberhard 2014) snagged at the junction of the insemination duct and spermathecal chamber I.

Sperm and additional material were transferred to the female during long insertions. Sperm introduced into the large, soft-walled chamber I of the spermatheca caused it to inflate. At least, some sperm in this chamber become decapsulated within an hour or so (Eberhard and Huber 1998a).

4.4.3.2 "Short" Insertions

Short insertions occurred in bursts of several short insertion movements of the conductor during a single period of cymbium–abdomen contact. In each insertion, the conductor tip was inserted into the atrium and the embolus base moved distally along the conductor. The conductor was withdrawn from the atrium after each insertion (it always came away smoothly, without a snap), and the embolus base withdrew to its resting position at the base of the conductor. Each time the palpal sclerites rotated again to bring the tips of the conductor and the embolus into contact with the epigynum, the base of the embolus gradually moved toward the tip of the conductor (as in a long insertion). Most inflations during short insertions resulted in a viscous white material with a consistency similar to that of toothpaste emerging from the tip of the palp during the period in which the base of the embolus moved distally. In most cases, the white plug material adhered only very poorly to the female. Sometimes, it came away still stuck to the male's palp when the embolus and conductor were withdrawn. Often when the tip of the conductor and the embolus were reinserted, they dislodged and removed a mass of material that had been deposited previously. During one copulation, for instance, the male more or less filled one side of the atrium with white material three different times, but each time eventually dislodged the accumulation during subsequent insertions. Most copulations with virgin females ended with the female still lacking a plug, even though the male had deposited material. In two cases, the plug material assumed a more liquid consistency and flowed into the atrium and presumably at least into the mouths of the insemination canals, where it condensed into a single, smooth mass that remained in place at the end of copulation.

A burst of short insertions lasted on average less than half as long as a long insertion (mean durations 128 and 359 s, respectively; Aisenberg 2009) and included only about one-fourth as many hematodochal inflations. Following a burst of short insertions, the cymbium was withdrawn, the palp was retracted, and the other palp was extended to the female's abdomen. The order of long and short insertions varied to some extent, and sometimes, a long insertion occurred after several short insertions had been performed on the same side of the epigynum.

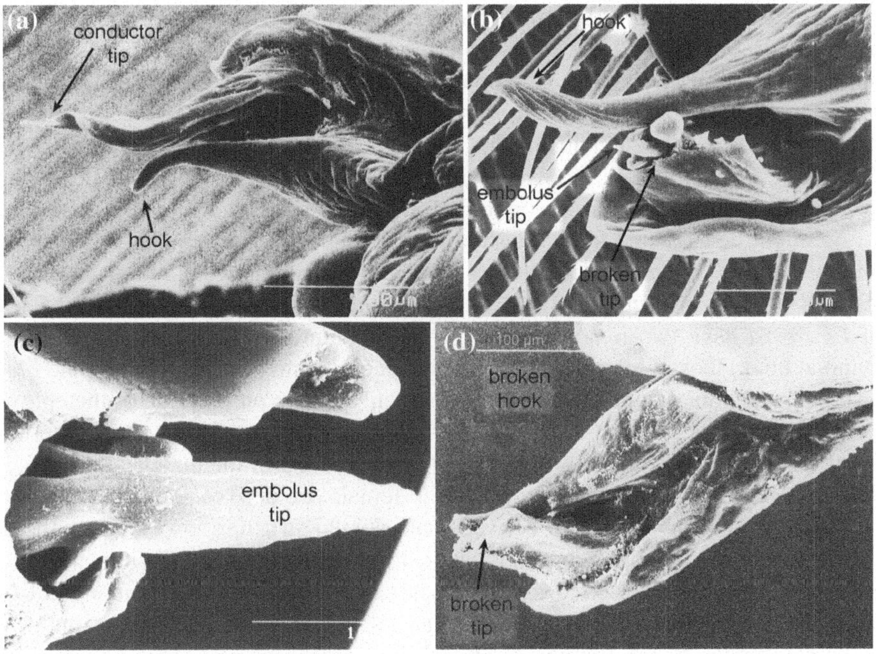

Fig. 4.4 Intact and modified palps of *L. mariana*. **a** Intact hook and conductor tip; **b** hook intact but tip of conductor removed and tip of embolus exposed; **c** triangular tip of embolus; and **d** both conductor tip and hook removed from Barrantes et al. 2013

4.4.3.3 "Flubs"

A third type of palpal contact consisted of what appeared to be failed attempts to insert the conductor ("flubs" in the terminology of Watson 1991). Inflation of the hematodochae caused the tips of the conductor and the embolus to scrape across the face of the epigynum without engaging it or only briefly engaging it, at an inappropriate site and then snap free. The male often repositioned his cymbium, lifting it briefly from the female's abdomen and then setting it down at a slightly different site after a flub. Flubs were more common later in copulation, when short insertions tended to occur.

4.4.4 L. argyra: Movements of Male Genitalia When the Female Lacked a Plug

Copulations in *L. argyra* differed in several ways. They included only long palpal insertions, and there were multiple hematodochal inflations after each palpal insertion; flubs were rare (Aisenberg and Barrantes 2011). Matings usually ended with a vigorous struggle, in which the female appeared to attempt to grasp the male's body or legs with her legs while maintaining her chelicerae locked. The male, for his part, frantically attempted to break away and escape. The female often attempted to wrap the male with silk during the struggle, as occurs during prey capture. Sexual cannibalism sometimes occurs (in 1 of 32 copulations of virgin females, and 4 of 13 in females with mating plugs) (Aisenberg and Barrantes 2011), both prior to and during copulation (it has never been seen in *L. mariana*).

As might be expected from the sharply different male and female genital morphology in this species, the copulatory behavior of the male genitalia of *L. argyra* differed in many aspects from that of *L. mariana* (Barrantes et al. 2013). Males of *L. argyra* have two major derived genital structures; one of them, the strong cymbial hook, locked his palp to the female's epigynum as follows. The male first inserted his cymbial hook into the laterally facing atrium on one side of the epigynum and then held it there while he inserted his conductor into the atrium on the opposite side (Figs. 4.5 and 4.6). Both the cymbial hook of the male and the large, widely flared, laterally oriented atria of the female are derived traits and may be unique to *L. argyra*. A second derived male genital structure, a curved tooth on the margin of the cymbium, probably served to lock the cymbial hook against the epigynum after it was inserted into the atrium. Both derived male structures of *L. argyra* may have evolved to stabilize the male's genitalia during intromission, perhaps in response to the frequently violent and dangerous resistance behavior of females of *L. argyra* during and at the end of copulation, or to perforate the especially strong mating plug (below). Clamping the female with the male's genitalia is not common in spiders (Eberhard and Huber 2010), in contrast to some other groups such as insects (e.g., Tuxen 1970; Scudder 1971). Possibly, the difference is due to spiders being unable to exert strong mechanical force with their genitalia because they lack in the palpal bulb muscles (Eberhard and Huber 2010).

4.4.5 Movements of Male Genitalia When the Female Has a Plug

In both species, some males which copulated with females that bore mating plugs were able to remove or penetrate the plug, but others failed to gain access to the insemination ducts despite sustained attempts and eventually left. Detailed observations of palp behavior are only available for *L. mariana*, in which the male succeeded in overcoming the plug in 68 % of 28 pairs (Méndez and Eberhard 2014) and we will describe them first. Copulations with unplugged non-virgins consisted mainly of short insertions and almost never included long insertions. Males employed three different mechanical ways to overcome plugs: snag the plug and pull it off; break it and then penetrate through it; and break its adhesion to the epigynum by injecting material under it. The genital bulb lacks muscles and

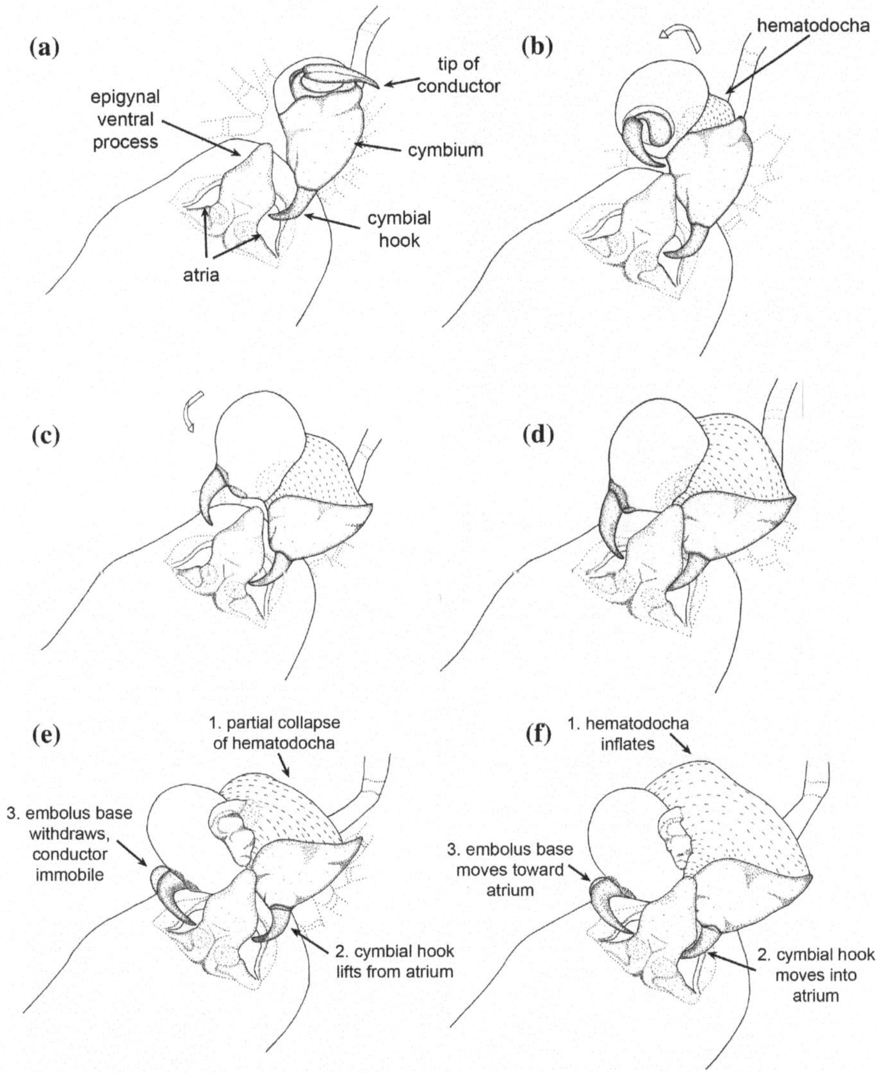

Fig. 4.5 Schematic drawings (in posterior–ventral views) that show how the male palp clasps the anterior portion of the female's epigynum by inserting the cymbial hook in one atrium and the conductor in the other atrium in *L. argyra* (his palp is partially hidden behind the ventral projection of the epigynum). Insertion of the conductor (**a**–**d**) follows insertion of the hook, as the basal hematodocha expands and drives the conductor into the atrium. Next, the hematodocha collapses partially and the conductor remains immobile, but the embolus base withdraws (**e**). Finally, the hematodocha inflates and the embolus base moves toward the atrium (**f**), presumably driving the embolus tip into the insemination duct (figure from Barrantes et al. 2013)

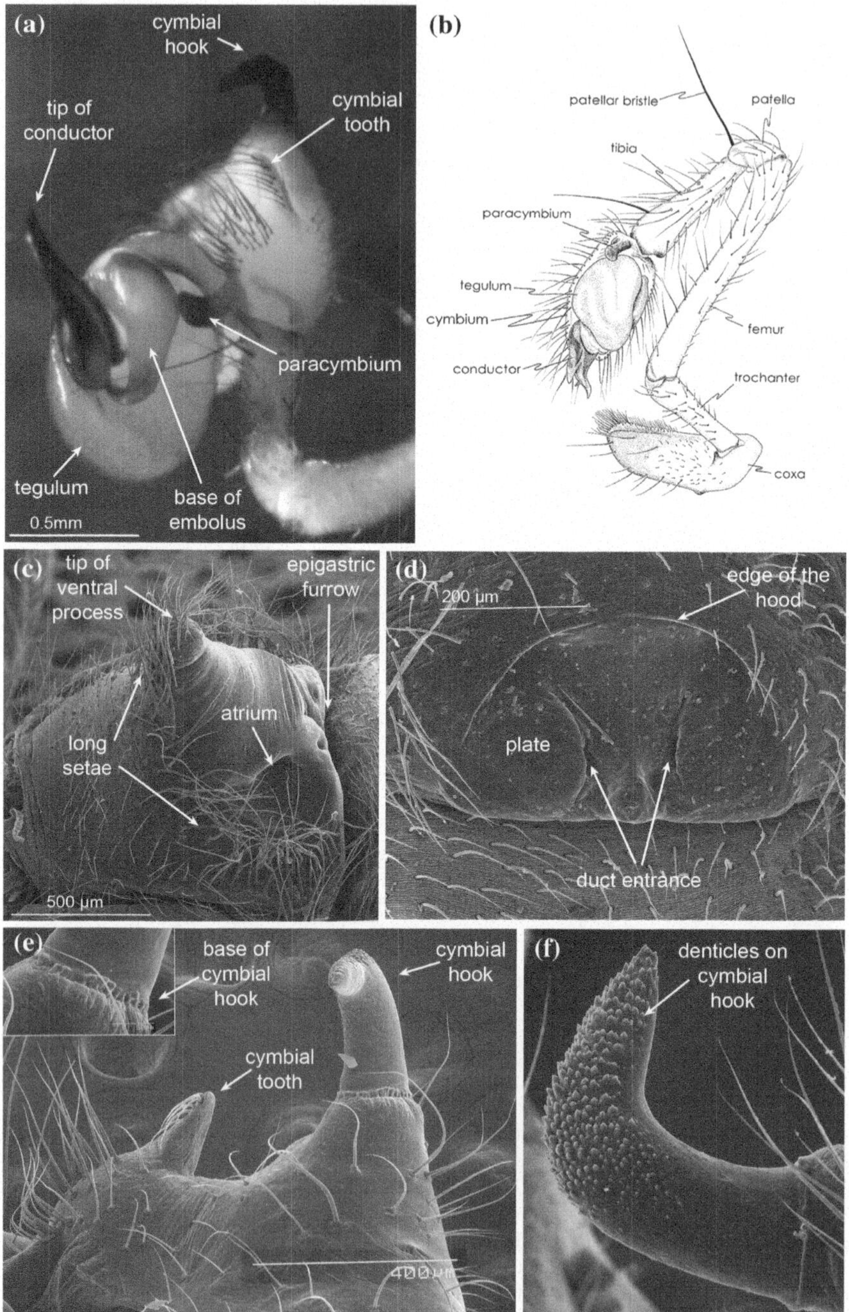
(a)
cymbial hook
tip of conductor
cymbial tooth
paracymbium
tegulum
base of embolus
0.5mm
(b)
patellar bristle
patella
tibia
paracymbium
tegulum
cymbium
femur
conductor
trochanter
coxa
(c)
tip of ventral process
epigastric furrow
atrium
long setae
500 µm
(d)
200 µm
edge of the hood
plate
duct entrance
(e)
base of cymbial hook
cymbial hook
cymbial tooth
400µm
(f)
denticles on cymbial hook

◀ **Fig. 4.6** Portraits of the male and female genitalia of *L. mariana* and *L. argyra* that document several derived traits in *L. argyra*. **a** The palp of the male of *L. argyra* has a long process on the cymbium (cymbial hook), with a smaller toothlike process (cymbial tooth) near its base on the border of the cymbium (**e, f**); **b** both are lacking in the palps of *L. mariana* (from Eberhard and Huber 1998a, b). **c** The epigynum of a *L. argyra* female has long, robust ventral process; the entrances of the two insemination ducts (the atria) are large and open laterally; and there are abundant long setae around the atria as well as on the anterior surface of the ventral process. None of these traits occurs in the epigynum of *L. mariana*. **d** Although there is no phylogeny of *Leucauge* currently available, the absence of all of the *L. argyra* traits in other *Leucauge* and other tetragnathids indicates clearly that they are all derived in the lineage of this species (Barrantes et al. 2013)

innervation, so the male's ability to guide these genital movements precisely seems likely to be limited (Eberhard and Huber 1998b, 2010). Neither movements nor sequences during plug removal showed any special traits that occurred when the female was plugged or that seemed especially designed for removing plugs. Further, more detailed analyses of palp behavior are needed, however (Méndez and Eberhard 2014). After overcoming a plug and making one or more insertions, the male deposited plug material of his own, as was described above.

Males of *L. argyra* also sometimes failed and sometimes succeeded in coupling their palps to the epigynum when it was plugged. The details of the palp movements have not been studied in detail, however, and it was usually not certain whether the conductor or the embolus entered the atrium. Typically, a male made repeated failed attempts to engage the cymbial hook with the epigynum, first with one palp and then the other, before succeeding in clasping the epigynum with the hook and then the conductor. The plugs in *L. argyra* are so hard and adhere so tightly to the epigynum that it is difficult to imagine a male being able to penetrate a plug unless it were somehow first softened. Nevertheless, males did sometimes penetrate them without removing them; in one case, the male broke the plug on one side free and then inserted the conductor along the side of the plug on the other side (Aisenberg and Barrantes 2011; A. Aisenberg, unpub.). More detailed observations of plug removal and penetration are needed.

4.4.6 Female Participation During Copulation

Because the events of copulation are played out on and inside the female's body, her morphology and behavior inevitably have important effects on copulation success. "Passive" female morphological participation is involved at all stages of copulation. The shape of her epigynum, the length of her spermathecal duct, the positions and shapes of her spermathecae (Eberhard and Huber 1998a; Triana and Quesada, in prep.), the shape and size of her chelicerae, and the presence of sense organs on her chelicerae, endites, and epigynum, etc., all have potentially important roles in determining the success of a copulation. Most of these female traits seem relatively free to evolve without having disadvantageous consequences under natural selection. Different forms could bias mating success with respect to different male traits (and thus exercise sexual selection on them), so this kind of female participation is not evolutionarily trivial.

Female *Leucauge* also participated actively in several aspects of copulation. Occasionally, a female *L. mariana* pushed away a male's palp that had been inserted in her epigynum with her leg IV. In both species, the female also sometimes terminated copulation itself by opening her chelicerae and pulling away. In addition, several lines of evidence showed that the females of both species also played active roles in the formation of mating plugs.

4.4.6.1 Copulatory Plugs

Copulatory plugs that adhere to female genitalia occur in various animal groups, including nematodes, insects, spiders, scorpions, reptiles, snakes, birds, and mammals (Drummond 1984; Andrade 1996; Gomiendo et al. 1998; Knoflach 1998; Simmons 2001; Wigby and Chapman 2004; Mattoni and Peretti 2004; Aisenberg and Eberhard 2009; Leonard and Córdoba-Aguilar 2010; Timmermeyer et al. 2010). Their composition varies and includes mucous, sperm, remains of spermatophores, parts of male genitalia that have broken off, and the male's entire body (Uhl et al. 2010). Traditionally, copulatory plugs have been considered male adaptations to gain advantage in sperm competition. By plugging the female after mating with her, a male imposes a physical barrier against other males, thus eliminating or reducing the chances that the female will receive sperm from additional males (Simmons 2001; Uhl et al. 2010).

Although mating plugs are widespread, their biological function is not always clear. The copulatory plugs of some spiders challenge the traditional interpretation because they are composed of a mixture of substances produced by the female as well as the male; those of still others are formed from substances produced exclusively by the female (Knoflach 1998; Aisenberg and Eberhard 2009; Uhl et al. 2010; Aisenberg and Barrantes 2011; Kuntner et al. 2012).

There are several additional possible functions for the plugs. Plugs could benefit females by preventing sperm desiccation or minimizing sperm loss (Boorman and Parker 1976; Huber 1995, 2005; Simmons 2001; Uhl et al. 2010). They could also act as mechanisms of CFC: By making a plug or allowing one to be made, a female could bias the chances of paternity for her offspring in favor of those future mates which are able to remove plugs. It is important to realize that a mating plug in *Leucauge* is probably of little significance from the female's point of view with respect to remating per se. This is because a female *Leucauge* need not rely on a plug to prevent additional copulations. Females can and do sometimes refuse males easily by failing to performing one or more of the several receptive responses that are necessary for copulation to occur (above).

The active female participation in plug formation in *Leucauge*, which has been documented in several ways, highlights the possible importance of females in determining the conditions under which remating occurs. Most importantly, we have made occasional direct observations (when viewing angles were favorable) of liquid surging up from the insemination duct of an unplugged female and hardening to form a plug in both species (Eberhard and Huber 1998a; Aisenberg and Barrantes 2011). In *L. mariana*, plug material in some females had acquired a clearer, less-white appearance and was still very liquid in appearance an hour after copulation ended. The prominent, lumpy pile of white material that had emerged from the male's palp and accumulated on the epigynum sometimes later acquired a more level, smooth surface; in one case, we observed that this transformation occurred when liquid welled up from the insemination duct (Eberhard and Huber 1998a). Similar smooth-surfaced masses were common in many field-collected *L. mariana* females.

Direct observations showed that female *L. argyra* had an even more exclusive role in plug formation. The male never deposited any material on the surface of the female's epigynum during copulation, as in *L. mariana*. A clear liquid was seen welling up from inside the insemination ducts up to an hour or more following some copulations (Aisenberg and Barrantes 2011). While the possibility that this liquid contained male products cannot be ruled out definitively, this seems unlikely. In only 5 % of examined plugs were there any sperm present (and then in only very small quantities) (Barrantes et al. 2013). Even if male products are included in the liquid, the movement of the liquid to the surface of the epigynum from inside the insemination ducts was presumably performed by the female. The timing of copulatory plug production *L. argyra* varied substantially; plug material was first seen at the beginning of mating, during palpal insertions, immediately after copulation, and many hours later in different pairs (Aisenberg and Barrantes 2011). The hardened plugs of *L. argrya* were more rigid and adhered more tightly to the epigynum than those of *L. mariana* (Méndez 2002; Barrantes et al. 2013; Méndez and Eberhard 2014).

4.4.6.2 Copulatory Plugs as Lethal Traps in *L. argyra*

An unusual feature of plugs in *L. argyra* is that while the male has his palp inserted, apparent plug material sometimes emerged from the female's genitalia and then hardened into a strongly adhesive state and caused the male's palp to adhere tenaciously to her epigynum (Aisenberg and Barrantes 2011). The male struggled frenetically to pull his palp free, pushing with his legs against the female. Because the palp bulb has no nerves or sense organs (Eberhard and Huber 1998b, 2010), it is not clear whether or not the male was able to sense that the liquid was beginning to accumulate before he attempted to pull his palp free. In some cases, the male succeeded in freeing his palp easily; in other cases, it was more difficult, and the rapidly hardening substance was pulled into a peak or thread during the process of pulling free and hardened in this form (Fig. 4.7). On some occasions, the male was unable to pull free, and he was then attacked and killed by the female. The danger to the male was greatest when plug liquid was more abundant and emerged while the male was still copulating. In addition, some

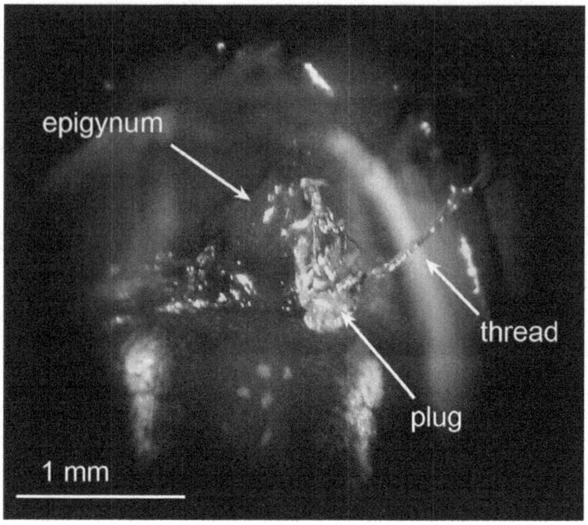

Fig. 4.7 This thread extension of the mating plug on the epigynum of a female *L. argyra* tells a story of female treachery. The female produced an abundant substance that adhered to the male's palp and made it difficult for him to pull out (from Barrantes et al. 2013)

plugs seemed to harden more rapidly; others remained liquid for a longer time. It thus appears that the male's chances of survival are affected by three female-controlled variables: the quantity of plug material; the timing of its emission onto the epigynum; and the rate of hardening. The frequency with which females in the field were found that had peaks or threads of plug material (21 % of 113) indicates that struggles by males to free themselves from plugs are not uncommon in nature (Aisenberg and Barrantes 2011).

In sum, active female cooperation is crucial in both species in determining copulation success at three stages: whether copulation would occur, when it would end, and whether a mating plug would be formed. In *L. argyra*, the female's behavior also affects the male's chances of survival.

4.5 Factors Affecting the Outcomes of Copulation

4.5.1 Correlations with Courtship Behavior

Higher numbers and durations of one type of copulatory courtship (male leg pushing) and greater numbers of short genital insertions during copulation were correlated in *L. mariana* with a higher probability that the female would form a mating plug (adding her liquid to the white substance deposited by the male) (Aisenberg and Eberhard 2009). In *L. argyra*, twanging during pre-copulatory courtship was also correlated with a greater tendency for females to form copulatory plugs (Aisenberg and Barrantes 2011).

4.5.2 Evidence from Experimental Manipulations

Data from experimental manipulations have an advantage over those from correlations because they allow cause and effect to be distinguished. However, many direct experimental tests of the functions of sexually dimorphic traits are incomplete; most experiments focus mainly on modifying one of the sexes (usually males) and recording the consequences of these modifications on female choice (Eberhard 2011). In addition, many studies fail to demonstrate unambiguously that it is stimulation from male genital or non-genital traits rather than some other associated trait that is under sexual selection through female choice, and do not eliminate the possibility that the behavior of morphologically modified animals remains unchanged (Eberhard 2011).

We have confronted these challenges by using experimental tests in *Leucauge* in which both male structures and female sensitivity to them are modified in two contexts. Experimental modification of some of the rigid cuticular structures involved in sexual interactions in *Leucauge* was relatively easy. The manipulations performed to date, however, have involved only relatively gross alterations; they do not test the consequences of particular designs of structures, but rather test the effects of the presence or absence of the structure.

4.5.2.1 Experiments with the Male's Cheliceral "Beard"

Aisenberg et al. (2015) tested the possibility in *L. mariana* that tactile stimuli from the male's dimorphic cheliceral setae (Fig. 4.3a) and/or the movements of his fangs during cheliceral clasps constitute courtship. We modified both the male stimulatory structures and the suspected female receptors (setae on her chelicerae and endites) and then checked whether these manipulations affected female sexual responses that could alter a male's chances of paternity. We shaved male cheliceral setae in one experimental group and shaved the setae on the female's chelicerae and endites which are often contacted by male's chelicerae on another group (Fig. 4.3). We analyzed the effects of these manipulations on the female's mating behavior (e.g., her tendency to interrupt copulation), her receptivity to remating, and the formation of copulatory plugs compared with controls.

Removing the male cheliceral setae resulted in a female response that seems likely to decrease the male's chances of paternity: The female was more likely accept a remating (Aisenberg et al. 2015). However, the effects were even more extreme when female cheliceral and endite setae were removed: Chelicerae disengagement was more frequent, and the female was more likely to accept remating and less likely to form a copulatory plug. These experiments were incomplete in that they did not include masking the setae on the female sternum that may also be implicated in female sexual responses to male stimulation (Fig. 4.2), nor did they include removal of her cheliceral setae without also removing her endite setae.

4.5.2.2 Experiments with Genitalia: Cutting Palpal Sclerites in *L. mariana*

In *L. mariana,* the tip of the male conductor, or the tip of the conductor and that of the nearby conductor hook were cut (Fig. 4.4) (Méndez and Eberhard 2014). In each case, the other palp was left intact as a control, and the male's ability to remove plugs and to inseminate the female with his two palps was observed. The effects of these manipulations, combined with direct observations of the morphology and behavior of these structures, indicated that the relatively flexible conductor tip is important in sperm transfer (its removal reduced sperm transfer). Plugs were snagged and penetrated, in contrast, by the combination of the protruding conductor tip and hook. Of the two, the hook was probably more important in removing plugs; it is much stiffer than the conductor tip, and the form of its pointed, hooked tip is appropriate to snag the edge of the plug or irregularities on its surface. In addition, experimental removal of the hook and the conductor tip reduced the palp's ability to break the plug, compared with the intact palp of the same male, while removal of only the conductor tip did not result in differences in plug breakage between the experimental and the control palp.

Interpretation of the results is complicated, however, for several reasons: The plugs were not bilaterally symmetrical in these tests; there was less than complete certainty whether insemination followed plug removal because we were not able to distinguish with certainty whether or not the embolus was inserted into the insemination duct when the tip of the conductor was in the atrium; and there was at least occasional uncertainty regarding which palp was the first to loosen a plug.

4.6 Discussion

4.6.1 The Non-passive Females of Leucauge

Perhaps one of the most striking aspects of the sexual biology of *L. mariana* and *L. argyra* is the pervasive need for the female to actively cooperate if a male's attempts to copulate and sire her offspring are to be successful. She performs six active responses prior to coupling, two during copulation, and one following copulation. Some of these same cooperative responses also occur in *L. regnyi* (Alayón 1979) and *L. venusta* (Castro 1995). One of these, the ventral flexion of the female abdomen to bring it into range of the male's palps appears to be widespread in Tetragnathidae (Levi 1981; Álvarez-Padilla and Hormiga 2011), as it is indeed, in many other spiders (Huber 1998).

Physical coercion of a female *Leucauge* by the male to oblige her to mate is simply not possible, and male *L. mariana* and *L. argyra* do not attempt to coerce females. Instead, they perform several types of behavior that are apparently designed to induce females to cooperate. Cooperation in female *L. argyra* is less complete, and females often attack and sometimes kill males either prior to or part way through a copulation.

4.6.2 Why Do Females Influence Copulation Outcomes? Functions of Male and Female Traits

The active female influences just mentioned are in accord with the possibility that CFC occurs in *L. mariana* and *L. argyra*. Several of the requirements for demonstrating CFC (Eberhard 1991, 1996) are fulfilled: Females are polyandrous; female copulatory and post-copulatory behaviors can determine male paternity success; and females decide to favor certain sexual partners over others based on both male genital (number and duration of short insertions) and non-genital (number and duration of leg pushes) copulatory behaviors. As discussed in Chap. 1, however, there are several alternative types of selection that need to be considered (for more detailed descriptions of these hypotheses, see Chap. 1).

4.6.2.1 Species Isolation

Although we have not attempted to make crosses of *L. mariana* and *L. argyra*, their genitalia seem likely to be exclude each other. The form of the epigynum of *L. mariana*, with its small atria that open medially rather than laterally, seems highly likely to preclude successful intromission by a male *L. argyra* (there is nowhere for the cymbial hook to be inserted). Epigynal designs similar in general form to that of *L. mariana* occur in many *Leucauge* species (Levi, unpub.). We doubt, however, whether selection favoring species isolation was responsible for the evolution of these mechanical incompatibilities, or for the many other traits that we have discussed which only come into play after a female has received various pre-copulatory male courtship stimuli. Terminating cross-specific sexual interactions earlier rather than later in an interaction would be advantageous for both males and females and would be expected to evolve. As expected from this argument, Castro (1995) found that in none of the 47 crosses among four species of *Leucauge* that she made in captivity did the male even perform clear pre-copulatory courtship behavior, much less approach and clasp chelicerae with the female. We thus suppose that most if not all of the traits we described above evolved due to sexual selection, rather than natural selection to isolate species.

4.6.2.2 Sperm Competition

One type of post-copulatory sexual selection that probably occurs in *Leucauge* is sperm competition. Removal of mating plugs deposited during previous copulations and thus gaining access to the female's insemination ducts is a male mechanism to win out in sperm competition. One genital trait in *L. mariana*, the conductor hook, appears to function in plug removal and is thus probably an adaptation to sperm competition (though some details remain to be resolved). The cymbial hook in *L. argyra* may also function in plug removal, but further observations are needed to test this speculation.

With respect to the deposition of mating plugs that are capable of excluding future males, however, it is the females of *Leucauge* rather than the males that appear to be in control, so the traits discussed here that are related to the formation or lack of formation of a sperm plug are not attributable to sperm competition. In fact, as argued above, they are not even likely to be attributable to selection on females to prevent future insemination, because females can easily prevent any interaction with an unwanted male from reaching the stage of contact with his genitalia. A female need not rely on a physical covering of her genitalia to avoid insemination. The most likely sexual selection advantage of plugs to females is as filters for potential mates to allow CFC (below).

Except for copulatory plugs, none of the other male traits that are discussed here seem to have any physical relation to competition involving deposition or removal of sperm.

4.6.2.3 Sexually Antagonistic Coevolution

Having eliminated species isolation and sperm competition as potential explanations for the evolution of male chelicerae in *L. mariana*, the male genitalia of *L. argyra*, and the female traits that interact with these structures, only two hypotheses remain to be tested: SAC (Arnqvist and Rowe 2005) and CFC (Eberhard 1996).

SAC can be classified into two basic types (Arnqvist and Rowe 2005; Chap. 1). One possibility is that SAC results from forceful, physical male manipulations of females (coercive manipulations) (Alexander et al. 1997; Arnqvist and Rowe 2005). This type of sexual selection is highly unlikely in either species of *Leucauge*, because there is no way males can physically oblige females to cooperate. As noted above, females must cooperate in four different ways for copulation to begin; once copulation begins, they can terminate it at any time by opening their chelicerae or pushing the male's palp away from the epigynum with their tarsi; and they are free to produce or not produce sperm plugs following copulation. Furthermore, the male traits that correlate with female cooperation, such as gentle leg pushing, short pedipalp insertions, and contact with setae on the female's chelicerae and endites, are not appropriately designed to physically force the female to cooperate.

Thus only an alternative, signaling version of SAC could work with *Leucauge* (a third, physiological type of male manipulations is not relevant to the traits discussed here). One possible objection to the signaling version of SAC is that we did not find female morphological defensive structures or devices to block and protect her sense organs in the areas contacted by male genitalia and cheliceral setae, as SAC might predict. This failure is suggestive, but not convincing, as evidence against SAC, however, because female resistance might be in her nervous system (lack of synapses between particular neurons, lower responsiveness of certain neurons, etc.) rather than in her external morphology.

A more convincing objection to SAC involves female sensitivity and responses to male stimuli. SAC supposes that males manipulate females via sensory traps. A sensory trap can occur when the male mimics a stimulus to which females have evolved under natural selection to respond in a way that would favor the male. For instance, the male *Calopteryx haemorrhoidalis* damselfly induces the female to discard sperm from previous males that are stored in her spermatheca, even though it may be in her best interests to conserve them. He uses a sensory trap (Córdoba-Aguilar 1999, 2002). He rubs his genitalia on receptors in the wall of the female's oviduct that evolved to sense the presence of an egg descending the oviduct; these receptors trigger the female to move sperm from her spermatheca to the oviduct and fertilize the egg. The male triggers this female sperm transport response and then removes the sperm from the oviduct with his genitalia. Sexual selection on the male (sperm competition) thus favors his use of a female response that evolved under natural selection, and continued natural selection on the female to sense and fertilize eggs in her oviduct prevents her from ceasing to respond to this stimulus in this way, even though her responses to males may reduce her own fitness.

Male *Leucauge* do not seem likely to be using such sensory traps. This is because it seems unlikely that the female sensitivities and responses to the male's stimuli are restrained by natural selection. Take, for instance, stimuli from the setae on the male's chelicerae, which increase the female's tendency to make a mating plug and to reject future mating attempts. Assume for the moment that there is conflict and that it is disadvantageous to the female to lose these mating opportunities. The female will lose nothing (via natural selection) from simply readjusting her threshold of response to the stimuli from the male's chelicerae (or eliminating her responses entirely), thus avoiding the cost of being manipulated. A similar argument could be made for the correlation between pre-copulatory male "twangs" during pre-copulatory courtship and the production of copulatory plugs. In this case, one might make the (strained) argument that sensitivity to the kinds of vibrations that a male produces with a twang might be under natural selection in another context, such as prey capture. But this neglects an additional necessity for the sensory trap argument: The female's sensitivity should be intrinsically linked, due to natural selection, to copulatory plug production. We see no logical reason to expect that such a linkage will occur under natural selection.

A limitation of these arguments is that they are not direct, and they involve theoretical predictions rather than empirical measurements (e.g., of the lack of costs to the females).

4.6.2.4 Cryptic Female Choice

The CFC hypothesis is favored as an explanation of several of the morphological and behavioral characteristics of male *L. mariana* and *L. argrya* that are employed during mating, both by the elimination of other alternatives and by the fact that several of the necessary prerequisites for CFC to occur are fulfilled (above). Our experimental manipulations of male signals and of female receptors demonstrated that stimulation from secondary sexual modifications of the male chelicerae (in *L. mariana*) and his courtship behavior (in both species) influence multiple female reproductive processes, including receptivity to remating (in *L. mariana*), female interruption of copulation (in *L. mariana*), copulatory plug formation (in *L. mariana*), and the consistency of plugs and the timing of their production (and thus both their probable effectiveness in resisting intromission attempts by future males and the possibility that the male will be cannibalized) (in *L. argyra*). In addition, some correlations suggested additional possible male effects on post-copulatory female reproductive processes: Two male copulatory courtship behavioral patterns correlate with greater plug formation (in *L. mariana*), and one pre-copulatory courtship behavioral pattern in males correlates with plug formation (in *L. argyra*) (Table 4.1).

The effects of experimentally reducing the female's sensitivity to stimuli from the male's chelicerae mirrored those of modifying the male chelicerae, confirming that stimulation itself rather than some other associated factor was responsible for the changes in female responses. If anything, modifying the female produced greater effects than modifying the male. This pattern, to obtain more dramatic effects from sensory "blinding" of the female than from modifying the male, is similar to those in studies of stimulation by male genitalia in tsetse flies (Briceño and Eberhard 2009a, b) and by species-specific male front leg clamping organs in sepsid flies (Eberhard 2001, 2002). The greater response to experimental blinding is logical, taking into account that the modifications of the females (complete or nearly complete elimination of signals from her sense organs) were probably more radical than those in the males (a change in the details but not complete elimination of the male stimuli).

The male and female genitalia and behavior of *L. argyra* are derived in several respects with respect to typical traits in *Leucauge* (Fig. 4.6). Two derived male genitalic clasping and clamping devices in *L. argyra* may be associated with the extremely aggressive female behavior in this species. In contrast to predictions from SAC theory, the female genitalia of this species have a derived trait (highly accessible, large atria) that facilitates rather than impedes male genital clamping. In contrast to predictions regarding lock-and-key mechanical fit between male and female, one prominent, derived female genital structure (the ventral projection) does not mesh physically with any male structure during copulation. Barrantes et al. (2013) speculated that it may instead function in sensing male genital movements.

4.6.3 Why Do Female *Leucauge* Make Mating Plugs?

Producing a mating plug is clearly to a male's advantage if it prevents at least some future males from being able to inseminate the female, but the advantage to a female *Leucauge* of making plugs is less obvious. In the first place, it is not reasonable to suppose that, from the female's point of view, the mating plugs function to prevent future males from inseminating her in either species (above). The active female participation in plug formation in both species is thus paradoxical in terms of blocking all future males.

One possible CFC explanation for female production of plugs is that the plugs serve as filters for males. In both species, some males but not all overcame or removed mating plugs. By forming a mating plug, a female could thus both favor the current male and filter future males by allowing only those able to remove the plug to inseminate her. Plug formation by females may favor males that are able to perform some types of both pre-copulatory and copulatory courtship (Table 4.1). These data are not completely convincing, however, because we were not able to confidently distinguish which plugs were sufficiently strong enough to exclude future males, as opposed to more fragile plugs, though in *L. argyra*, only relatively thick plugs that covered the atria completely were counted. In fact, the existence of some very fragile plugs in *L. mariana* (Méndez 2002; Méndez and Eberhard 2014) argues that they have additional, as yet undetermined functions (see Uhl 2010 for further discussion).

A second possibility relates to cannibalism of *L. argyra* males. In some spider species, females determine insemination rates and copulation durations by timing the attack on their sexual partners (Elgar et al. 2000; Schneider and Elgar 2001; Fromhage et al. 2003) (see also Andrade and MacLeod, this volume). Female *L. argyra*could exert mate choice through the timing and sizes of plugs that they form, which is in turn affected by the number of bursts of male twanging during pre-copulatory courtship via CFC (Aisenberg and Barrantes 2011). The data on this point are only preliminary, however, because only plug formation per se rather the effectiveness in excluding future males was shown to correlate with pre-copulatory twanging. The general questions of why *L. argyra* females appear to attempt to use plugs to capture some males but not others and why they attack some males more intense than others during pre-copulatory courtship and at the end of copulation (while *L. mariana* females do not kill males) are still unanswered. As far as we know, the mating plug of *L. argyra* is the first plug known to function as a sticky mortal trap for males, and it opens multiple doors for future research.

Females also participate in producing mating plugs in some other spiders, such as the theridiid *Therion varians* (Knoflach 1998) and *Nephila pilipes* (Kuntner et al. 2012). Not enough is known in these cases to evaluate whether some of the same or other explanations also apply to their formation.

4.7 Directions for Future Research

The results described here, particularly the unusually detailed descriptions of the behavior of different portions of the male genitalia, shed new light on the general pattern in spiders of rapid evolutionary diversification of male genital morphology. The movements of the male genitalia are basically quite different in *L. mariana* and *L. argyra*. If, as in these species, differences in morphology in other *Leucauge* species are accompanied by differences in behavior, and if the movements of male palpal sclerites are controlled by differences in the form, elasticity, and twisting of the flexible membranes (hematodochae), then it seems likely that there is an entire character system in the forms and extensibility and folding abilities of hematodochae that could be used to discriminate closely related species. However, this possibility has not yet been exploited by taxonomists.

Although there are reasons to suppose that neither the first nor the last male to mate with a female achieves complete sperm precedence (above), the preference of male *L. mariana* to associate with penultimate rather than mature females (Eberhard et al. 1993) predicts that the first male to mate with a female enjoys a paternity bias; quantitative paternity analyses when a female has mated with more than one male are needed. An additional important aspect of paternity is the numbers of sires of clutches laid by females in the field. The data available to date generally concern only matings with only two males and younger females; it is possible that expanding the range of ages and sexual experience of females would reveal further important details. More complete reproductive histories of females (visits by males, matings, and durability of plugs) in the field would also be useful to put present data into context.

The discovery of sexually dimorphic setae on the sternum of female *L. mariana* that may be stimulated by the male's chelicerae during copulation opens the exciting possibility of a female specialization whose function is to screen males on the basis of copulatory courtship stimulation and needs to be followed up by checking for effects of experimental modifications of these setae. Further experimental modifications of the tall epigynal process and the male cymbial hook in *L. argyra* are also needed to clear up the mystery still surrounding these derived structures. Even the use of an observation technique (e.g., mirrors) that would change in the observer angle of viewing copulation might help determine whether the male's palps ever deflect the setae on this female process. Finer modifications of male and/or female setae could permit finer conclusions regarding the functions of particular designs. It is still a mystery whether male *L. argyra* are able to penetrate the extremely hard mating plugs of this species, and if so, how they accomplish this feat.

Of course, there are many species of *Leucauge* yet to be observed. Some have mysterious structures (e.g., the pits on the epigynum of *L. venusta*, Barrantes et al. 2013), while others have interesting variant behavior (e.g., the lack of a cheliceral clasp in at least one *Leucauge* species, Barrantes et al. in press). A phylogeny of the genus would be of great help in interpreting further studies. It would also be useful to extend Castro's (1995) tests to see whether cross-specific pairs consistently fail to court and attempt to mate, in order to further test our conclusion that species isolation is an unlikely explanation for the rapid divergence in morphological and behavioral traits in *Leucauge*.

Acknowledgments We thank Rosannette Quesada, Catalina Sánchez, Emilia Triana, and Vivian Méndez for providing access to unpublished data; Jairo Moya-Ramirez for photographs; Rafael Lucas Rodriguez and A.V. Peretti for useful suggestions on the manuscript; and the Smithsonian Tropical Research Institute (AA, WE), Agencia Nacional de Investigación e Innovación (AA), Programa de Desarrollo de Ciencias Básicas (AA), and the Universidad de Costa Rica (GB, WE) for financial support.

References

Aisenberg A (2009) Male performance and body size affect female re-mating occurrence in the orb web spider *Leucauge mariana* (Araneae, Tetragnathidae). Ethology 115:1127–1136

Aisenberg A, Barrantes G (2011) Sexual behavior, cannibalism and mating plugs as sticky traps in the orb weaver spider *Leucauge argyra* (Tetragnathidae). Naturwissenschaften 98:605–613

Aisenberg A, Eberhard WG (2009) Possible cryptic female choice in a spider: female cooperation in making a copulatory plug depends on male copulatory courtship. Behav Ecol 20:1236–1241

Aisenberg A, Barrantes G, Eberhard WG (2015) Hairy kisses: tactile cheliceral courtship affects female mating decisions in *Leucauge mariana* (Araneae, Tetragnathidae). Behav Ecol Sociobiol. 69(2):313–323

Alayón G (1979) Procesos copulatorios de *Leucauge regnyi* Simon (Araneae: Tetragnathidae). Misc Zool Acad Ciencias Cuba 4:2–3

Alexander RD, Marshall DC, Cooley JR (1997) Evolutionary perspectives on insect mating. In: Choe JC, Crespi BJ (eds) Mating systems in insects and arachnids. Cambridge University Press, Cambridge, pp 4–31

Álvarez-Padilla F, Hormiga G (2011) Morphological and phylogenetic atlas of the orbweaving spider family Tetragnathidae (Araneae: Araneoidea). Zool J Linn Soc 162:713–879

Álvarez-Padilla F, Dimitrov D, Giribet G, Hormiga G (2009) Phylogenetic relationships of the spider family Tetragnathidae (Araneae, Araneoidea) based on morphological and DNA sequence data. Cladistics 25:109–146

Andrade MCB (1996) Sexual selection for male sacrifice in the Australian redback spider. Science 240:70–72

Arnqvist G, Rowe L (2005) Sexual conflict. Princeton University Press, Princeton

Barrantes G, Aisenberg A, Eberhard WG (2013) Functional aspects of genital differences in *Leucauge argyra* and *L. mariana* (Araneae, Tetragnathidae). J Arachnol 41:59–69

Barrantes G, Sánchez-Quirós C, Aisenberg A, Leitch K, Eberhard WG (in press) Loss of cheliceral clasping in *Leucauge* sp. (Araneae, Tetragnathidae). Arachnology

Boorman E, Parker GA (1976) Sperm (ejaculate) competition in *Drosophila melanogaster*, and the reproductive value of female of remating to male in relation to female age and mating status. Ecol Entomol 1:145–155

Briceño RD, Eberhard WG (2009a) Experimental modifications imply a stimulatory function for male tsetse fly genitalia, supporting cryptic female choice theory. J Evol Biol 22:1516–1525

Briceño RD, Eberhard WG (2009b) Experimental demonstration of possible cryptic female choice on male tsetse fly genitalia. J Insect Physiol 55:989–996

Bristowe WS (1929) The mating habits of spiders, with special reference to the problems surrounding sex dimorphism. Proc Zool Soc 21:309–358

Castro TJ (1995) Estudio comparativo del comportamiento reproductor, en las arañas del género *Leucauge* (Araneae: Tetragnathidae), del Soconusco, Chiapas. Graduate Thesis, Universidad de Ciencias y Artes del Estado de Chiapas, México

Córdoba-Aguilar A (1999) Male copulatory sensory stimulation induces female ejection of rival sperm in a damselfly. Proc R Soc Lond B 266:779–784

Córdoba-Aguilar A (2002) Sensory trap as the mechanism of sexual selection in a damselfly genitalic trait (Insecta: Calopterygidae). Amer Nat 160:594–601

Drummond BA (1984) Multiple mating and sperm competition in the Lepidoptera. In: Smith RA (ed) Sperm competition and the evolution of animal mating systems. Academic Press, New York, pp 291–371

Eberhard WG (1991) Copulatory courtship and cryptic female choice in insects. Biol Rev 66(1):1–31

Eberhard WG (1996) Female control: sexual selection by cryptic female choice. Princeton University Press, Princeton

Eberhard WG (2001) Species-specific genitalic copulatory courtship in sepsid flies (Diptera, Sepsidae, Microsepsis) and theories of genitalic evolution. Evolution 55(1):93–102

Eberhard WG (2002) Physical restraint or stimulation? The function (s) of the modified front legs of male *Archisepsis diversiformis* (Diptera, Sepsidae). J Insect Behav 15(6):831–850

Eberhard WG (2011) Experiments with genitalia: a commentary. Trends Ecol Evol 26(1):17–21

Eberhard WG, Huber BA (1998a) Courtship, copulation and sperm transfer in *Leucauge mariana* (Araneae, Tetragnathidae) with implications for higher classification. J Arachnol 26:342–368

Eberhard WG, Huber BA (1998b) Possible links between embryology, lack of innervation, and the evolution of male genitalia in spiders. Bull British Arachnol Soc 11(2):73–80
Eberhard WG, Huber BA (2010) Spider Genitalia. In: Leonard J, Córdoba A (eds) The evolution of primary sexual characters in animals, pp 249–284
Eberhard WG, Guzmán-Gómez S, Catley KM (1993) Correlation between genitalic morphology and mating systems in spiders. J Linn Soc 50:197–209
Elgar MA, Schneider JM, Herberstein ME (2000) Female control of paternity in the sexually cannibalistic spider *Argiope keyserlingi*. Proc R Soc Lond B 267:2439–2443
Fromhage L, Uhl G, Schneider JM (2003) Fitness consequences of sexual cannibalism in female *Argiope bruennichi*. Behav Ecol Sociobiol 55:60–64
Gomiendo M, Harcourt AH, Roldán ERS (1998) Sperm competition in mammals. In: Møller AP, Birkhead TR (eds) Sperm competition and sexual selection. Academic Press, London, pp 667–756
Huber BA (1995) The retrolateral tibial apophysis in spiders—shaped by sexual selection? Zool J Linn Soc 113:151–163
Huber BA (1998) Spider reproductive behaviour: a review of Gerhardt's work from 1911–1933, with implications for sexual selection. Bull Brit Arachnol Soc 11:81–91
Huber BA (2005) Sexual selection research on spiders: progress and biases. Biol Rev 80:363–385
Knoflach B (1998) Mating in *Theridion varians* Hahn and related species (Araneae: Theridiidae). J Nat Hist 32:545–604
Kraus O (1984) Male spider genitalia: evolutionary changes in structure and function. Verh Naturwiss Ver Hamburg 27:373–382
Kuntner M, Gregoric M, Zhang S, Kralj-Fišer S, Li D (2012) Mating plugs in polyandrous giants: which sex produces them, when, how and why? PLoS ONE 7(7):e40939
Leonard J, Córdoba-Aguilar A (2010) The evolution of primary sexual characters in animals. Oxford University Press, Oxford
Levi HW (1981) The American orb-weaver genera *Dolichognatha* and *Tetragnatha* North of Mexico (Araneae: Araneidae, Tetragnathinae). Bull Mus Comp Zool 149:271–318
Masumoto T (1993) The effect of the copulatory plug in the funnel-web spider, *Agelena limbata* (Araneae: Agelenidae). J Arachnol 21:55–59
Mattoni CI, Peretti AV (2004) The giant and complex genital plug of the asper group of *Bothriurus* (Scorpiones, Bothriuridae): morphology and comparison with other scorpion genital plugs. Zool Anz 243:75–84
Méndez V (2002) Comportamiento sexual y dinámica de población de *Leucauge mariana* (Araneae: Tetragnathidae). MSc thesis, Universidad de Costa Rica, Costa Rica
Méndez, V., Eberhard WG (2014) Removal of genital plugs and insemination by males with normal and experimentally modified palps in *Leucauge mariana* (Araneae, Tetragnathidae). J Arachnol 42(3):284–292
Schneider JM, Elgar MA (2001) Sexual cannibalism and sperm competition in the golden orb-web spider *Nephila plumipes* (Araneoidea): female and male perspectives. Behav Ecol 12:552–574
Scudder G (1971) Comparative morphology of insect genitalia. Ann Rev Entomol 16:379–406
Simmons LW (2001) Sperm competition and its evolutionary consequences in the insects. Princeton University Press, Princeton
Timmermeyer N, Gerlach T, Guempel C, Knoche J, Pfann JF, Schliessmann D, Michiels NK (2010) The function of copulatory plugs in *Caenorhabditis remanei*: hints for female benefits. Front Zool. doi:10.1186/1742-9994-7-28
Tuxen SL (1970) Taxonomist's glossary of genitalia in insects. CON: S-H Service Agency, Darien
Uhl G, Nessler SH, Schneider JM (2010) Securing paternity in spiders? A review on occurrence and effects of mating plugs and male genital mutilation. Genetica 138:75–104
Watson PJ (1991) Multiple paternity as genetic bet-hedging in female sierra dome spiders, *Linyphia litigiosa* (Linyphiidae). Anim Behav 41:343–360
Wigby S, Chapman T (2004) Sperm competition. Curr Biol 14:100–103

Chapter 5
Copulatory and Post-copulatory Sexual Selection in Haplogyne Spiders, with Emphasis on Pholcidae and Oonopidae

Lucía Calbacho-Rosa and Alfredo V. Peretti

Abstract Cryptic female choice (CFC) in spiders may involve several mechanisms to bias paternity including early termination of copulation, remating likelihood, and sperm dumping. In Pholcidae, these mechanisms seem to be very common and will be examined in the present chapter. In the Pholcidae *Physocyclus globosus*, sperm dumping involves an active role of the female. In contrast, in the Pholcidae *Holocnemus pluchei*, sperm mass ejection during copulation is mainly under male control. In another haplogyne spider, the Oonopidae *Opopaea fosuma*, females are able to influence male's chances of rearing their offspring by also exerting CFC by sperm dumping. Among pholcids, rhythmic genitalic movements of the pedipalps (squeezes) during copulation have been interpreted as genitalic copulatory courtship. Additionally, recent studies have evaluated the possibility that the outcome of male–female copulatory communication affects paternity. Future attention to the behavior of both female and male, and to the possible dialogues during copulation, promises to be a valuable tool in understanding sexual interactions in these spiders.

5.1 Introduction

5.1.1 Overview on Post-copulatory Sexual Selection

Afterward Parker's (1970) paper on sperm competition, the study of sexual selection were not only processes prior to copulation (male–male competition and female choice) but also processes that act during and after copulation. Sperm

L. Calbacho-Rosa (✉) · A.V. Peretti
Instituto de Diversidad y Ecología Animal, CONICET - Universidad Nacional de Córdoba, Vélez Sarsfield 299 (5000), Córdoba, Argentina
e-mail: luciacalbacho@gmail.com

A.V. Peretti
e-mail: aperetti@com.uncor.edu

A.V. Peretti and A. Aisenberg (eds.), *Cryptic Female Choice in Arthropods*,
DOI 10.1007/978-3-319-17894-3_5

competition, the competition between the ejaculates from two or more males over fertilization of the female's eggs, is currently viewed as a selective pressure driving the evolution of numerous behavioral, physiological, and anatomical male traits (Birkhead and Møller 1998; Smith 1984). Sperm competition creates selection pressure in males, which leads to an increase in the fertilization success of a given male (Parker 1970). Selection should favor the ability to outcompete the sperm of a previous male, and at the same time, it should favor resistance against sperm competition by subsequent males.

For some animal species, it has been observed that after a female has mated, some males choose to guard her against a possible further insemination, whereas other males seek to copulate with another female (Alcock 1994; Birkhead and Møller 1992; Parker 1970; Smith 1984; Thornhill and Alcock 1983; Waage 1979). This is referred to as post-copulatory guarding and often occurs when there is a high degree of sperm precedence for the last male to mate. There is a number of different behavioral and physiological ways in which a male may guard a female against a second insemination (Alcock 1994; Parker 1970). These adaptations include (a) producing a mating plug, which is transferred to the female genital tract after insemination and prevents the future transfer of the sperm by another male, consequently preventing sperm competition (e.g., Masumoto 1993; Schneider et al. 2001, see also Chap. 4). In addition, (b) a prolonged copulation has a similar function in that it temporary prevents the access to the females by a potential rival male (Elgar 1995). There are several mechanisms to explain why copulation may be longer than is necessary to transfer sperm including the occurrence of contact and non-contact guarding strategies. In the contact guarding, the male remains attached to the female after copulation (e.g. in Odonata, Parker 1970). In contrast, in the non-contact guarding, the male remains near the female (e.g. pholcid spiders, Calbacho-Rosa et al. 2010). (Pholcidae). Males may have evolved these strategies given that females may remain sexually receptive throughout their adult lives (Shillington and Verrell 1997; West and Toft 1999), and despite their ability to store sufficient sperm for several clutches, females commonly mate more than once (Austad 1984; Bukowski and Christenson 1997; Drengsgaard and Toft 1999; Schneider and Lubin 1998; Watson 1990).

Female choice during or after mating is less obvious compared to male–male competition, and this is especially true for the female counterpart of sperm competition: cryptic female choice (CFC). CFC is a pattern of non-random post-copulatory fertilization success among male phenotypes that is caused by a trait in females (Thornhill 1983; Eberhard 1996). CFC is expressed in many ways, for example, premature interruption of copulation, lack of sperm transport to storage and/or fertilization sites within the female, discharge or digestion of the current male's sperm or those of previous or subsequent males, lack of ovulation, lack of preparation of the uterus for implantation of embryos, abortion, lack of oviposition, rejection or removal of mating plugs, prevention of removal of plugs by subsequent males, and selective use of stored sperm (Eberhard 1996, Chap. 1). Also, CFC traits can be broadly categorized into morphological, behavioural, and physiological CFC traits (Arnqvist 2014). However, why females perform post-copulatory sexual selection by CFC? Although copulation has been commonly used as the criterion for final acceptance in female choice, if CFC occurs, a male that copulates with a female may fail to sire her offspring (Eberhard 1996). Just as sperm competition extends the possibility of male–male competition past the beginning of copulation, so the idea of CFC extends the possibility of female choice past the initiation of copulation (Eberhard 1996). Ultimately, the complexity of male and female behaviors and processes that occur before, during, and after copulation necessarily leads to large variation in male fertilization success (Eberhard et al. 1993).

5.1.2 The Need for Fine-Scaled Studies on Cryptic Female Choice in Different Organisms

Spiders are useful model organisms to uncover patterns of copulatory or post-copulatory sexual selection on traits, such as genitalic movements for two main reasons. First, sperm are encapsulated when transferred, and thus, it is feasible to see which sex produces movements that allow sperm transport during their transfer (Eberhard 2004). Second, most copulatory movements are relatively stereotyped and usually occur on the female's external surface, so it is possible to observe male genitalic movements. Such male movements include the rhythmic movements, twisting, vibrating, and repeated insertions of pedipalps (Huber 1998a).

Additionally, the reproductive biology of spiders is unusual, that make them compelling aims for studies on sperm competition and CFC (Eberhard 2004). The highly complex female genital morphology of some species suggests that females play active roles in processes associated with receiving, storing, and dumping sperm (Burger 2007; Burger et al. 2003, 2006a). In addition, mating behavior observations in combination with a thorough investigation of the genital morphology in spiders show large potential consequences for understanding the reproductive biology of particular species (Burger 2007; Huber 1994, 1995; Huber and Eberhard 1997; Uhl et al. 1995). In particular, fine genital morphological details help to understand the evolution of the genitalia and to explain their function in the context of sexual selection (Burger 2007; Burger et al. 2003, 2006a; Eberhard 1985, 1996).

5.2 Haplogyne and Entelegyne Spiders

According to Simon (1893), araneomorph spiders are classified into two groups, Haplogynae and Entelegynae, based on the complexity of their external genitalia. Females without an epigynum (a hardened plate on the underside of the abdomen surrounding the gonopore) and males with comparatively simple copulatory palpal organs are grouped under Haplogynae. Wiehle (1967) gave more precise definition by considering the structure of the internal female genitalia, where there is a single genital cavity in the Haplogynae, which serves also for fertilization. The spermathecae of haplogyne spiders have only a single duct, through which sperm both enters and exits the receptacle (the "cul-de-sac spermathecae"), a morphological design that in other arthropods is associated with a reduced advantage for the first male (Austad 1984; Parker 1984; Ridley 1989; Walker 1980). Haplogynes are traditionally considered to have simple female genitalia (Austad 1984; Uhl 2002; Wiehle 1967), compared with entelegynes. Entelegyne spiders have separate insemination ducts that connect each spermatheca with the outside and are used to introduce sperm into the spermatheca. The emergence of sperm from the spermatheca is done through separate fertilization ducts, running from the receptacle to the uterus, which are used to transfer sperm to the eggs (Wiehle 1967). Austad (1984) called this two-duct arrangement as "conduit spermathecae" and proposed that haplogyne and entelegyne female morphology may influence sperm precedence patterns. For entelegyne spiders, a first male sperm precedence pattern is predicted (Austad 1984; Snow and Andrade 2005; Uhl 2000).

Nonetheless, recent studies showed that the internal genitalia of some haplogyne spiders are very complex and do not correspond to the conventional type of a haplogyne vulva described by Wiehle (1967), for example, complex accessory structures (Brignoli 1978), with some sort of epigynum (Huber and Eberhard 1997), or with two distinctly different sperm storage organs (Uhl 2000), with complicated asymmetric genitalia (Huber 2006), and with the entelegyne vulval type (Huber 1997). Such complex morphology could function to lock one area of the female's reproductive tract, to pack a male's ejaculate in a secretion preventing sperm mixing, and eject it from her body as a single mass (Burger 2007, 2008; Burger et al. 2006b). All of these strategies may enable females to influence the fate of transferred sperm (Burger 2007; Burger et al. 2003, 2006a, c; Eberhard 1985, 1996).

However, it turned out that general confirmations of these patterns are limited (Eberhard et al. 1993; Elgar 1998; Uhl 2002; Uhl and Vollrath 1998) as several other factors beneath the genital morphology can influence a sperm precedence pattern (Austad 1984; Uhl 2002; Watson 1991). For example, in a haplogyne spider (*Pholcus phalangioides*, Pholcidae), experiments demonstrated that in matings with two males, 88 % of offspring are sired by the second male to copulate, as predicted by the Austad's hypothesis (Schäfer and Uhl 2002). There is, however, considerable variation in the number of offspring sired, and evidence suggests that this variation is controlled by the behavior of the female after mating (Schäfer et al. 2008). Current evidence suggests that last-male sperm priority may be augmented by mechanisms promoting sperm removal by males (see below) and sperm dumping by females (see below) (Burger 2007, 2010; Burger et al. 2006c; Schäfer and Uhl 2002). Furthermore, females of different species show behavioral, physiological, or morphological mechanisms that favor sperm of certain males against others during or after copulation (CFC in the sense of Eberhard and Cordero 1995, Chap. 1). In this way, they are able to influence sperm priority patterns

(Burger et al. 2003; Eberhard 1985, 1996, 2004; Hellriegel and Ward 1998). The morphological differences between haplogyne and entelegyne have been questioned by studies on various haplogyne taxa, which have revealed a complexity in female genital morphology (Burger 2007, 2008, 2009, 2010; Burger et al. 2003, 2006a, b), and by the genital morphology defined for haplogyne in species of several Entelegynae families (Dimitrov et al. 2007; Griswold et al. 1998; Platnick and Forster 1989). These species are considered secondarily haplogyne. For example, *Cyrtognatha keyserling* (Tetragnathidae) shows female genital morphology with reduced spermathecae and an unpaired sac that holds the sperm (Dimitrov et al. 2007; Dimitrov and Hormiga 2009). Another species that is considered secondarily haplogyne is *Tetragnatha extensa* (Tetragnathidae), where the patterns of hatching success are in accordance with the hypothesis that last-male priority is the predominant pattern of sperm use in haplogyne species (West and Toft 1999). However, due to the particular morphology and sexual behaviors of this mostly understudied group, our emphasis in this chapter will be on Haplogynae.

5.2.1 Haplogyne Spiders: Reproductive Sexual Morphology and Sexual Behavior

5.2.1.1 Sexual Morphology

Detailed morphological studies on genitalia have large potential value for understanding the reproductive biology of species, (Burger 2009; Burger et al. 2006c; Eberhard et al. 1993; Michalik et al. 2005; Uhl 2002). Furthermore, such details shed insights into the evolution of the genitalia and their function in the context of sexual selection (Alberti and Michalik 2004; Burger and Kropf 2007; Burger et al. 2003; Eberhard 1985, 1996, 2004, 2006; Huber 2003a, 2006). Several studies revealed that the internal female genitalia of some haplogynes are in fact highly complex (see above). Below, we consider some species of Pholcidae and Oonopidae relating their reproductive morphology and sexual behavior (also see Table 5.1). In Pholcidae, rotation of the pedipalps of 180° before intromission seems to be as usual as pedipalp movements and opisthosoma-vibrations during copulation (Gerhardt 1921, 1923, 1924, 1927; Montgomery 1903). Male pedipalps in Pholcidae are highly derived and easily distinguishable from any other spider pedipalp. The sclerites of the genital bulb are fused (Kraus 1984), and the presence of an apophysis of the distal pedipalpal segment (the procursus) is unique among spiders. We will focus on four genera, *Pholcus*, *Physocyclus*, *Holocnemus*, and *Psilochorus*, in order to show similarities and differences in both reproductive morphology and sexual behavior. We will also focus on three male structures, genital bulb, pedipalps, and chelicerae, as these structures in males are in contact with the female during copulation, and then discuss the similarities and differences in the copulatory organ of female. *Pholcus phalangiodes* (Uhl et al. 1995) and *P. opilionoides* (Huber 1995) males have a bulb provided with several protrusions.

Table 5.1 External reproductive morphology of some Pholcidae in which sexual behavior was studied

Species	Male genitalia	Chelicerae	Female genitalia	References
Holocnemus pluchei	Genital bulb with two bulbal appendages, dorsal and ventral apophysis, without forming an embolus.	Frontal apophysis.	The uterus externus is bordered by bulges anteriorly and posteriorly.	Huber (1995)
Pholcus opilionoides	With bulbal appendages (uncus, appendix, embolus).	Frontal apophysis with two cones, lateral apophysis and a frontal prominence.	Triangular plate, highly sclerotized.	Huber (1995)
Pholcus phalangiodes	Wth bulbal appendages (uncus, sclerotized denticles, appendix, embolus).	Two apophyses that are partially surrounded by robust bristles and carry sensilla at their ápices.	With sclerotized hook.	Uhl et al. (1995)
Physocyclus dugesi	No published data	Frontal area with tooth-like processes and a large toothlike process on the lateral basal part of each chelicera.	Epigynum with a large flattened and sclerotized apophysis that is bifurcated on its posterior end.	Rodríguez-Márquez and Peretti (2010)
Physocyclus globosus	Procursus with tuft of hairs and cymbium, and small denticles on the embolic sclerite.	With paired cavities and tooth-like processes.	With the bifurcated apophysis on the anterior end of the epigynum and heavily sclerotized bulges.	Huber and Eberhard (1997)
Psilochorus simoni	Genital bulb with dorsally there are two prominences, ventrally with a sclerotized pouch.	A pair of simple apophyses at their anterior faces.	Simple transversal slit that is bordered by lightly sclerotized epigyneal plates, and two pouches anterior to these plates.	Huber (1994)

Surrounding the bulb runs a dark-colored ribbon, the uncus (Fig. 5.1). However, in *Physocyclus globosus*, the genital bulbs have cuticular structures and apophyses, and in *H. pluchei*, the genital bulbs are situated distally on the pedipalpal tarsi (Huber 1995). These genital bulbs have two bulbal appendages, the dorsal and the ventral bulbal apophyses, and the sperm reservoir opens between them without forming an embolus (Fig. 5.1). Finally, in *Psilochorus simoni*, the genital bulb is situated in the prolateral side of the pedipalpal tarsus (Huber 1994). Dorsally, the species has two further prominences, with numerous cuticular denticles, whereas ventrally, the genital bulb is provided with a sclerotized pouch.

Another interesting structure is the male pedipalp, a massive structure that extends far from the prosoma (Fig. 5.1). In the resting posture, it is U-shaped, with trochanter, femur, and patella forming one axis, tarsus with procursus and genital bulb forming the other axis, and tibia being the intermediate (Uhl et al. 1995). In *P. phalangiodes*, male pedipalp consists of various structures including a small trochanter, only connected with the coxa by means of membranes (Uhl et al. 1995). In *P. opilionoides*, only the pedipalp trochanter is further modified, bearing a long apophysis (Huber 1995). On the other hand, in *P. simoni*, the pedipalpal femora are provided with two apophyses each, one located basally with two long setae and one sclerotized apophysis located distally, whereas *H. pluchei* has a spine at the distal area (Huber 1994, 1995).

Other non-pedipalpal male structures that are modified in the context of copulatory mechanics in pholcids are the male chelicerae. In *P. opilionoides*, each chelicera bears a distal and frontal apophysis (Huber 1995). However, in *Physocyclus dugesi*, the frontal area of the chelicerae possesses many small toothlike processes and a large toothlike process on the lateral basal part of each chelicerae

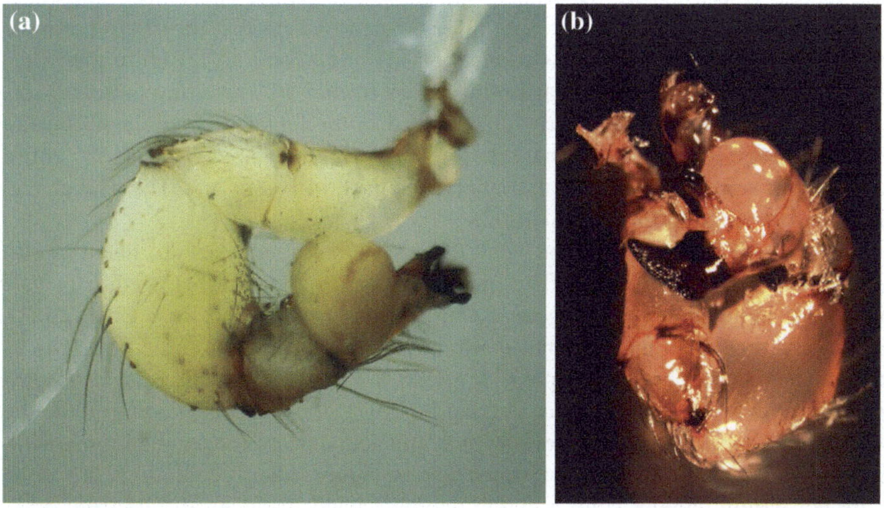

Fig. 5.1 Male genitalia of two Pholcidae spiders **a** pedipalps of *Holocnemus pluchei* and **b** pedipalps of *Pholcus phalangiodes*

(Rodríguez-Márquez and Peretti 2010). Male of *P. globosus* has chelicerae with their frontal armature of spines, teeth, and modified hairs (Huber and Eberhard 1997). Finally, in *P. simoni* and *H. pluchei*, only a frontal cheliceral apophysis is present (Huber 1994, 1995).

Regarding females, there are also differences in their sexual structures. For example, in *P. opilionoides*, female copulatory organ is marked externally by a highly sclerotized, triangular plate which bears an anterior sclerotized hook (Huber 1995). The uterus externus is characterized by a pair of dorsal pore plates through which the vulval glands discharge their products (Uhl 1994). A complex valvular apparatus separates the uterus externus from the uterus internus. On the contrary, in *H. pluchei*, the transversal slit that opens into the uterus externus is bordered by conspicuous bulges (Huber 1995; Uhl 1994), and the epigynum of *P. dugesi* is constituted by a large flattened and sclerotized apophysis that is bifurcated on its posterior end (Rodríguez-Márquez and Peretti 2010). Finally, the female copulatory organ of *P. simoni* has a transversal slit that is bordered by lightly sclerotized epigyneal plates and two pouches located anterior to these plates (Huber 1994).

5.2.1.2 Sexual Behavior

In Pholcidae, the male courts the female in her web by performing bursts of quick dorsoventral vibrations with the opisthosoma when moving toward her (Huber 1994, 1995; Huber and Eberhard 1997; Calbacho-Rosa et al. 2013). Females that allow copulation respond with tapping movements and move a short distance toward the approaching male. Correct positioning of the male relative to the female is assisted by the frontal cheliceral apophyses, and female epigyneal structures usually appear either neutral or cooperative (in that they help the male to lodge his chelicerae and thus to position his body correctly). Before the insertion, the pedipalps are twisted again through a further 90°, resulting in a total torsion of 180°. Both pedipalps are kept inserted in the female genital cavity during copulation and the male performed rhythmic twisting movements during the entire copulation, simultaneously or alternately (Calbacho-Rosa et al. 2013; Huber and Eberhard 1997; Schäfer and Uhl 2002).

In *H. pluchei*, the dorsal and ventral male bulbal apophyses are inserted into corresponding female cavities. The frequency of movement of male pedipalps varies along copulation, being higher at the beginning (Calbacho-Rosa et al. 2013), and copulations with mated females are longer than those with virgin ones. Males use cheliceral apophyses and palps to grasp the epigynum (Fig. 5.2). In *P. globosus*, according to Huber and Eberhard (1997), the male chelicerae hook the female epigynum, but copulations with virgin females are longer than those with mated females, and the rate of squeezing is higher when copulations are done with non-virgin females (Fig. 5.2). On the other hand, *P. dugesi* differs from other pholcids in that an apparent sexual conflict over mating occurs. Indeed, females are very reluctant to copulate more than once prior to the first oviposition, and

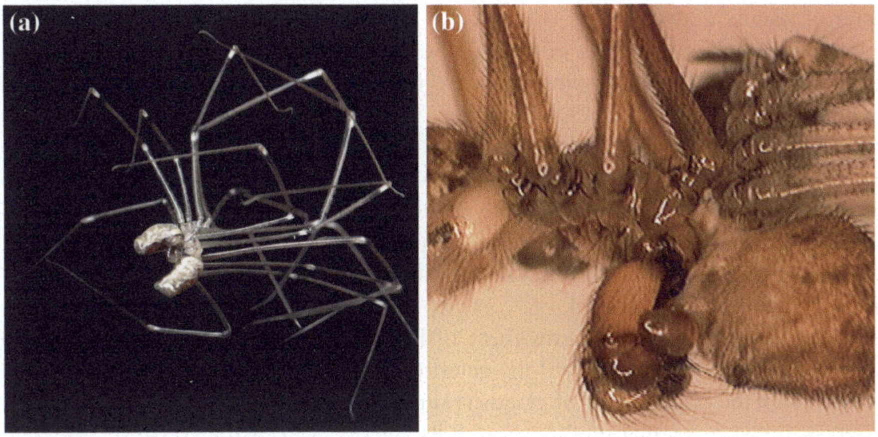

Fig. 5.2 **a** Male and female *Holocnemus pluchei* (Pholcidae) in copulation; **b** copulatory posture in *Physocyclus globulosus*, lateral view. Male on the left side. Both palps are inserted into the female's genital opening

precopulatory sexual cannibalism by both males and females often appears (Peretti and Rodríguez-Márquez, unpub. data). In *P. simoni*, intensive leg movements characterize the first minutes of copulation (Huber 1994). Then, the movements decrease in intensity, and during the last minutes, the male gently vibrates his opisthosoma and occasionally moves his pedipalps sideways, either simultaneously or alternately.

5.2.1.3 Oonopidae Morphology

Oonopidae was included in our chapter for being the only other Haplogynae where there have been studies of CFC, conducted by Mathias Burger and his collaborators.

In this family, studies have particularly focused on the description of reproductive morphology, rather than in sexual behaviors (see below). Oonopidae are haplogyne spiders that have incredibly complex genital structures that contradict the mentioned "simplicity" of the haplogynes and deserve to be described in detail. The subfamily Gamasomorphinae comprises all oonopids with abdominal scuta. These spiders are most abundant in the tropics where they show an enormous species diversity (Burger et al. 2003). In this chapter, we will focus on two commonly studied species: *Opopaea fosuma* and *Silhouettella loricatula*. In *O. fosuma*, females have complex genital structures and muscles, allowing females to perform sperm dumping, indicating that CFC may occur in these spiders (Burger et al. 2003). The opisthosoma is covered by both a dorsal and a ventral scutum. The ventral scutum is separated by the presumed epigastric furrow into an anterior and a posterior plate. There are two pairs of oval openings. The anterior pair is connected by two internal sclerotized ridges. They are separated by a furrow, which expands in the middle to the copulatory orifice. This furrow extends between two large sclerites, while the presumed epigastric furrow extends to the lateral edges of the ventral scutum. The two sclerites that function as muscle attachments originate at the sclerotized ridges and are directed backward. The posterior pair of oval openings is connected by a thin sclerotized ridge. Then, two sclerites directing backward originate at the posterior oval openings. The copulatory orifice leads directly to the longish spermatheca, which extends posteriorly. The distal part of its posterior wall appears only chitinized, whereas the upper parts are distinctly sclerotized.

The anterior wall of the spermatheca of female *O. fusuma* is strongly sclerotized and enormously extended toward the anterior. Ventrally, the wall of the spermatheca is fused with the ridge. A pair of glands is situated in a cavity on each side. The uterus externus shows a peculiar modification in its anterior wall: A transverse sclerite with a nail-like structure lies in a thickened chitinized area. A sclerotized stalk extends distally to a fold that is directed forward. The uterus externus extends through the muscles and opens in the presumed epigastric furrow. The posterior wall of the uterus borders directly on the strongly sclerotized anterior wall of the spermatheca, forming a second fold directing backward. In the males, there is a single large ventral scutum and only the posterior pair of the oval openings is connected by an internal sclerotized ridge. The ridges, the two sclerites, and the transverse sclerite that bears a nail-like structure are missing. A nail-like structure and an opposite fit for it are also present in the four Gamasomorphinae spiders, *O. lena*, *O.* c.f. *cornuta*, *Xyphinus* sp., and *Gamasomorpha loricatula.* The enlarged patella of the male palp is characteristic of the genus *Opopaea.* The long, oval genital bulb extends into a slender embolus with a complex tip. There is a striking furrow at the distal end of the bulb. The cymbium and bulb are only partly fused (Burger et al. 2002). In some cases, an embolus is visible in the spermatheca of the female. The embolus has to be pushed through the genital orifice into the spermatheca. The furrow of the genital bulb fits exactly into the ridge which is little bit compressed in its soft chitinized region. This coupling mechanism attains a full locking of the male bulb into the vulva of the female.

In the other Oonopidae species, *S. loricatula*, the sperm storage site of female is an almost entirely sclerotized, single receptaculum (Burger 2007). Only the ventral half of its posterior wall appears chitinized. The muscles connect the stalk-like parts of the sclerites with the posterior scutum. A curved slit took its course through the ventral part of a sclerite, which is situated anteriorly to the receptaculum. Gland cells extend around pores in the dorsal wall of the receptaculum. All mated females have a sac in their receptaculum. The sac consists of a secretion and fills the lumen of the receptaculum almost entirely. All sacs contain spermatozoa and globular secretion (Burger 2007). The globular palpal bulb of *S. loricatula* continues into a slender embolus consisting of a sclerotized proximal part with four lateral extensions and a membranous distal part (Burger 2007). The embolus is strongly furrowed on its retrolateral side. The palpal bulb is filled with seminal fluid consisting of a globular secretion in which spermatozoa are embedded. An extensive glandular epithelium with secretory vesicles is evident within the palpal bulb. During copulation, the male inserts the pedipalps simultaneously and moves them rhythmically. Males move the pedipalps in circles and/or back and forth. The pedipalps start to move in tandem.

5.2.2 The Value of Haplogyne Spiders to Evaluate CFC

Conflict between the sexes over control of copulation may drive the coevolution of complex genitalia (Arnqvist and Rowe 1995; Holland and Rice 1999). This coevolution is predicted to result in the appearance of traits for resistance by females and for coercion by males (Arnqvist and Rowe 2002; Chapman 2006; Chapman et al. 2003; Parker 2006). Sexual coercion during mating can select for male structures to grasp the partner. Females can use anticlasper devices or resistance behavior to avoid or minimize the costs imposed by forced copulation (Arnqvist and Rowe 1995; Peretti and Willemart 2007). In contrast to the sexual conflict hypothesis, the female choice hypothesis suggests a "selective cooperation" from females toward males before and during mating (Cordero and Eberhard 2003, 2005; Eberhard 1996).

In many arthropods, the data to support either of these two hypotheses have mainly come from comparative morphological studies focused on the male grasping structures and the corresponding anchored body region of the female (Eberhard 2006; Huber 1998a, 1999). In addition, experimental manipulations of these structures have been proved useful in determining the function of particular structures (Arnqvist and Rowe 1995; Eberhard 2001, 2002).

Therefore, what can the haplogyne spiders offer on CFC? As we saw above, haplogyne spiders also show conspicuous and interesting traits to investigate the sexual coercion and sexual cooperation hypotheses. A practical advantage is that a large proportion of the male's genitalia remain visible during copulation, and the male's stereotyped, rhythmic, and sustained squeezing movements can be observed directly (Huber and Eberhard 1997). Rodríguez-Márquez and Peretti (2010) tested the effect of manipulation of clasping mechanism on behavior in the pholcid *P. dugesi*. In females, they covered the epigynum apophyses and in males, they covered the cheliceral apophyses with a plaster. The increase in courtship duration between experimental and control groups occurred because males could not insert their palps into the female genitalia. In relation to this difficulty during clasping, males performed more stimulatory patterns in the experimental groups in comparison with the control group. Regarding copulatory behavior, Rodríguez-Márquez and Peretti (2010) found that in manipulated groups, the duration of copulation was shorter than that of the control group. Because of this decrease in copulation duration, the absolute number of squeezes was higher in control group than in the group of manipulated females. Females cooperated with courting males by keeping the genital aperture open and moving the forelegs laterally in order to facilitate the body contact, while males increased their stimulatory patterns (Rodríguez-Márquez and Peretti 2010). Females of *P. dugesi* with modified epigynum did not affect the occurrence of copulation; it resulted in shorter mating duration, suggesting that the epigyneal apophyses are cooperative structures that allow a long-lasting male clasping (Rodríguez-Márquez and Peretti 2010).

In another *Physocyclus* species, *P. globosus*, the anterior tip of the epigynum is small and fits into a groove that the male presents on the upper medial region of his chelicerae (Huber and Eberhard 1997). Preliminary data on experimental manipulation of these areas in *P. globosus* also demonstrate the existence of inter-sexual cooperation during clasping (A. Peretti, unpublished data).

5.3 Evaluating CFC and Other Alternatives Step by Step

5.3.1 Genitalic and Non-genitalic Male Copulatory Courtship

Several studies have shown that copulation duration in arthropods is longer than expected, given that the basic function of copulation is the transfer of sperm from males to females (Eberhard 2009). Thus, a long copulation suggests that functions other than sperm transfer may be operating. When females mate multiply, a fertilization success may be accomplished by three processes: an increase in ejaculate volume (Kelly and Jennions 2011; Pilastro et al. 2002; Wada et al. 2010); direct removal of previously stored rival sperm (Michiels 1989; Naud et al. 2004; Ono et al. 1989; Waage 1979; Wada et al. 2006, 2010); and female stimulation that promotes ejection of rival sperm by the female (Burger 2007, 2010; Burger et al. 2003, 2006; Huber and Eberhard 1997; Schäfer and Uhl 2002). This last mechanism has been interpreted as a form of copulatory courtship (Eberhard 1991, 2009, 2011). Genitalic copulatory courtship has also been documented in several taxa and often occurs during or following copulation (Eberhard 1994, 2001, 2009; Otronen 1990).

Additionally, males of hundreds of species of animals perform non-genital behavior during copulation that appears to be courtship. Such behaviors include biting, tapping, rubbing, squeezing, shaking, vibrating, and singing to and feeding the female (Eberhard 1991, 1994, 1996). These behaviors can result in a decreased mobility of the female during copulation (Humphries 1967), but also to increase dumping of sperm from previous males (Otronen 1990; Otronen and Siva-Jothy 1991), increase copulation duration (Hoikkala and Crossley 2000), increase sperm precedence (Edvardsson and Arnqvist 2000), relaxation of bursal muscles allowing the male to penetrate deeper and to transfer sperm (Tallamy et al. 2003), increase resistance to subsequent mating (King and Fischer 2005), and increase oviposition (Barbosa 2009).

5.3.1.1 Pholcids

In several species of Pholcidae, males court females during copulations with non-genitalic structures, as abdomen vibrations, leg shaking, and stridulation (Table 5.2). For example, *H. pluchei* males engaged in longer mating events and

Table 5.2 Male copulatory courtship (CC), paternity and sperm output of some Pholcidae and Oonopidae. A) non-genitalic CC, B) genitalic CC: male's rhythmic genitalic movements of the pedipalps (squeezing in Pholcidae)

	A			B			
Pholcidae	abdomen vibrations	leg shaking	stridulation		Paternity	Sperm output	References
Anopsicus zeteki	Yes	Yes	No	yes (simultaneously and symmetrically)	??	??	Huber (1998b)
Hedypsilus culicinus	??	??	No	Yes	??	??	Huber 1997b (Note: possible gustatorial courtship)
Holocnemus pluchei	Yes	Yes	No	Yes (simultaneous and alternate)	Second male	Yes (sperm removal)	Huber (1995), Calbacho-Rosa et al. (2013)
Metagonia rica	??	??	No	Yes	??	??	Huber 1997a
Pholcus phalangiodes	Yes	Yes	No	Yes	Second male	Yes	Uhl (1994), Uhl et al. (1995), Schäfer and Uhl (2002)
Physocyclus globosus	Yes	Yes	Yes (female)	Yes	Equal	Yes (sperm dumping)	Huber and Eberhard (1997)
Psilochorus simoni	Yes	Yes	No	Yes	??	??	Huber (1994)
Spermophora senoculata	Yes	Yes	No	Yes	??	??	Huber (2002a)
Oonopidae							
Silhouettella loricatula	??	??	No	Yes	??	Yes (sperm dumping)	Burger (2007), Burger et al. (2006a)
Orchestina sp.	Yes	Yes	No	Yes	??	??	Burger et al. (2010)
Opopaea fosuma	??	??	No	Yes	??	Yes (sperm dumping)	Burger et al. (2003)

more abdominal movements when mating with mated females than with virgin females (Calbacho-Rosa et al. 2013). Males could stimulate mated females using abdominal vibratory movements as a form of copulatory courtship. Such stimulation may induce the female to use the current male's sperm for fertilization instead of previous mating males' sperm. In *P. globosus*, males vibrated their abdomen in short bursts of rapid dorsoventral flexions at the pedicel (Huber and Eberhard 1997). The frequencies of the bursts of abdomen vibrations decreased as copulation proceeded. In addition, *Psilochorus simoni* and *Anopsicus zeteki* males rhythmically moved their abdomens during copulation (Huber 1994, 1998b).

Leg shaking is another non-genitalic behavior. For example, in *P. simoni*, the first minutes of copulation are characterized by intensive leg movements, in which the male seems to stroke back the fidgeting female legs with his legs (Huber 1994). Similarly, males of *P. globosus* (Huber and Eberhard 1997) and *H. pluchei* (Calbacho-Rosa et al. 2013) during copulation performed occasional leg vibrations. Other less studied cases in haplogynes include *Sicarius* sp. (Sicariidae) and *Misionella mendensis* (Filistatidae). *Sicarius* sp. males gently contact the dorsal abdomen of the female with his long legs. Then, with a movement, the male turns the female to mating position, clamps her down with his legs, and inserts both pedipalps (Levi 1967). In *M. mendensis*, the male rubs the tarsi of his legs on the basal segments of the female's legs and on the lateral surface of her cephalothorax. In addition, male abdomen vibrations seem to reduce female aggressiveness (Barrantes and Ramírez 2013).

Another non-genitalic behavior is stridulation; for example, *P. globosus* females use stridulation as a communicatory signal to modulate the male copulatory movements. A strong relationship between female stridulation and male papal squeezing during copulation has been found (Peretti et al. 2006, see further description in the next section).

Finally, a special case of non-genitalic contact structure is the frontal lobe in the pholcid male *Modisimus culicinus*. Clypeal glands open at the lobe, and during copulation, the female mouth is in contact with the lobe, suggesting gustatorial courtship (Huber 1997).

These strategies recently found in spiders have been also demonstrated in other arthropods (Andrés and Cordero-Rivera 2000; Marcotte et al. 2005; Solensky and Oberhauser 2009). Males engage in longer mating events and more complex pedipalp movements when copulating with mated females than when copulating with virgin females. Males can stimulate females using pedipalp movements as a form of copulatory courtship. In *H. pluchei*, female reproductive status influenced male mating behavior, specially the time spent in copulation and the frequency and duration of pedipalp movements (Calbacho-Rosa et al. 2013). Copulation duration is also influenced by the frequency and duration of palpal movements, and different types of these movements are associated with different copulatory functions. Simultaneous pedipalp movements involve strong squeezing and twisting movements. One function for simultaneous pedipalp movements is sperm transfer. However, although these movements have a higher frequency of occurrence and longer absolute duration in copulations with mated females compared with copulations with virgin females, the final sperm volume in twice-mated females does not differ from single-mated females. In addition, second male transfers less sperm than the first male since we commonly observed that a part of the previous male's sperm mass remained in the female even after the sperm removal (see below) phase ended. We suggest that longer and more frequent simultaneous pedipalp movements not only function to transfer more sperm but also as a form of genitalic copulatory courtship.

Another example of males that can stimulate females using pedipalp movements, as a form of copulatory courtship, is *P. globosus* (Huber and Eberhard 1997), where the pattern of pedipalp movements gradually changed during copulation. Early in the copulation, twists are more frequent and then the inward twists become longer. A large area surrounding the epigynum is deformed by the male pedipalpal movements, which could cause stimulation of putative stretch receptors (Huber and Eberhard 1997). In *P. phalangioides*, copulatory behavior differs depending on the mating order (Uhl 1994; Uhl et al. 1995). First copulations are much longer than second copulations, and the rate of pedipalp movements does not remain constant throughout the copulation. In the first copulations, the frequency of pedipalp movements is typically high during the first minutes, and then it decreases. Second matings take only 5 min, during which the frequency of pedipalp movements is always high. The total number of pedipalp movements is correlated with copulation duration. This variation in pedipalp movements is associated with variation in fertilization success: The higher the total number of pedipalp movements, the higher the paternity achieved by a given male (Schäfer and Uhl 2002). The procursus with its various apophyses, lamellar processes, and the conspicuous zone that becomes inflated during intromission might play an important role in stimulating the female (Uhl et al. 1995).

Spermophora senoculata male starts copulation by moving his palps rhythmically. During the "*in*" phase, the female opisthosoma is considerably deformed, but the male movements become weaker later, and at the end of the copulation, no movements are observed (Huber 2002a). Finally, there are other examples of Pholcidae where pedipalp movements also occur. The male *Anopsicus zeteki* performs simultaneous movements (Huber, unpublished data), from 2 to about 10 s by cycle. Several species of *Modisimus* make rhythmic pedipalpal movements (Huber, unpublished data). Male *P. simoni* repeatedly changed between simultaneous and alternate rhythmic pedipalp movements (Huber 1994). Males of *Metagonia rica* first move its procursi non-rhythmically in the female genital cavity and then changed its movements to a rhythmic pattern (Huber 1997). Therefore, multiple observations indicate that rhythmic pedipalpal movements during copulation are both common and diverse in pholcids (Huber and Eberhard 1997).

The most conclusive proof supporting that a given behavior pattern functions as courtship is to demonstrate that females respond to the behavior in ways that increase male's reproductive success. Only some authors tested the possibility that copulatory courtship affects paternity. In pholcid spiders, paternity has been estimated in only three species: In *H. pluchei* (P_2, a measure of 74 %, Calbacho-Rosa et al. 2010; Kaster and Jakob 1997) and *P. phalangioides* (P_2, a measure of 88.7 %, Schäfer and Uhl 2002), analysis reveals second-male sperm priority, whereas in *P. globosus* (P_2, a measure of 50 %, Eberhard et al. 1993), results suggest sperm mixing.

Since the procursus is not connected to the sperm duct, it can be a form of genital stimulation (Eberhard 1996; Huber and Eberhard 1997; Peretti et al. 2006; Peretti and Eberhard 2010; Schäfer and Uhl 2002).

Another less studied process is the para-genital courtship, where the bristles of males touch the female during copulation. For example, in *Mesabolivar delclaroi*, the bristles located in the genital male palpal cymbia rub the female's abdomen whenever she moves (Stefani et al. 2012). Also, in *M. globulosus* and *H. pluchei*, the bristles of the male touch females during copulation (Peretti and Calbacho-Rosa, obs. pers.). This interesting type of putative copulatory courtship remains unexplored for other species.

5.3.1.2 Oonopids

Pedipalp movements are considered as a possible copulatory courtship in some species of the family Oonopidae (Table 5.2). For instance, in *S. loricatula* (Burger 2007), males insert both pedipalps simultaneously and move them rhythmically during the entire copulation. The pedipalps start to move in tandem. The movements performed by one or both pedipalps together change irregularly. The male may perform copulatory courtship in order to stimulate the female. The extensions and the deeply furrowed proximal parts of the emboli further suggest that they may be used as stimulatory courtship devices (Burger 2007). The existence of various different muscles in the female genital region suggests that females play active roles in receiving, storing, and dumping sperm (Burger et al. 2006a) and that the sac is not dispelled by the male alone. Males continue moving their pedipalps after the ejection of a secretory sac, indicating that these movements have an additional function.

As in *S. loricatula*, the male palps of *Myrmopopaea* sp., *Grymeus* sp., and *Lionneta* sp. show furrows and spines on the emboli, suggesting their function as copulatory courtship devices and potentially removing the secretory sac from the female receptaculum during copulation (Burger 2010). In *Orchestina* sp., male palps are inserted simultaneously and moved alternately during the entire copulation (Burger et al. 2010).

Eberhard (1985) argued that male genitalia may function as courtship devices and therefore are under sexual selection. However, in some spiders, the female cuticle contacted by the male genitalia seems to lack receptors, casting doubt on the importance of stimulation (Huber 1993a, b, 1995). This potential problem was discussed by Eberhard (1996), pointing the fact that "stimulation … might also occur on the basis of overall fit, perhaps signaled by pressing or twisting the entire epigynum or the female's entire abdomen …" Due to the morphological and behavioral evidence (see above), this process might be happening in *S. senoculata* (Huber 2002a).

5.3.1.3 Dysderids

An extreme example where it seems that male courtship is absent is the case of *Harpactea sadistica* (Dysderidae). In this species, the male penetrates the female's body wall at some location away from her genitalia, which is known as traumatic insemination (Rezác 2008). The genitalia of both the male and the female of *H. sadistica* are modified in ways that are compatible with traumatic insemination. The needlelike structure of the male's embolus functions in penetrating the female's body wall, and the sperm storage organs are atrophied in the female. These males seem to have reversed the last-male sperm priority pattern (Rezác 2014).

5.3.2 Female Copulatory Behavior: Resistance Versus Copulatory Dialogues with the Male

The great diversity of biological definitions of communication as well as controversies about the role of information in studies of animal communication are a result of many different approaches (Carazo and Font 2010; Rendall et al. 2009; Seyfarth et al. 2010). Reproductive biology and in particular sexual behavior is a field in which the main components of communication (signals, information, signaler, and receiver) have been widely evaluated (Searcy and Nowicki 2005). Many examples of intra- and intersexual communication have been described in the classic work of Darwin (1871), as well as in reviews on sexual selection (Andersson 1994; Arnqvist and Rowe 2005; Andersson and Simmons 2006; Eberhard 1996). Traditionally, intersexual communication has been considered essential for specific identification and as a mechanism for mate choice (Peretti and Aisenberg 2011).

The costs incurred by polyandrous females during their reproductive life as well as the different strategies that they use to counteract these costs have been the focus of many studies and discussion along the last decades (Arnqvist and Rowe 2005; Chapman 2006; Chapman et al. 2003; Cordero and Eberhard 2003; Martin and Hosken 2004). Although a basic strategy used by females to avoid unwanted insemination is direct rejection, females also exhibit a wide diversity of behaviors to communicate sexual reluctance (Blanckenhorn et al. 2000; González and Costa 2008; Hosken et al. 2003; Peretti and Willemart 2007; Rowe 1992).

Some behavioral patterns related to female sexual reluctance such as "body shaking" or "hitting the male" have been frequently interpreted as indiscriminate female tactics to avoid male coercion (Arnqvist and Rowe 1995, 2005). Although this could be the case in some animals, fine behavioral observations have shown in other animals that certain patterns of female resistance should be interpreted as flexible, non-coercive signals that communicate lack of sexual receptivity (Baena and Eberhard 2007; Peretti and Córdoba-Aguilar 2007).

Acoustic intersexual communication by stridulation occurs in many insects (e.g., Hemiptera, Manrique and Lazzari 1994; Heteroptera, Rodríguez 1998; Orthoptera, Zuk and Simmons 1997). In some arachnids such as spiders, the male also produces sexual acoustic signals (Huber and Eberhard 1997; Uetz and Stratton 1982). Stridulation is produced by friction between two rigid structures (Legendre 1963) and has been found in 31 spider families (Uhl and Schimtt 1996), which all use a variety (20 types have been described) of stridulatory organs (Legendre 1963; Uetz and Stratton 1982). Huber (1995) indicates that in Pholcidae, stridulatory organs can be found in at least 11 genera, and according to the classification of Legendre (1963), they are all of type "d" (pedipalp rubs against chelicera), or type "a" (opisthosoma rubs against prosoma).

As in many other pholcids, *Holocnemus pluchei* have stridulatory organs that are composed of a line of cuticular ridges on the chelicera and a scraper near the basis of the pedipalpal femur (Huber 1995). Huber (1995) described the courtship and copulation of *H. pluchei* and observed that the male stridulation is used during courtship as a luring behavior. In contrast, his observations of female stridulatory behavior suggest a defensive or even aggressive function, because females that readily accepted the male never rub their pedipalps against the chelicerae, while females that seemed less sexually receptive often rub the pedipalps against the chelicerae (Huber 1995). In *H. pluchei*, female stridulation occurs during courtship but never during copulation (Dutto et al. 2011; Huber 1995). Therefore, stridulation by females conveys a negative message both in intrasexual and in intersexual contexts (Dutto et al. 2011). The strength of the "negative" message seems to be defined by the intensity of stridulation. In addition, the negative meaning involved in female stridulation is in agreement with its usual occurrence during agonistic interactions in another pholcid, *P. phalangioides* (Dutto 2006).

There are two possible non-exclusive advantages to the female of demonstrating reluctance: (1) She could avoid potential costs associated with injuries, increased predation, or energetic costs because of long courtships performed by persistent males or (2) she could filter males based on their persistence in such signals, thereby obtaining more persistent male offspring. This second advantage could explain the paradoxical signals of non-receptivity produced by virgin females that eventually mated. For a male, recognizing the female signals of lack of receptivity could be advantageous if he avoids wasting time and risking injury or predation in fruitless attempts to induce receptivity in a non-receptive female (Dutto et al. 2011).

In addition, females perform courtship behavior during copulation in some species (Rodríguez 1998), but its significance remains unknown. Females probably often have a large degree of control over paternity (Eberhard 1996), so female signals may affect the ultimate outcomes of sexual interactions (Peretti et al. 2006).

Female signaling during copulation is probably more common than that has been appreciated (Eberhard 1994). There are reports of possible female signaling during copulation in insects and spiders (Arnqvist et al. 2003; Cheng et al. 2004; Cowan 1991; Crudgington and Siva-Jothy 2000; Eberhard 2005; Huber 1994; Kotrba 1996; O'Neill 2001), but little has been established regarding their functions (Peretti et al. 2006). One function of possible female signals during copulation is that they serve to induce changes in male behavior (Crudgington and Siva-Jothy 2000; Knoflach and van Harten 2000; Ridsdill-Smith 1970; Rodríguez 1998).

An interesting example comes from females of the spider *P. globosus*. Females stridulate during copulation (Huber and Eberhard 1997), a behavior that is not physically coercive. The female often moves her pedipalps during copulation in short bursts of rapid, alternating dorsoventral movements. Sound recordings show that each burst of movements by the female's pedipalps is associated with a burst of sound, resembling squeaking leather. Female stridulation could presumably be transmitted to the male through the air, through the silk web lines through which both animals are gripped or through body contact. In *P. globosus*, the female uses stridulation as a communicatory signal to modulate the male copulatory movements, and there is a strong relationship between female stridulation and male papal squeezing during copulation. Moreover, the relative fertilization success of a male depends partially on the type of responses given to the female stridulatory request, demonstrating a dialogue between male and female during copulation (Peretti et al. 2006).

Nonetheless, it is unclear whether female copulatory stridulation is widespread in other pholcids. Females of the congeneric *P. dugesi* lack stridulatory ridges and do not perform stridulatory movements with their palps during mating (A. Peretti, unpublished data).

5.3.3 Pholcid Genitalia and Post-copulatory Sexual Selection

5.3.3.1 CFC by Mechanical Fit Hypothesis and CFC by Genitalic Copulatory Courtship

Eberhard (1985) suggested species specificity of genitalia to be a result of female choice and proposed two mechanisms to this process: female choice by mechanical fit and female choice by stimulations.

From the male's evolutionary perspective, variations in genital morphology that enable him to solve mechanical challenges of copulation could confer advantages over other males. In contrast, from the evolutionary perspective of females, the mechanical problems experienced by males that lack sense organs in their genitalia could lead to selection on females to discriminate against those males least able to achieve effective genitalic alignment, either through the stimuli received or via changes in morphology that bias male abilities to fit mechanically (Huber and Eberhard 2010). Such selection to discriminate among male designs could favor changes in female morphology that would make her genitalia more selective, facilitating given male's chances of getting his sperm into her spermathecae only if his genitalia have certain mechanical properties. The female would thus be exercising sexual selection by CFC with respect to the male's ability to adjust mechanically to her genitalic morphology (Huber and Eberhard 2010).

Female choice by mechanical fit does not depend on the presence of mechanoreceptors. The prediction is that the contact zones of females should be highly sclerotized because membranous pouches cannot discriminate between variants in the male genitalia. This might be the only plausible mechanism of genitalic discrimination when there are no neurons and stimuli involved (Huber 1993a). As an example, the sheet-web spider *Neriene montana* (Lyniiphidae) may serve to answer these questions, where van Helsdingen (1969, 1972) shows that the spirally wound male terminal apophysis is inserted into the female genital cavity during copulation. In accordance with van Helsdingen, histological sections reveal that the male apophysis almost exactly closes with the open spiral groove in the female genital cavity that is highly sclerotized (Huber 1993a). Females of few species of the pholcid genus *Mesabolivar* have exaggerated external genitalia, and those exaggerations are functionally correlated with extravagant male cheliceral morphology (Huber et al. 2005). In *M. yuruani*, the males have the procursus about six times longer as usual in the family, and this exaggeration is paralleled in the female internal genitalia (Huber 2006).

On the other hand, when females choose males by their stimulations, there should be selection toward optimized reception of these stimuli. It should not be enough to sense the mere presence of the male genital organ, but it should be crucial for the female to evaluate the minimal morphological variants that exist within a population (Huber 1993b). The only way to sense this is by mechanoreceptors. The prediction is that at the contact zones, mechanoreceptors should be found in high density (Huber 1993b). A study of these contact zones using scanning electron microscope revealed that there are no sensory hairs in these areas (Huber 1993b). However, it is yet to be determined that the occurrence of genitalic stimulation in spiders could not be ruled out with current available information. There might be undiscovered internal receptors in certain membranous parts, and females could sense the male genital organs with slit sensilla far away from the copulatory organ (Huber 1993b). Another possibility was found in *P. globosus*, where a large area around the epigynum is deformed by the male pedipalpal movements (Huber and Eberhard 1997), which could cause stimulation of stretch receptors in this area.

Preliminary measurements suggest that the right male palp of *Metagonia mariguitarensis* may acquire a more stimulatory function, while the left male palp may have taken over the sperm transfer function, and the significant differences in pedipalp size indicate that there is indeed a tendency toward functional segregation between right and left male palps (Huber 2004). The procursus in *P. phalangioides* have various apophyses, lamellar processes, and the conspicuous zone that becomes inflated, which inflates during intromission, and these structures could play the most important part in stimulating the female (Uhl et al. 2005).

5.3.3.2 Selective Cooperation During Genitalic Coupling

Pholcids are unusual among spiders by showing a functional correlation between the male chelicerae and the female external genitalia ("epigynum"). At the onset of copulation, the male contacts the epigynum with his sexually modified chelicerae, and during copulation, parts of the female genitalia are firmly locked between male chelicerae and palps (Huber et al. 2005). This correlation has been shown both by direct observations (Huber 1995, 1997, 1998a, 2002; Huber and Eberhard 1997; Uhl et al. 1995) and by comparative analyses (Huber 1999, 2003b; Kraus 1984). Comparative analyses are possible in cases where the female develops specific pockets or cavities for engagement of specific male apophyses. Such cooperative structures have evolved several times convergently within Pholcidae (Huber 2000, 2003a, b).

In several species, male and female morphologies seem to reflect arms races (Huber et al. 2005b). This process predicts evolutionary changes in female morphology that are mirrored in corresponding changes in male morphology. In some cases, this does not seem to involve any significant cost to the female, for example, when the only change in females is the position of a pair of pockets on the epigynum (Huber 2000; Huber et al. 2005). For males, in contrast, a simple shift of the female pockets may select for exaggerated apophyses whose length reflects their need to reach the female pockets (Huber 2000, 2003a, b; Huber et al. 2005).

Evidences in pholcids suggest a functional correlation between male cheliceral apophyses and projections and pockets of female genitalia. First, the length of male chelicerae corresponds tightly with the length of the female epigyneal projections (Huber et al. 2005). All direct observations of copulation in pholcids have found a functional correlation between male chelicerae and female epigynum (Huber 1995, 1997, 1998a, 2002; Huber and Eberhard 1997; Uhl et al. 1995).

The female morphology of the two species (*Mesabolivar samatiaguassu* and *Mesabolivar cuarassu*) shows that females offer a cooperative structure to the male (a median pocket), but a pair of projections confer difficult access to it (Huber et al. 2005). In spiders in general, males typically lure rather than coerce females to mate, and female cooperation is usually a prerequisite for successful genital coupling (Huber 1998b). Female epigyneal structures in Pholcidae usually appear either neutral (flat plates) or even cooperative (hoods, grooves, pits, scapes) in the sense that they help the male to place his chelicerae and position his body correctly. In *H. pluchei*, correct positioning of the male relative to the female is achieved with the help of the frontal cheliceral apophyses; they are brought into contact with a specially sculptured area anterior to the epigyneal bulges (Huber 1995). Also, in *Spermophora senoculata* and genus *Belisana*, females have structures as epigyneal and abdominal pits and males have cheliceral and bulbal apophyses (Huber 2002, 2005).

The CFC hypothesis could explain the prevalence of "selective cooperative" female designs as evidence against the sexually antagonistic coevolution hypothesis (Chap. 1). However, CFC might seem unable to explain why male genitalia and female genitalia would change rapidly. Once the males of a species evolved a genitalic design that fits with the corresponding structures of conspecific females, further changes in either males or females would seem to be disadvantageous. A male with variant genitalia could have a disadvantage because he would have a poor couple with females (Huber and Eberhard 2010).

There is evidence that the intraspecific variation in genitalia found in spiders has been selectively important. The genitalia of both male and female spiders resemble those of insects in showing negative static allometry (relatively large genitalia in smaller individuals and relatively small genitalia in large individuals of the same species) (Eberhard et al. 1998, 2009). These low allometric values probably represent special evolutionary adjustments to reduce the amount of difference in genital size between males and females, allowing the male to fit effectively with the most common size of female (Eberhard et al. 2009).

The one-size-fits-all hypothesis (Eberhard et al. 1998) attempted to explain the adaptive significance of this pattern of low allometric slopes in genitalia. It is proposed that the low slopes might be due to the advantage of a male having genital sizes that are appropriately adjusted to the most common size of females in the population and that such size adjustments might improve male's abilities to stimulate the female and thus succeed under sexual selection by female choice (Eberhard et al. 1998). An additional, non-exclusive possible advantage (which emphasizes the possible role of natural selection rather than sexual selection) is that these adjustments may facilitate precise mechanical fits between male and female genitalia that are needed to carry out sperm transfer (House and Simmons 2003).

In *H. pluchei*, the male genital structures and female genital structures show no significant allometric values, i.e., body size does not go hand in hand with the genital structures (L. Calbacho-Rosa, PhD Dissertation Thesis). In addition, in *P. globosus*, both males and females show lower values of allometry of genital structures than those of no-genital structures (Huber 1996). In the same way, a study by Eberhard et al. (1998) found lower values of allometry in 20 species of insects and spiders.

Eberhard et al. (1998) attempted to explain the adaptive significance of this pattern of low slopes allometric in genitalia. They proposed that the low slopes might be due to the advantage to a male of having genital sizes that are adjusted to the most common size of females and that adjustments might improve male abilities to stimulate the female.

A currently used hypothesis for explaining rapid divergent evolution in sexual traits, such as genitalia, is sexually antagonistic coevolution (SAC) of males and females (Chap. 1). SAC hypothesis assumes that because interests of males and females are different, conflict between the sexes over control of copulation will lead to coevolutionary races between "aggressive" male traits that enhance the male's control over copulation and "defensive" female traits that enhance the female's control and thus reduce the costs (Huber and Eberhard 2010).

One SAC prediction is that female morphology should tend to coevolve with male morphology. This prediction is partially supported in spiders. However, one aspect of this prediction is not fulfilled in spiders: If genitalic diversification is due to an arms race between males and females for control of copulation, female genitalia should often have recognizably "defensive" designs, for excluding male genitalia. Instead, many of the traits of female spider genitalia are most easily understood as being "selectively cooperative" structures (see above). Another prediction of SAC is that rapid divergent genitalic evolution should be associated with only certain types of male–female precopulatory interactions (Alexander et al.

1997). Coevolutionary races are most likely to occur in groups in which males are more able to physically coerce or sexually harass unreceptive females (Alexander et al. 1997). In contrast, such races are less likely to occur in groups in which males are not able to physically coerce females into copulating and only interact with females that are receptive ("luring" interactions) (Alexander et al. 1997). A major review of spider mating behavior in more than 150 species (Huber 1998a) showed that interactions preceding copulation are typically of the luring type; nevertheless, in contradiction to the SAC prediction, spider genitalia show a sustained and rapid divergent evolution (Huber and Eberhard 2010, Chap. 1).

This failure of SAC predictions extends to the fine details of the physical coupling between male and female genitalia. It is clear in a number of spider groups that tiny movements of the female can easily disrupt the difficult process of alignment of the male. This goes against the likelihood that adjustments of the morphology of the female genitalia are needed as defenses against males and thus against the idea that such morphological differences in females function in this context (Huber and Eberhard 2010).

5.3.4 Sperm Dumping and Its Role to Bias Paternity

5.3.4.1 What Is a "Dumping"?

In animals where females are promiscuous, the competition between sperm of different males for the fertilization of an ovum ("sperm competition," Parker 1970) represents a strong evolutionary force that has determined a wide range of adaptations in sexual behavior (Birkhead and Møller 1998). One mechanism that can considerably influence male sperm competitiveness and paternity assurance is the ejection of sperm transferred by the previous or current male from the female's sperm storage site during or after copulation. The phenomenon of sperm dumping is reported in a wide variety of taxa, ranging from birds (Pizzari and Birkhead 2000), insects (Snook and Hosken 2004), and round worms (Barker 1994; Eberhard 1996). Often, sperm dumping is associated with movements of the male genitalia during copulation interpreted as copulatory courtship (Eberhard 1985, 1994, 2009, 2011; Rodríguez 1995). Nowadays, it is widely accepted that sperm dumping has evolved under sexual selection by CFC (Córdoba-Aguilar 2006; Eberhard 1996). For example, males of the beetle *Chelimorpha alternans* perform rhythmic movements with their genitalia during copulation that are sometimes followed by the ejection of previous sperm by the female (Eberhard 1996; Rodríguez 1995). In the fly *Dryomyza anilis*, the male taps the female's external genitalia after copulation with his genitalic claspers and presses her abdomen. Before oviposition, the female ejects a droplet containing mostly sperm of the previous male, thus favoring last-male sperm precedence (Eberhard 1996; Otronen 1990, 1997; Otronen and Siva-Jothy 1991). However, in spiders, sperm dumping has only been reported in three species: the two pholcid *P. phalangioides* by Uhl et al. (1995) and Schäfer and Uhl (2002), the pholcid *P. globosus* by Eberhard (1996) and Huber and Eberhard (1997), and the tetragnathid *Pachygnatha clercki* by Gerhardt (1923). Based on the morphology of the female genital organs, sperm dumping is also suggested for the oonopids *O. fosuma* and *S. loricatula* by Burger et al. (2003, 2006b) (please see Sect. 5.2.1.3).

5.3.4.2 Morphological and Behavioral Evidences that Support Female Control

The evolution of complex reproductive by means of sexual selection by CFC is currently widely accepted (Eberhard 1985, 1996; Thornhill 1983). CFC, in the sense of Eberhard and Cordero (1995), means that female behavior, physiology, or morphology that favors certain males against others during or after copulation is operating. In many animal groups, females have a physiologically and morphologically complex reproductive system, which enables them to control or directly influence the fate of stored sperm (Hellriegel and Ward 1998). Frequently, the sperm storage site and the fertilization site are separated, allowing females to eject sperm with the same genital mechanisms that move stored sperm from the storage site to the fertilization site (Eberhard 1996). Eberhard (1994) reported emission from the female during or immediately following copulation in 25 % of 53 species of insects and spiders.

If sperm dumping is simply due to an excessive ejaculate volume respect to the capacity of the female's reproductive tract, as suggested by Rehfeld and Sudhaus (1985) and Hass (1990), it may not have reproductive consequences for the male. Alternatively, sperm dumping may result in a reduction in the male's chances of paternity, if his success in fertilization depends on the amount of his sperm that remains inside the female. Most of the previous studies of sperm dumping have assumed that there is a negative effect of sperm dumping on paternity (Otronen and Siva-Jothy 1991; Rodríguez et al. 2004; Snook and Hosken 2004). If sperm dumping affects paternity, it can only exercise sexual selection by CFC if two additional conditions are fulfilled: The proportion of sperm discarded must be correlated with a male trait (such as copulatory courtship), and females must normally mate with more than a single male.

Convincing positive associations of sperm dumping with particular male traits have been established in few species. The traits include male social dominance in birds (Pizzari and Birkhead 2000) and the sizes and movements of genital sclerites in insects (Córdoba-Aguilar 2006; Rodríguez et al. 2004).

Females of *P. globosus* (Pholcidae) often emit a mass of sperm in an irregular white mass from the female's genitalia near the end of the copulation, or just following copulation (Huber and Eberhard 1997) (Fig. 5.3). Discarded masses in first copulations seemed to usually have a more liquid consistency than those in second copulations (Peretti and Eberhard 2010). A study of genital morphology and copulatory behavior indicated that sperm dumping is not simply removal by the male and instead almost certainly involves the active role of the female (Huber and Eberhard 1997). In fact, the positions and movements of the male's

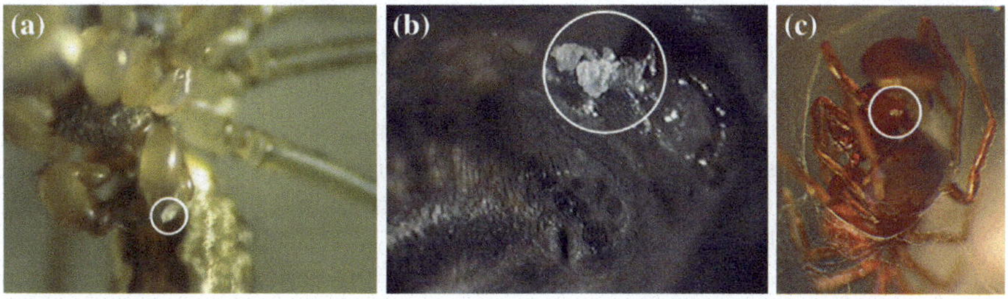

Fig. 5.3 **a** Sperm removal from the female gonopore during phase I of copulation in *Holocnemus pluchei* (Pholcidae); **b** ventral surface of the abdomen of female *Physocyclus globosus*, showing the sperm mass that was dumped from her gonopore (Pholcidae); **c** *Silhouettella loricatula* (Oonopidae) in copulation. The circle shows the dumped secretory sac

genitalia within the female with respect to the site in the female where sperm are stored are not appropriate to remove stored sperm. Moreover, sperm is ejected in copulations with virgin females. Sperm ejection involves active expulsing movements of the female's genital opening and sometimes occurs after copulation has ended. Removal of sperm by the female is morphologically feasible because the male apparently deposits his sperm in a liquid matrix (Huber and Eberhard 1997). Contractions of the vaginal walls similar to those presumably used during oviposition could expel this liquid or viscous mass through the female's gonopore.

When a female of *P. globosus* mates with two males, paternity averages were approximately 50 % for each male (Eberhard et al. 1993), but percentages varied widely. Two biases in sperm use have been found: Males that relaxed their palpal squeezes more consistently in response to female stridulation achieve about three times greater paternity, and greater numbers of palpal squeezes also correlate with greater paternity (Peretti et al. 2006). The pattern of larger discarded masses is weakly associated with reduced paternity for the first male. Two male characteristics of first copulations that are associated with paternity success also have strong negative correlations with the volumes of discarded sperm masses: more bursts of abdomen vibration and a higher frequency of palpal squeezes (Peretti and Eberhard 2010). Abdomen vibration and palpal squeezing are under male rather than female control and thus are appropriately taken to be favoured male traits. It is possible that female stridulation or other unobserved female responses, such as events in her internal genitalia, could induce male responses in these behavioral patterns and that the male trait under selection involves his responses to the female. Male abdominal vibration is not a physically coercive behavior, and its effects on sperm dumping seem likely to result from stimulation of the female rather than physical force. It is likely that the selection favouring male abdomen vibration is CFC rather than SAC (Peretti and Eberhard 2010). In another Pholcidae, *P. phalangioides*, the movements of male pedipalps result in extrusion of spermatozoa from the female genital cavity (Uhl et al. 1995).

Sperm displacement could be suspected if the extrusion occurred exclusively in previously mated females. Nevertheless, the ejection of sperm from the genital cavity is sometimes observed in virgin females (Uhl 1994; Schäfer and Uhl 2002; Schäfer et al. 2008), suggesting that females would have an active participation. Copulation duration varies depending on the reproductive history of the female. Second males do not copulate for longer than five minutes (Uhl 1993); thus, variation in fertilization success can be explained by variation in pedipalp movements, where the higher number of pedipalp movements determines higher paternity.

The higher number of pedipalp movements in first mating compared to second one may have evolved either in the context of numerical sperm competition or in the context of CFC (Schäfer and Uhl 2002). Preliminary data show that in *M. globulosus* (Peretti and Calbacho-Rosa, unpublished data), sperm output occurs during and after copulation, suggesting the occurrence of sperm dumping.

In Oonopidae, sperm is dumped in *S. loricatula* and *O. fosuma*. In *S. loricatula*, females use muscles attached to the receptaculum to dump sperm during copulation (Burger 2007; Burger et al. 2006b). Females of *S. loricatula* manage to process sperm in a very unusual and previously unknown way (Burger 2007). The entire ejaculate of the male is enclosed in a discrete package or sac, consisting of a secretion inside the female's receptaculum (Fig. 5.3). The secretion enclosing sperm in a sac is produced by glands adjoining the receptaculum, which is in accordance with the findings of Burger et al. (2006a). The enclosure of sperm from the current male takes place during or immediately after copulation. Burger et al. (2006a) found microorganisms inside the receptacula and proposed that one function of the secretory sac could be to protect the sperm from infectious agents. Another function of the sac could be to avoid the mixing of spermatozoa from different males inside the receptaculum, which should prevent sperm competition. Burger (2007) showed that females of *S. loricatula* are able to dump the secretory sac and therewith the entire ejaculate of a male during copulation. Dumping sperm of a previous male during the next copulation may allow females to bias sperm precedence. It is suggested that sperm dumping in *S. loricatula* is the main reason for the process of sperm enclosure in a sac (Burger 2007). In this way, the secretory sac gets into the genital opening and is dumped (Burger 2007; Burger et al. 2006a). Males move their pedipalps rhythmically during copulation and possibly use them as copulatory courtship. The sac dumping is proposed to occur in cooperation with the male, which moves the pedipalps during the entire copulation (Burger 2007).

In the same way, in *O. fosuma*, Burger et al. (2003) describe the complexity of the female genital structures and their allied muscles, also suggesting the possibility of sperm dumping. If the interpretations of Burger et al. (2003) are correct, female genital organs in *Opopaea*, *Xyphinus*, and *Gamasomorpha* are able to influence strongly male's chances of fathering their offspring by exerting CFC.

5.3.4.3 Other Types of Sperm Expelling

Competition among sperm from different males to fertilize the eggs of a single female has important implications for the mating strategies of both sexes (Parker

1970). Priority advantage is often a function of the order of mating, so that either first or last males to mate fertilize a disproportionately greater number of eggs. Priority advantage may arise through numerous mechanisms, including preferential sperm use by the female, stratification of the sperm within the sperm storage, site of the female, or displacement of sperm by a later male. In certain insects, the male genitalia are even specialized to remove active sperm from the female's sperm storage organs, as earwig *Euborellia plebeja* (Kamimura 2000, 2005) or in several odonates (Córdoba-Aguilar 1999; Córdoba-Aguilar et al. 2003; Eberhard 1996; González-Soriano and Córdoba-Aguilar 2003; Waage 1979).

In the pholcid spider *Holocnemus pluchei*, female remains receptive after copulation and the mate guarding strategy of males seems effective to impede competitors' access to recently mated females (Calbacho-Rosa et al. 2010). The duration of post-copulatory mate guarding seems adaptive, as it is adjusted to that time when females would be more likely to remate. Mating duration of the last male is significantly longer than that of the first male. In addition, Kaster and Jakob (1997) observed a last-male sperm precedence of 74 % in this species, and one possible explanation for second-male sperm priority is male removal ability.

Calbacho-Rosa et al. (2013) found that males of *H. pluchei* perform two types of pedipalp movements, simultaneous pedipalp movements and alternating pedipalp movements, and interpret these movements as operating in different sperm competition-related functions. In the simultaneous pedipalp movements, males moved both pedipalps in a rhythmic and simultaneous manner. Both pedipalps contracted and relaxed in a regular fashion. Simultaneous movements are commonly observed during copulations with both virgin and mated females. The male introduced both the procursus and the two bulbal apophyses into the gonopore. Thus, at least one function for simultaneous pedipalp movements is sperm transfer sensu stricto (Calbacho-Rosa et al. 2013). In pholcids, simultaneous pedipalp movements have been linked to a male stimulatory role (Huber and Eberhard 1997; Peretti and Eberhard 2010; Schäfer and Uhl 2002) (see above). The other type of behavior, alternating pedipalp movements are quickly alternated, almost superficial and disordered, coincides with sperm mass ejection from the female's genital opening. Calbacho-Rosa et al. (2013) suggested that the function of this alternating movement is to remove previously stored, rival sperm (Fig. 5.3). This hypothesis is based on (a) the time of copulation in which sperm ejection occurs (at the beginning of the second copulation), (b) the specific region in which procursus of each pedipalps is inserted (the internal and dorsal side of the epigyne that contains stored sperm), and (c) the fact that sperm are often attached to the male procursus. During alternating pedipalp movements, the two apophyses of the genital bulb remain completely outside the female. Therefore, it is not possible that the male transfers its own sperm during alternating pedipalp movements (Calbacho-Rosa et al. 2013). Removal of rival sperm is a widespread trait and can be mechanically achieved using sexual as well as non-sexual morphological structures (Leonard and Córdoba-Aguilar 2010).

In *Pholcus phalangioides*, despite much shorter duration of second copulations, last-male sperm precedence could be also explained by a sperm removal mechanism (Schäfer and Uhl 2002). The current knowledge on the male and female genital morphology (Uhl 1994; Uhl et al. 1995) and the role of pedipalp movements for fertilization success—as described in the present study—provides insights into the mechanisms of sperm precedence. Access of male genital structures to the female sperm storage site (Uhl 1994) and the fact that sperm mass emerges in the center of the female epigynum coinciding with pedipalp movements strongly suggest a sperm removal mechanism in this species (Schäfer and Uhl 2002).

However, we cannot discard a possible mechanism of CFC, as has been hypothesized for other arthropods in cases where the female participates in sperm ejection (e.g., Burger 2007; Córdoba-Aguilar 1999; Eberhard 1994, 1996; Peretti and Eberhard 2010; Snook and Hosken 2004). Finally, one possibility is that stored sperm is acted upon by both male-driven sperm removal processes and female-driven active ejection.

5.4 Conclusions and Prospects

Haplogyne spiders are a good model to study copulatory and post-copulatory sexual selection. The fact that during copulations males maintain much of their palps outside females allows direct observation of structures. The haplogyne spiders, due to the morphology of the female reproductive tract, mainly indicate last-male sperm priority. In agreement, male guarding of penultimate females has not been observed in these species. Conversely, several behaviors have been observed during copulation and post-copulation, which could favor the reproductive success of males.

In this chapter, we showed several examples of two diverse families of haplogyne spiders: Pholcidae and Oonopidae. There is great diversity in morphological and behavioral male traits associated to copulation. This allows the evaluation of different mechanisms by means of which males try to increase their reproductive success. The duration of copulations varies among species and according to the previous reproductive status of the females, virgins and non-virgins. In copulations, males stimulate females through copulatory courtship, which can be genital or non-genital. In the latter, behaviors such as abdomen vibrations, leg shaking, and stridulation are included, whereas genital copulatory courtship stimulation of females is achieved by male's rhythmic genitalic movements of the pedipalps. Genital courtship has paid particular interest, and it was found that different species exhibit variations in the types of movements and their frequencies. Males used these movements, in addition to the sperm transfer function, to cause sperm dumping through the stimulation of the female or perform active removal of sperm from already mated females.

However, to date, there have been very few studies that determine paternity in haplogyne spiders. For example, last-male sperm priority was found only in *H. puchei* and *P. phalangiodes*. Despite general predictions, in *P. globosus* the sperm priority for the last male was not observed. Therefore, not only the morphology of the reproductive tract of females is what determines fertilization success of males: Different degrees of fertilization success may be also the mixed result of multiple mechanisms of post-copulatory sexual selection, including sperm competition and CFC.

Acknowledgments We thank Bernhard Huber and Mathias Burger for photographs, Anita Aisenberg, Ignacio Escalante, and Fernando G. Costa for useful suggestions on the manuscript, and the Consejo Nacional de Investigaciones Científicas y Técnicas (CONICET), FONCYT, and SECYT-UNC for financial support.

References

Alberti G, Michalik P (2004) Feinstrukturelle Aspekte der Fortpflanzungssysteme von Spinnentieren (Arachnida). In: Thaler K (ed) Diversit¨at und Biologie von Webspinnen, Skorpionen und anderen Spinnentieren, vol 12. Denisia, pp 1–62

Alcock J (1994) Post-insemination associations between male and female insects: the mate guarding hypothesis. Annu Rev Entomol 39:1–21

Alexander RD, Marshall D, Cooley J (1997) Evolutionary perspectives on insect mating. In: Choe JC, Crespi BJ (eds) The evolution of mating systems in insects and arachnids. Cambridge University Press, Cambridge, pp 4–31

Andersson M (1994) Sexual selection. Princeton University Press, Princeton, NJ

Andersson M, Simmons L (2006) Sexual selection and mate choice. Trends Ecol Evol 21(6):296–302

Andrés J, Cordero Rivera A (2000) Copulation duration and fertilization success in a damselfly: an example of cryptic female choice? Anim Behav 59:695–703

Arnqvist G (2014) Cryptic female choice. In: Shuker DM, Simmons LW (eds) The evolution of insect mating systems. Oxford University Press, Oxford

Arnqvist G, Rowe L (1995) Sexual conflict and arms races between the sexes: a morphological adaptation for control of mating in a female insect. Proc R Soc Lond Ser B 261:123–127

Arnqvist G, Rowe L (2002) Antagonistic coevolution between the sexes in a group of insects. Nature 415:787–789

Arnqvist G, Rowe L (2005) Sexual conflict. Princeton University Press, Princeton

Arnqvist G, Jones T, Elgar M (2003) Reversal of sex roles in nuptial feeding. Nature 424:387

Austad S (1984) Evolution of sperm priority patterns in spiders. In: Smith RL (ed) Sperm competition and the evolution of animal mating systems. Academic Press, San Diego, pp 223–249

Baena ML, Eberhard WG (2007) Appearances deceive: female "resistance" in a sepsid fly is not a test of male ability to hold on. Ethol Ecol Evol 19:27–50

Barbosa F (2009) Cryptic female choice by female control of oviposition timing in a soldier fly. Behav Ecol 20:957–960

Barker DM (1994) Copulatory plugs and paternity assurance in the nematode *Caenorhabditis elegans*. Anim Behav 48:147–156

Barrantes G, Ramírez MJ (2013) Courtship, egg sac construction, and maternal care in *Kukulcania hibernalis*, with information on the courtship of *Misionella mendensis* (Araneae, Filistatidae). Arachnology 16(2):72–80

Birkhead TR, Møller AP (1992) Sperm competition in birds: evolutionary causes and consequences. Academic Press, London

Birkhead TR, Møller AP (1998) Sperm competition and sexual selection. Academic Press, London

Blanckenhorn WU, Muhlhauser C, Morf C, Reusch T, Reuter M (2000) Female choice, female reluctance to mate and sexual selection on body size in the dung fly *Sepsis cynipsea*. Ethology 95:466–482

Brignoli PM (1978) Some remarks on the relationships between the Haplogynae, the Semientelegynae and the Cribellatae (Araneae). Symp Zool Soc Lond 42:285–292
Bukowski TC, Christenson TE (1997) Mating in the orbweaving spider *Micrathena gracilis*—II. Factors influencing copulation and sperm release and storage. Anim Behav 53:381–395
Burger M (2007) Sperm dumping in a haplogyne spider. J Zool 273:74–81
Burger M (2008) Functional genital morphology of armored spiders (Arachnida: Araneae: Tetrablemmidae). J Morphol 269:1073–1094
Burger M (2009) Female genitalia of goblin spiders (Arachnida: Araneae: Oonopidae): a morphological study with functional implications. Invertebr Biol 128:340–358
Burger M (2010) Complex female genitalia indicate sperm dumping in armored goblin spiders (Arachnida, Araneae, Oonopidae). Zoology 113:19–32
Burger M, Kropf C (2007) Genital morphology of the haplogyne spider *Harpactea lepida* (Arachnida, Araneae, Dysderidae). Zoomorphology 126:45–52
Burger M, Nentwig W, Kropf C (2002) *Opopaea fosuma*, n. sp. from Sumatra, Indonesia (Araneae, Oonopidae). Bull Br Arachnol Soc 12:244–248
Burger M, Nentwig W, Kropf C (2003) Complex genital structures indicate cryptic female choice in a haplogyne spider (Arachnida, Araneae, Oonopidae, Gamasomorphinae). J Morphol 255:80–93
Burger M, Jacob A, Kropf C (2006a) Copulatory behavior and web of *Indicoblemma lannaianum* from Thailand (Arachnida, Araneae, Tetrablemmidae). J Arachnol 34:176–185
Burger M, Graber W, Michalik P, Kropf C (2006b) Silhouettella loricatula (Arachnida, Araneae, Oonopidae): a haplogyne spider with complex female genitalia. J Morphol 267:663–677
Burger M, Michalik P, Graber W, Jacob A, Nentwig W, Kropf C (2006c) Complex genital system of a haplogyne spider (Arachnida, Araneae, Tetrablemmidae) indicates internal fertilization and full female control over transferred sperm. J Morphol 267:166–186
Burger M, Izquierdo M, Carrera P (2010) Female genital morphology and mating behavior of *Orchestina* (Arachnida: Araneae: Oonopidae). Zoology 113:100–109
Calbacho-Rosa L (2012) L. Calbacho-Rosa, PhD Dissertation Thesis: Patrones de Selección Sexual en arañas Pholcidae: comportamiento sexual y procesos asociados. Tesis doctoral. Universidad Nacional de Córdoba, Argentina
Calbacho-Rosa L, Córdoba-Aguilar A, Peretti A (2010) Occurrence and duration of post-copulatory mate guarding in a spider with last sperm precedence. Behaviour 147:1267–1283
Calbacho-Rosa L, Galicia-Mendoza I, Dutto MS, Córdoba-Aguilar A, Peretti AV (2013) Copulatory behavior in a pholcid spider: males use specialized genitalic movements for sperm removal and copulatory courtship. Naturwissenschaften 100:407–416
Carazo P, Font E (2010) Putting information back into biological communication. J Evol Biol 23:661–669
Chapman T (2006) Evolutionary conflicts of interest between males and females. Curr Biol 16:744–754
Chapman T, Arnquist G, Bangham J, Rowe L (2003) Sexual conflict. Trends Ecol Evol 18:41–47
Cheng LI, Howard RW, Campbell JG, Charlton RE, Nechols JR, Ramaswamy SB (2004) Mating behavior of *Cephalonomia tarsalis* (Ashmead) (Hymenoptera: Bethylidae) and the effect of female mating frequency on offspring production. J Insect Behav 17:227–246
Cordero C, Eberhard WG (2003) Female choice of sexually antagonistic male adaptations: a critical review of some current research. J Evol Biol 16:1–6
Cordero C, Eberhard WG (2005) Interaction between sexually antagonistic selection and mate choice in the evolution of female response to male traits. Evol Ecol 19:111–122
Córdoba-Aguilar A (1999) Male copulatory sensory stimulation induces female ejection of rival sperm in a damselfly. Proc R Soc Lond 266:779–784
Córdoba-Aguilar A (2006) Sperm ejection as a possible cryptic female choice mechanism in Odonata (Insecta). Physiol Entomol 31:146–153
Córdoba-Aguilar A, Uhía E, Cordero Rivera A (2003) Sperm competition in Odonata (Insecta): the evolution of female sperm storage and rivals' sperm displacement. J Zool Lond 261:381–398
Cowan D (1991) The solitary and presocial Vespidae. In: Ross KG, Matthews RW (eds) The social biology of wasps. Comstock, Ithaca, pp 33–73

Crudgington HS, Siva-Jothy MT (2000) Genital damage, kicking and early death. Nature 407:855–856
Darwin C (1871) The descent of man and selection in relation to sex (reprinted). Random House, New York
Dimitrov D, Hormiga G (2009) Revision and cladistic analysis of the orbweaving spider genus *Cyrtognatha Keyserling*, 1881 (Araneae, Tetragnathidae). Bull Am Mus Nat Hist 317:1–140
Dimitrov D, Alvarez-Padilla F, Hormiga G (2007) The female genital morphology of the orb weaving spider genus *Agriognatha* (Araneae, Tetragnathidae). J Morphol 268:758–770
Drengsgaard IL, Toft S (1999) Sperm competition in a nuptial feeding spider, *Pisaura mirabilis*. Behaviour 136:877–987
Dutto MS (2006) Análisis funcional del comportamiento de estridulación en *Holocnemus pluchei* (Scopoldi) (Araneae: Pholcidae) femenino. Graduation thesis, Facultad de Ciencias Exactas, Físicas y Naturales, Universidad Nacional de Córdoba, Argentina, p 56
Dutto MS, Calbacho-Rosa L, Peretti A (2011) Signalling and sexual conflict: female spiders use stridulation to inform males of sexual receptivity. Ethology 117:1040–1049
Eberhard W (2001) Species-specific genitalic copulatory courtship in sepsid flies (Diptera, Sepsidae, Microsepsis) and theories of genitalic evolution. Evolution 55(1):93–102
Eberhard WG (1985) Sexual selection and animal genitalia. Harvard University Press, Cambridge
Eberhard WG (1991) Copulatory courtship and cryptic female choice in insects. Biol Rev 66:1–31
Eberhard WG (1994) Evidence for widespread courtship during copulation in 131 species of insects and spiders, and implications for cryptic female choice. Evolution 48:711–733
Eberhard WG (1996) Female control: sexual selection by cryptic female choice. Princeton University Press, Princeton
Eberhard WG (2002) The function of female resistance behavior: intromission by male coercion vs. female cooperation in sepsid flies (Diptera: Sepsidae). Rev Biol Trop 50:485–505
Eberhard WG (2004) Rapid divergent evolution of sexual morphology: comparative tests of antagonistic coevolution and traditional female choice. Evolution 58:1947–1970
Eberhard WG (2005) Sexually reversed copulatory courtship roles and possible nuptial feeding in the soldier beetle *Ditemnus acantholabus* (Coleoptera, Cantharidae). J Kans Entomol Soc 79:13–22
Eberhard WG (2006) Sexually antagonistic coevolution in insects is associated with only limited morphological diversity. J Evol Biol 19:657–681
Eberhard WG (2009) Post-copulatory sexual selection: Darwin's omission and its consequences. Proc Natl Acad Sci USA 106:10025–10032
Eberhard WG (2011) Experiments with genitalia: a commentary. Trends Ecol Evol 26:17–21
Eberhard WG, Cordero C (1995) Sexual selection by cryptic female choice on male seminal products—a new bridge between sexual selection and reproductive physiology. Trends Ecol Evol 10:493–496
Eberhard WG, Huber BA (2010) Spider genitalia: precise maneuvers with a numb structure in a complex lock.In: Leonard JL, Cordoba-Aguilar A (eds) The evolution of primary sexual characters in animals. Oxford University Press, Oxford, pp 249–284
Eberhard WG, Guzmán-Gómez S, Catley KM (1993) Correlation between spermathecal morphology and mating systems in spiders. Biol J Linn Soc 50:197–209
Eberhard WG, Bernhard B, Huber BA, Rodríguez R, Briceno D, Salas I, Rodríguez V (1998) One size fits all? Relationships between the size and degree of variation in genitalia and other body parts in twenty species of insects and spiders. Evolution 52:415–431
Eberhard W, Rodríguez RL, Polihronakis M (2009) Pitfalls in understanding the functional significance of genital allometry. J Evol Biol 22:435–445
Edvardsson M, Arnqvist G (2000) Copulatory courtship and cryptic female choice in red flour beetles *Tribolium castaneum*. Proc R Soc Lond B 267:559–563
Elgar MA (1995) Copulation duration in spiders: comparative patterns. Rec. WA Mus 51:1–11

Elgar MA (1998) Sperm competition and sexual selection in spiders and other arachnids. In: Birkhead TR, Møller AP (eds) Sperm competition and sexual selection. Academic Press, London, pp 307–337
Gerhardt U (1921) Vergleichende Studien über die Morphologie des männlichen Tasters und die Biologie der Kopulation der Spinnen. Arch. Naturgesc 87(A, 4):78–247
Gerhardt U (1923) Weitere sexualbiologische Untersuchungen an Spinnen. Arch. Naturgesellschaft 90:85–192
Gerhardt U (1924) Weitere Studien über die Biologie der Spinnen. Archiv für. Naturgeschichte 90:85–192
Gerhardt U (1927) Neue biologische Untersuchungen an einheimischen und ausländischen Spinnen. Z Morphol Oekol Tiere 8:96–186
González M, Costa F (2008) Persistence of sexual reluctance in mated females and the importance of regular copulation in a wolf spider. Ethol Ecol Evol 20:115–124
Gónzalez-Soriano E, Córdoba-Aguilar A (2003) Sexual behaviour in Paraphlebia quinta Calvert: male dimorphism and a possible example of female control (Zygoptera: Megapodagrionidae). Odonatologica 32:345–353
Griswold CE, Coddington JA, Hormiga G, Scharff N (1998) Phylogeny of the orb web building spiders (Araneomorphae, Orbiculariae: Deinopoidea, Araneoidea). Zool J Linn Soc 123:1–99
Hass B (1990) A quantitative study of insemination and gamete efficiency in different species of the Rhabditis strongyloides group (Nematoda). Invertebr Reprod Dev 18:205–208
Hellriegel B, Ward PI (1998) Complex female reproductive tract morphology: its possible use in postcopulatory female choice. J Theor Biol 190:179–186
Hoikkala A, Crossley S (2000) Copulatory courtship in Drosophila: behavior and songs of *D. birchii* and *D. serrata*. J Insect Behav 13:71–86
Holland B, Rice WR (1999) Experimental removal of sexual selection reverses intersexual antagonistic coevolution and removes a reproductive load. Proc Nat Acad Sci USA. 96:5083–5088
Hosken D, Stockley P (2003) Sexual selection and genital evolution. Trends Ecol Evol 19:87–93
House CM, Simmons LW (2003) Genital morphology and fertilization success in the dung beetle *Onthophagus taurus*: an example of sexually selected male genitalia. Proc Roy Soc B 278:447–455
Huber BA (1993a) Genital mechanics and sexual selection in the spider *Nesticus cellulanus* (Araneae: Nesticidae). Can J Zool 71:2437–2447
Huber BA (1993b) Female choice and spider genitalia. Bolletino dell' Accademia Gioenia di Scienze Naturali (Catania) 26:209–214
Huber BA (1994) Genital morphology, copulatory mechanism and reproductive biology in *Psilochorus simoni* (Berland, 1911) (Pholcidae; Araneae). Neth J Zool 44:85–99
Huber BA (1995) Copulatory mechanism in *Holocnemus pluchei* and *Pholcus opilionoides*, with notes on male cheliceral apophyses and stridulatory organs in Pholcidae (Araneae). Acta Zool. Stockholm 76:291–300
Huber BA (1996) Genitalia, fluctuating asymmetry, and patterns of sexual selection in *Physocyclus globosus* (Araneae, Pholcidae). Rev Suisse Zool (Suppl.):289–284
Huber BA (1997) Evidence for gustatorial courtship in a haplogyne spider (*Hedypsilus culicinus:* Pholcidae: Araneae). Neth J Zool 47(1):95–98
Huber BA (1998a) Spider reproductive behaviour: a review of Gerhardt's work from 1911-1933, with implications for sexual selection. Bull Br arachnol Soc 11(3):81–91
Huber BA (1998b) Genital mechanics in some Neotropical pholcid spiders (Araneae: Pholcidae), with implications for systematics. J Zool London 244:587–599
Huber BA (1999) Sexual selection in pholcid spiders (Araneae, Pholcidae): artful chelicerae and forceful genitalia. J Arachnol 27:135–141
Huber BA (2000) New World pholcid spiders (Araneae: Pholcidae): a revision at generic level. Bull Am Mus Nat Hist 254:1–348

Huber BA (2002a) Functional morphology of the genitalia in the spider *Spermophora senoculata* (Pholcidae, Araneae). Zool Anz 241:105–116
Huber BA (2002b) Functional Morphology of the Genitalia in the Spider *Spermophora senoculata* (Pholcidae, Araneae). Zool Anz 241:105–166
Huber BA (2003a) Rapid evolution and species-specificity of arthropod genitalia: fact or artifact? Org Divers Evol 3:63–71
Huber BA (2003b) Southern African pholcid spiders: revision and cladistic analysis of *Quamtana* gen. Nov. and *Spermophora* Hentz (Araneae: Pholcidae), with notes on male-female covariation. Zool J Linn Soc 139:477–527
Huber BA (2004) Evidence for functional segregation in the directionally asymmetric male genitalia of the spider *Metagonia mariguitarensis* (González-Sponga) (Pholcidae: Araneae). J Zool Lond 262:317–326
Huber BA (2005) Sexual selection research on spiders: progress and biases. Biol Rev 80:363–385
Huber BA (2006) Cryptic female exaggeration: the asymmetric female internal genitalia of *Kaliana yuruani* (Araneae: Pholcidae). J Morphol 267:705–712
Huber BA, Eberhard WG (1997) Courtship, copulation, and genital mechanics in *Physocyclus globosus* (Araneae, Pholcidae). Can J Zool 75:905–918
Huber BA, Brescovit AD, Rheims CA (2005) Exaggerated female genitalia in two new spider species (Araneae: Pholcidae), with comments on genital evolution by female choice versus antagonistic coevolution. Insect Syst Evol 36:285–292
Humphries DA (1967) The mating of the hen flea *Certophyllus gallinae* (Schrank) (Siphonaptera: Insecta). Anim Behav 15:82–90
Kamimura Y (2000) Possible removal of rival sperm by the elongated genitalia of the earwig, *Euborellia plebeja*. Zool Sci 17:667–672
Kamimura Y (2005) Last-male paternity of *Euborellia plebeja*, an earwig with elongated genitalia and sperm removal behavior. J Ethol 23:35–41
Kaster JL, Jakob EM (1997) Last-male sperm priority in a haplogyne spider (Araneae: Pholcidae): correlations between female morphology and patterns of sperm usage. Ann Entomol Soc Am 90:254–259
Kelly CD, Jennions MD (2011) Sexual selection and sperm quantity: meta-analyses of strategic ejaculation. Biol Rev 86:863–884
King BH, Fischer CR (2005) Males mate guard in absentia through extended effects of postcopulatory courtship in a parasitoid wasp *Spalangia endius* (Hymenoptera: Pteromalidae). J Insect Physiol 51:1340–1345
Knoflach B, van Harten A (2000) Palpal loss, single palp copulation and obligatory mate consumption in *Tidarren cuneolatum*(Tullgren, 1910) (Araneae, Theridiidae). J Nat Hist 34:1639–1659
Kotrba M (1996) Sperm transfer by spermatophore in Diptera: new results from the Diopsidae. Zool J Linn Soc 117:305–323
Kraus O (1984) Male spider genitalia: evolutionary changes in structure and function. Verh Naturwiss Ver Hamburg 27:373–382
Legendre R (1963) L'audition et l'e'mision de sons chez les Araneides. Ann Biol 2:371–390
Leonard J, Córdoba-Aguilar A (2010) The evolution of primary characters in animals. Oxford University Press, Oxford
Levi HW (1967) Predatory and sexual behavior of the spider Sicarius (Araneae: Sicariidae). Museum of Comparative Zoology, Harvard University, Canbridge
Manrique G, Lazzari CR (1994) Sexual behaviour and stridulation during mating in *Triatoma infestans* (Hemiptera: Reduviidae). Mem Inst Oswaldo Cruz 89:629–633
Marcotte M, Delisle J, McNeil JN (2005) Impact of male mating history on the temporal sperm dynamics of *Choristoneura rosaceana* and *C. fumiferana* females. J Insect Physiol 51:537–544
Martin OY, Hosken DJ (2004) Copulation reduces male but not female longevity in *Saltella sphondylli* (Diptera: Sepsidae). J Evol Biol 17:357–362
Masumoto T (1993) The effect of the copulatory plug in the funnel-web spider, *Agelena limbata* (Araneae: Agelenidae). J Arachnol 21:55–59

Michalik P, Reiher W, Tintelnot-Suhm M, Coyle FA, Alberti G (2005) Female genital system of the folding trapdoor spider Antrodiaetus unicolor (Hentz, 1842) (Antrodiaetidae, Araneae): ultrastructural study of form and function with notes on reproductive biology of spiders. J Morphol 263:284–309

Michiels NK (1989) Morphology of male and female genitalia in *Sympetrum danae* (Sulzer), with special reference to the mechanism of sperm removal during copulation (Anisoptera: Libellulidae). Odonatologica 18:21–31

Montgomery TH (1903) Studies on the habits of spiders, particularly those of the mating period. Proc Acad Nat Sci Phila 55:59–151

Naud MJ, Hanlon RT, Hall KC, Shaw PW, Havenhand JN (2004) Behavioural and genetic assessment of reproductive success in a spawning aggregation of the Australian giant cuttlefish, *Sepia apama*. Anim Behav 67:1043–1050

O'Neill K (2001) Solitary wasps behavior and natural history. Comstock, Ithaca

Ono T, Siva-Jothy MT, Kato A (1989) Removal and subsequent ingestion of rivals' semen during copulation in a tree cricket. Physiol Entomol 14:195–202

Otronen M (1990) Mating behavior and sperm competition in the fly, *Dryomyza anilis*. Behav Ecol Socbiol 26:349–356

Otronen M (1997) Variation in sperm precedence during mating in male flies, *Dryomyza anilis*. Anim Behav 53:1233–1240

Otronen M, Siva-Jothy MT (1991) The effect of postcopulatory male behaviour on ejaculate distribution with in the female sperm storage organs of the fly, *Dryomyza anilis* (Diptera: Dryomyzidae). Behav Ecol Sociobiol 29:33–37

Parker GA (1970) Sperm competition and its evolutionary consequences in the insects. Biol Rev 45:525–567

Parker GA (1984) Sperm competition and the evolution of animal mating strategies. In: Smith RL (ed) Sperm competition and the evolution of animal mating systems. Academic Press, London, pp 1–59

Parker GA (2006) Sexual conflict over mating and fertilization: an overview. Philos Trans R Soc Lond B Biol Sci 361:235–259

Peretti AV, Aisenberg A (2011) Communication under sexual selection hypotheses: challenging prospects for future studies under extreme sexual conflict. Acta ethol. doi:10.1007/s10211-011-0099-4

Peretti AV, Córdoba-Aguilar A (2007) Sexual conflict over mating: on the value of fine-scaled behavioral observations for studies on sexual coercion. Ethol Ecol Evol 19:77–86

Peretti AV, Eberhard WG (2010) Cryptic female choice via sperm dumping favours male copulatory courtship in a spider. J Evol Biol 23:271–281

Peretti AV, Willemart RH (2007) Sexual coercion does not exclude luring behavior in the climbing camel-spider *Oltacola chacoensis* (Arachnida, Solifugae, Ammotrechidae). J Ethol 25:29–39

Peretti AV, Eberhard WG, Briceño D (2006) Copulatory dialogue: female spiders sing during copulation to influence male genitalic movements. Anim Behav 72:413–421

Pilastro A, Scaggiante M, Rasotto MB (2002) Individual adjustment of sperm expenditure accords with sperm competition theory. Proc Natl Acad Sci USA 99:9913–9915

Pizzari T, Birkhead TR (2000) Female feral fowl eject sperm of subdominant males. Nature 405:787–789

Platnick NI, Forster RR (1989) A revision of the South American and Australasian spiders of the family Anapidae (Araneae, Araneoidea). Bull Am Mus Nat Hist 190:1–139

Rehfeld VK, Sudhaus W (1985) Comparative studies of sexual behavior of two sibling species of Rhabditis (Nematoda). Zool Jb Syst 112:435–454

Rendall D, Owren MJ, Ryan MJ (2009) What do animal signals mean? Anim Behav 78:233–240

Rezác M (2008) Description of *Harpactea sadistica* n. sp. (Araneae: Dysderidae) a haplogyne spider with reduced female genitalia. Zootaxa 1698:65–68

Rezác M (2014) The spider *Harpactea sadistica*: co-evolution of traumatic insemination and complex female genital morphology in spiders. Proc R Soc B 276:2697–2701

Ridley M (1989) The incidence of sperm displacement in insects: four conjectures, one corroboration. Biol J Linn Soc 30:349–367

Ridsdill Smith TJ (1970) The behaviour of *Hemithynnus hyalinatus* (Hymenoptera: Tiphiidae), with notes on some other Thynnine. J Aust Entomol Soc 9:196–208

Rodriguez V (1995) Relation of flagellum length to reproductive success in male *Chelymorpha alternans* Boheman (Coleoptera: Chrysomelidae: Cassidinae). Coleopt Bull 49(3):201–205

Rodríguez RL (1998) Possible female choice during copulation in *Ozophora baranowskii* (Heteroptera: Lygaeidae): female behavior, multiple copulations, and sperm transfer. J Insect Behav 11:725–741

Rodríguez V, Windsor DM, Eberhard WG (2004) Tortoise beetle genitalia and demonstration of a sexually selected advantage for flagellum length in *Chelymorpha alternans* (Chrysomelidae, Cassidini, Stolaini). In: Jolivet P, Santiago-Blay JA, Schmitt M (eds) New developments in the biology of chrysomelidae. SPB Academic Publishing, The Hague, pp 739–748

Rodríguez-Márquez I, Peretti AV (2010) Intersexual cooperation during male clasping of external female genitalia in the spider *Physocyclus dugesi* (Araneae, Pholcidae). J Ethol 28:153–163

Rowe L (1992) Convenience polyandry in a water strider: foraging conflicts and female control of copulation frequency and guarding duration. Anim Behav 44:189–202

Schäfer MA, Uhl G (2002) Determinants of paternity success in the spider *Pholcus phalangioides* (Pholcidae: Araneae): the role of male and female mating behaviour. Behav Ecol Sociobiol 51:368–377

Schäfer MA, Misof B, Uhl G (2008) Effects of body size of both sexes and female mating history on male behaviour and paternity success in a spider. Anim Behav 76:75–86

Schneider JM, Lubin Y (1998) Intersexual conflict in spiders. Oikos 83:496–506

Schneider JM, Thomas ML, Elgar MA (2001) Ectomised conductors in the golden orb-web spider, *Nephila plumipes* (Araneoidea): a male adaptation to sexual conflict? Behav Ecol Sociobiol 49:410–415

Searcy WA, Nowicki S (2005) The evolution of animal communication: reliability and deception in signaling systems. Princeton University Press, Princeton

Seyfarth RM, Cheney DL, Bergman T, Fischer J, Zuberbühler K, Hammerschmidt K (2010) The central importance of information in studies of animal communication. Anim Behav 80:3–8

Shillington C, Verrell P (1997) Sexual strategies of a North American 'tarantula' (Araneae: Theraphosidae). Ethology 103:588–598

Simon E (1893) Histoire Naturelledes Araigne' es, vol.1, part 2. Roret, Paris, pp 257–488

Smith RL (1984) Sperm competition and the evolution of animal mating systems. Academic Press, New York

Snook RR, Hosken DJ (2004) Sperm death and dumping in *Drosophila*. Nature 428:939–941

Snow LSE, Andrade MCB (2005) Multiple sperm storage organs facilitate female control of paternity. Proc R Soc Lond B 272:1139–1144

Solensky MJ, Oberhauser KS (2009) Male monarch butterflies, *Danaus plexippus*, adjust ejaculates in response to intensity of sperm competition. Anim Behav 77:465–472

Stefania V, Garciaa K, Vecchiaa C, Silva LA, Guimarãesa B, Tizo-Pedrosoc E, Machado de EO, Brescovit AD, Del-Clarof K (2012) Mating behaviour, nympho-imaginal development and description of a new *Mesabolivar* species (Araneae: Pholcidae) from the Brazilian dry forest. J Nat Hist 1–13

Tallamy DW, Darlington MB, Pesk JP, Powell BE (2003) Copulatory courtship signals male genetic quality in cucumber beetles. Proc R Soc Lond Ser B 270:77–82

Thornhill R (1983) Cryptic female choice and its implications in the scorpionfly *Hylobittacus nigriceps*. Am Nat 122:765–788

Thornhill R, Alcock J (1983) The evolution of insect mating systems. Harvard University Press, Cambridge

Uetz GW, Stratton GE (1982) Acoustic communication and reproductive isolation in spiders. In: Witt PN, Rovner JS (eds) Spider communication: mechanisms and ecological significance. Princeton University Press, Princeton, pp 123–158

Uhl G (1993) Mating behaviour and female sperm storage in *Pholcus phalangioides* (Fuesslin) (Araneae). Mem Queens Mus 33:667–674

Uhl G (1994) Genital morphology and sperm storage in *Pholcus phalangioides* (Fuesslin, 1775) (Pholcidae; Araneae). Acta Zool Stockholm 75:1–12

Uhl G (2000) Two distinctly different sperm storage organs in female *Dysdera erythrina* (Araneae: Dysderidae). Arthropod Struct Dev 29:163–169

Uhl G (2002) Female genital morphology and sperm priority patterns in spiders (Araneae). In: Toft S, Scharff N (eds) European arachnology 2000. Aarhus University Press, Aarhus, pp 145–156

Uhl G, Schimtt M (1996) Stridulation in *Palpimanus gibbulus* Dufour (Araneae: Palpimanidae). Rev Suisse Zool (Suppl):649–660

Uhl G, Schmitt S, Schäfer M (2005) Fitness benefits of multiple mating versus female mate choice in the cellar spider (*Pholcus phalangioides*). Behav Ecol Sociobiol 59:69–76

Uhl G, Vollrath F (1998) Genital morphology of *Nephila edulis*: implications for sperm competition in spiders. Can J Zool 76:39–47

Uhl G, Huber BA, Rose W (1995) Male pedipalp morphology and copulatory mechanism in *Pholcus phalangioides* (Fuesslin, 1775) (Araneae, Pholcidae). Bull Br Arachnol Soc 10:1–9

van Helsdingen PJ (1969) A reclassification of the species of *Linyphia* Latreille based on the functioning of the genitalia (Araneida, Linyphiidae). Part I. *Linyphia* Latreille and *Neriene* Blackwell. Zool Verh 105:1–303

van Helsdingen PJ (1972) The function of genitalia as a useful taxonomic character. Arachn Congr Int V Brno 1971:123–128

Waage JK (1979) Dual function of the damselfly penis: sperm removal and transfer. Science 203:916–918

Wada T, Takegaki T, Mori T, Natsukari Y (2006) Reproductive behavior of the Japanese spineless cuttlefish *Sepiella japonica*. Venus 65:221–228

Wada T, Takegaki T, Mori T, Natsukari Y (2010) Sperm removal, ejaculation and their behavioral interaction in male cuttlefish in response to female mating history. Anim Behav 79:613–619

Walker WF (1980) Sperm utilization strategies in nonsocial insects. Am Nat 115:78–799

Watson PJ (1990) Female-enhanced male competition determines the first mate and principal sire in the spider *Linyphia litigiosa*. Behav Ecol Sociobiol 26:77–90

Watson PJ (1991) Multiple paternity as genetic bet-hedging in female sierra dome spiders, *Linyphia litigiosa* (Linyphiidae). Anim Behav 41:343–360

West HP, Toft S (1999) Last-male sperm priority and the mating system of the haplogyne spider *Tetragnatha extensa* (Araneae: Tetragnathidae). J Insect Behav 12:433–450

Wiehle H (1967) Metaeine semi entelegyne Gattung der Araneae. Senckenbergiana Biol 48:183–196

Zuk M, Simmons LW (1997) Reproductive strategies of the crickets (Orthoptera: Gryllidae). In: Choe JC, Crespi BJ (eds) The evolution of mating strategies in insects and arachnids. Cambridge University Press, Cambridge, pp 89–109

Chapter 6
Cryptic Female Choice and Nuptial Prey Gifts in a Spider Model

Luiz Ernesto Costa-Schmidt

Abstract Reproduction is composed by a series of strategies performed by each sex in order to gain control over the outcome of the entire process. Males' attempts to fertilize the females may deal with a hidden factor: the cryptic female choice, which is the females' ability to control fertilization based on males' sexual performance. A particular kind of reproductive strategy adopted by males of several invertebrate orders involves the transfer of substances or materials during the courtship and/or copulation, also known as nuptial gifts (NG). At least nine spider families have representatives that rely on the transfer of NG, all of them classified as oral gifts. Given its appeal in broad areas of behavioral ecology, NG received great attention in terms of their role in determining the outcome of sexual interactions, but little effort was applied toward the integrate analysis of the entire courtship/copulatory process. NGs are obviously just a stage of the entire mate choice process, and the empirical data available supports the idea of sexual conflict. However, I defend that only an integrated analysis will allow us to understand if NG does play some role as a source of information for cryptic female choice, or if it still remains as a way that males found to mitigate the sexual conflict issues.

L.E. Costa-Schmidt (✉)
Instituto de Diversidad y Ecología Animal/CONICET, Universidad Nacional de Córdoba, Córdoba, Argentina
e-mail: luizernesto@gmail.com

L.E. Costa-Schmidt
Programa de Pós-Graduação em Biologia, Universidade do Vale do Rio dos Sinos/UNISINOS, São Leopoldo, Brazil

A.V. Peretti and A. Aisenberg (eds.), *Cryptic Female Choice in Arthropods*,
DOI 10.1007/978-3-319-17894-3_6

6.1 Introduction

> Female morphology, physiology, and behavior determine the playing field on which males must compete, and many of the rules by which they must play. (Eberhard 1996, p. 70)

The second half of the XX century proved to be really exciting in terms of improvement of our understanding of the reproductive process. After the seminal works of Parker (1970) and Eberhard (1985, 1996), reproduction became to be interpreted beyond the limits of mating success, by including post-copulatory sexual selection processes. Since then, fertilization control became the main focus of most reproductive studies, especially for polyandrous species (Eberhard 1994; Schneider and Andrade 2011, p. 241).

There is an inherent difficulty in tracking the fate of seminal products transferred via ejaculate inside females' genital tracts, leading to a great amount of discussion about the mechanisms behind fertilization control. According to Eberhard (2000, p. 1049), "*(...) Questions about which sex controls processes that result in differences in paternity will have different answers, depending on the level of analysis*," and he is probably right. The cryptic female choice hypothesis (CFC) provides an elegant explanation for the differential fertilization pattern observed in polyandrous species in terms of the use of the available sperm pool (fertilization set sensu Parker et al. 1990). The main argument behind CFC hypothesis is that the fertilization process involves active participation by the females, even in the presence of males' fertilization adaptations to monopolize the amount of fertilized ova (male manipulation) (Vahed 2007). It is important to notice that both post-copulatory sexual selective forces (CFC and sperm competition) may occur together in a single organism, something turning the processes even more complex.

From the first appearance of the term CFC (Thornhill 1983), through a formal definition provided in the early 1990s (Eberhard 1996), until a modern understanding and interpretation of the process (Schneider and Andrade 2011, p. 242; Manier et al. 2013; Eberhard, Chap. 1), CFC poses as an elegant hypothesis, but one that is hard to demonstrate. The intrinsic difficulty in raising evidences supporting CFC is that several of its mechanisms take action inside the female's body and are therefore difficult to evaluate directly. The fact that males' adaptations may also be involved in the processes brings more complexity to the subject. A precise definition of how male and female factors interact with each other inside the female body may be not only difficult but may be futile, where CFC and male adaptations to sperm competition could be seen as two sides of the same coin (Vahed 2007).

When Randy Thornhill presented the term "cryptic female choice" in his 1983 paper, the limits of mate choice were extended beyond the copula itself. By doing that, CFC considers that females would perform mate choice along the entire reproductive process, comprising a multi-step process starting in pre-mating, continuing to mating, to finally reach a post-mating step for mate choice. Besides its theoretical simplicity, direct evidence supporting CFC is difficult to obtain even at the present time, when experimental manipulations may control several key factors, like sperm precedence patterns, for example. Probably, the greater difficulty is not in terms of demonstration, but the search for an overall explanation for the observed fertilization patterns.

It is quite surprising that several studies claiming to show evidences demonstrating CFC in action completely ignore the participation of pre-copulatory steps (see commentaries in Eberhard, this volume). Such absence of an integrative study considering all the decisions involved during the reproductive process may mask important females' strategies with important consequences in terms of differential reproductive allocation. In the words of Eberhard (1996, p. 81): "*Many processes in animal behavior (…) involve a sequence or hierarchy of decisions, and female reproductive behavior is no exception.*" This idea is reinforced later in the same book, where Eberhard (1996) not only affirms that "*(…) Precopulatory female choice probably also occurs along with cryptic choice in many species (…)*" (p. 202), but also call to attention that "*(…) cryptic female choice could be based entirely on courtship prior to intromission*, which could also serve to induce post-*intromission female processes*" (p. 204). This is by far the most neglected idea regarding the mechanisms involved along CFC, and only recently has the *in concert* action of distinct sexual selection pressures been considered (Sbilordo and Martin 2014). Most of the existing papers take into consideration only a subset of the entire range of stimulatory signals that reach the female, simplifying the processes of sexual communication that is well known to be driven by multimodal sensorial mechanisms (Partan and Marler 1999, 2005; Partan 2004). If the main improvement in terms of evolutionary explanation associated with the CFC hypothesis is the ability to consider subsequent selective steps inside the female reproductive tract, most of the assumed cases that corroborate CFC in wild populations do not consider the effects of copulatory steps. CFC is the final cut of a complex evolutionary play, where each scene contributes for the definition of further interaction outcomes.

The main goal of this chapter is to present the available evidence connecting nuptial prey gifts to any mechanistic source for CFC occurrence. I will use a broad definition of CFC as being non-random paternity biases resulting from a multitude of behavioral, physiological, and/or morphological mechanisms under female control (a similar definition can be found in Mainer et al. 2013). By saying that, I am considering mechanisms where females have a direct control and/or indirect participation as well. I will restrain my analysis to spider models, with the intention that some of my arguments could be applied to any biological model that relies upon nuptial gifts as a mating strategy.

6.2 What Is This Chapter About?

Eberhard (1996) described at least 20 possible mechanisms by which females may perform CFC. Of several mechanisms by which CFC may occur, some of them demand a complete knowledge on the biomechanical properties of how the individuals perform stimulatory movements that will determine the fertilization outcome (grabbing, tapping, rubbing, biting, etc.; Eberhard 1991), something way beyond the actual knowledge for spiders, even for better studied animal groups (Manier et al. 2013). So, my entire rationale will be restricted to the subset of mechanisms acting over female-mediated process influencing intromission, sperm transfer, and sperm storage, the major processes toward the fertilization set formation (Mainer et al. 2013). Most of my arguments will be certainly speculative given the absence of any source of evidence, either direct or indirect, for the proposed mechanism. I hope that the following comments could at least bring new discussion for a topic traditionally centered in a small portion of its possibilities.

6.3 Nuptial Gifts and CFC: Beyond the Males' World

Nuptial gifts are a widespread reproductive strategy (Vahed 1998, 2007; Lewis and South 2012) and may be defined as "*(...) materials (beyond the obligatory gametes) provided by a donor to a recipient during courtship or copulation in order to improve donor fitness*" (Lewis et al. 2014). Given this loose definition, some cases traditionally interpreted under different theoretical rationale may be included as a nuptial gift as well. For example, copulatory cannibalism was interpreted as a maladaptation resulted from a spillover effect of female aggressiveness (Anrqvist and Henriksson 1997; for a recent debate on the subject: Johnson 2013; Kralj-Fišer et al. 2013; Pruitt and Keiser 2013). However, a closer look to some biological models indicates that the males' fate does also represent an ultimate reproductive effort with positive fitness consequences for males (Buskirk et al. 1984; Schneider and Elgar 2001), where sexual cannibalism is sometimes better described as a male's self-donation as a nuptial feeding (see Andrade and MacLeod, Chap. 2). Such a perspective improved our understanding of this extreme strategy, by adding an extra and interacting element to act with the female aggression for a proper description of the entire processes.

As with any emerging field, the urge for the description of basic information regarding the natural history of the species comes along the search for peculiarities about their reproductive behavior. Therefore, much attention has being paid to nuptial gifts as an isolated behavioral trait, and little has been said about their integrative role along the entire reproductive processes. Even those works looking for some connection between nuptial gifts and CFC represent an initial approach to a complex subject (Albo et al. 2013; Albo and Peretti in press), bringing interesting indirect evidence into discussion.

The traditional approach to interpret nuptial gifts as sexually selected traits is that they are used as a source of information for mate choice (Austad and Thornhill 1986), especially when transferred before and/or during copulation. Nuptial gifts have also an intimate association with polyandry, reinforced by direct non-genetic benefits provided by them (Arnqvist and Nilsson 2000). In such a situation, CFC may emerge as a "side effect" of polyandry: multiple mated females improve their energetic budget by acquiring resources from nuptial gifts (Arnqvist and Nilsson 2000; Boggs 1990), resulting in an opportunity to choose among those males that compose the fertilization set. Beyond the direct benefits involved during nuptial gift assimilation, CFC may take its course by the set of indirect genetic benefits provided through the nuptial gift. In this situation, it will be expected that females would rely on another set of information to perform CFC than those provided from the quality of the resources that composes the nuptial gift, since nuptial gift quality signals may be deceptive (LeBas and Hockham 2005; Albo et al. 2011; Ghislandi et al. 2014). Empirical evidences indicate that the resources transferred via nuptial gift by the males are correlated with males' condition or to the compatibility of the mating pair (Albo et al. 2012). Therefore, indirect benefits may be an important aspect guiding the mechanisms of CFC in several gift-giving species (Warwick et al. 2009; Albo et al. 2012). For males, deception may be an evolutionary path worth following, especially if such a strategy allows them to get further into the mate choice process where other sources of stimulation will take place.

It is "easy" to hypothesize about the adaptive role of nuptial gifts in terms of traditional mate choice, but few works explicitly suggest their role during CFC (Thornhill 1983; Albo et al. 2013; Albo and Peretti in press). As mentioned before, it is complex to build a general framework for such a diverse strategy, and apparently, there is no single explanation that could connect nuptial gifts to CFC in a general sense. Even though the available evidence suggests that each case must be analyzed individually, direct and indirect benefits must be balanced against the involved costs, favoring that nuptial gift may have a primary and/or a secondary association with several mechanisms that promote CFC (see below). For example, theoretical evidence supports the emergence of CFC when direct benefits are involved along the reproductive processes of a hypothetical species (Knowlton and Greenwell 1984). These authors investigated the combination of direct benefits and male manipulation in a game theory model, where late manipulative behavior in copulation (or quick re-mate by the males) followed by little investment into the female and into the offspring resulted in a female control of the fertilization processes through premature termination of copulation (Eberhard 1996, p. 44).

6.4 A Sexual Strategy Shaped in Concert

In terms of the evolution and maintenance of nuptial gift as a sexually selected strategy, both male and female perspectives most probably have been acted *in concert*. Several interpretations of the observed patterns available in the literature present an overemphasis toward the males' perspective, which could easily be changed for alternative explanations rooted in the females' perspective. For example, that "males manipulate female position while copulating" may be a consequence of females' help once stimulated to assume such position; that "males have their overall fitness increased by offering a nuptial gift" may be a consequence of females' CFC mechanisms by biasing the fertilization of her eggs favoring an specific male; and so on. I am not diminishing here the importance of the males' perspective to understand a representative fraction of the life history of a species, such as the way that nuptial gifts are shaped in order to better explore females' sensory system, for example (Warwick et al. 2009; Brum et al. 2012). The point is that much emphasis has been applied only toward the male's perspective, viewing the females as just passive players in this complex and amazing interaction, a criticism raised by Eberhard (1996) and several others.

It is also evident that the intrinsic sexual conflict between males and females acts as a sexual selection force, which defines much of the evolutionary scenario on how CFC mechanisms rely upon nuptial gifts. Moreover, it is also expected that from time to time there is an inversion in the control of the process between males and females, or even depending on the level of analysis (Eberhard 1996, 2000). In most cases, nuptial gifts represent a variable investment from males with a potential manipulative behavior (Warwick et al. 2009; Brum et al. 2012). This hypothesis of manipulation considers that nuptial gifts constitute a male sexual strategy that exploits how females perform their choice, which may lead the females to a sub-optimal fitness condition. This scenario is an open road for the emergence of a chase-away evolutionary process of mate choice (Holland and Rice 1998), where females would then become more resistant to males' manipulation along the generations, forcing males to improve their manipulation capabilities in exploring females' sensory systems, which eventually results in an evolutionary arms race. There is a straightforward association between males' manipulation and indirect benefits, since sons of a manipulated female will inherit such ability to manipulate females.

We cannot forget that some classes of nuptial gifts do provide direct benefits that would be a significant fraction of the female's energetic budget available for reproduction (Gwynne 2008). In such scenario, even under the presence of a manipulative trait, female reproductive effort will not necessarily reach a sub-optimal level (Arnqvist and Rowe 2005, p. 102).

6.5 Nuptial Gifts and CFC in Spiders

6.5.1 Conditions for CFC Occurrence via Nuptial Gifts Within Spiders

Spiders are great study models to investigate the effects of CFC, not only because most species are polyandrous (Schneider and Andrade 2011, p. 233), but also because the resulting paternity bias emerges as a combination of male and female behavioral mechanisms (e.g., courtship). These are also associated with morphological variation within the females' sperm storage organs (Uhl 2002; Eberhard 2007; Schneider and Andrade 2011, p. 215). Classical generic predictions about the influence of the female genital tract morphology on how the fertilization set is used proved to be inadequate (Austad 1984; Uhl 2002; Eberhard 2007; Elias et al. 2011; Herberstein et al. 2011). This lack of generalization leads to an emerging pattern where each case must be taken separately, reinforcing the underlying complexity of sperm use bias within the female genital tract. To make clear predictions about the patterns of fertilization under the CFC rationale, it is necessary to understand how the internal structure of the female genital tract is shaped and how male's genitals interact with it. In the absence of such information, any prediction may be classified as speculative, a situation that still holds true for most spider species used as biological models in reproductive studies.

Nuptial gifts may vary the amount of information provided for females. We can take the premise that the more modified or complex the nuptial gift, the more information it contains. Let us take as an example a special class of nuptial gifts within spiders, the *prey gifts* (see *Prey gifts* section for a definition): It may consist of nothing but the prey, or may be also covered by a layer of silk that may be used as a source of information between males and females, or at least as a substrate for information exchange (Costa-Schmidt et al. 2008; Bilde et al. 2007; Brum et al. 2012). Certainly, the chance that this character will be used as a source of information for CFC is higher the more complex it is. Moreover, to have access to such information content, the individuals must be capable of perceiving that information. This leads to an important aspect that may promote nuptial gift to become traits used for CFC: the more sensory channels involved, the more variance among the available nuptial gifts in the population will be perceived, and such variance will be the raw data for selection to operate.

Following the same information criteria linking nuptial gifts to CFC, the timing of the transfer is of special interest. The moment of the transfer will be directly correlated to the amount of time that females would have to extract information from the nuptial gift. First, if such transfer occurs before insemination, females may be able to reject a male in terms of the quality of the nuptial gift. In this situation, pre-copulatory female choice is the main selective process associated to the nuptial gift, and a possible consequence is that the rejected male may improve the quality of the nuptial gift and then present himself for another round of evaluation. Especially during the pre-insemination and insemination stages, females may walk away with the transferred nuptial gift, and this may also have crucial consequences for males if they cannot rescue at least some of the energetic investment provided within the nuptial gift. Second, if the nuptial gift is transferred during insemination, females may be able to abort copulation prematurely based on the quality of the nuptial gift, resulting in drastic consequences for male's fitness both in terms of CFC and sperm competition. Spider males transfer their sperm by inserting their first appendage pairs, the pedipalps, inside the female genital tract. Long-lasting pedipalp insertions may be crucial for males to perform deep intromissions that would result in better placement of sperm inside female's genital tract, thus allow for a greater amount of sperm transfer as well. Finally, if the nuptial gift is transferred after insemination, evidences of the occurrence of CFC become harder to obtain, since the mechanisms involved are less conspicuous than those form other insemination stages. As stated by Manier et al. (2013, p. 552): "*(…) the closer we get to fertilization, the less we understand about the mechanisms of cryptic female choice*."

Along all those stages when nuptial gift transfer may occur, there is a potential that male manipulation strategies take part of the process, which may influence post-copulatory females' decisions. Of course, the descriptions provided here do not consider the entire range of influence of nuptial gifts over CFC along the reproductive processes, but provide us an idea of the whole evolutionary picture (not to say "evolutionary potential"). Male copulatory courtship may be recognized by the occurrence of repetitive behaviors with no association to any other non-reproductive function during copulation (Eberhard 1996, 2007), and the participation of nuptial gifts during courtship stages seems to be a rule among the known spider models.

Finally, it is important to emphasize that the nuptial gift participation in CFC is a processes that depends on the degree of mating rates of both sexes. Beyond the obvious obligatory occurrence of polyandry, variation on the degree of male mating rates will define the evolutionary scenario for the emergence of males' strategies. This inevitably leads to the emergence of sexual conflict, and so the emergence of deceptive or manipulative strategies is intuitively expected (Albo et al. 2011, 2014b).

6.5.2 Classes of Nuptial Gifts in Spiders

The study of nuptial gifts among spiders became more active after the description of several models' species where nuptial gifts have an important role in their reproductive behavior. There is a single classification for spiders' nuptial gifts proposed by Albo et al. (2014a) (see also: Vahed 2007; Lewis and South 2012), which considered only the occurrence of *oral gifts* as recognized forms of nuptial gift in spiders. However, the emphasis over the oral gift pattern within spiders may be the consequence of interpretative bias by not considering other male strategies that could fit the nuptial gift definition. It would be interesting to seek along the specialized literature for extra classes of nuptial gifts beyond those considered in Albo et al. (2014a), such as *seminal gifts* and *structural gifts* (Aisenberg et al. 2007 for structural gifts). Such inclusions would be justified if previous works could provide evidences connecting the reproductive behavior of the involved species with the transfer of a nuptial gift, which obviously would not be explicitly defined as nuptial gifts in the original papers.

Most described cases of nuptial gifts in spiders entail the exchange of *oral gifts*, both endogenous and exogenous (Albo et al. 2014a), that can be divided in three classes: *prey gifts*, *body gifts*, and *glandular gifts*. Spiders are known as being voracious predators, and the available evidence supports the hypothesis that oral gifts have emerged from males' ability to exploit the foraging motivation of females (prey gifts: Bilde et al. 2007; Brum et al. 2012; body gifts: Andrade 1996; Miller 2007; Wilder et al. 2009; glandular gifts: Michalik and Uhl 2011; Kunz et al. 2013). Such exploitation allows males to not only securely approach an adult female (Schneider and Andrade 2011, p. 216), but also to induce her to assume an adequate copulation posture, which may have substantial influence over CFC.

From all aspects associated with oral gifts in spiders that may have a direct influence over CFC, insemination duration is one that shows the most impressive pattern. The quality of the oral gift appears to be the main source of information for females to proceed with CFC. For all classes of oral gifts (see above), longer insemination provides at least three extra advantages for the males in terms of: (1) sperm competition, by the transference of a large amount of sperm; (2) by allowing males to reach with their genitals deeper regions of females' genital tract; and (3) increasing the amount of copulatory courtship that may stimulate the female to place the received sperm in a better position for fertilization.

We cannot exclude the possible role of oral gift assimilation in influencing females' physiological and biochemical conditions, especially for those cases involving endogenous oral gifts. As stated by Eberhard (1996, p. 305), "*(…) the nutritional value of male products may be relatively insignificant, and their action stem from hormone-like triggering effects on the female.*" Therefore, the composition of endogenous oral gifts probably has evolved through a coevolutionary process rooted in the sexual conflict for the control of fertilization. Unfortunately, nothing is known about the internal fate of oral gifts within spiders, and the investigation of the effects of their assimilation in CFC is very promising.

6.5.3 Spiders' Prey Gifts: Structure and Multimodal Communication Issues

Prey gifts are the best-studied case of nuptial gift among spiders. They are defined as prey captured by the male that is offered to the female during the initial stages of courtship (but see: Cobbold and Su 2010; Uetz et al. 2010). Besides being mostly composed of an exogenous material, the prey handling by the male involves the inclusion of endogenous substances through a silk layer that most of the times surrounds the entire prey, which have strong influence over several stages of the reproductive process. Therefore, the prey gift constitutes a sexually selected signal, composed by two components (external silk layer and the inner content) with different optimized multimodal responses depending on the mating stage.

The main representative species of this nuptial gift class is by far the Palearctic *Pisaura mirabilis* (Clerck 1757) (Pisauridae), whose reproductive behavior was first described by Van Hasselt in 1884 (Austad and Thornhill 1986). For a long time, prey gift behavior was considered an exclusive strategy for *Pi. mirabilis*, but a closer look toward the Pisauridae family resulted in a list of species incorporating this same behavioral strategy into their courtship repertoire (Nitzsche 1988; Itakura 1993, 1998; Bastos-Pereira 2009; Nitzsche 2011). Since the description of some new records for the Trechaleidae family (Costa-Schmidt et al. 2008; Silva and Lise 2009), prey gifts turned into a common reproductive strategy within these two families from the Lycosoidea clade. Within the Trechaleidae family, *Paratrechalea ornata*

(Mello-Leitão 1943) (Carico 2005) is being the subject of intensive evolutionary studies regarding the adaptive interpretation of prey gift as a sexually selected trait.

It is inevitable to make a direct comparison between *Pi. mirabilis* and *Pa. ornata*. The similarities between their reproductive behaviors suggest a strong phylogenetic influence in their evolutionary path (Costa-Schmidt et al. 2008; Albo et al. 2014a), although such interpretation lacks a formal analytical demonstration. The standard mating behavior for these two species is briefly described in the diagram provided in (Table 6.1).

The evolution of the prey gift structure is clearly the result of several selective pressures acting over each component separately, even though is the in concert action of those components that defines the adaptive maintenance of prey gifts as a reproductive trait in a population. Separately, each component has its own (and important) role, which varies along the different stages of the reproductive process (pre-copulatory, copulatory, and post-copulatory), where their relative adaptive outcome may be redundant or non-redundant to each other (Table 6.1; Partan and Marler 1999, 2005). However, maybe the application of a multimodal interpretation of prey gifts may bring us an extra source of evidences that may explain some subtleties about prey gifts as a sexually selected trait and its intrinsic variation.

The actual interpretation of the role of the external silk layer and the inner content is summarized in Table 6.1, which was filled following the terminology proposed by Partan and Marler (1999). During the pre-copulatory stage, both prey gift components have being associated to attract females (*Pi. mirabilis*: Bilde et al. 2007; Prokop and Maxwell 2009; Tuni and Bilde 2010; Albo et al. 2011, 2014b; *Pa.* ornata: Albo and Costa 2010; Brum et al. 2012; Trillo et al. 2014), a redundant role leading toward a multimodal response of increasing their effects in the same direction (enhancement). However, since the inner content of prey gifts may represent a source of deceptive signal, the information contained within the external silk layer may sometimes overcome the associated information within the inner content (Albo et al. 2011), a situation that defines a dominance condition of

Table 6.1 Multimodal signal interpretation for prey gifts, following the classification provided in Partan and Marler (1999). The description provided for separate components is supported by evidences, which are quoted along the text. Interpretations followed by a question mark refer to possible mechanisms lacking any source of evidence for the studied models so far. Enhancement: increase in the intensity of a signal when occurring with a redundant signal; Dominance: when the effect of one component overshadows the effect of the other; Modulation: when the effect of one component changes the intensity of the effect of the other component; and Independence: when the effects of each component do not interact

Mating stage	Separate components		Multimodal response
	External silk layer	Inner content	
Pre-copulatory	Female attraction	Female attraction	Enhancement[a] Dominance[b]
Copulatory	Copula control	Copula duration	Modulation
Post-copulatory	Refractoriness (?) Hormone-like trigger (?)	Nutritional value	Dominance (?) Independence (?)

[a]Honest prey gifts
[b]Deceptive prey gifts

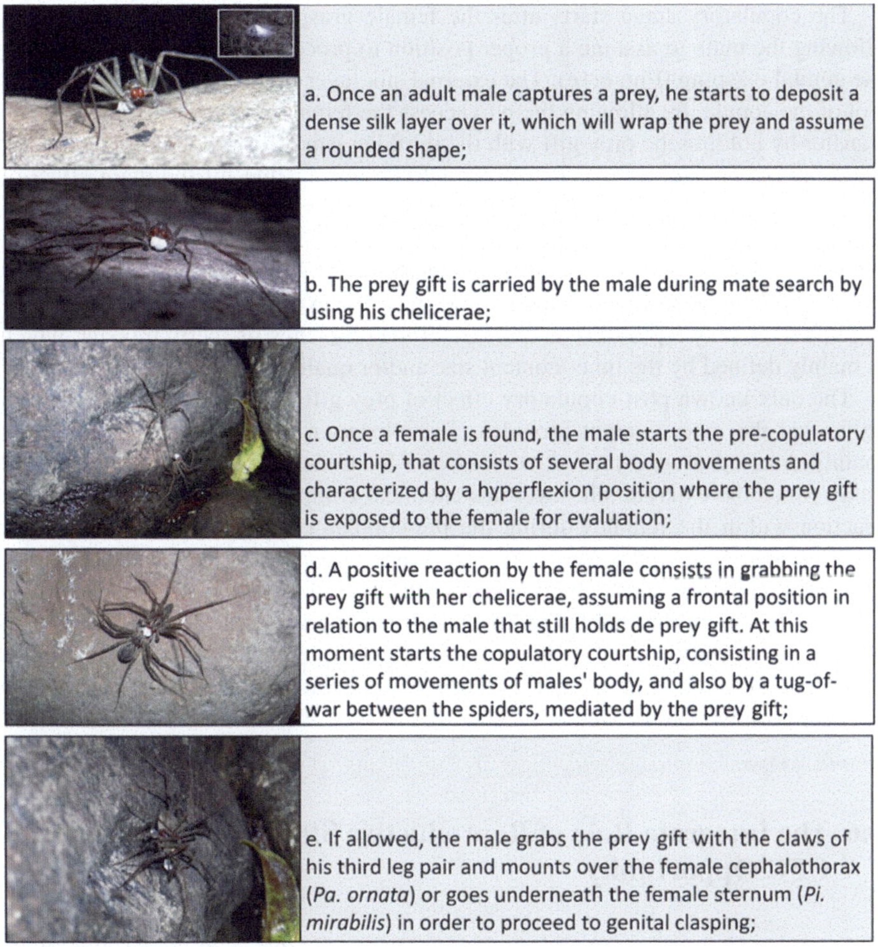

Fig. 6.1 Description of the main steps that compose the sexual behavior of the prey gift-giving spiders *Paratrechalea ornata* and *Pisaura mirabilis*. The former species was taken as reference for the pictures, and except when mentioned, the behavioral pattern is similar for the other species. Palpal insertion occurs alternately, and between two insertions, the male returns to the frontal position, grabs the prey gift with his chelicerae, stops holding the prey gift with the third pair of legs, and resumes the same copulatory courtship position described in "**d**". The sequence "**d–e**" is repeated until the end of the copulation that almost always is determined by the female

the external silk layer over the inner content. It is important to notice that most of the initial communication during the pre-copulatory stage occurs without the contact of the partners. In such situation, the external silk layer is the main source of information, both chemically (Brum et al. 2012) and visually (Trillo et al. 2014). Only after the female acceptance of the prey gift (prey gift grasping: Fig. 6.1d), the inner content can be accessed and properly evaluated.

The copulatory stage starts after the female grasps the prey gift (Fig. 6.1d), allowing the male to assume a proper position to proceed to sperm transfer during the genital clasping (Fig. 6.1e). The external silk layer offers to males a better control of the copula, by allowing them to access the female reaction during the sperm transfer by holding the prey gift with the third leg pair (Costa-Schmidt et al. 2008; Nitzsche 2011). The inner content seems to be responsible for the main effect of prey gifts is in terms of copula duration, which seems to be associated to both the quality and the size of the inner content (Bruun et al. 2004; Albo and Costa 2010; Albo et al. 2011; Klein et al. 2014). The multimodal interpretation for those components' effects would be of modulation, since the extra control of the males during this stage may represents an increase of the efficiency of sperm transfer, which is mainly defined by the inner content size and/or quality.

The only known post-copulatory effect of prey gifts is related to the nutritional value that the inner content provides. The allocation of such nutritional input is mainly directed toward reproduction (Costa-Schmidt and Machado, unpublished data). Since the external silk layer has chemical components that activate specific reaction within the females during the pre-copulatory stage (Brum et al. 2012), it is hypothesized if the prey gift chemical components could also act during the post-copulatory stage. Such effect could be expressed in the induction of a refractoriness period to the females right after eating the prey gift (but see Prokop and Maxwell 2009), or even in terms of triggering specific biochemical routes based in a hormone-like trigger. Besides its speculative nature, such hypotheses are very promising to be studied.

6.6 The Integrate Role of Reproductive Stages Defining CFC Opportunity

As mentioned before, several authors considered that CFC decisions were influenced exclusively by the action of copulatory and post-copulatory courtships. An alternative way to look for the CFC process is to consider that it is the result of the sum of information gathered along all phases of the sexual behavior of a species (Sbilordo and Martin 2014). This will also include the mate choice stage, since to be cryptically selected, the male should have been explicitly selected. This is particularly true for prey gift giving models, where coercion does not seem to take part of the process and where a prey gift does represent a nutritional transfer (Austad and Thornhill 1986, but see deceptive offerings below). Evidence indicates a huge participation of prey gifts during all courtship stages involved (Table 6.1).

The variety of possible direct interactions between prey gifts and CFC (from now on referred to as primary effects) is reinforced during the pre-copulatory and copulatory stages, when related sexual selection processes have an indirect participation with prey gifts (secondary effects).

6.6.1 Prey Gift and Mate Choice

Prey gifts were traditionally interpreted as source of information for mate choice (Austad and Thornhill 1986), especially when transferred before or during copulation. This hypothesis states that female mate choice is partially defined by the intrinsic variation in the quality of the prey gifts offered by the available males in a population. Female mate choice defines the assemblage of the sperm pool on which CFC will be performed, and for this reason, if prey gift quality confers some advantage in terms of mating success, it is essential to understand its role at this important stage.

Another important selective pressure associated with prey gifts during female mate choice is the occurrence of prey gift theft by females (Prokop and Maxwell 2009). Prey gift stealing in the female perspective is a win-or-win situation: If females are successful in stealing a prey gift, they obtain a free meal; if they are not successful in stealing, they may have found a skillful male in avoiding stealing. Such male will then proceed to further steps of the sexual interaction. There is much information exchange during prey gift grasping, and the strength and/or resistance of a male could easily be evaluated during such interaction and be a source of selection for indirect benefits.

Prey gifts are not the only sexually selected trait involved during mate choice, especially within spiders, where male pre-copulatory courtship usually involves multiple sensory channels in a concomitant way (Hebets and Papaj 2005; Wilgers and Hebets 2011; Hebets 2011). So, the list of factors influencing mate choice in prey gift giving spiders would include not only the presence and the quality of a prey gift, but also additional courtship traits/units performed by the males (e.g., Costa-Schmidt et al. 2008). There is a widespread correlation between mating success and male condition, which will reflect in the intensity of courtship and also in the degree of investment in prey gift quality (Lang 1996; Albo et al. 2014b). As an emerging pattern from the available evidences, it is possible to set a hierarchy of the factors that influence mate choice in prey gift giving spiders, starting with the presence of a prey gift, followed by the male condition (independent of the prey gift), and ending with the quality of the prey gift.

The first step in this hierarchical set list is the presence of a prey gift, whatever its quality. Several studies at least mention the positive causal effect of the presence of a prey gift for mating success in *Pi. mirabilis* (Stålhandske 2001; Prokop and Maxwell 2009; Albo et al. 2011; Nitzsche 2011) and *Pa. ornata* (Costa-Schmidt et al. 2008; Albo and Costa 2010), and the overall conclusion was that prey gifts are a facultative trait, but with several and strong benefits for males in terms of frequency and latency until acceptance by the females. This pattern of prey gifts being determinant traits for mating success goes along with the hypothesis that prey gifts exploit the female foraging motivation through chemical signals found within the external silk layer (*Pi. mirabilis*: Lang 1996; Bilde et al. 2007; *Pa. ornata*: Brum et al. 2012). As mentioned before, the external silk layer functions as an advertisement signal and seems to be evolved to include some sort of attractive chemical signal that permits the males to proceed further to the next step of the mate choice process (Brum et al. 2012).

Another role of prey gifts in this initial phase is to place both sexes in an adequate pre-mating position, allowing an exchange of multimodal information mediated by the prey gift. This moment is crucial for CFC, because communication via the prey gift helps males and females to synchronize their movements, which eventually will influence the adjustments in body position during mating itself. In the case of spiders' prey gifts, such pre-mating position is represented by the frontal position adopted by males and females which allows a close contact between the sexes, where another set of courtship movements will take place. At this moment, females are able to access the inner content of the prey gift and so decide to continue or to abort the entire mating process. However, indirect evidences derived from an experimental study with *Pi. mirabilis* suggest that at this stage, male condition overcomes any information obtained via prey gifts (Albo et al. 2012), i.e., other channels that comprise the male courtship are more informative for females to take any decision. Given the once believed rarity of prey gifts in spiders, much attention was paid for prey gift roles in the courtship, and consequently, little attention was paid to the extra amount of courtship traits involved.

Finally, once females have access to the inner content of the prey gift, they are able to make any judgment based on its quality. If we are able to infer quality to a trait, it is expected that there exists some sort of variation on it, and consequently, it is expected that sexual selection operates over it. In the case of spiders' prey gifts, variation may be found in their inner content and/or in the external silk layer that is used to wrap the gift. Inner content quality depends not only on the nutritional value of the prey used, but also on the amount of substances that males have consumed from it before offering it to females. The inner content is hardly accessed by the females during early stages of the pre-copulatory courtship, since they need to grasp the prey gift in order to make any inference on its contents (Albo et al. 2011). On the other hand, the external silk layer serves as an advertisement signal, allowing males to hide the inner content of the prey gift, and also helping them to adjust their bodies for a proper pre-mating position.

Deceptive prey gifts are now seen as a possible evolutionary path adopted by the males to overcome several energetic costs involved in prey gift investment (Albo et al. 2011, 2014b; Ghislandi et al. 2014; but see Prokop and Maxwell 2012). Such widespread occurrence leads to an emergent conclusion that maybe the inner content again is not the main information source of the prey gift during mate choice, but the external silk layer may provide other source of benefits to promote male mating success. Besides the existence of a strong correlation between male condition and the investment in the quality of the prey gift, the variation found in the prey gift silk layer quality is not used as a source of information during the initial steps of the mating processes (Albo et al. 2012).

It is curious that most of the explanations for nuptial gift adaptive maintenance relies in classical mate choice hypothesis (Albo et al. 2011, 2012, 2013, 2014b; Trillo et al. 2014; Albo and Peretti in press), such as runaway process or even Zahavian processes, even though this is a system based on sensory exploitation and, therefore, more prone to evolution by sexually antagonistic coevolution.

6.6.2 Copulatory Courtship

Male copulatory courtship may be recognized by the occurrence of repetitive behaviors with no association to any other non-reproductive function during copulation (Eberhard 1996, 2007). In the case of prey gifts, they have at least two sets of primary and secondary effects over CFC during this mating stage. The first set is described by the primary role of prey gifts in maintaining the mating position (Table 6.1). This allows the male to reach appropriately the female genital tract, where two secondary processes will proceed: genital stimulation and sperm transfer inside the female genital tract. The second set of prey gift effects over CFC is related to female prey gift evaluation while holding it with her chelicerae. Differently from the previous phase, prey gift evaluation now is performed over its inner content, where both the prey gift nutritional value and its size (interpreted as the amount of energy available) will exert an indirect influence over copula duration (Bruun et al. 2004), which is directly associated to the length of the insemination process. Besides occurring simultaneously, these sets of effects probably interact with each other.

Mating position is directly influenced by the presence and the shape of the prey gift (Andersen et al. 2008; Costa-Schmidt et al. 2008). Several lines of indirect evidence support the hypothesis that prey gifts may serve as a channel that transmits information between the mating pair (Costa-Schmidt et al. 2008; Brum et al. 2012). Males do have mating success without prey gift offering, but they pay several costs in terms of courtship investment and efficiency along subsequent steps of the entire process. The quality of the external silk layer also influences male mating performance, where poorly wrapped gifts result in less control over the copulatory movements involved (Andersen et al. 2008).

Linked to the mating position, genital stimulation comes as an underestimated ubiquitous action in terms of sexual selection pressures within prey gift giving spiders, with a great potential for CFC. Males' unsuccessful intromission attempts (genital flubs) during the palp insertion are among those copulatory movements that may play an important role to prepare the female's genital tract for insemination (Huber 2005). At least for the Trechaleidae family, genital flubs are usually interpreted as a misfit during male's genital clasping into the female's epigynum. Males' flubs during insertion attempts are frequent for *Pa. ornata* (Albo and Peretti in press; Luiz Ernesto Costa-Schmidt, personal observation), and after consecutive series of flubs, the male returns to the frontal position, where some sort of information exchange between the sexes seems to be facilitated through the prey gift. Moreover, the timing of the point at which the male stops trying to insert the flubbing palp may be a male reaction to subtle movements performed by the female that indicates her intention to interrupt copula, i.e., the prey gift mediates an important source of information favoring males' ability to avoid a precocious end to copulation by the female. This would be a situation where the prey gift may transport a cue to the male about the female intention to interrupt copulation, leading him to assume a previous step of the courtship process in order to proceed to other copulation attempts.

Copula position also reflects the males' ability to properly insert their palps into the female's genital tract, where insertions may vary in their efficiency in placing the sperm in a strategic position inside the female genital tract. Little is known of the mechanics of genital clasping for both prey gift species analyzed here, and a very promising venue for further investigations is the comparison of the genital mechanical fitting between males and females when prey gifts are present or absent. Even if prey gifts do not influence genital coupling, it may influence other copulatory courtship traits that depend on this structure for information exchange.

Assuming that a male has been selected through the initial steps of mate choice, then he reaches to one of the most important steps of the whole process: insemination. As mentioned before, the general copulation pattern of *Pi. mirabilis* and *Pa. ornata* consists in several palpal insertions that occur alternately. Each round of insemination leads to an unknown amount of sperm transfer, and we assume that several insertions also leave the male in a better condition in terms of sperm competition (Albo et al. 2013; Albo and Peretti in press). Given the alternate insemination pattern in relation to the bilateral symmetry of the female genitalia, a male that copulates twice will fill both sides of the female genital tract with his sperm. This seems to be under a strong sexual selected pressure, since males frequently follow this alternate pattern of insemination even when having complications in clasping their palp into the female epigynum in one of the sides (personal observation). From the female's perspective, such alternate insemination represents an important mechanism of CFC, by controlling the assembling of the fertilization set in her spermatheca. Females may be able to control the access of a male to her genital tract by simply not cooperating with a proper body positioning for genital clasping, and this may occur every time that a male resumes the frontal position and courts the female for another insemination round. This scenario gives a huge control for the female over the assembling of the fertilization set, where she may allow a first male to access both sides of her genital tract and restrict one of the sides for a second male that is not so attractive, or the other way round, by allowing a second, attractive male to perform extra inseminations in order to dilute and/or displace the sperm of the first male. There are numerous possibilities by which CFC may take part of this process, especially considering the alternate insemination order, multiple inseminations per copulation event, and male control of sperm transfer rate, which is something completely unknown for the species analyzed here (Fig. 6.2).

Mate choice and insemination occur in an intermittent way in both *Pi. mirabilis* and *Pa.* ornata (Table 6.1). After female acceptance, both sexes remain in the frontal position, from where the male will proceed to another sequence of movements in order to inseminate the female. After each event of insemination, that is composed by male and female genital clamping with further male ejaculation, the male returns to the frontal position for another round of mate choice. At this moment, the female has already been inseminated by the male, so the next round of mate choice will have another meaning for the interacting couple. From the male's perspective, the first insemination round is the most important, since in that way, he will make part of the fertilization set. A second round could mean that the male

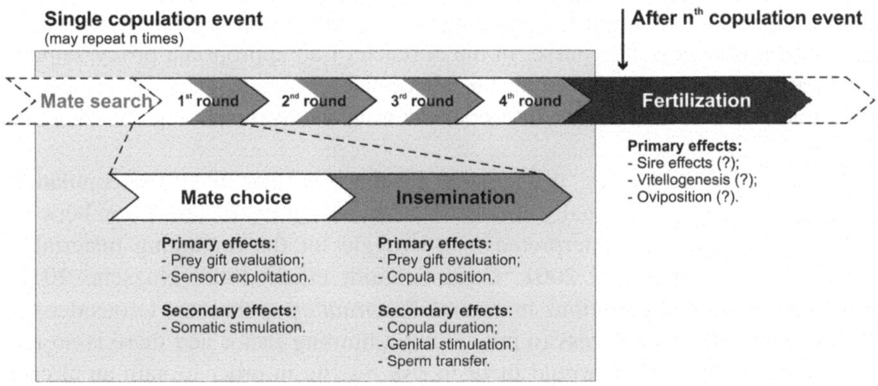

Fig. 6.2 Main reproductive steps of a prey gift giving species reproduction event, with a list of primary and secondary effects of the prey gift over cryptic female choice (CFC). Arrow color indicates different sets of CFC mechanisms. During a single copulation event, males could pass through several rounds of mate choice and insemination (up to four rounds in *Paratrechalea ornata*). Fertilization is supposedly under the female control and will occur after nth copulation event, which may be with the same male or with different males. White arrow: pre-copulatory; light gray: copulatory; dark gray: post-copulatory

will increase his competitiveness for fertilization since he will have his sperm inserted in both sides of female's genital tract, so the effort to succeed in this round could be similar as to the first insemination. From then on, any extra insemination may be interpreted as a further male investment in sperm competition, in terms of dilution of the assembled fertilization set. Such investment is probably variable, depending mostly on the female reproductive status.

It is also important to take into account that the observed fertilization patterns may hide other complex mechanisms of differential sexual investment (Herberstein et al. 2011), especially if the inference of CFC is based on the indirect evidence of the amount of sperm transferred during insemination. A strong assumption of most studies involving prey gift giving spiders is that it is assumed that there is a constant rate of sperm transfer while there is genital clasping (Albo et al. 2013; Albo and Peretti in press). By assuming this, we are also assuming that the longer the insertion, the more sperm is transferred. However, it is already known that males are able to control not only the quality of their sperm, but also they are capable of performing subtle changes in the way sperm are transferred. As stated by Schneider and Andrade (2011, p. 249): "*(…) the duration of copulation may not equal sperm transfer, which may not equal sperm storage, which may not equal sperm used for fertilisation (…) it is problematic to consider individual processes such as copulation duration or number of sperm transferred in isolation because cryptic processes that occur between sperm transfer and fertilisation may affect relative paternity (…).*"

The thing is that it is hard to buy the assumption of a constant sperm transfer rate for a species where copulatory courtship is so complex, possibly triggering internal reactions inside the female genital tract, and where cryptic male choice[1] may also be in action. Of course, in the absence of an appropriate proxy, copulation time holds as the best option, but alternative explanations considering such inferential bias must be taken into consideration while discussing the observed patterns.

In both prey gift species, the females are the ones that finish the copulation. The end of copulation is usually characterized by an agonistic interaction between the mating pair, usually interpreted as a struggle for the remaining material of the prey gift (Stålhandske 2001; Costa-Schmidt et al. 2008; Nitzsche 2011). Based on personal observations in several *Pa. ornata* populations, I consider this a weak argument, since access to prey is not a limiting factor and there is no reason to believe that a male would incur to risk his life in order to gain an already consumed prey gift. An extra source of evidence supports my conclusion: Sexual cannibalism favoring females does occur after such agonistic interactions in small frequencies (Austad and Thornhill 1986; Drengsgaard and Toft 1999; Stålhandske 2001; Nitzsche 2011; Costa-Schmidt and Machado 2012), representing a selective pressure over males for not fighting for a prey gift in this final stage of the reproductive process. But why do fights happen? An alternative interpretation would be that those struggles at the end of the copulatory process are maintained as another female choice opportunity that would favor those males that are capable to escape from a female attack. It is important to notice that a refused and attacked male may resume all the courtship process toward the same female (personal observation).

The rate of consumption of the prey gift during insemination process may be a selective force in direction to optimize prey gift size and sperm transfer. Even though expected, there is little evidence supporting such hypothesis. The single study that presents this information was conducted by Prokop and Maxwell (2009), where they recorded that in a laboratory experimental set, 60 % of the trials (38/63) involving *Pi. mirabilis* had the prey gifts completely consumed before the termination of insemination. No such data are available for *Pa. ornata*.

6.6.3 Post-copulatory CFC

It is during the post-copulatory processes that most of the hidden CFC mechanisms occur. Our knowledge about the processes inside the female genital tract is scarce, and in the case of prey gift giving spider species, post-copulatory mechanisms are completely unknown. The focus of this stage of the reproductive process poses as a very promising research program for the near future. Considering what might happen after male's departure from the copulatory process, we can expect

[1]Cryptic male choice definition would be analogous to CFC, but in relation to strategic sperm allocation by the male in each mating event.

prey gift effects during the assimilation process in the form of direct benefits in terms of assimilation of nutrients dedicated to fecundity and also by indirect benefits through the assimilation of manipulative substances that enhances fecundity.

Our understanding about spiders' prey consumption has received an impressive improvement with the emergence of a theoretical framework provided by Nutritional Ecology (Boggs 2009; Morehouse et al. 2010; Wilder 2011) that considers the influence of an individual diet over its reproductive success. For example, the integrative approach proposed by emergent research programs like Nutritional Ecology indicates a very promising source of evidence for the near future. Even with the occurrence of deceptive strategies in prey gift offering, most of the time prey gifts represent a nutritional donation that may be used as an important source for synthesizing eggs or simply as a trigger for their maturation (Costa-Schmidt and Machado, unpublished data (see also Albo et al. 2014b for an extreme case of high percentage of deceptive gifts in nature)).

It is most probably that nuptial gifts may have an important contribution for CFC during the post-copulatory phase, especially for oral gifts since "*(…) the possibility that nutritional contributions from males may trigger female responses only indirectly was not considered*" (Eberhard 1996, p. 306). Furthermore, Eberhard also says "*(…) a common test of the nourishment hypothesis (…) is to ascertain whether the effect of male donations on female fecundity varies with the female's nutritional status*" (Eberhard 1996, p. 307).

The post-copulatory phase is also where male manipulation can be performed also in a cryptic way. Therefore, separating what would be a CFC effect from male manipulation becomes a difficult task even in experimental terms. Some of the effects that a prey gift would impose over female reproductive decisions can be represented by the induction of an extended inter-copula refractory interval (but see Prokop and Maxwell 2009), which would be an advantage for a male in terms of assuming a partial control over females' polyandric decisions. Such induction could be interpreted as a reproductive cost for females, since they would lose their control over the assemblage of the fertilization set. The simplest way that refractory induction may be triggered would be through the transfer of some substance together with the prey gift that would delay prey gift assimilation without compromising the physiological demands for fertilization.

6.7 Concluding Remarks

Along this chapter, I tried to bring to our attention the possible roles of prey gifts during CFC. Off course, the entire chapter may be classified as speculative, but such classification is directly related to our ignorance over fundamental mechanisms that nuptial gifts takes a major role. Probably a major effort to be worth, it is the investigation of the occurrence of chemical signals associated to nuptial gift construction in several cases, and their role as a trigger of a female physiological state that leads to CFC.

Several aspects were not considered in this first approach for such a complex theme. For example, certainly, the importance of each of those mentioned CFC mechanisms will vary among the population of a single species. Ecological factors are not homogeneous, there are interspecific interactions occurring in natural populations, and reproductive interference may take its influence over the process, to name a few. Physiological factors, such as the male's and/or female's physiological condition, most probably have its share in the variation among populations. However, it is important to note that nuptial gifts have a tremendous potential to participate as a main trait for CFC.

Acknowledgments Thanks to the editors for their invitation to contribute to the volume, and to Maria Jose Albo and Karim Vahed for valuable comments on the chapter. The writing of this chapter was supported by the Fondo para la Investigación Científica y Tecnológica—FONCyT (Argentina, FONCYT 2010-0906), and by the Coordenação de Aperfeiçoamento de Pessoal de Nível Superior—CAPES (Brazil, PNPD 20132737).

References

Aisenberg A, Viera C, Costa FG (2007) Daring females, devoted males, and reversed sexual size dimorphism in the sand-dwelling spider *Allocosa brasiliensis* (Araneae, Lycosidae). Behav Ecol Sociobiol 62:29–35

Albo MJ, Peretti AV (In press) Worthless and nutritive nuptial gifts: mating duration, sperm stored and potential female decisions in spiders. PLos One

Albo MJ, Costa FG (2010) Nuptial gift-giving behaviour and male mating effort in the Neotropical spider *Paratrechalea ornata* (Trechaleidae). Anim Behav 79:1031–1036

Albo MJ, Winther G, Tuni C, Toft S, Bilde T (2011) Worthless donations: male deception and female counter play in a nuptial gift-giving spider. BMC Evol Biol 11:329

Albo MJ, Toft S, Bilde T (2012) Female spiders ignore condition-dependent information from nuptial gift wrapping when choosing mates. Anim Behav 84:907–912

Albo MJ, Bilde T, Uhl G (2013) Sperm storage mediated by cryptic female choice for nuptial gifts. Proc R Soc B 280:20131735

Albo MJ, Toft S, Bilde T (2014a) Sexual selection, ecology, and evolution of nuptial gifts in spiders. In: Macedo R, Machado G (eds), Sexual selection: perspectives and models from the neotropics. Academic Press, pp 183–200

Albo MJ, Melo-González V, Carballo M, Baldenegro F, Trillo MC, Costa FG (2014b) Evolution of worthless gifts is favoured by male condition and prey access in spiders. Anim Behav 92:25–31

Andersen T, Bollerup K, Toft S, Bilde T (2008) Why do males of the spider *Pisaura mirabilis* wrap their nuptial gifts in silk: female preference or male control? Ethology 114:775–781

Andrade MCB (1996) Sexual selection for male sacrifice in the australian redback spider. Science 271:70–72

Arnqvist G, Henriksson S (1997) Sexual cannibalism in the fishing spider and a model for the evolution of sexual cannibalism based on genetic constraints. Evol Ecol 11:255–273

Arnqvist G, Nilsson T (2000) The evolution of polyandry: multiple mating and female fitness in insects. Anim Behav 60:145–164

Arnqvist G, Rowe L (2005) Sexual conflict. Princeton University Press, Princeton

Austad SN (1984) Evolution of sperm priority patterns in spiders. In: Smith RL (ed), Sperm competition and the evolution of mating systems. Harvard University Press, pp 223–249

Austad SN, Thornhill R (1986) Female reproductive variation in a nuptial-feeding spider, *Pisaura mirabilis*. Bull Br Arachnol Soc 7:48–52

Bastos-Pereira R (2009) Comportamento sexual e a importância do presente nupcial na reprodução de *Thaumasia* sp. nov. (Araneae, Pisauridae). Masters Dissertation, University of Brasilia. Available at http://bdtd.bce.unb.br/tedesimplificado/tde_busca/arquivo.php?codArquivo=5380

Bilde T, Tuni C, Elsayed R, Pekar S, Toft S (2007) Nuptial gifts of male spiders: sensory exploitation of the female's maternal care instinct or foraging motivation? Anim Behav 73:267–273

Boggs CL (1990) A general model of the role of male-donated nutrients in female insects' reproduction. Am Nat 136:598–617

Boggs CL (2009) Understanding insect life histories and senescence through a resource allocation lens. Func Ecol 23:27–37

Brum PED, Costa-Schmidt LE, Araújo AM (2012) It is a matter of taste: chemical signals mediate nuptial gift acceptance in a neotropical spider. Behav Ecol 23:442–447

Bruun LE, Michaelsen KR, Sørensen A, Nielsen MH, Toft S (2004) Mating duration of *Pisaura mirabilis* (Aranea: Pisauridae) depends on size of the nuptial gift and not on male size. Atrhopoda Selecta 1(Special Issue):35–39

Buskirk RE, Frohlich C, Ross KG (1984) The natural selection of sexual cannibalism. Am Nat 123:612–625

Carico JE (2005) Descriptions of two new spider genera of Trechaleidae (Araneae, Lycosoidea) from South America. J Arachnol 33:797–812

Clerck C (1757) Svenska spindlar, uti sina hufvud-slågter indelte samt under några och sextio särskildte arter beskrefne och med illuminerade figurer uplyste. Stockholmiae, 154

Cobbold SM, Su Y (2010) The host becomes dinner: possible use of *Cyclosa* as a nuptial gift by *Argyrodes* in a colonial web. J Arachnol 38:132–134

Costa-Schmidt LE, Machado G (2012) Reproductive interference between two sibling species of gift-giving spiders. Anim Behav 84:1201–1211

Costa-Schmidt LE, Carico JE, Araujo AM (2008) Nuptial gifts and sexual behavior in two species of spider (Araneae, Trechaleidae, *Paratrechalea*). Naturwissenschaften 95:731–739

Drengsgaard IL, Toft S (1999) Sperm competition in a nuptial feeding spider, *Pisaura mirabilis*. Behaviour 136:877–897

Eberhard WG (1985) Sexual selection and animal genitalia. Harvard University Press, Cambridge

Eberhard WG (1991) Copulatory courtship and cryptic female choice in insects. Biol Rev 66:1–31

Eberhard WG (1994) Evidence for widespread courtship during copulation in 131 species of insects and spiders, and implications for cryptic female choice. Evolution 48:711–733

Eberhard WG (1996) Female control: sexual selection by cryptic female choice. Princeton University Press, Princeton

Eberhard WG (2000) Criteria for demonstrating postcopulatory female choice. Evolution 54:1047–1050

Eberhard WG (2007) Escolha críptica pela fêmea e fenômenos associados em aranhas. In: Gonzaga MO, Santos AJ, Japyassú HF (eds) Ecologia e comportamento de aranhas. Editora Interciência, Rio de Janeiro, pp 115–136

Elias DO, Andrade MCB, Kasumovic MM (2011) Dynamic population structure and the evolution of spider mating systems. Adv Insect Physiol 41:65–114

Ghislandi PG, Albo MJ, Tuni C, Bilde T (2014) Evolution of deceit by worthless donations in a nuptial gift-giving spider. Curr Zool 60:43–51

Gwynne D (2008) Sexual conflict over nuptial gifts in insects. Annu Rev Entomol 53:83–101

Hebets E (2011) Current status and future directions of research in complex signaling. Curr Zool 57:i–v

Hebets E, Papaj DR (2005) Complex signal function: developing a framework of testable hypotheses. Behav Ecol Sociobiol 57:197–214

Herberstein ME, Schneider JM, Uhl G, Michalikc P (2011) Sperm dynamics in spiders. Behav Ecol 22:692–695

Holland B, Rice WR (1998) Chase-away sexual selection: antagonistic seduction versus resistance. Evolution 52:1–7

Huber BA (2005) Sexual selection research on spiders: progress and biases. Biol Rev 80:363–385
Itakura Y (1993) The life history and nuptial feeding of a nursery web spider, *Pisaura lama*. The Insectarium 30:4–9
Itakura Y (1998) Discovery of nuptial feeding in the spider, *Perenethis fascigera* (Araneae: Pisauridae). Acta Arachnol 47:173–175
Johnson JC (2013) DEBATES: challenging a recent challenge to the aggressive spillover hypothesis. Ethology 119:811–813
Klein AL, Trillo MC, Costa FG, Albo MJ (2014) Nuptial gift size, mating duration and remating success in a Neotropical spider. Ethol Ecol Evol 26:29–39
Knowlton N, Greenwell SR (1984) Male sperm competition avoidance mechanisms: the influence of female interests. In: Smith RL (ed) Sperm competition and the evolution of animal mating systems. Academic Press, Orlando, pp 61–84
Kralj-Fišer S, Schneider JM, Kuntner M (2013) Challenging the aggressive spillover hypothesis: is pre-copulatory cannibalism a part of a behavioral syndrome? Ethology 119:615–623
Kunz K, Michalik P, Uhl G (2013) Cephalic secretion release in the male dwarf spider *Oedothorax retusus* (Linyphiidae: Erigoninae): an ultrastructural analysis. Arthropod Struct Dev 42:477–482
Lang A (1996) Silk investment in gifts by males of the nuptial feeding spider *Pisaura mirabilis* (Araneae: Pisauridae). Behaviour 133:697–716
LeBas NR, Hockham LR (2005) An invasion of cheats: the evolution of worthless nuptial gifts. Curr Biol 15:64–67
Lewis S, South A (2012) The evolution of animal nuptial gifts. Adv Study Behav 44:53–97
Lewis S, Vahed K, Koene JM, Engqvist L, Bussière LF, Perry JC, Gwynne D, Lehmann GUC (2014) Emerging issues in the evolution of animal nuptial gifts. Biol Lett 10:20140336
Manier MK, Lüpold S, Pitnick S, Starmer WT (2013) An analytical framework for estimating fertilization bias and the fertilization set from multiple sperm-storage organs. Am Nat 182:552–561
Mello-Leitão CF (1943) Arañas nuevas de Mendoza, La Rioja y Córdoba colectadas por el Professor Max Birabén. Revista del Museo de La Plata (N.S., Zool.) 3:101–121
Michalik P, Uhl G (2011) Cephalic modifications in dimorphic dwarf spiders of the genus *Oedothorax* (Erigoninae, Linyphiidae, Araneae) and their evolutionary implications. J Morphol 272:814–832
Miller JA (2007) Repeated evolution of male sacrifice behavior in spiders correlated with genital mutilation. Evolution 61:1301–1315
Morehouse NI, Nakazawa T, Booher CM, Jeyasingh PD, Hall MD (2010) Sex in a material world: why the study of sexual reproduction and sex-specific traits should become more nutritionally-explicit. Oikos 119:766–778
Nitzsche ROM (1988) "Brautgeschenk" und Umspinnen der Beute bei *Pisaura mirabilis*, *Dolomedes fimbriatus* und *Thaumasia uncata* (Arachnida, Araneida, Pisauridae). Verh Naturwiss Ver Hamburg 30:353–393
Nitzsche ROM (2011) Courtship, mating and agonistic behaviour in *Pisaura mirabilis* (Clerck, 1757). Bull Br Arachnol Soc 15:93–120
Parker GA (1970) Sperm competition and its evolutionary consequences in the insects. Biol Rev 45:525–567
Parker GA, Simmons LW, Kirk H (1990) Analysing sperm competition data: simple models for predicting mechanisms. Behav Ecol Soc 27:55–65
Partan SR (2004) Multisensory animal communication. In: Calvert G, Spence C, Stein BE (eds) The handbook of multisensory processes. MIT Press, Cambridge, pp 225–240
Partan SR, Marler P (1999) Communication goes multimodal. Science 283:1272–1273
Partan SR, Marler P (2005) Issues in the classification of multimodal communication signals. Am Nat 166:231–245
Prokop P, Maxwell MR (2009) Female feeding regime and polyandry in the nuptially feeding nursery web spider, *Pisaura mirabilis*. Naturwissenschaften 96:259–265

Prokop P, Maxwell MR (2012) Gift carrying in the spider *Pisaura mirabilis*: nuptial gift contents in nature and effects on male running speed and fighting success. Anim Behav 83:1395–1399

Pruitt JN, Keiser CN (2013) DEBATES: The aggressive spillover hypothesis: existing ailments and putative remedies. Ethology 119:807–810

Sbilordo SH, Martin OY (2014) Pre- and postcopulatory sexual selection act in concert to determine male reproductive success in *Tribolium castaneum*. Biol J Linn Soc 112:67–75

Schneider J, Andrade M (2011) Mating behavior and sexual selection. In: Herberstein ME (ed) Spider behavior: flexibility and versatility. Cambridge University Press, Cambridge, pp 215–274

Schneider JM, Elgar MA (2001) Sexual cannibalism and sperm competition in the golden orb-web spider *Nephila plumipes* (Araneoidea): female and male perspectives. Behav Ecol 12:547–552

Silva ELC, Lise AA (2009) New record of nuptial gift observed in *Trechalea amazonica* (Araneae, Lycosoidea, Trechaleidae). Rev Peru Biol 16:119–120

Stålhandske P (2001) Nuptial gift in the spider *Pisaura mirabilis* maintained by sexual selection. Behav Ecol 12:691–697

Thornhill R (1983) Cryptic female choice and its implications in the scorpionfly *Harpobittacus nigriceps*. Am Nat 122:765–788

Trillo MC, Melo-González V, Albo MJ (2014) Silk wrapping of nuptial gifts as visual signal for female attraction in a crepuscular spider. Naturwissenschaften 101:123–130

Tuni C, Bilde T (2010) No preference for novel mating partners in the polyandrous nuptial-feeding spider *Pisaura mirabilis* (Araneae: Pisauridae). Anim Behav 80:435–442

Uetz GW, McCrate A, Hieber CS (2010) Stealing for love? Apparent nuptial gift behavior in a kleptoparasitic spider. J Arachnol 38:128–131

Uhl G (2002) Female genital morphology and sperm priority patterns in spiders (Araneae). In: Toft S, Scharff N (eds) European arachnology 2000. Aarhus University Press, Aarhus, pp 145–156

Vahed K (1998) The function of nuptial feeding in insects: a review of empirical studies. Biol Rev 73:43–78

Vahed K (2007) All that glisters is not gold: sensory bias, sexual conflict and nuptial feeding in insects and spiders. Ethology 113:105–127

Van Hasselt AWM (1884) Waarnemingen omtrent anomaliën van de geslachtsdrift bij spinnenmares. Tijd Ent 27:197–206

Warwick S, Vahed K, Raubenheimer D, Simpson SJ (2009) Free amino acids as phagostimulants in cricket nuptial gifts: support for the 'Candymaker' hypothesis. Biol Lett 5:194–196

Wilder SM (2011) Spider nutrition: an integrative perspective. Advances in Insect Physiology 40:87–136

Wilder SM, Rypstra AL, Elgar MA (2009) The importance of ecological and phylogenetic conditions for the occurrence and frequency of sexual cannibalism. Annu Rev Ecol Evol Syst 40:21–39

Wilgers DJ, Hebets E (2011) Complex courtship displays facilitate male reproductive success and plasticity in signaling across variable environments. Current Zoology 57:175–186

Chapter 7
Male and Female Mate Choice in Harvestmen: General Patterns and Inferences on the Underlying Processes

Glauco Machado, Gustavo S. Requena, Carlos Toscano Gadea, Estefanía Stanley and Rogelio Macías-Ordóñez

Abstract Harvestmen belong to the order Opiliones, and, unlike other arachnids, they are highly polygynandrous, with both males and females mating multiply throughout the breeding season. In this chapter, we review the current information on sexual selection in the group, focusing mostly on intersexual interactions. Particularly, we provide an overview of harvestman mating systems, examine different temporal phases of male–female sexual interactions, and explore cases of sex role reversal. Several traits in harvestmen make them unique in the context of most previous studies of sexual selection. First, they have evolved an intromittent organ independently of other well-studied taxa, such as insects, spiders, and mammals. Second, the lack of long-range perception mechanisms reduces the window of opportunity for males and females to exchange information during the very short period between the first contact and intromission. In some cases, however,

G. Machado (✉)
Departamento de Ecologia, Instituto de Biociências, Universidade de São Paulo,
Rua do Matão, trav. 14, no 321, Cidade Universitária, São Paulo, SP 05508-900, Brazil
e-mail: glaucom@ib.usp.br

G.S. Requena
Department of Ecology and Evolutionary Biology, Yale University,
165 Prospect Street, New Haven, CT 06520, USA
e-mail: gs.requena@gmail.com

C. Toscano-Gadea · E. Stanley
Laboratorio de Etología, Ecología y Evolución, Instituto de Investigaciones Biológicas
Clemente Estable, Av. Italia 3318, C. Postal 11.600 Montevideo, Uruguay
e-mail: ctoscanogadea@gmail.com

E. Stanley
e-mail: estefaniastanley@gmail.com

R. Macías-Ordóñez
Red de Biología Evolutiva, Instituto de Ecología, A.C., Carretera antigua a Coatepec 351,
91070 Xalapa, Veracruz, México
e-mail: rogeliomaciasordonez@gmail.com

A.V. Peretti and A. Aisenberg (eds.), *Cryptic Female Choice in Arthropods*,
DOI 10.1007/978-3-319-17894-3_7

acceptance or rejection of a mate may be based on information gathered before contact, such as the quality of the male territory or the presence of eggs in his nest. Regardless of the role of pre-copulatory interactions, actual fertilization success is likely to be strongly dependent on the outcome of copulatory and post-copulatory processes. In this sense, the fact that most species are highly promiscuous and have sperm cells that lack flagella, which are stored near to the tip of the ovipositor and used to fertilize eggs immediately before oviposition, renders Opiliones a fertile ground to study the role of cryptic female choice, sperm competition, and sexual conflict.

7.1 Introduction

The order Opiliones is a major group of arachnids that comprises nearly 6,500 species distributed in all continents, except for Antarctica (Machado et al. 2007; Kury 2012). The order is divided into four living suborders (Fig. 7.1a), whose species show marked differences in morphology and behavior. The suborder Cyphophthalmi is composed of nearly 190 small-bodied (1–3 mm in body length), short-legged species (Giribet 2007; Kury 2012; Fig. 7.1b), in which sperm transfer occurs via spermatophore (Karaman 2005). The remaining three living suborders form a clade called Phalangida, defined among other characters by the presence of intromittent male genitalia (Shultz and Pinto-da-Rocha 2007). The suborder Eupnoi comprises nearly 1,800 species and includes the forms widely known as daddy longlegs (Cokendolpher et al. 2007; Kury 2012; Fig. 7.1c). The suborder Dyspnoi comprises nearly 350 species exhibiting great diversity of body plans (Gruber 2007; Kury 2012; Fig. 7.1d). Finally, the suborder Laniatores is the most diverse lineage, including nearly 4,200 species, typically bearing spiny pedipalps and legs that are sexually dimorphic (Kury 2007, 2012; Buzatto and Machado 2014; Fig. 7.1e).

The book Harvestmen: The Biology of Opiliones, published in 2007, was a landmark in our understanding about the morphology, systematics, behavior, and ecology of the group, synthesizing in a single volume all available information on harvestmen that was scattered in thousands of papers published in several idioms. Although the publication of the book is relatively recent, some chapters are already outdated as a consequence of fast advances in our knowledge in recent years. The subject "reproduction" is perhaps the most emblematic of such a fast advance. From 2007 to now, the existence of male dimorphism and alternative reproductive tactics has been discovered in many species (Buzatto and Machado 2014), several new cases of both maternal and paternal care have been described (Requena et al. 2013; Buzatto et al. 2013), correlates of male mating success have been investigated (e.g., Buzatto and Machado 2008; Nazareth and Machado 2010; Fowler-Finn et al. 2014), and the use of phylogenetic information to infer the evolution of sexually selected traits has increased (e.g., Burns et al. 2013; Buzatto et al. 2014).

Fig. 7.1 **a** Two concurrent hypotheses for the relationship among the living suborders of Opiliones: *on the left side* Eupnoi + Dyspnoi form a clade called Palpatores, and *on the right side* Dyspnoi + Laniatores form a clade called Dyspnolaniatores (Giribet and Kury 2007). The two topologies recognize a clade called Phalangida that includes Eupnoi, Dyspnoi, and Laniatores. **b** Representative of Cyphophthalmi belonging to the family Sironidae (*photograph* Marshal Hedin). **c** Representative of Eupnoi belonging to the family Phalangiidae (*photograph* Daniel Proud). **d** Representative of Dyspnoi belonging to the family Ischyropsalididae (*photograph* Gonzalo Giribet). **e** Representative of Laniatores belonging to the family Stygnidae (*photograph* Glauco Machado)

Despite the great leap forward in the last seven years, there are still many gaps in our knowledge, and for the great majority of harvestman species, the only available information is the taxonomic description. Moreover, there is great geographic bias in our knowledge about harvestman behavior. Contrary to nearly all other animal groups, the reproductive behavior of harvestmen is better studied in the neotropics and southern temperate regions than in northern temperate regions (Buzatto et al. 2013). Considering that the diversity of the suborders is not evenly distributed in the world (Kury 2012), the geographic bias also implies strong taxonomic bias. Finally, regardless of the region or taxonomic group, most of the studies on sexual selection in harvestmen conducted to date are devoted to describe patterns rather than to investigate mechanisms. Therefore, hard data or experimental evidence on cryptic female choice (CFC), sexual antagonistic coevolution, and sperm competition, for instance, are virtually absent.

If there is no such information on harvestmen, what should readers expect to find in this chapter? We will review the current information on sexual selection in harvestmen, focusing mostly on intersexual interactions (for an updated account on intrasexual selection, see Buzatto and Machado 2014). As we intend to show here, harvestmen are an interesting group to study sexual selection in general and CFC in particular, because they have evolved an intromittent organ independently of other taxa, such as spiders, insects, and mammals (Dunlop 2007; Macías-Ordóñez et al. 2010). Moreover, contrary to most species of these well-studied groups, in which long-range sexual information is exchanged in the form of exaggerated structures, colored ornaments, elaborate songs, or potent airborne pheromones, the great majority of harvestman species seems incapable of forming images, only perceiving changes in light intensity (Curtis 1970; Willemart et al. 2009). Although there is increasing evidence that harvestmen rub body parts against the substrate (Fernandes and Willemart 2014), probably leaving chemicals behind, they seem to be unable to detect long-range, vibratory or airborne chemical information (Edgar 1963; Willemart and Chelini 2007). However, their legs, especially the first two pairs, and sometimes their pedipalps, are equipped with sensitive contact chemo- and mechanoreceptors (Willemart et al. 2009). The complexity of their sensory mechanisms can be exemplified by a recent study that identified nine different sensory structures on the tegument of both male and females of *Dicranopalpus ramosus* (Eupnoi), with clear sexually dimorphic distribution throughout their legs, pedipalps, and chelicerae (Wijnhoven 2013). Sexual communication in animals that rely mainly on tactile or short-range chemical cues has been greatly overlooked, and here, we will explore the implications of the unique sensory system of harvestmen in a sexual selection context.

The chapter is divided into six major sections. In the first section, we provide an overview of harvestmen mating systems, which we consider the scenario for the sexual interactions we describe afterward. The following three sections explore the temporal sequence of male–female interactions, starting with a pre-copulatory phase, passing through the copulatory phase in which we focus on genital interactions, and finishing with the post-copulatory phase (Fig. 7.2). Then, we will explore cases of partial or total sex role reversal, focusing on male mate choice and female courtship behavior. Finally, we will conclude indicating the points that make harvestmen special for the study of sexual selection, with special emphasis on CFC, and will also suggest potentially promising research questions.

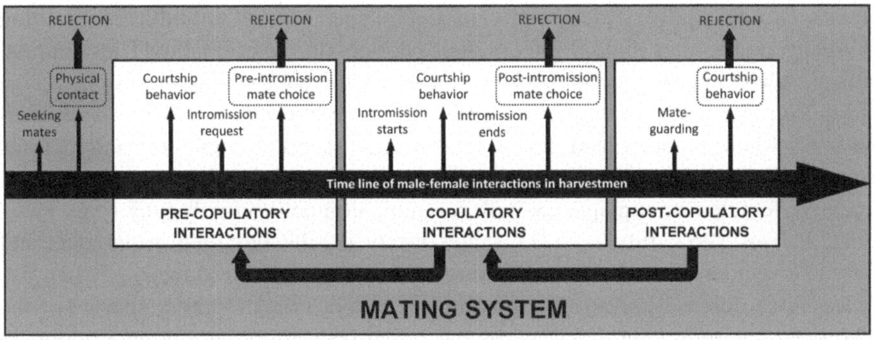

Fig. 7.2 Scheme of the chronological order of events during male–female interactions in harvestmen. Ecological and historical characteristics of populations determine the mating systems (*gray area*), which sets the specific rules of how and by which sex the reproductive behaviors are performed. Due to the harvestman sensory mechanisms, mating interactions only start after individuals find each other and establish physical contact. The whole mating process can be divided into three stages (depicted as *white boxes*): pre-copulatory, copulatory, and post-copulatory. Rejection of the mating partner may occur at different moments, which are highlighted by the *dotted line*. Multiple intromissions are common in species of the suborder Eupnoi, and this behavioral pattern is represented here by the *black arrow* connecting the pre- and copulatory stages at the bottom of the scheme. In some species of the suborder Laniatores, remating frequently occurs during mate guarding, and this behavioral pattern is represented here by the *black arrow* connecting the post- and copulatory stages at the bottom of the scheme

7.2 Mating Systems

7.2.1 General Characterization

Describing mating systems is important to address sexual selection in any group because they describe the set of reproductive strategies present in a given population (Emlen and Oring 1977) and thus the social context in which males and females can exert reproductive decisions (Fig. 7.2). There are several studies devoted to harvestmen reproductive biology (review in Machado and Macías-Ordóñez 2007), but only a few accounts provide details on their mating systems. Buzatto et al. (2013) provide a review on what it is currently known about harvestman mating systems, and here, we only summarize their main findings. Most harvestman species reproduce sexually, and, unlike other arachnids, they are highly polygynandrous, i.e., both males and females engage in copulation with multiple mates throughout the breeding season (Machado and Macías-Ordóñez 2007). The most likely mating system of species belonging to the suborders Cyphophthalmi and Dyspnoi, as well as many species of the suborder Eupnoi, is scramble competition polygyny (Buzatto et al. 2013), which involves multiple mating events for both males and females. In these groups, females generally lay eggs on sites that cannot be profitably monopolized by males, such as the bark of trees, leaf litter, cracks on rock walls, and empty snail shells (Machado and

Macías-Ordóñez 2007). Additionally, in several species with scramble competition polygyny, males offer nuptial gifts to their partners during courtship. These nuptial gifts are glandular secretions produced either on the base of the male's chelicerae and delivered directly to the female's mouth before intromission or on the base of the penis where females feed during intromission (see Pre-copulatory interactions below). Although there are some records of male–male fights for the possession of receptive females among species with scramble competition polygyny (e.g., Pabst 1953; Parisot 1962; Edgar 1971), males rarely exhibit enlarged armaments and there is no evidence of alternative mating tactics (Buzatto and Machado 2014).

Resource defense polygyny is a common mating system among species of the suborder Laniatores, but a few cases have also been reported among Eupnoi. In both suborders, males fight each other for the possession of reproductive territories that are visited by females looking for particular oviposition sites. These reproductive territories include natural cavities on trunks and riverside banks (Machado et al. 2004; Nazareth and Machado 2010), rocks (Macías-Ordóñez 1997, 2000; Zatz et al. 2011), specific host plants (Buzatto and Machado 2008), or mud nests built by males (Rodríguez and Guerrero 1976). In some species, females remain inside the male's territory after oviposition, forming harems that may include as much as ten females (Buzatto and Machado 2008; Zatz et al. 2011). In other species, females abandon the male's territory after oviposition, leaving their eggs hidden either inside small crevices (Macías-Ordóñez 1997, 2000; Wijnhoven 2011) or under the male's guard (Mora 1990; Nazareth and Machado 2010). Post-copulatory association between male and female is widespread in the order (see Post-copulatory interactions below), but it seems to be particularly common among species exhibiting resource defense polygyny (Buzatto et al. 2013) and/or exclusive paternal care (Requena et al. 2013).

Another feature that seems to be widespread among species exhibiting resource defense polygyny is the presence of alternative reproductive tactics, often associated with male dimorphism (Buzatto and Machado 2014). Large males (commonly referred to as majors or territorials) fight for the possession of reproductive territories or harems using elaborate weaponry. At least among gonyleptids, these weapons frequently include apophyses growing from the carapace or from one or more segments of their legs, especially the fourth pair. In some species, however, the whole leg is extremely elongated and functions as a whip in male–male contests. On the other hand, small males (commonly referred to as minors or sneakers) have reduced or completely absent weaponry and usually sneak copulations invading majors' territories (Buzatto and Machado 2014). The presence of two male morphs with different mating tactics in the population has profound implications for sexual selection (Kvarnemo and Simmons 2013), influencing the level of sperm competition (Muniz et al. 2014) and the level of sexual conflict within and between sexes (Buzatto and Machado 2014). However, these subjects have been poorly explored in harvestmen.

7.2.2 Potential for Female Mate Choice

Among polygynous populations, the freedom females have to perform mate choice is partially determined by the type of male–female associations. When females acquire resources on their own, they are free to choose among available males based on phenotypic differences (Borgia 1979). This situation occurs in male dominance polygyny (such as leks) and in some types of scramble competition polygyny. As far as we know, there is no case of lek among harvestmen, probably because the great majority of the species of the order is unable to form images and is very limited in the use of long-range chemical, vibratory, and acoustic stimuli (Willemart et al. 2009). On the other hand, no study conducted so far has investigated female choice in harvestman species exhibiting scramble competition polygyny, despite the fact that this is probably the most common mating system in the order (Buzatto et al. 2013). In the family Trogulidae (Dyspnoi), for instance, males search for receptive females and pre-copulatory and copulatory interactions involve mutual cheliceral rubbing and male leg tapping on the female's body (Pabst 1953). Given that trogulids, as well as many other harvestman species exhibiting scramble competition polygyny, are easily maintained in captivity (where females copulate and lay eggs), future studies should investigate whether specific morphological and behavioral traits in males influence their mating and fertilization success.

When males control reproductive resources, such as oviposition sites, females only have access to these resources by mating with the males controlling them (Borgia 1979). In this situation, typical of resource defense polygyny (associated or not with exclusive paternal care), females may actively choose mates based on resource quality and/or male phenotypic attributes, such as body size, weapon size, conspicuousness of ornaments, concentration of pheromones, or rates of displays. Alternatively, females may choose mates passively, which occurs when any female trait (morphological or behavioral) promotes or intensifies male–male competition, increasing females' chance of mating with a successful competitor (Wiley and Poston 1996). Passive female choice may be common among harvestman species exhibiting site or resource defense polygyny. Females of the harvestman *Zygopachylus albomarginis* (Laniatores), for instance, copulate exclusively with males associated with cup-like mud nests used as oviposition sites (Mora 1990). Therefore, by choosing where to lay their eggs, females of *Z. albomarginis* may passively select traits that indicate male ability to build, maintain, or even take over nests.

Females of the harvestman *Serracutisoma proximum* (Laniatores) lay eggs on the vegetation at river margins, showing marked preference for certain plant species. Preferred plants are predictable resources searched by females at the time of breeding, so that males benefit from defending and monopolizing territories containing these plants as a means of acquiring mates (Buzatto and Machado 2008). Since males use legs II as weapons to resolve contests for the possession of territories, the reported positive correlation between male leg length and number of females in his harem may have emerged as a consequence of two non-mutually exclusive processes. First, since males owning territories are likely to be under intense intrasexual competition, females that mate with these males may benefit from producing offspring fathered by the best competitors. The hypothesis of passive mate choice would require a further correlation between harem quality and female choice, because the higher the quality of oviposition sites, the more intense male–male contests should be. However, variation in harem quality in *S. proximum* does not explain variation in the number of females found across harems (Buzatto and Machado 2008), and thus, this hypothesis has no empirical support so far. Second, females may actively choose males with longer legs II if this trait is genetically correlated with males' viability (good genes hypothesis) or increases the mating success of their sons due to the hereditary covariance between male trait and female preference (sexy sons hypothesis). Although it is hard to disentangle these two possibilities (Kotiaho and Puurtinen 2007), information on the correlation between male leg length and the fitness of his offspring, as well as on the genetic covariance between female preference and male leg length, may indicate the relative importance of these two different processes for the maintenance of the observed patterns.

When males directly monopolize females instead of territories, females may choose males actively, using phenotypic traits, or passively, if many receptive females aggregate in particular sites and males that better defend harems against competitors accumulate in the owner position (Borgia 1979; Wiley and Poston 1996). As far as we know, this mating system, known as female defense polygyny, has been reported for a single harvestman species, *S. proximum*. After the arrival of several females in a harem, territorial males concentrate their patrolling activity mostly on egg-guarding females. At this stage, the mating system seems to shift from resource defense to female defense polygyny (Buzatto and Machado 2008), and the most likely explanation for this shift relies in the gonadotrophic cycle of *S. proximum* females. Although females lay nearly 80–90 % of the eggs in the first 24 h after copulation, they may take up to 14 days to complete oviposition. This asynchronous egg deposition increases the possibility of polyandry, which could bring benefits for females, such as increased genetic variability in the offspring (Arnqvist and Nilsson 2000). However, by directly associating with egg-guarding females after the first oviposition bout, territorial males can increase their chance of fertilizing additional eggs retained in the females' ovaries and can also reduce the copulation success of males of a small morph (sneakers) that invade their territory (Buzatto et al. 2011; see also Post-copulatory interactions below). The extent to which mate-guarding behavior can influence female fitness and the potential conflict over promiscuity level between the sexes is an interesting area for future studies.

Finally, when males provide food resources to females in exchange for copulation or when males care for the offspring alone, females can choose males in terms of resource quality and/or direct or indirect benefits to the offspring (Borgia 1979; Hoelzer 1989). In harvestmen of the genera *Ischyropsalis* and *Paranemastoma* (Dyspnoi), for instance, males offer glandular secretions as a nuptial gift to females before intromission, but the role of this secretion for female mate choice is still unknown (Martens 1969; Meijer 1972). Harvestman species exhibiting exclusive paternal care, in turn, have received increasing attention in recent years, and some information on female mate choice is now available (Requena et al. 2013). Given that paternal care minimizes the foraging costs related to egg guarding by females and may also provide information about the quality of paternal care, males providing care should be chosen by females and obtain a greater number of copulations than males that are not associated with eggs (Hoelzer 1989; Tallamy 2000, 2001). In fact, there is experimental evidence showing that females of *Magnispina neptunus* (Laniatores) consistently prefer to lay eggs inside nests that already contain eggs, regardless of individual male traits. Not surprisingly, males that usurp a nest containing eggs usually protect these eggs against predation (Nazareth and Machado 2010), and this egg adoption behavior probably increases the chance of acquiring their own eggs later.

Considering that the production of nuptial gifts (Vahed 1998; Gwynne 2008) and paternal care in arthropods (Requena et al. 2013) can be energetically expensive, passive female choice based on intense male–male competition is unlikely because any investment in weaponry may trade off the investment in food resource or parental behavior, reducing the direct benefits of female choice (Price et al. 1993). On the other hand, mate preferences based on the presence of eggs under a male's guard are the equivalent of passive female choice based on copying the decision of other females (sensu Dugatkin 1992) and may additionally benefit females by reducing the costs of mate search and assessment (Trumbo 1996). However, given that consistency and reliability of male care quality signals are crucial to determine female decisions (e.g., Hoelzer 1989; Price et al. 1993; Wagner 2011), the relative importance of active and passive female choice on males' reproductive success should respond to ecological and social factors affecting the costs paid by parental males (Requena et al. 2013). Considering that exclusive paternal care has evolved several times independently in harvestmen, the group offers a unique opportunity to test this hypothesis using a comparative approach.

7.3 Pre-copulatory Interactions

7.3.1 General Characterization

Courtship in most arachnids is a process that generally requires careful approach from males and long distance, elaborated, highly stereotyped, and species-specific visual or vibratory displays because females may attack and cannibalize approaching partners before copulation (Thomas and Zeh 1984). However, pre-copulatory cannibalism has never been recorded in harvestmen (Acosta and Machado 2007), and in most species, males seem unable to detect females until direct physical contact is established (e.g., Willemart et al. 2006; Fowler-Finn et al. 2014). Once male and female detect each other, pre-copulatory interactions are generally brief and involve mostly tactile and close range chemical stimuli (Machado and Macías-Ordóñez 2007; Fig. 7.2). After the first contact, the mating pair generally adopts a face-to-face position, and males of many species of Eupnoi and Laniatores grasp the female using their pedipalps. In some species of Eupnoi, the male hooks his long, sexually dimorphic pedipalps at the base of female's second pair of legs (e.g., Edgar 1971; Macías-Ordóñez 1997; Burns et al. 2013; Fowler-Finn et al. 2014; Fig. 7.3a). In Laniatores, males grasp the females' pedipalps (Fig. 7.4a), but sexual dimorphism in the length or armature of the pedipalps is rare in species of this suborder (Macías-Ordóñez et al. 2010). Pedipalpal grasping has not been described for any species of Dyspnoi (Machado and Macías-Ordóñez 2007).

An active role of the male genitalia in male–female interactions in the pre-copulatory phase, prior to actual genital contact, is another unique feature of mating interactions in harvestmen. In some species of *Leiobunum* (Eupnoi), for instance, the pre-copulatory phase is characterized by the transference of secretions produced in glands located on the penis, in a region that is contacted by the female's mouth before and during intromission (Willemart et al. 2006; Macías-Ordóñez et al. 2010; Burns et al. 2013; Fowler-Finn et al. 2014). Therefore, we will use the term eversion to designate genital exposure of the penis prior to entering the female's genital opening and intromission as the actual insertion through such structure, but not necessarily through the female's vagina placed at the tip of the female ovipositor (for more details on genital interactions, see Copulatory interactions below).

7.3.2 *Tactile Courtship and Nuptial Gifts*

Sexual selection theory predicts that, unless males have some investment at stake in terms of material resources and/or time (see Male mate choice below), they should seek intromission soon after the first contact. Females, on the other hand, should carefully evaluate and reject some partners before intromission, unless they have full control of the fate of the sperm they receive (Andersson 1994). Most reports of pre-copulatory interactions in harvestmen indeed mention lack of evident courtship by males and some female reluctance before intromission (e.g., Roters 1944; Cloudsley-Thompson 1948; Parisot 1962; Edgar 1971; Macías-Ordóñez 1997; Fig. 7.2). Females of many species of Eupnoi and Laniatores seem to resist male advances by fleeing away or lowering their frontal end and placing the genital operculum close or in contact with the substrate, which makes intromission impossible (Machado and Macías-Ordóñez 2007; Fig. 7.3b). Nevertheless, in many species of these two suborders, the male taps different parts of the female's body using his front legs, while pedipalpal grasping is maintained (Willemart et al. 2006; Machado and Macías-Ordóñez 2007; Nazareth and Machado 2009; Fowler-Finn et al. 2014). Although it is short, the period between the first contact and intromission may be a phase of intense courtship because once a male and a female establish physical contact, the sensitive chemo- and

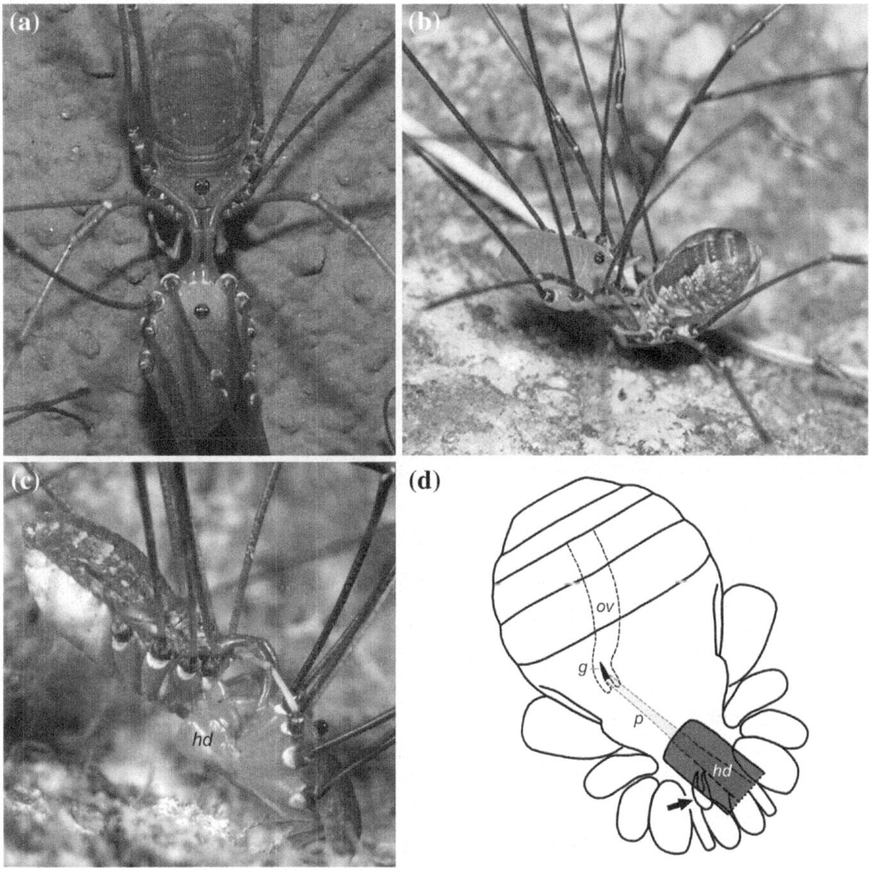

Fig. 7.3 **a** Mating pair of an unidentified Sclerosomatidae (Eupnoi) showing the male (*below*) grasping the female's first pair of legs using his pedipalps (photograph: Jerry Armstrong). **b** Female of *Leiobunum* (*right*) rejecting a male lowering the frontal end of her body so that intromission is impossible (photograph: Michael Marlow). **c** Lateral view of a mating pair of *Leiobunum* during intromission; the male is at the right, and *hd* indicates the inflated hematodocha, whose dorsal surface forms a conduit from basal nuptial glands to the female's mouth (*photograph* Joseph G. Warfel). **d** Schematic representation of the genital interaction in the same mating pair illustrated in **c**. The body of the female (*ventral view*) is outlined in *black* and her ovipositor (ov), which is retracted inside her body, is depicted by a *dashed black line*. Only the male's penis is outlined, with its tip (glans, g) depicted in *black* and the inflated hematodocha (hd), in *dark gray*. Note that (1) the penis only reaches the tip of the ovipositor below the vagina in the region where seminal receptacles are located, and (2) the female chelicerae are in close contact with the hematodocha (*black arrow*), probably feeding on glandular secretions

mechanoreceptors on their legs probably allow abundant flow of information by means of chemical and tactile stimuli. Moreover, mutual assessment of male and/or female relative body size may play an important role during pedipalpal grasping and pre-copulatory struggles (see below).

Fig. 7.4 **a** Mating pair of an unidentified Tithaeidae (Laniatores) in a forest in Singapore (photograph: Melvyn Yeo). Note that the male (*right*) uses his pedipalps to grasp the female's pedipalps during intromission. The *arrow* indicates the penis. **b** Male of *Serracutiosma proximum* (Laniatores) mate guarding an ovipositing female (photograph: Bruno A. Buzatto). The male (*above*) maintains his second pair of legs extended toward her and occasionally touches her legs or dorsum

A recent captivity study with *Leiobunum vittatum* (Eupnoi) investigated the influence of multiple male traits on the outcome of staged mating interactions (Fowler-Finn et al. 2014). Shortly after contacting the female, the male vigorously attempted to secure her in a mating embrace, typically wrapping the distal portion of his third pair of legs on the femur of her second pair of legs. During these initial pre-copulatory interactions, females were able to resist male mating attempts in more than half of the mating trials, even when males engaged in pedipalpal grasping (Fowler-Finn et al. 2014). Males that succeeded in starting the mating embrace had shorter pedipalps than unsuccessful males, and the authors suggest that the length of this appendage is decisive to the outcome of the male–female struggle: short pedipalps would provide greater mechanical advantage to overcome female resistance. While pedipalp length did not correlate with male body size, successful males that were larger compared to females were able to achieve intromission in just a few seconds, while smaller males spent much longer in this stage (Fowler-Finn et al. 2014). The relative importance of specific male traits (e.g., pedipalp length and body size) in overcoming female resistance and/or in signaling male quality, however, is still an open question that deserves future attention (Machado and Macías-Ordóñez 2007).

The total time males and females remain in contact before intromission corresponds to the window of opportunity for pre-copulatory courtship and thus female evaluation. In a well-studied clade of North American Eupnoi, the duration of pre-copulatory interactions shows great differences between sacculate and non-sacculate species (Burns et al. 2013). Sacculate species are those in which the penis has a subterminal pair of cuticular sacs used for delivering a glandular secretion to the females before intromission. Females spend a few seconds or minutes feeding on this secretion, after which they may accept or reject intromission (Macías-Ordóñez et al. 2010). In non-sacculate species, little, if any, glandular secretion is transferred before intromission, and male pedipalps seem to be selected for strongly grasping the female. Moreover, in non-sacculate species, the pregenital opening of females is sclerotized and apparently serves as a barrier against forced intromission (Burns et al. 2013). The duration of the pre-copulatory phase in the non-sacculate species can last for up to 1 h, with periods of wrestling in which the male seems to forcefully penetrate the female's pregenital opening (Macías-Ordóñez et al. 2010; Burns et al. 2013). Therefore, males of non-sacculate species seem to rely more on a powerful grasping to negotiate with the female and less on pre-copulatory courtship, including nuptial gifts. We must highlight, however, that females of both sacculate and non-sacculate species do not commit to fertilization of their gametes by merely accepting copulation (see Copulatory interactions below).

Results of a phylogenetic comparative study show that loss of penile sacs and elaboration of male pedipalps are correlated in the clade of Eupnoi mentioned above, which includes species of the genera *Leiobunum*, *Eumesosoma*, and *Hadrobunus*. Moreover, the loss of penile sacs is also correlated with the gain of pregenital barriers in females (Burns et al. 2013), suggesting that male and female morphology coevolved in the group, probably in response to sexual selection during pre-copulatory interactions. A similar result has also been found in Japanese species of the curvipalpe group of *Leiobunum*. Males of species in which females may reproduce parthenogenetically (*L. manubriatum* and *L. globosum*) have larger and more powerful pedipalps than species with exclusive sexual reproduction, presumably as a result of selection to counteract female possibilities to reproduce asexually (Tsurusaki 2004). In the concluding remarks, we are going to explore the role of female choice, male–male competition, and intersexual antagonism for the emergence of the coevolutionary patterns described here.

As we stated before, pre-copulatory interactions in some species of Dyspnoi also include the transference of secretions, but in this case, they are produced in a pair of glands located dorsally on the first segment of the male chelicerae, which are either offered or somewhat forced into the female's mouth before intromission (Martens 1969; see also Fig. 12.3a in Machado and Macías-Ordóñez 2007). Although the composition, amount, and quality of the secretion offered by males in the suborders Eupnoi and Dyspnoi may influence the donors mating success and paternity, no study has investigated these questions in nuptial gift-giving harvestmen so far. An unpublished record under natural conditions shows two males of *L. vittatum* exhibiting pedipalpal grasping, one of them seemingly mimicking female stimulation on the other male's genitalia and obtaining the secreted nuptial gift, obviously without intromission involved (R. Macías-Ordóñez, unpub. data). This isolated observation, though anecdotal, suggests that glandular secretions in Eupnoi are valuable and/or nutritive. If these glandular secretions were only exploiting a sensory bias in females, other males should not be expected to cheat in order to feed on this secretion.

Recent observations with *Discocyrtus prospicuus* (Laniatores) show complex and still poorly understood pre-copulatory interactions (Stanley and Toscano-Gadea 2011). In this species, males approach females slowly waving the second pair of legs in front of their bodies. When individuals establish physical contact, the male slightly raises the front of his body and begins to evert and retract his penis. The female touches the penis with her first pair of legs for up to 1 min. No contact between the penis and the female's mouth has ever been recorded, and the first pair of female's legs is not inserted in her mouth after contacting the penis. Therefore, if there is some chemical signal during male–female interactions, it does not fit the definition of a nuptial gift, which implies the transference of food items or inedible tokens from males to females prior to or during copulation (sensu LeBas and Hockham 2005). After touching the penis of the potential mating partner, the female generally lowers her pedipalps allowing the male to perform cheliceral and pedipalpal grasping. All cases of male rejection (n = 10 observations) occurred when he attempted to grab the female without allowing her to touch the penis. In these cases of rejection, the behavior was similar to other harvestman species, with the female lowering the frontal end and approaching her venter to the substrate, preventing male access to her genital opening. It seems, therefore, that females use information acquired when touching the penis to evaluate their partners, but the type of information used remains to be investigated.

7.4 Copulatory Interactions

7.4.1 General Characterization

As far as we know, all harvestman species require direct male–female contact to transfer sperm. If there is an exception to this pattern, it should be found in the Cyphophthalmi, in which males of some species might leave spermatophores on the substrate for the female to find them, as occurs with some pseudoscorpions and collembolans (Proctor 1998). However, although only a few spermatophores have actually been recorded for species of Cyphophthalmi, the shape of the male genitalia and the way the spermatophores have been found attached to the female strongly suggest direct participation of the male during sperm transfer (Macías-Ordóñez et al. 2010). Moreover, results from a recent phylogeny of Opiliones including fossil species (Garwood et al. 2014) suggest that the presence of a spermatophore may be derived in Opiliones, which would contrast with all other major groups of arachnid, in which the spermatophore is plesiomorphic (Proctor 1998). Regardless of the evolutionary history of sperm transfer in the order, genitalic intromission is the rule in the suborders Eupnoi, Dyspnoi, and Laniatores, and direct interaction during intromission is well documented for a few species of these groups.

In species belonging to the clade Phalangida, both males and females may actively interact during intromission by touching, rubbing, tapping, grasping, pushing, and pulling their partners in many ways with legs, pedipalps, chelicerae, and mouthparts (Willemart et al. 2006; Nazareth and Machado 2009; Macías-Ordóñez et al. 2010; Fowler-Finn et al. 2014; Fig. 7.2). As in the pre-copulatory phase, abundant chemical and tactile information probably flows during intromission, but we can do little but speculate about its meaning. Repeated intromissions while male pedipalpal grasping is maintained have been reported for many species of Eupnoi, but seem to be uncommon among representatives of Dyspnoi and Laniatores (Machado and Macías-Ordóñez 2007; Fig. 7.2). Along with scarce behavioral records, genital structure offers the basis to infer the potential processes that take place during male–female interactions during copulation. In the following topics, we summarize the available information on genital morphology and genital interactions in harvestmen. Our synthesis is largely based on Shultz and Pinto-da-Rocha (2007) and Macías-Ordóñez et al. (2010), to which the readers should refer for illustrations and further information.

7.4.2 Genital Morphology

The eversible genitalia of male Cyphophthalmi is called spermatopositor because it is much shorter than that of other harvestmen and appears to be used in the transfer of spermatophores rather than direct copulation (Van der Hammen 1985; Karaman 2005). The penis of the harvestmen belonging to the clade Phalangida is an intromittent organ typically divided into two main parts: pars basalis, which corresponds to most of the long shaft called truncus, and pars distalis, which contains the distal end of the truncus and the terminal or subterminal glans (Fig. 7.5). The pars distalis is the part that interacts with the ovipositor (Fig. 7.3c, d) and is often equipped with spines, sensilla, and other projections (Fig. 7.5). The glans is the most variable structure of the penis and contains the opening of the ejaculatory duct, located at the end of the stylus. Typically in Eupnoi and Dyspnoi, the pars distalis is composed almost exclusively of a relatively simple glans with an apical stylus, with the glans being only slightly differentiated from the truncus (Fig. 7.5a). In most species of Laniatores, however, the glans is very complex and can be divided into two main parts, capsula interna and capsula externa. The capsula interna is formed by sclerites associated with the distal end of the ejaculatory duct that are surrounded totally or partially by the capsula externa, formed simply by a soft sac called follis or by highly modified, sclerotized structures (Fig. 7.5b–h).

In muscular penises, the movement of the glans relative to the truncus is provided by one or two muscles that originate from the shaft and insert on a cuticular tendon that ends at the base of the glans. This muscular type of penis occurs in Eupnoi and Dyspnoi, as well as in two superfamilies of Laniatores, namely Travunioidea and Triaenonychoidea (Macías-Ordóñez et al. 2010). Representatives of the remaining superfamilies of Laniatores have hydraulic penises, i.e., the muscles are absent and the glans is apparently operated by

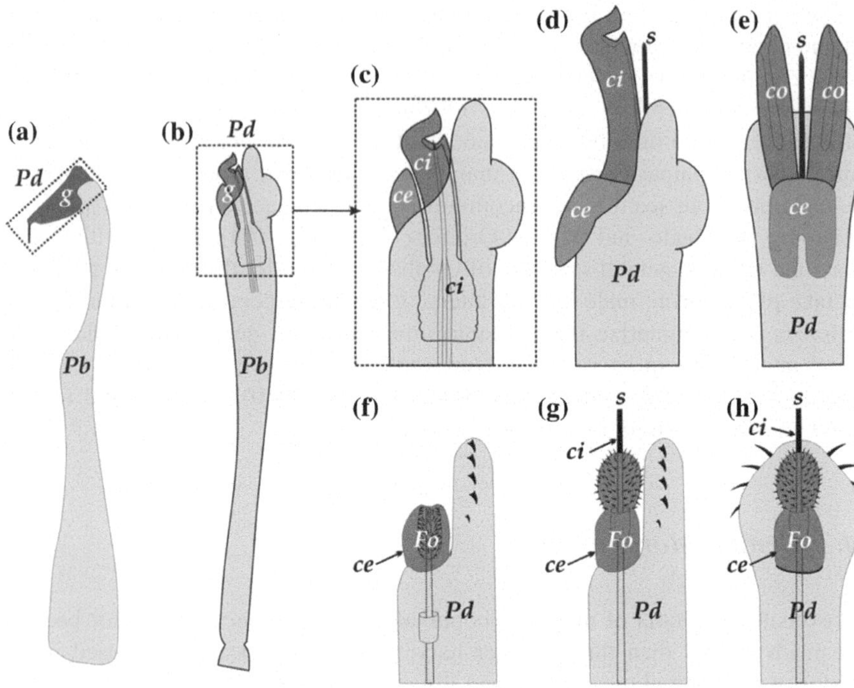

Fig. 7.5 Schematic representation of male genitalia in **a** Eupnoi and **b** Laniatores. The pars basalis (Pb) is a long shaft in both suborders, but the par distalis (Pd), especially the glans (g), is more complex in Laniatores. **c** Detail of the pars distalis in Biantidae (Laniatores) showing the capsula interna (ci) and capsula externa (ce). The capsula interna is everted by hydraulic pressure exposing the conductors (co), which are probably used to open the lumen of the ovipositor, and the stylus (s), in the tip of which is the opening of the ejaculatory duct: **d** lateral view and **e** frontal view. **f** Detail of the pars distalis in Assamiidae (Laniatores) showing the stylus retracted inside the capsula externa. When the follis (Fo) is everted by hydraulic pressure, the stylus is probably exposed inside the open lumen of the ovipositor: **g** lateral view and **h** frontal view. Modified from Macías-Ordóñez et al. (2010)

internal hemolymph pressure. It is important to note, however, that the terms muscular and hydraulic refer to the operation of the glans only. Eversion and inversion of the entire penis in all harvestmen are achieved by a combination of muscular and hydraulic mechanisms (Shultz and Pinto-da-Rocha 2007). Seminal products are apparently pushed through the long ejaculatory duct by a muscular propulsive organ located at the base of the penis; this organ is absent in Cyphophthalmi (Macías-Ordóñez et al. 2010).

The ovipositor in Cyphophthalmi and most Eupnoi has a shaft (sometimes as long as or longer than the female body) composed of cuticular rings connected by segmentally arranged muscles. It ends in paired bilateral processes derived from one or more rings, and the genital opening is located basally between these processes. Each process generally bears a tuft of sense organs on the latero-subdistal surface. A pair of sclerotized seminal receptacles is found inside the genital opening, sometimes associated with glands (Martens 1986; see also Fig. 13.5 in Macías-Ordóñez et al. 2010). Lack of seminal receptacles is often associated with parthenogenesis (Shultz and Pinto-da-Rocha 2007). The ovipositors of Dyspnoi and Laniatores differ markedly from those of the other suborders. Their ovipositors are always much shorter than the female body and show marked structural differences such as circular muscles around the vagina, X-shaped lumen in cross section, and one or more short and sclerotized seminal receptacles in each of their four lobes (see Fig. 13.5 in Macías-Ordóñez et al. 2010).

7.4.3 Genital Interaction

Based on morphological and behavioral evidence, male and female genitalia interactions are likely to be greatly diverse in harvestmen. However, there are no detailed descriptions of genitalic interaction published for any species of Opiliones. Unpublished studies of *L. verrucosum* (Eupnoi) show that the stylus does not go beyond the first third of the ovipositor, where the openings of the seminal receptacles are located (J.W. Shultz, pers. comm.). Therefore, although a large portion of the penis shaft goes through the female's genital operculum, only its tip reaches the female reproductive tract since the ovipositor is retracted (Fig. 7.3c, d). This seems to be the general pattern for all harvestmen belonging to the clade Phalangida, and it is markedly different from other animal groups with internal fertilization, such as insects, birds, and mammals, in which the entrance to the female reproductive tract is continuous with the apparent external genital opening. Once inside the ovipositor, the glans in species of Eupnoi seems to have the appropriate shape and angle to enter the seminal receptacles after the ovipositor atrium, where it could release the sperm. With the stylus inserted in the seminal receptacles, the basal portion of the glans would lay near the tip of the ovipositor, seemingly in contact with the region containing abundant sensilla. This contact while the penis is reaching the seminal receptacle may suggest that the glans either stimulates females as a form of copulatory courtship or exploits female sensory bias by seductively stimulating the sensilla, used by females to probe optimal sites for oviposition (Macías-Ordóñez et al. 2010; see also concluding remarks below).

The penis in Dyspnoi and Eupnoi are fairly similar, but their ovipositors are strikingly different. The Dyspnoi ovipositor is shorter, and the seminal receptacles are smaller and highly variable in number. Detailed records of copulation for two species of *Ischyropsalis* (Dyspnoi) show a few short intromissions, while the female grasps and apparently feeds from the male's cheliceral glands (Martens 1969). The genitalic interaction during copulation, however, has never been described. Nevertheless, it seems that the Eupnoi-like penis with its stylus does not fit seminal receptacles in this suborder. The Dyspnoi penis seemingly leaves the sperm in the lumen of the ovipositor, still far from the sperm receptacles, where the female would need to somehow transport sperm to the seminal receptacles or directly to the eggs. Moreover, the Dyspnoi ovipositor has fewer sensilla than the Eupnoi ovipositor and thus may be less sensitive to the stimulation of the penis.

Among the families of Laniatores, the modified morphology of the pars distalis shows very different arrangements of structures (Martens 1976, 1986), which apparently play three main roles during genital interactions (Macías-Ordóñez et al. 2010). First, they fasten the par distalis at the distal end of the ovipositor, where the seminal receptacles are located. Second, they seem to promote intromission of the penis in the ovipositor. Third, they may open the X-shaped vagina by hydraulic pressure and expose the stylus, which will release sperm in the lumen. The ringed muscles of the vagina in Laniatores may constrict the lumen so that the sperm transferred fill up the multiple seminal receptacles when it closes, after penis retraction. Such a muscular system might also enable females to reject sperm if the entrance to the seminal receptacles is obstructed. These hypotheses have never been tested, and an ongoing project with *Pachyloides throrelli* (Laniatores) may shed some light on sperm use and ejection in harvestmen (A. Pérez-González, pers. comm.). The ovipositors of Laniatores show fewer sensilla than other suborders, and the male genitalia possess a unique set of more proximal and highly variable and complex structures that seem in an ideal position to interact with the few sensilla. In contrast to the Eupnoi, in which males constantly move the body and the penis during multiple intromissions, in many species of Laniatores, the male's body remains motionless during the single intromission (Mora 1990; Buzatto and Machado 2008; Nazareth and Machado 2009, 2010; Requena and Machado 2014). Moreover, unlike Eupnoi males, which need to find the opening of the seminal receptacle to release sperm, Laniatores males may simply leave sperm in the lumen, passively entering into the multiple seminal receptacles through muscular contraction of the lumen (Macías-Ordóñez et al. 2010).

As explained above, the Laniatores penis does not go far inside the female reproductive tract, and sperm deposition is restricted to the tip of the ovipositor. It has been hypothesized that the function of a specific structure of the glans, known as ventral process (see Fig. 13.9 in Macías-Ordóñez et al. 2010), would be the removal of sperm from previous mates. The ventral process would penetrate the lumen of the ovipositor "brushing" the inner walls going in, but scraping off the same surface going out (Macías-Ordóñez et al. 2010). One important assumption of this hypothesis is that the sperm stored in the seminal receptacles should be transferred to the ovipositor lumen because the ventral process almost certainly does not enter these receptacles (A. Pérez González, pers. comm.). Moreover, because sperm removal is hypothesized to be a mechanical process, males would not discriminate between self and non-self ejaculates. Therefore, males should perform multiple intromissions, removing first the competitors' sperm and then transferring their own ejaculate later on. As far as we know, however, there is no obvious case of multiple intromissions in Laniatores. We argue, therefore, that a better understanding of sperm transfer and storage, as well as male–female genital interactions during intromission, is necessary to make stronger inferences on role of the ventral process and on the possibility of sperm removal in Laniatores.

Finally, lack of flagella in harvestmen sperm has important implications for post-copulatory processes in this group: their movement, if any, must be mostly controlled by female action. Once inside the seminal receptacles, being so close to the vagina at the tip of the ovipositor, their only movement may be restricted to exit seminal receptacles either by the flushing action of new sperm or by female control squeezing the seminal receptacles by contracting the ovipositor muscles. The later could potentially be done in the absence of any fertilizable egg as a means of spermatic rejection, or as mature eggs travel through the ovipositor and may come in contact with sperm just prior to crossing the vagina during oviposition, thus allowing fertilization (Macías-Ordóñez et al. 2010).

7.5 Post-copulatory Interactions

7.5.1 General Characterization

Male behavior after copulation shows great variation across Opiliones suborders. Anecdotal observations with a single species of Cyphophthalmi suggest that the male remains attached to the female after sperm transference, but additional information is necessary to understand this behavior (Schwendinger and Giribet 2005). Mating pairs in Dyspnoi generally do not show any post-copulatory interaction, and in some species, male and female may even flee from each other after copulation (Immel 1954). In Eupnoi and Laniatores, males of several species actively grasp or remain close to females until oviposition (Machado and Macías-Ordóñez 2007; Fig. 7.2). In *L. vittatum* (Eupnoi), for instance, the male wraps the female's legs using his own first pair of legs and follows her while she wanders inside his territory, apparently inspecting rock crevices for oviposition (see Fig. 12.2c in Machado and Macías-Ordóñez 2007). The male does not attempt to copulate again with the female during this association period and aggressively attacks any other approaching male. The post-copulatory association can last as much as 2 h and only ends when the female abandons the male's territory (Macías-Ordóñez 1997, 2000). In many species of Laniatores, females lay eggs immediately after copulation and males stay close to their mates during oviposition (Fig. 7.4b). Due to the great diversity of parental care forms in the suborder, a mating pair may remain together until the female abandons the clutch under the male's protection (Requena et al. 2013) or until the male returns to his activities as a territorial owner, leaving the eggs under female's protection (e.g., Machado and Oliveira 1998; Buzatto and Machado 2008). In both situations, post-copulatory associations may last more than 24 h, during which the male often attempts (and occasionally succeeds) to remate with the same female (Machado and Macías-Ordóñez 2007; Macías-Ordóñez et al. 2010).

The post-copulatory associations described here are not exclusive to harvestmen, and in many arthropod species, the mating pair remains together for periods longer than the necessary to transfer sufficient sperm to fertilize the eggs.

Alcock (1994) categorized the post-copulatory interactions of insects based on the extent of contact between male and female. According to this categorization, paired individuals may (1) keep genital contact for longer than needed for insemination; (2) exchange mating plugs, with males transferring secretions or even parts of his body that remain attached to female's genital opening; (3) keep body contact, usually with males grasping females using jaws, claspers, or legs; or (4) keep close proximity, with males apparently monitoring females without physical contact after copulation. Regarding category (1), intromission in harvestmen can last from few seconds in some Eupnoi to more than 10 min in some Dyspnoi and Laniatores (Macías-Ordóñez et al. 2010), but based on the scarce available data, it is not possible to say whether male and female keep genital contact for longer than needed for insemination. Category (2) does not apply to harvestmen, since there is no description of mating plug in the order. Nevertheless, mating pairs in Eupnoi and Laniatores may remain together after copulation, with males either grasping females with their front legs or guarding females at close proximity (Machado and Macías-Ordóñez 2007; Fig. 7.4b). In some species, males can also block the entrance of their nests, preventing females from leaving and other males from entering (Nazareth and Machado 2010). Therefore, we will explore categories (3) and (4), which are intersexual associations that may prevent females from remating and avoid sperm competition (i.e., mate guarding). These post-copulatory interactions may also prolong courtship signals that provide additional information to females (i.e., CFC) and protect female and/or offspring against natural enemies (i.e., direct benefits). All these possibilities are non-mutually exclusive, and we argue that researchers should consider multiple alternative predictions when designing experiments to understand the meaning of post-copulatory associations for both males and females (see below).

7.5.2 Mate Guarding

Prolonging association after sperm transference may benefit males by reducing the chances of the female accepting copulation with rival males, which would increase sperm competition due to the presence of additional ejaculates in the female's reproductive tract and potentially the removal of his sperm by female's subsequent mates (Alcock 1994; Simmons 2001). The key prediction of this hypothesis is that males exhibiting mate guarding should fertilize relatively more eggs than males that abandon their mates after copulation. To date, no experiment has been conducted with harvestmen to estimate males' fertilization success. Furthermore, molecular markers developed for a couple of species of Laniatores (AFLP: B.A. Buzatto, pers. comm.; microsatellites: G.S. Requena, unpub. data.) have shown surprisingly low levels of polymorphism within populations, preventing any reliable estimation of paternity. Therefore, advances in the techniques used to properly assign parentage are still needed for species of the order Opiliones.

A more complete analysis of the evolution and maintenance of mate guarding should consider the balance between male fitness benefits and costs associated with this behavior (Alcock 1994). Last male sperm precedence has been argued to be the main determinant of the benefits of mate guarding (Alcock 1994; Simmons 2001; Harts and Kokko 2013). The proportion of offspring sired by the last male to mate (usually referred to as P2) is a measure commonly used to infer fertilization mechanisms. Unfortunately, the same observed P2 pattern may be achieved by several alternative mechanisms biasing fertilization (see Simmons and Siva-Jothy 1998; Simmons 2001). For instance, Simmons (2001) lists six different processes that may increase P2, claiming that the underlying mechanisms are poorly understood and refined information about sperm dynamics during and after copulation is lacking for most species. Although direct quantification of P2 seems technically problematic in harvestmen, questions regarding mechanisms associated with the fertilization pattern of last male sperm precedence may be answered by characteristics of sperm biology. If post-copulatory association in harvestmen has evolved or has been maintained by sperm competition, the presence and the intensity of mate-guarding behavior should be related, for instance, to patterns of sperm longevity (e.g., Lessels and Birkhead 1990; Greeff and Parker 2000), sperm displacement (e.g., Parker and Simmons 1991; Parker et al. 2010), and sperm use and storage between reproductive events (Requena and Alonzo 2014).

Remaining with a female after sperm transference may also impose costs to males in terms of injury when repelling rival males or limited time available to invest in additional mates and territorial defense (Alcock 1994; Simmons 2001). Males of *L. vittatum* guarding females after copulation ignore other females and fight more vigorously than non-guarding males and commonly defeat approaching males (usually in less than 10 s), regardless of differences in body size or number of legs between contestants. Although the male never abandons his mate for additional mating opportunities with unattended females within his territory, the frequency of such additional encounters is low compared to additional mating opportunities after the mate-guarding period under natural conditions (Macías-Ordóñez 1997). Therefore, male mate guarding in *L. vittatum* may have been favored by a combination of low risks of injury (since agonistic interactions with conspecific males are quickly resolved), potentially low energetic costs of guarding (since males just follow females within their territories), and low reproductive costs due to small loss of mating opportunities.

In *S. proximum* (Laniatores), although territorial fights are common for the establishment of harems at the beginning of the breeding season, aggressive contests between males during the mate-guarding period have never been reported, and sneaker males quickly retreat if detected by territorial males (Buzatto et al. 2011). Therefore, the risk of injury of males during mate guarding in this species is probably very low as well. However, given that a territorial male can guard only one female at a time (Fig. 7.4b), as the number of females in his harem increases, it becomes more difficult to successfully defend all of them from sneakers. When the territorial male is copulating or mate guarding one female of his harem, sneakers can seize the opportunity to copulate with other unguarded females (Buzatto et al. 2011). Considering that social factors, such as adult sex ratio, density, and relative frequency of sneakers, show great variation across populations of *S. proximum* (Munguía-Steyer et al. 2012), the costs-to-benefits ratio of mate guarding should also vary in different populations. In this sense, the mate-guarding hypothesis provides additional predictions on the relationship between the intensity of mate guarding and ecological pressures on males regarding the costs and benefits of this behavior (Alcock 1994). As far as we know, those predictions have never been tested, but harvestmen are an ideal group of organisms to study interpopulation variations in male post-copulatory behavior.

One last possibility, particular to some species in which males establish their nests and take care of the offspring inside natural cavities, is that males may block nest entrance after copulation so that females are prevented from leaving. This behavior has been frequently observed in captivity conditions for *M. neptunus* (Laniatores) (Nazareth and Machado 2010), but it may also exist in *Gonyleptes saprophilus* and *Neosodocus* sp. (Laniatores) (Machado et al. 2004; Requena et al. 2013). While the behavior of blocking nest entrance can result in paternity protection, by preventing females to mate with additional males before oviposition, males can also prevent females from leaving their nest, prolonging their permanence inside and perhaps increasing the number of eggs they lay. This situation creates the possibility of sexual conflict, depending on how male behavior may affect female fitness. If attempts of nest takeovers by competitor males or nest abandonment by the ovipositing female are likely, successfully sustaining the blocking behavior could provide additional information to females about male quality. On the other hand, if remaining inside the nest for long periods limits female foraging activity and/or decreases the benefits derived from polyandry (Kvarnemo and Simmons 2013), blocking behavior could impose fitness costs to females. How sensitive those costs and benefits are to variation in ecological conditions is an open question that still deserves attention.

7.5.3 Cryptic Female Choice

An additional explanation for remaining with the female after copulation is to prolong behavioral interactions between the mating pair, so that females would be able to assess complementary male traits during post-copulatory courtship stimulation and "decide" about sperm use (Eberhard 1996; Fig. 7.2). The key prediction of this hypothesis is that male reproductive success should be directly related to his behavioral performance during the post-copulatory period. In several species of Eupnoi and Laniatores, males grasp or tap females' body and legs after intromission (Machado and Macías-Ordóñez 2007). For instance, *L. vittatum* males tap the female's dorsum while holding her within his territory (Macías-Ordóñez 1997). Evidence for several species of Goniosomatinae (Laniatores) exhibiting maternal care shows that territorial males remain with females until they complete oviposition, waving their second pair of legs and occasionally touching females' legs and dorsum (Gnaspini 1995; Machado and Oliveira 1998; Willemart and Gnaspini 2004; Buzatto and Machado 2008).

Despite the observational evidence of post-copulatory interactions between male and female in harvestmen, there is no study linking the post-copulatory performance of the males to the number of eggs sired by them. As discussed before, this gap in our knowledge may be explained in part by the challenge of estimating paternity in species of the order. Male post-copulatory performance, however, may also affect other components of his success that may provide circumstantial evidence of CFC. For instance, females may respond to intense male courtship behavior by reducing the interval between copulation and oviposition, increasing the number of eggs laid after the copulation, allocating additional resources into the eggs sired by the courting male, or simply staying for longer near that mating partner. Females of many harvestman species copulate and lay eggs in the laboratory, and it is relatively easy to manipulate the intensity of male courtship behavior by immobilizing their second pairs of legs and preventing them from touching their partners both during and after intromission. Therefore, these species are good models to investigate female behavioral response to varying intensities of male courtship both during and after intromission.

In species exhibiting exclusive paternal care, the post-copulatory association between the mated couple lasts until the end of oviposition and it is strictly necessary for males to provide care. Exclusive paternal care has independently evolved in at least nine lineages of Opiliones, all of them belonging to the suborder Laniatores. In all those lineages, females mature eggs continuously throughout the breeding season, and male care has been demonstrated to improve the survival of their clutches, which usually contain eggs laid by multiple females (Requena et al. 2013). The post-copulatory behavior of parental males varies across species, from sporadic touches on females' legs with the second pair of legs (Requena and Machado 2014) to active tapping on females' dorsum and venter with the first and second pair of legs (Mora 1990; Nazareth and Machado 2009, 2010). Moreover, females are also predicted to evaluate the prospective survival chances of the offspring to decide where to lay their own eggs (Sargent 1988; Hoelzer 1989; Tallamy 2001). For instance, the hygienic condition of a clutch of *Z. albomarginis* (Laniatores) may indicate the microclimatic quality of the oviposition site or the male ability to clean the eggs (Mora 1990). Additionally, ovipositing in clutches containing a large number of eggs may also benefit a female because her offspring would be protected by both the dilution effect during a predation attack (Sargent 1988) and the high quality of male protection (Hoelzer 1989; Tallamy 2001). Finally, when paternal activities are condition dependent, females should benefit by evaluating accurately and directly the current condition of guarding males (Requena et al. 2013). Therefore, depending on the context, characteristics of the broods themselves and/or directly of the males are expected to affect post-copulatory female decisions, which may be expressed not only in terms of biases in paternity, but also in the number of eggs laid or resource allocation.

7.5.4 Direct Benefits

The last explanation we explore for post-copulatory association in harvestmen considers the delivery of direct benefits from males to females. Under this perspective, by protecting females from predators or from the sexual harassment of rival conspecifics, a male would benefit from increasing the chances of his mate surviving to lay eggs sired by him (Alcock 1994). After copulation, females of *L. vittatum* may spend as much as 2 h in the male's territory inserting her ovipositor under the moss layer and inside rock crevices, probably selecting specific oviposition sites based on microclimatic conditions (Macías-Ordóñez 1997). Because territory owners maintain close contact with females while they are wandering in the territory and aggressively attack approaching males, female protection from sexual harassment during rock inspection and oviposition may confer direct benefits to the females (Machado and Macías-Ordóñez 2007). From the males' perspective, mate guarding may additionally protect their paternity by avoiding sperm competition, as discussed above.

Post-copulatory association between the mated couple may also improve males' reproductive success by directly increasing offspring survival due to helping females to provide parental care (Alcock 1994). Although there is no record of strict biparental care in any harvestmen, species with polygynous mating systems associated with maternal care can provide an equivalent scenario for male post-copulatory decisions. Territorial males of *S. proximum* patrol females that remain inside their harems after oviposition, in a period during which maternal care takes place (Buzatto and Machado 2008). When females abandon their eggs, either naturally or experimentally, some territorial males have been observed to care temporarily for unattended clutches within their harems (Buzatto and Machado 2009). Temporary male care in this species may be important for offspring survival (and hence female fitness) since egg predators usually consume entire clutches in just a few hours or days (Buzatto et al. 2007). This association, however, can be costly as it limits male patrolling behavior toward additional females in his harem and may compromise his ability to prevent sneaker males' invasion and copulation. As mentioned before, non-mutually exclusive mechanisms for post-copulatory association may be operating simultaneously and the balance between the costs and benefits to males of such association should be taken into account for a more complete appreciation of the processes driving male behavior evolution and maintenance.

7.6 Male Mate Choice

For a long time, the so-called Darwin–Bateman paradigm was the cornerstone of sexual selection theory, leading to three general predictions (Dewsbury 2005): (1) male reproductive success should vary more than that of females; (2) male reproductive success should be more influenced by the number of mating events than that of females; and (3) males should mate indiscriminately, while females should be discriminating. However, in the last two decades, it has become clear that sex roles are dynamic and variable, with an increasing number of empirical works reporting male choosiness and female–female competition (Bonduriansky 2001; Edward and Chapman 2011). Moreover, recent studies have built up the theoretical framework to investigate the evolution of male mate choice (e.g., Servedio and Lande 2006; Barry and Kokko 2010; Edward and Chapman 2011). In this sense, one essential condition favoring male choosiness is the cost associated with male reproductive decisions. For instance, relatively inexpensive mate search and assessment would allow males to sample and evaluate a great number of receptive females before effectively mating. On the other hand, great investment in parental activities or nuptial gifts by males is likely to energetically constrain their allocation to mate with many females and/or to produce sufficient ejaculate, which is predicted to favor male discriminatory behavior. Other condition favoring the evolution of male mate choice is related to the benefits of choosiness. Great variation in viability or fecundity among females is predicted to increase the fitness advantages of male selectiveness in terms of the potential quality and quantity of offspring to be sired. The balance between the costs and the benefits of male mate choice should determine the net advantage of performing such behavior.

Although there are several nuptial gift-giving species in harvestmen (see Precopulatory interactions above), there is no reported case of sex role reversal, as those described for some orthopterans (Vahed 1998; Gwynne 2008). However, at least, some harvestman species exhibiting exclusive paternal care show evidence of male mate choice and we will explore these cases in more detail in the following topics. A comprehensive review of sex role reversal in harvestmen and other arthropods exhibiting exclusive paternal care is presented in Requena et al. (2013).

7.6.1 Males Repelling Females

Females of *Iporangaia pustulosa* (Laniatores) lay eggs on the vegetation, and males take care of the offspring during the entire period of embryonic development (Machado et al. 2004). Paternal care, however, does not seem to constrain mate acquisition: caring males can mate with as much as 15 females and take care of all of their eggs simultaneously in multiple clutches (Requena et al. 2012). Mate search is exclusively accomplished by females, who may visit several males over the course of the breeding season. Field observations show that caring males aggressively repel some females upon arrival, even before copulation (Requena and Machado 2014). One possible explanation for this behavior is that the high mating frequency of some males and the low food intake during the caring period (Requena et al. 2012) may jointly compromise sperm production and replenishment (Requena and Machado 2014), negatively affecting future mating opportunities (e.g., Härdling et al. 2008). Under this circumstance, males would benefit from evaluating the quality of the arriving females, which can be accomplished using close range volatile chemicals or contact hydrocarbons during the brief pre-copulatory interactions (Requena and Machado 2014). Evaluating variation in traits signaling female fertility and sperm competition risk between rejected and accepted females, as well as sperm load in the seminal vesicle of males that accept and reject visiting females, is crucial to understand male selectivity in *I. pustulosa*.

Females of *Z. albomarginis* and *M. neptunus* (Laniatores) exclusively lay eggs inside nests that consist of cuplike structures built by males in the first case (Mora 1990) and natural cavities on roadside banks in the latter case (Nazareth and Machado 2010). Males of both species protect their nests against the invasion of conspecific males as well as egg predators, and behavioral observations show that they also attack some visiting females, biting their legs and chasing them out of the nest (Mora 1990; Nazareth and Machado 2010). Contrary to the pattern described for *I. pustulosa*, attacks in both species usually take place after males and females interact or even copulate. Additionally, territorial males of *Z. albomarginis* and *M. neptunus* use their second pair of legs to touch the venter of visiting females during such interactions (Nazareth and Machado 2010; G.S. Requena, unpub. data). Given that oviposition in Laniatores usually happens immediately after copulation (see Post-copulatory interactions above), males may touch females' venter to assess whether they everted the ovipositor and started oviposition. Considering that cannibalistic activities from conspecific females constitute an important source of egg mortality in harvestmen with paternal care (Mora 1990; Requena et al. 2009; Nazareth and Machado 2010), any delay in everting the ovipositor may indicate females' predatory intentions. In fact, cannibalistic females of *M. neptunus* first copulate with the guarding males and then attempt to eat some of the eggs while they are being courted by the guarding male (Nazareth and Machado 2010). Therefore, male attack toward females may represent a parental protective behavior, and manipulations of female feeding condition should help elucidating whether male aggression is preferentially directed to cannibalistic females.

7.6.2 Female Courtship and Aggressive Behaviors

Although *L. vittatum* (Eupnoi) shows conventional sex roles, females seem to court males during grasping by rubbing, pinching, and grasping the male genital opening and the genitalia using their chelicerae and pedipalps, prior to and during penis eversion and nuptial feeding. Furthermore, some sexual interactions may involve grasping, but not intromission, even when females do not seem reluctant to mate (R. Macías-Ordóñez, unpub. data). Males in this, and probably other nuptial gift-giving harvestman species, may have a choice of whether to offer glandular secretions and proceed with the copulatory sequence or not. Males may also modulate how much secretion they offer, if they are able to assess female quality. If so, some form of female courtship may be expected, but it may be really subtle and very detailed observations of the sexual interaction must be carried out to discover it. This scenario would also explain the possibility of cheater males mimicking female stimulation in order to obtain nuptial gifts as described above (see Pre-copulatory interactions).

In *Z. albomarginis*, nesting males remain most of the time inside their territories and females actively visit nests over the course of the breeding season. After approaching a nest, a female enters and taps the walls and the floor of the nest with the first pair of legs, and either abandons the nest or initiates courtship. In this second scenario, she taps the partner using her second pair of legs, moves around, and faces the male until he starts tapping her with his first legs, or biting and chasing her off the nest (Mora 1990). It seems that mate interactions in this species allow not only females to evaluate nests and males' qualities (see Post-copulatory interactions above), but also males to assess potential mates (Mora 1990). Although females usually abandon the clutch after oviposition (Rodríguez and Guerrero 1976), they sometimes spend several days in the vicinity of one or two nests, where they may engage into aggressive behaviors toward newcomer females that approach one of these nests (Mora 1990). Given that conspecific females are the main egg predators in *Z. albomarginis* (Mora 1990), the risk of cannibalism may also explain female–female aggression. Therefore, the partial sex role reversal of *Z. albomarginis* offers a remarkable opportunity to investigate not only the criteria used during mutual mate choice, but also the conditions favoring female strategies of male monopolization, a mating strategy extremely rare in nature.

7.7 Concluding Remarks

Several traits in harvestmen make them unique in the context of most previous studies of sexual selection. The lack of long-range perception mechanisms reduces the window of opportunity for males and females to exchange information during the very short period of close proximity or physical contact prior to intromission (Fig. 7.2). In some cases, however, acceptance or rejection of a mate may be based on information gathered even before contact. Although empirical data are scarce, we argue that this would be the case in harvestman species where: (a) individuals of one sex leave chemical cues on the substrate that may be used by individuals of the other sex to evaluate mate quality; (b) males defend specific oviposition sites, and females' mating decisions are based either on the direct assessment of territory quality or on the passive mate choice; or (c) males suffer from sperm depletion, and the acceptance or rejection of a female is primarily based on the amount of ejaculate males have in their seminal vesicles. Regardless of the role of pre-copulatory interactions, actual fertilization success is likely to be strongly dependent on the outcome of copulatory and post-copulatory processes. In this sense, the fact that most species are highly promiscuous and have sperm cells that lack flagella, which are stored near to the tip of the ovipositor and used to fertilize eggs immediately before oviposition (Macías-Ordóñez et al. 2010), renders Opiliones a fertile ground to study the role of CFC, sperm competition, and sexual conflict.

In the context of CFC, it is important to define when hypotheses are made on the original (ancestral) function of a trait involved in male–female interactions or on its current function, which may not only differ, but to some degree are expected to change over time along the evolutionary history of a lineage (Eberhard 2004). The use of phylogenies to infer evolutionary changes in the function of traits under sexual selection is a powerful tool that has only recently been applied in Opiliones. For instance, the study by Burns et al. (2013) with a clade of Nearctic Eupnoi clearly shows that, in species in which males deliver a nuptial gift to the females before intromission, sexual dimorphism in the pedipalps is absent and females lack physical barriers in their genital operculum that seem to interfere with intromission in other species. On the other hand, in species in which males lack gift-giving glands, sexual dimorphism in the length of the pedipalps is well marked and females have structures that seem to function as barriers against intromission. The latter scenario constitutes the derived state and has evolved independently at least four times in the clade. Based on these results, the authors argue that female choice is the ancestral state and sexual conflict is a repeatedly evolved derived state. Furthermore, they hypothesize that major ecological factors, such as the length of the reproductive season, may trigger the change of state through three non-mutually exclusive mechanisms: (1) natural selection for male economy in nuptial gift production in extreme latitudes with shorter reproductive seasons, (2) a change of target trait of female choice from production of nuptial gifts to mechanical or stimulatory attributes of potential mates, and (3) stronger intrasexual selection among males under shorter mating seasons would result in selection for behavioral (e.g., mate guarding) or morphological (e.g., size or weapons) traits that could also be used on female coercion.

In conclusion, a male sexual trait may originate in a sexual conflict context, promoting evolutionary response of a wide array of female sexual traits and falling later in a female-screening evolutionary process (e.g., Macías-García and Ramirez 2005). Conversely, as suggested by Burns et al. (2013), a male sexual trait may have originally had a female courting function and then switched to a coercive one. Testing for these possibilities is a highly attractive research line, which requires a robust phylogenetic hypothesis on the study group, detailed studies on form and function, as well as detailed data on behavior in wild populations. Besides the uniqueness of Opiliones in the terms stated above, all these conditions have been developing in recent years, transforming the order in one of the most exciting and promising ones in terms of research opportunities in reproductive evolutionary ecology.

Acknowledgments We thank Alfredo Peretti and Anita Aisenberg for the invitation to write this chapter, and Bruno A. Buzatto, Rodrigo H. Willemart, and Abel Pérez González for comments on an early version of the manuscript. We are also grateful to all colleagues that kindly provided photographs to the plates we present here, namely Jerry Armstrong (Fig. 7.3a), Bruno A. Buzatto (Fig. 7.4b), Gonzalo Giribet (Fig. 7.1d), Marshal Hedin (Fig. 7.1b), Michael Marlow (Fig. 7.3b), Daniel Proud (Fig. 7.1c), Joseph G. Warfel (Fig. 7.3c), and Melvyn Yeo (Fig. 7.4a). GM is funded by a research grant from National Counsel of Technological and Scientific Development (CNPq) and by a project grant from São Paulo Research Foundation (FAPESP: 2012/50229-1); GSR is supported by funds from Yale University and Coordenação de Aperfeiçoamento de Pessoal de Nível Superior (Capes) under a "Science without Borders" postdoctoral fellowship; RMO is supported by INECOL.

References

Acosta L, Machado G (2007) Diet and foraging. In: Pinto-da-Rocha R, Machado G, Giribet G (eds) Harvestmen: the biology of Opiliones. Harvard University Press, Cambridge, pp 309–338

Alcock J (1994) Postinsemination associations between males and females in insects: the mate-guarding hypothesis. Ann Rev Entomol 39:1–21

Andersson M (1994) Sexual selection. Princeton University Press, Princeton, NJ

Arnqvist G, Nilsson T (2000) The evolution of polyandry: multiple mating and female fitness in insects. Anim Behav 60:145–164

Barry KL, Kokko H (2010) Male mate choice: why sequential choice can make its evolution difficult. Anim Behav 80:163–169

Bonduriansky R (2001) The evolution of male mate choice in insects: a synthesis of ideas and evidence. Biol Rev 76:305–339

Borgia G (1979) Sexual selection and the evolution of mating systems. In: Blum MS, Blum NA (eds) Sexual selection and reproductive competition in insects. Academic Press, New York, pp 19–80

Burns MM, Hedin M, Shultz JW (2013) Comparative analyses of reproductive structures in harvestmen (Opiliones) reveal multiple transitions from courtship to precopulatory antagonism. PLoS ONE 8:e66767

Buzatto BA, Machado G (2008) Resource defense polygyny shifts to female defense polygyny over the course of the reproductive season of a Neotropical harvestman. Behav Ecol Sociobiol 63:85–94

Buzatto BA, Machado G (2009) Amphisexual care in *Acutisoma proximum* (Arachnida, Opiliones), a Neotropical harvestman with exclusive maternal care. Insectes Soc 56:106–108

Buzatto BA, Machado G (2014) Male dimorphism and alternative reproductive tactics in harvestmen (Arachnida: Opiliones). Behav Proc 109:2–13

Buzatto BA, Requena GS, Martins EG, Machado G (2007) Effects of maternal care on the lifetime reproductive success of females in a Neotropical harvestman. J Anim Ecol 76:937–945

Buzatto BA, Requena GS, Lourenço RS, Machado G (2011) Conditional male dimorphism and alternative reproductive tactics in a Neotropical arachnid (Opiliones). Evol Ecol 25:331–349

Buzatto BA, Macías-Ordóñez R, Machado G (2013) Macroecology of harvestman mating systems. In: Macedo RH, Machado G (eds) Sexual selection: perspectives and models from the neotropics. Elsevier, Amsterdam, pp 115–162

Buzatto BA, Tomkins JL, Simmons LW, Machado G (2014) Correlated evolution of sexual dimorphism and male dimorphism in a clade of Neotropical harvestmen. Evolution 68:1671–1686

Cloudsley-Thompson JL (1948) Notes on arachnida. 4. Courtship behavior of the harvestmen *Mitopus morio*. Anm Mag Nat Hist 11:809–810

Cokendolpher JC, Tsurusaki N, Tourinho AL, Taylor CK, Gruber J, Pinto-da-Rocha R (2007) Taxonomy. Eupnoi. In: Pinto-da-Rocha R, Machado G, Giribet G (eds) Harvestmen: the biology of Opiliones. Harvard University Press, Cambridge, pp 108–131

Curtis DJ (1970) Comparative aspects of the fine structure of the eyes of Phalangida (Arachnida) and certain correlations with habitat. J Zool 160:231–265

Dewsbury DA (2005) The Darwin-Bateman paradigm in historical context. Integr Comp Biol 45:831–837

Dugatkin LA (1992) Sexual selection and imitation: females copy the mate choice of others. Am Nat 139:1384–1389

Dunlop JA (2007) Paleontology. In: Pinto-da-Rocha R, Machado G, Giribet G (eds) Harvestmen: the biology of Opiliones. Harvard University Press, Cambridge, pp 247–265

Eberhard WG (1996) Female control: sexual selection by cryptic female choice. Princeton University Press, Princeton

Eberhard WG (2004) Rapid divergent evolution of sexual morphology: comparative tests of antagonistic coevolution and traditional female choice. Evolution 58:1947–1970

Edgar AL (1963) Proprioception in the legs of phalangids. Biol Bull 124:262–267

Edgar AL (1971) Studies on the biology and ecology of Michigan Phalangida (Opiliones). Misc Pub Mus Zool, Univ Mich 144:1–64
Edward DA, Chapman T (2011) The evolution and significance of male mate choice. Trends Ecol Evol 26:647–654
Emlen ST, Oring LW (1977) Ecology, sexual selection, and evolution of mating systems. Science 197:215–223
Fernandes N, Willemart RH (2014) Neotropical harvestmen (Arachnida, Opiliones) use sexually dimorphic glands to spread chemicals in the environment. C R Biol 337:69–77
Fowler-Finn KD, Triana E, Miller OG (2014) Mating in the harvestman *Leiobunum vittatum* (Arachnida: Opiliones): from premating struggles to solicitous tactile engagement. Behaviour 151(12–13):1663–1686
Garwood RJ, Sharma PP, Dunlop JA, Giribet G (2014) A paleozoic stem group to mite harvestmen revealed through integration of phylogenetics and development. Curr Biol 24:1–7
Giribet G (2007) Taxonomy: Cyphophthalmi. In: Pinto-da-Rocha R, Machado G, Giribet G (eds) Harvestmen: the biology of Opiliones. Harvard University Press, Cambridge, pp 89–108
Giribet G, Kury AB (2007) Phylogeny and biogeography. In: Pinto-da-Rocha R, Machado G, Giribet G (eds) Harvestmen: the biology of Opiliones. Harvard University Press, Cambridge, pp 62–87
Gnaspini P (1995) Reproduction and postembryonic development of *Goniosoma spelaeum*, a cavernicolous harvestman from southeastern Brazil (Arachnida: Opiliones: Gonyleptidae). Invertebr Reprod Dev 28:137–151
Greeff JM, Parker GA (2000) Spermicide by females; what should males do? Proc Roy Soc Lond B 267:1759–1763
Gruber G (2007) Taxonomy: Dyspnoi. In: Pinto-da-Rocha R, Machado G, Giribet G (eds) Harvestmen: the biology of Opiliones. Harvard University Press, Cambridge, pp 131–159
Gwynne DT (2008) Sexual conflict over nuptial gifts in insects. Annu Rev Entomol 53:83–101
Härdling R, Gosden T, Aguilée R (2008) Male mating constraints affect mutual mate choice: prudent male courting and sperm-limited females. Am Nat 172:259–271
Harts AM, Kokko H (2013) Understanding promiscuity: when is seeking additional mates better than guarding an already found one? Evolution 67:2838–2848
Hoelzer GA (1989) The good parent process of sexual selection. Anim Behav 38:1067–1078
Immel V (1954) Zur Biologie und Physiologie von *Nemastoma quadripunctatum* (Opiliones, Dyspnoi). Zool Jb Abt Syst 83:475–484
Karaman IM (2005) Evidence of spermatophores in Cyphophthalmi (Arachnida, Opiliones). Rev Suisse Zool 112:3–11
Kotiaho JS, Puurtinen M (2007) Mate choice for indirect genetic benefits: scrutiny of the current paradigm. Func Ecol 21:638–644
Kury AB (2007) Taxonomy: Laniatores. In: Pinto-da-Rocha R, Machado G, Giribet G (eds) Harvestmen: the biology of Opiliones. Harvard University Press, Cambridge, pp 159–246
Kury AB (2012) A synopsis of catalogs and checklists of harvestmen (Arachnida, Opiliones). Zootaxa 3184:35–58
Kvarnemo C, Simmons LW (2013) Polyandry as a mediator of sexual selection before and after mating. Phil Trans R Soc B 368:20120042
LeBas NR, Hockham LR (2005) An invasion of cheats: the evolution of worthless nuptial gifts. Current Biol 15:64–67
Lessells CM, Birkhead TR (1990) Mechanisms of sperm competition in birds: mathematical models. Behav Ecol Sociobiol 27:325–337
Machado G, Macías-Ordóñez R (2007) Reproduction. In: Pinto-da-Rocha R, Machado G, Giribet G (eds) Harvestmen: the biology of Opiliones. Harvard University Press, Cambridge, pp 414–454
Machado G, Oliveira PS (1998) Reproductive biology of the neotropical harvestman (Goniosoma longipes) (Arachnida, Opiliones: Gonyleptidae): mating and oviposition behaviour, brood mortality and parental care. J Zool 246:359–367

Machado G, Requena GS, Buzatto BA, Osses F, Rossetto LM (2004) Five new cases of paternal care in harvestmen (Arachnida: Opiliones): implications for the evolution of male guarding in the Neotropical family Gonyleptidae. Sociobiol 44:577–598
Machado G, Pinto-da-Rocha R, Giribet G (2007) What are harvestmen? In: Pinto da Rocha R, Machado G, Giribet G (eds) Harvestmen: the biology of Opiliones. Harvard University Press, Cambridge, pp 1–13
Macías-García C, Ramírez E (2005) Evidence that sensory traps can evolve into honest signals. Nature 434:501–505
Macías-Ordóñez R (1997) The mating system of *Leiobunum vittatum* Say 1821 (Arachnida: Opiliones: Palpatores): resource defense polygyny in the striped harvestman. PhD thesis, Lehigh University, USA
Macías-Ordóñez R (2000) Touchy harvestmen. Nat Hist 109:58–67
Macías-Ordóñez R, Machado G, Pérez-González A, Shultz JW (2010) Genitalic evolution in Opiliones. In: Leonard J, Córdoba-Aguilar A (eds) The evolution of primary sexual characters in animals. Oxford University Press, New York, pp 285–306
Martens J (1969) Die Sekretdarbietung während des Paarungsverhaltens von Ischyropsalis C. L. Koch (Opiliones). Z Tierpsychol 26:513–523
Martens J (1976) Genitalmorphologie, System und Phylogenie der Weberknechte (Arachnida: Opiliones). Entomol German 3:51–68
Martens J (1986) Die Grossgliederung der Opiliones und die Evolution der Ordnung (Arachnida). In: Barrientos A (ed) Proceedings of the 10th international congress of arachnology. Instituto Pirenaico de Ecología and Grupo de Aracnología, Barcelona, pp 289–310
Meijer J (1972) Some data on the phenology and the activity patterns of *Nemastoma lugubre* (Müller) and *Mitostoma chrysomelas* (Herman) (Nemastomatidae: Opilionida: Arachnida). Neth J Zool 22:105–118
Mora G (1990) Parental care in a Neotropical harvestman, *Zygopachylus albomarginis* (Arachnida: Gonyleptidae). Anim Behav 39:582–593
Munguía-Steyer R, Buzatto BA, Machado G (2012) Male dimorphism of a neotropical arachnid: harem size, sneaker opportunities, and gonadal investment. Behav Ecol 23:827–835
Muniz GM, Guimarães Jr PR, Buzatto BA, Machado G (2014) A sexual network approach to sperm competition in a species with alternative mating tactics. Behav Ecol. doi:10.1093/beheco/aru166
Nazareth TM, Machado G (2009) Reproductive behavior of *Chavesincola inexpectabilis* (Opiliones, Gonyleptidae) with description of a new and independently evolved case of paternal care in harvestmen. J Arachnol 37:127–134
Nazareth TM, Machado G (2010) Mating system and exclusive postzygotic paternal care in a Neotropical harvestman (Arachnida: Opiliones). Anim Behav 79:547–554
Pabst W (1953) Zur Biologie der mitteleuropäischen Troguliden. Zool Jb Abt Syst Ökol. u. Geog Tiere 82:1–156
Parisot C (1962) Étude de quelques opilions de Lorraine. Vie et Millieu 13:179–197
Parker GA, Simmons LW (1991) A model of constant random sperm displacement during mating: evidence from Scatophaga. Proc Roy Soc Lond B 246:107–115
Parker GA, Immler S, Pitnick S, Birkhead TR (2010) Sperm competition games: sperm size (mass) and number under raffle and displacement, and the evolution of P2. J Theor Biol 264:1003–1023
Price T, Schluter D, Heckman NE (1993) Sexual selection when the female directly benefits. Biol J Linn Soc 48:187–211
Proctor HC (1998) Indirect sperm transfer in arthropods: behavioral and evolutionary trends. Ann Rev of Entomol 43:153–174
Requena GS, Alonzo SH (2014) Female sperm use and storage between fertilization events drive sperm competition and male ejaculate allocation. Evolution 68(12):3433–3444
Requena GS, Machado G (2014) Mating behavior of a Neotropical arachnid with exclusive paternal care. Acta Ethol 17:23–30
Requena GS, Buzatto BA, Munguía-Steyer R, Machado G (2009) Efficiency of uniparental male and female care against egg predators in two closely related syntopic harvestmen. Anim Behav 78:1169–1176

Requena GS, Buzatto BA, Martins EG, Machado G (2012) Paternal care decreases foraging activity and body condition, but does not impose survival costs to caring males in a Neotropical arachnid. PLoS ONE 7:e46701

Requena GS, Munguía-Steyer R, Machado G (2013) Paternal care and sexual selection in arthropods. In: Macedo RH, Machado G (eds) Sexual selection: perspectives and models from the Neotropics. Elsevier, Amsterdam, pp 201–234

Rodríguez CA, Guerrero S (1976) La historia natural y el comportamiento de *Zygopachylus albomarginis* (Chamberlin) (Arachnida, Opiliones: Gonyleptidae). Biotropica 8:242–247

Roters M (1944) Observations on British harvestmen. J Quekett Microsc Club Ser 4(2):23–25

Sargent RC (1988) Paternal care and egg survival both increase with clutch size in the fathead minnow, *Pimephales promelas*. Behav Ecol Sociobiol 23:33–37

Schwendinger PJ, Giribet G (2005) The systematics of the south-east Asian genus Fangensis Rambla (Opiliones: Cyphophthalmi: Stylocellidae). Invertebr Syst 19:297–323

Servedio MR, Lande R (2006) Population genetic models of male and mutual mate choice. Evolution 60:674–685

Shultz JW, Pinto-da-Rocha R (2007) Morphology and functional anatomy. In: Pinto-da-Rocha R, Machado G, Giribet G (eds) Harvestmen: the biology of Opiliones. Harvard University Press, Cambridge, pp 14–61

Simmons LW (2001) Sperm competition and its evolutionary consequences in the insects. Princeton University Press, Princeton

Simmons LW, Siva-Jothy MT (1998) Sperm competition in insects: mechanisms and the potential for selection. In: Birkhead TR, Møller AP (eds) Sperm competition and sexual selection. Academic Press, New York, pp 341–434

Stanley E, Toscano-Gadea CA (2011) Comportamiento sexual en *Discocyrtus prospicuus* (Opiliones, Gonyleptidae): cuanto mas conocemos mas nos preguntamos. In: III Congreso Latinoamericano de Aracnología, Quindío, Colombia, p 176

Tallamy DW (2000) Sexual selection and evolution of exclusive paternal care in arthropods. Anim Behav 60:559–567

Tallamy DW (2001) Evolution of exclusive paternal care in arthropods. Ann Rev Entomol 46:139–165

Thomas RH, Zeh DW (1984) Sperm transfer and utilization strategies in arachnids: ecological and morphological constraints. In: Smith RL (ed) Sperm competition and evolution of animal mating systems. Academic, Orlando, pp 180–221

Trumbo ST (1996) Parental care in invertebrates. Adv Stud Behav 25:3–52

Tsurusaki N (2004) Are parthenogenetic females willing to mate with males? Implications of exaggerated male palpi in two thelytokous harvestmen. In: 16th International congress of arachnology. Ghent, Belgium, p 163

Vahed K (1998) The function of nuptial feeding in insects: a review of empirical studies. Biol Rev 73:43–78

van der Hammen L (1985) Comparative studies in Chelicerata. III. Opilionida. Zool Verhand 220:1–60

Wagner WE Jr (2011) Direct benefits and the evolution of female mating preferences: conceptual problems, potential solutions, and a field cricket. Adv Stud Behav 43:273–319

Wijnhoven H (2011) Notes on the biology of the unidentified invasive harvestman *Leiobunum* sp. (Arachnida: Opiliones). Arachnol Mitt 41:17–30

Wijnhoven H (2013) Sensory structures and sexual dimorphism in the harvestman *Dicranopalpus ramosus* (Arachnida: Opiliones). Arachnol Mitt 46:27–46

Wiley RH, Poston J (1996) Indirect mate choice, competition for mates, and coevolution of the sexes. Evolution 50:1371–1381

Willemart RH, Chelini MC (2007) Experimental demonstration of close-range olfaction and contact in the Brazilian harvestman, Iporangaia pustulosa. Ent Exp Appl 123:73–79

Willemart RH, Gnaspini P (2004) Spatial distribution, mobility, gregariousness, and defensive behaviour in a Brazilian cave harvestman *Goniosoma albiscriptum* (Arachnida, Opiliones, Gonyleptidae). Anim Biol 54:221–235

Willemart RH, Farine JP, Peretti AV, Gnaspini P (2006) Behavioral roles of the sexually dimorphic structures in the male harvestman *Phalangium opilio* (Opiliones, Phalangiidae). Can J Zool 84:1763–1774

Willemart RH, Farine JP, Gnaspini P (2009) Sensory biology of Phalangida harvestmen (Arachnida, Opiliones): a review, with new morphological data on 18 species. Acta Zool 90:209–227

Zatz C, Werneck RM, Macías-Ordóñez R, Machado G (2011) Alternative mating tactics in dimorphic males of the harvestman *Longiperna concolor* (Arachnida: Opiliones). Behav Ecol Sociobiol 65:995–1005

Chapter 8
Cryptic Female Choice in Crustaceans

Stefan Dennenmoser and Martin Thiel

Abstract Cryptic female choice may be common among crustaceans, but few studies have thoroughly examined it in this diverse taxonomic group. Herein, we summarize current genetic evidence for multiple paternity and skewed offspring ratios in crustaceans, and document observations that could suggest cryptic female choice. Behaviors indicative of cryptic female choice have been reported from numerous crustacean taxa (e.g., crayfish, hermit crabs, isopods), showcasing a diverse array of behavioral mechanisms such as failed copulations, spermatophore removal, selective sperm passage, chemical signaling, adjusting duration of receptivity, delayed copulation, or discriminative reproductive investment in favor of preferred males. We highlight a few case studies, in which a suite of different cryptic behaviors permits females to maintain control over fertilizations. The possibility of selective sperm–egg interactions is briefly discussed, and parallels to other aquatic invertebrates are drawn revealing similar cryptic choice mechanisms. The disparity of body forms and reproductive strategies found in crustaceans and the fact that they inhabit many different habitats with variable selective environments makes them an ideal model taxon for future studies on cryptic female choice.

S. Dennenmoser (✉)
Max-Planck Institute for Evolutionary Biology, August Thienemann Strasse 2, 24306 Plön, Germany
e-mail: dennenmoser@evolbio.mpg.de

M. Thiel
Facultad Ciencias del Mar, Universidad Católica del Norte, Larrondo 1281, Coquimbo, Chile
e-mail: thiel@ucn.cl

M. Thiel
MN Ecology and Sustainable Management of Oceanic Island (ESMOI), Coquimbo, Chile

M. Thiel
Centro de Estudios Avanzados en Zonas Áridas (CEAZA), Coquimbo, Chile

A.V. Peretti and A. Aisenberg (eds.), *Cryptic Female Choice in Arthropods*,
DOI 10.1007/978-3-319-17894-3_8

8.1 Introduction

Cryptic female choice, which can be broadly defined as female-controlled mechanisms during and following copulation allowing females to bias fertilization success in favor of males with certain traits (Eberhard 1996), has long been neglected as a possible driver of sexual selection in crustaceans. Although the last decade has seen a recurrent interest in female control of the insemination process (e.g., Thiel and Hinojosa 2003; Galeotti et al. 2006; Aquiloni and Gherardi 2008), cryptic female choice remains challenging to study because the underlying mechanisms are usually subtle, difficult to observe, and closely interact with sperm competition and male-driven processes.

Both cryptic female choice and sperm competition are widely recognized to play crucial roles in postcopulatory sexual selection, evolutionary diversification of reproductive traits, and the evolution of polyandry (Birkhead and Pizzari 2002). However, the effects of postcopulatory selection on female traits have seldom been addressed, and no coherent framework exists that incorporates cryptic female choice to combine both pre- and postcopulatory mechanisms in sexual selection theory (Kvarnemo and Simmons 2013; Lüpold et al. 2014). It has been repeatedly suggested that cryptic female choice is favored when precopulatory choice is costly and that postcopulatory mechanisms are better suited to avoid genetic incompatibilities (e.g., Eberhard 1998; Jennions and Petrie 2000; Birkhead and Pizzari 2002; Slatyer et al. 2011). As a result of the increasing availability of genetic markers, the occurrence of multiple paternity has been confirmed for several taxa, and tests of genetic compatibilities underlying biased fertilization success are becoming more common (e.g., Simmons et al. 2006; Løvlie et al. 2013).

Among the least understood mechanisms underlying cryptic female choice are sperm–egg interactions, which may promote selective fertilization success (e.g., Zeh and Zeh 1997; Simmons 2005; Evans and Sherman 2013) and could play an important role in biasing paternity, particularly in crustacean species with external fertilization in which females lack internal sperm storage structures. While some cryptic female choice mechanisms can be directly observed, including sperm dumping, interruption of copulation, differential abortion or changes in oviposition rates (Eberhard 2000), a close inspection of processes occurring in the female reproductive tract and at the egg–sperm interface may be necessary to unveil the numerous features suited to cause cryptic female choice, and to demonstrate the outcome of paternity biases in favor of males bearing preferred traits (Eberhard 1996).

Herein, we suggest that crustaceans represent a diverse group of promising study organisms to address how cryptic female choice affects postcopulatory sexual selection. Multiple mating is common across species with a variety of mating systems, which may allow investigation of how pre- and postcopulatory mate choice processes interact while promoting sexual selection. The existence of both external and internal fertilization systems could help to disentangle male- and female-mediated effects on postcopulatory selection, as has been suggested for hermit crabs (Contreras-Garduño and Córdoba-Aguilar 2006). In most species, females are receptive for a limited but variable time (minutes to days) after molting, and species with internal fertilization differ considerably in mechanisms and duration of sperm storage. The resulting variation in the strength of pre- and postcopulatory sexual selection likely creates numerous opportunities and limitations for cryptic female choice to evolve.

Crustaceans have already served as valuable study organisms for investigating patterns of sperm competition (Duffy and Thiel 2007). Detailed descriptions of female reproductive tracts (e.g., Becker et al. 2011) and egg–sperm interactions (e.g., Dupré et al. 2012) in some species may facilitate the detection of female control mechanisms that allow for differential sperm use. Finally, many crustacean species are excellent model organisms for sexual selection studies because they are easy to manipulate under both natural and laboratory conditions (aided by small body sizes and short generation times). They often develop conspicuous sexual dimorphisms indicating a strong role of sexual selection, and genetic resources are becoming increasingly available to investigate patterns of biased paternity or the role of genetic compatibility in fertilization success. Herein, we aim to highlight indicators for cryptic female choice in crustaceans, summarize molecular evidence for multiple paternity, and exemplify the importance of manipulative female behaviors during and after copulation in selected case studies. We hope this brief review will contribute to postcopulatory sexual selection theory by identifying gaps in our present knowledge, as well as future research avenues for cryptic female choice in crustaceans.

8.2 A Brief Introduction to Crustacean Mating Patterns

8.2.1 Mating Systems

Mating patterns in crustaceans are extraordinarily diverse, including hermaphroditism, monogamy, polygamy, or the queen-centered mating system of eusocial snapping shrimps (Subramoniam 2013). Polyandry (females mating with multiple males) appears to be particularly widespread in decapods (e.g., crayfish, lobsters, crabs, shrimps), but is also known from other groups such as isopods, amphipods, and some copepods (e.g., Johnson 1982; Moreau et al. 2002; Todd et al. 2005).

Consequences of polyandry have been studied mainly from the perspective of males, which may respond to sperm competition for example by monitoring female receptivity, pre- and postcopulatory mate guarding, sealing off the female's reproductive tract with sperm plugs, diluting sperm from competitors with increased sperm supplies, or partially removing the sperm from previous males (Duffy and Thiel 2007; Galeotti et al. 2008). The often-conspicuous mating behavior of males has traditionally been used to categorize mating systems, which differ in the extent of male–male contest, monopolization of females and resources, and mate searching behavior (Christy 1987; Bauer 1996; Shuster 2007;

Baeza and Thiel 2007; Asakura 2009). Accordingly, mating systems in caridean shrimps have been classified as "monogamy," "pure searching," "neighborhoods of dominance," and "search and attend" (Correa and Thiel 2003), and similar classifications have been applied to other crustacean groups (Duffy and Thiel 2007).

Reproduction in crustaceans is synchronized with the molt cycle in many crustacean taxa, because females of such groups are receptive only for a limited time after molting until the shell hardens (e.g., Hartnoll 1969; Raviv et al. 2008). However, timing and number of reproductive cycles vary, and some species are capable to mate during the intermolt interval (Hartnoll 2000; Raviv et al. 2008; Espinoza-Fuenzalida et al. 2012). Females approaching their reproductive molt are often guarded by males (precopulatory mate guarding), which is commonly observed in decapods, amphipods, and isopods and can continue for prolonged times following copulation (postcopulatory mate guarding) (Duffy and Thiel 2007).

8.2.2 Reproductive Biology

In decapod crabs, spermatozoa are non-flagellate and non-motile and usually densely packed into spermatophores, which are highly diverse in shape and often transferred together with some amount of seminal fluids (Krol et al. 1992; Subramoniam 1993). Seminal fluids are usually assumed to provide nutrients to spermatozoa (likely for supporting long-term storage), but have also been proposed to play a role in sperm plug formation, displacement of sperm from previous males, or antimicrobial protection (e.g., Subramoniam 1993; Moyano et al. 2010). A special case of potential nutrition provisioning to the zygote has been suggested for the giant sperm of ostracods (Cypridoidea), which can make up 0.4–1.3 % of the egg volume and even supply paternal mitochondrial derivates (Matzke-Karasz 2005). In species with external fertilization (e.g., caridean shrimps, lobsters, hermit crabs), males attach their spermatophores beneath the female's posterior–ventral thorax or anterior abdomen, and females initiate spawning soon after sperm transfer. In amphipods, males transfer sperm into the female's marsupium ("brood pouch"), and females release their eggs shortly after (e.g., Clark and Caudill 2001). In species with internal fertilization (e.g., brachyuran crabs, isopods) sperm may be used right after copulation, or stored in sperm storage organs for some time (weeks to months) until the next oviposition or across multiple reproductive cycles. Female sperm storage can be located either in internal structures (spermatheca) suitable for long-time storage, or at external pocket-like structures in the genital area (thelycum) used for spermatophore attachment (Hartnoll 1969; Diesel 1991; López-Greco 2013). Spermathecae in Brachyura have been categorized into two major types, which differ in the positioning of the vagina relative to the oviduct (Diesel 1991). In the ventral type, vagina and oviduct open into the spermatheca in close proximity to each other.

Consequently, the sperm of each male is pushed away from the oviduct by the sperm of each subsequent male, promoting a higher fertilization chance for the last male (last-male precedence) particularly if sperm mixing within the spermatheca is prevented by layers of male-deposited "sperm gel" (sometimes referred to as "internal sperm plug") (e.g., Diesel 1991; Pardo et al. 2013). In the dorsal type, oviduct and vagina are well separated from each other, which promotes first-male precedence because sperm of the first male should reach the oviduct first. Regardless of the spermatheca type, mating plugs deposited into the vaginal region by males during copulation promote first-male precedence (e.g., Hartnoll 1969; Jensen et al. 1996). Such mating plugs are commonly found in brachyuran crabs and thought to ensure paternity (Hartnoll 1969; Christy 1987; Jensen et al. 1996), although they may not always completely prevent subsequent inseminations (e.g., Gosselin et al. 2005; Todd et al. 2005).

8.3 Cryptic Female Choice in Crustaceans

Since Eberhard's (1996) influential book on possible female control mechanisms enabling postcopulatory mate choice, crustaceans still remain largely underrepresented in studies of cryptic female choice. One reason for the dearth of crustacean examples could be the common occurrence of external fertilization in crustaceans, which may limit cryptic female choice mechanisms to sperm–egg interactions. However, internal fertilization systems can be frequently found among brachyuran crustaceans, which have been widely studied for patterns of sperm storage and mating behaviors (Henmi and Murai 1999; Duffy and Thiel 2007). If multiple matings occur, cryptic female choice mechanisms within the female reproductive tract could be common in such species. A strong indicator for cryptic female choice is when multiple paternity leads to skewed offspring ratios, which is increasingly discovered in crustacean species by the use of molecular genetic markers (e.g., Walker et al. 2002; Yue et al. 2010). However, thorough examinations of cryptic female choice mechanisms are largely absent, and no systematic investigation has been conducted to disentangle the relative importance of sperm competition and cryptic female choice, to convincingly demonstrate the biological significance of cryptic female choice in crustaceans. In this chapter, we aimed to briefly show the common occurrence of multiple matings in different crustacean species, summarize indicators and possible mechanisms for cryptic female choice in crustaceans, and outline the potential role of cryptic female choice in selected case studies.

8.3.1 Multiple Matings: Behavioral and Molecular Evidence

Multiple mating is a prerequisite for cryptic female choice to evolve and is commonly found in animals (e.g., Jennions and Petrie 2000; Simmons 2005). In crustaceans, a plethora of studies has accumulated evidence for multiple mating over the past decades (e.g., Sassaman 1978; Johnson 1982; Christy and Salmon 1984; Shuster 1989a; Bauer 1996; González-Gurriarán et al. 1998; MacDiarmid and Butler 1999; Franke 2000; Hartnoll 2000; Clark and Caudill 2001; Zimmer 2001; Moreau et al. 2002; Jensen and Bentzen 2012; Bailie et al. 2014). Most studies rely on direct behavioral observations, but multiple matings do not necessarily translate into multiple paternity or biased offspring ratios (Eberhard 1996). Hence, the use of genetic markers to assess paternity has proven particularly useful to confirm multiple paternity and skewed offspring ratios in an increasing number of crustacean species (summarized in Table 8.1).

Multiple paternity rates can be as low as 3 % of all broods with more than one father (*Tetraclita rubescens*, Kelly et al. 2012), or up to 100 % (*Caridina ensifera*, Yue and Chang 2010), and are generally highly variable (Table 8.1). Geographic variation within species has been found in species such as the American lobster *Homarus americanus* or the barnacle *T. rubescens*, in which rates of multiple paternity apparently depend on lobster exploitation rates and barnacle density, respectively (Gosselin et al. 2005; Kelly et al. 2012). Skewed offspring ratios among fathers appear to be particularly common when rates of multiple paternity are high: in fourteen out of seventeen species with documented skewed offspring ratios, multiple paternity was found in >50 % of broods (Table 8.1). However, more studies are needed to assess patterns and possible mechanisms underlying biased paternity. While molecular paternity analyses have advanced from the use of single allozymes (Nelson and Hedgecock 1977; Sassaman 1978) toward applying multiple microsatellite loci (e.g., Mathews 2007; Jossart et al. 2014; Bailie et al. 2014), the detection of multiple paternity can still be hampered by small sample sizes or low incidences of multiple fertilized egg clutches. For example, Urbani et al. (1998) found no evidence for multiple paternity in the snow crab *Chionoecetes opilio* using a sample size of seven broods, whereas two subsequent studies using larger sample sizes of 79 and 20 females found that 3.2 and 12.5 % were on average sired by multiple fathers, respectively (Roy 2003; Sainte-Marie et al. 2008). Similarly, the finding of single paternity in 18 broods of *Cancer pagurus* does not exclude multiple paternity for this species, but instead could be a consequence of rare occurrences of multiple fertilizations and be supported by effective cryptic female choice or last-male sperm precedence (McKeown and Shaw 2008). Taken together, broods sired by multiple fathers have been found in a diverse number of crustacean taxa, and genetic studies have revealed the common occurrence of skewed offspring ratios and a high variability in the rates of multiple paternity (Table 8.1).

8.3.2 Indicators and Mechanisms of CFC in Crustaceans

Cryptic female choice in crustaceans has been largely suggested based on behavioral observations such as failed copulations (Ra'anan and Sagi 1985; Diesel 1990; Bauer 1992), adjusted duration of receptivity, or delayed oviposition (e.g., Thiel

Table 8.1 Molecular genetic studies assessing multiple paternity in crustaceans

Species	Genetic marker	Outcome	Reference
Astacidea (Decapoda)			
Homarus americanus (American lobster)	4 microsatellites	Multiple paternity (2–3 males) in 14 out of 108 females (13 %); skewed offspring ratios in half of the broods	Gosselin et al. (2005)
Homarus gammarus (European lobster)	3–6 microsatellites	Multiple paternity in 13 out of 73 females (17.8 %)	Sørdalen (2012)
Nephrops norvegicus (Norway lobster)	2 microsatellites	Multiple paternity (2–3 males) in 6 out of 11 broods (54 %). Skewed reproductive success among males	Streiff et al. (2004)
Orconectes placidus (Crayfish)	3 microsatellites	Females mated multiply in 60 % of 15 broods (usually 2 males). Males contributed highly skewed numbers of offspring	Walker et al. (2002)
Orconectes sanbornii, *O. obscurus* (Crayfish)	4 microsatellites	Multiple paternity (2–3 males) in 5 O. *sanbornii* and 3 *O. obscurus* broods (100 %). Skewed paternity ratios	Kahrl et al. (2014)
Procambarus clarkii (Red swamp crayfish)	4 microsatellites	Multiple paternity (2–4 males) in 29 out of 30 females (97 %). Skewed paternity ratios	Yue et al. (2010)
Brachyura (Decapoda)			
Cancer pagurus (Brown crab)	1 allozyme locus; 3 microsatellites	No evidence for multiple paternity, but female long-term sperm storage makes it a possibility	Burfitt (1980), McKeown and Shaw (2008)
Chionoecetes opilio (Snow crab)	2 microsatellites	Single paternity in 7 broods. Multiple paternity (3.8 %) in 79 broods. Multiple paternity (12.5 %) in 20 broods, with two fathers and skewed offspring ratios	Urbani et al. (1998), Roy (2003), Sainte-Marie et al. (2008)

(continued)

Table 8.1 (continued)

Species	Genetic marker	Outcome	Reference
Dissodactylus primitivus (Pea crab)	4 microsatellites	Multiple paternity in 11–12 out of 18 clutches (>60 %). Up to six fathers. Skewed offspring ratios	Jossart et al. (2014)
Metacarcinus magister (Dungeness crab)	3 microsatellites	Multiple paternity (2–3 males) in 4 out of 10 wild-caught females (40 %). Skewed offspring ratios	Jensen and Bentzen (2012)
Metacarcinus edwardsii	8 microsatellites	Single paternity in five females	Rojas-Hernandez et al. (2014)
Ucides cordatus (Mangrove land crab)	6 microsatellites	Multiple paternity in 4 out of 10 females (40 %, 2 males)	Baggio et al. (2011)
Uca mjoebergi (Fiddler crab)	2–3 microsatellites	Burrow-mated females with multiple paternity (at least 2 males) in 5 out of 9 clutches (56 %). Highly skewed offspring ratios	Reaney et al. (2012)
Anomura (Decapoda)			
Munida rugosa, *M. sarsi* (Squat lobsters)	3 microsatellites	Multiple paternity (2–4 males) in both species (86 % in 25 females of *M. rugosa*, 100 % in 5 females of *M. sarsi*). Offspring ratios skewed toward a single male	Bailie et al. (2011)
Petrolisthes cinctipes (Porcelain crab)	2 microsatellites	Multiple paternity (2–3 males) in 8 out of 10 broods (80 %)	Toonen (2004)
Thalassinidea (Decapoda)			
Callichirus islagrande (Ghost shrimp)	2 microsatellites	Multiple paternity (2–3 males) in 8 out of 40 females (20 %)	Bilodeau et al. (2005)
Caridea (Decapoda)			
Acanthephyra pelagica (Deep-sea shrimp)	4 microsatellites	Multiple paternity (2–4 males) in 19 out of 19 females (100 %). Skewed offspring ratios	Paegelow (2014)

(continued)

Table 8.1 (continued)

Species	Genetic marker	Outcome	Reference
Alpheus angulosus (Snapping shrimp)	5 microsatellites	Multiple paternity (2 males) in 17 out of 53 clutches (31 %). Skewed offspring ratios	Mathews (2007)
Caridina ensifera (Freshwater shrimp)	4 microsatellites	Multiple paternity (2–11 males) in all 20 tested broods (100 %). Skewed offspring numbers (lower contribution of genetically similar males)	Yue and Chang (2010)
Palaemonetes pugio (Grass shrimp)	2 microsatellites	Multiple paternity in 8 out of 10 broods (80 %)	Baragona et al. (2000)
Rhynchocinetes typus (Marine rock shrimp)	3 microsatellites	Multiple paternity (2–4 males) in 11 out of 15 broods (73.3 %). Skewed paternities	Bailie et al. (2014)
Isopoda			
Porcellio scaber (Woodlouse)	1 allozyme locus	Multiple paternity in >80 % of 20 broods. Skewed paternities	Sassaman (1978)
Sphaeroma rugicauda (Isopod)	1 allozyme locus	Multiple mating (2 males) in 37 out of 211 broods (17.5 %)	Heath et al. (1990)
Thermosphaeroma thermophilum (Socorro isopod)	5 allozyme loci	Weak support for multiple paternity. In four broods, one might have had two fathers	Jormalainen et al. (1999)
Copepoda			
Lepeophtheirus salmonis (Salmon louse)	2 microsatellites	Multiple paternity (2 males) in 5 out of 10 females (50 %), despite effective blockage of copulatory ducts by first male	Todd et al. (2005)
Cirripedia			
Pollicepes polymerus (Gooseneck barnacle)	16 SNPs	Multiple paternity (mostly 2 males) in 17 out of 130 broods (13 %); spermcast mating	Barazandeh et al. (2013)

(continued)

Table 8.1 (continued)

Species	Genetic marker	Outcome	Reference
Pollicipes elegans (Pacific gooseneck barnacle)	3 microsatellites	Multiple paternity (2–5 males) in 11 out of 14 broods (79 %). Skewed offspring ratios	Plough et al. (2014)
Tetraclita rubescens (Eastern Pacific volcano barnacle)	4 microsatellites	Rates of multiple paternity were 3 % (in 74 broods) in low-density, and 25 % (in 56 broods) in high-density patches	Kelly et al. (2012)

and Hinojosa 2003; Brockerhoff and McLay 2005; Table 8.2). While providing valuable information, such observations are not sufficient to demonstrate cryptic female choice in favor of certain male traits, or to reveal benefits for females such as avoidance of genetic incompatibilities. The same applies to observations of male stimulation during and after copulation ("copulatory courtship," Eberhard 1994), which is often discussed to play a role in cryptic female choice (for example, see Rodriguez, Chap. 18) and appears to be observed in many crustacean taxa including both decapods and non-decapods (e.g., Bauer 1992; Titelman et al. 2007; Almerão et al. 2010; Tanaka 2013). While it remains unclear whether or not females use male stimulation cues for mate choice, the possibility of females stimulating the male remains largely unexplored. As an exception, studies on hermit crabs revealed that spermatophore transfer does not start unless the female touches the male mouthparts with her chelipeds or legs (Contreras-Garduño et al. 2007), which could represent a form of precopulatory mate choice.

More direct support for cryptic female choice in crustaceans is provided by studies reporting biased reproductive effort in favor of certain male traits such as body and chela size (Galeotti et al. 2006; Aquiloni and Gherardi 2008), which are not always convincingly explainable by sperm competition alone. Similarly, biased offspring ratios can indicate a role for cryptic female choice (e.g., Gosselin et al. 2005; Bailie et al. 2011; Jossart et al. 2014), particularly if accompanied by additional indicators for cryptic female choice such as sperm removal or selective sperm passage through the female reproductive tract (see Table 8.2). Strong support for cryptic female choice would also be provided by demonstrating "good gene" and "compatible gene" benefits for females, which may include, for example, decreased costs of inbreeding, increased viability of offspring, avoidance of genetic incompatibilities, or increased immunity-gene (MHC) diversity (e.g., Jennions and Petrie 2000; Neff and Pitcher 2005). In crustaceans, a study on the freshwater shrimp *C. ensifera* found that offspring ratios were skewed in favor of genetically dissimilar males, which appears consistent with inbreeding avoidance (Yue and Chang 2010; Fig. 8.1). In contrast, males of higher genetic similarity

Table 8.2 Indicators for cryptic female choice in crustaceans; O indicates observed and S indicates suggested mechanisms that could underlie or facilitate cryptic female choice

Species	Failed copulations	Sperm removal	Sperm selection	Biased reproductive investment	Chemical signaling	Delayed receptivity or copulation	Cryptic female choice?	References
Astacidea (Decapoda)								
Austropotamobius italicus (White-clawed crayfish)	O			O			Biased reproductive investment	Galeotti et al. (2006)
Pacifastacus leniusculus (Signal crayfish)		O			S		Female urine signaling triggers male searching and courtship behavior. Urine release during fights by both sexes. Suggested to favor stronger males as mating partners	Stebbing et al. (2003), Berry and Breithaupt (2010)
Procambarus clarkii (Red swamp crayfish)				O			Discriminate reproductive investment against small males	Aquiloni and Gherardi (2008)
Homarus americanus (American lobster)						O	Females delayed molting until mated with a dominant male	Cowan and Atema (1990)

(continued)

Table 8.2 (continued)

Species	Failed copulations	Sperm removal	Sperm selection	Biased reproductive investment	Chemical signaling	Delayed receptivity or copulation	Cryptic female choice?	References
Brachyura (Decapoda)								
Chionoecetes opilio (Snow crab)	O		S		O	S	Suggested possibility of selective sperm use mediated by spermathecal epithelium. Males increase searching and clasping behavior when exposed to water from a tank containing a female	Sainte-Marie and Lovrich (1994), Bouchard et al. (1996), Sainte-Marie et al. (2000)
Hemigrapsus sexdentatus (Purple rock crab)						O	Females isolated from males delayed oviposition and stayed receptive longer than in the presence of males	Brockerhoff and McLay (2005)
Inachus phalangium (Leach spider crab)			S				Multiple matings, long-term sperm storage; suggestion that velum structure might control amount of sperm used	Diesel (1988, 1989)

(continued)

Table 8.2 (continued)

Species	Failed copulations	Sperm removal	Sperm selection	Biased reproductive investment	Chemical signaling	Delayed receptivity or copulation	Cryptic female choice?	References
Libinia spinosa (Spider crab)			S				Sperm storage, multiple matings; suggestion that velum might control amount of sperm used during spawning	Moyano et al. (2010)
Neohelice granulata (Semiterrestrial crab)					S	S	Suggested to adjust duration of receptivity, and control moment of fertilization. Female mate choice (for large unmated males) suggested to be aided by male chemical signals	Moyano et al. (2012, 2013)
Anomura (Decapoda)								
Pagurus filholi (Hermit crab)					O	O	Delaying copulation could facilitate exploitation of male–male competition in favor of dominant males. Indirect female choice by sex pheromones that intensify male–male combat	Yamanoi et al. (2006), Okamura and Goshima (2010)

(continued)

Table 8.2 (continued)

Species	Failed copulations	Sperm removal	Sperm selection	Biased reproductive investment	Chemical signaling	Delayed receptivity or copulation	Cryptic female choice?	References
Aegla platensis (Freshwater crab)	O						No release of oocytes following copulation (one female)	Almerão et al. (2010)
Caridea (Decapoda)								
Macrobrachium rosenbergii (Giant river prawn)	O						Failed fertilization more common with subordinates "small males" and "orange-claw" males (>40 %) compared to dominant "blue-claw" males (3.3 %)	Ra'anan and Sagi (1985)
Rhynchocinetes typus (Marine rock shrimp)		O				O	Females delay spawning when held by subordinate males, but not with the preferred dominant males, and remove spermatophores from subordinate males	Thiel and Hinojosa (2003)

(continued)

Table 8.2 (continued)

Species	Failed copulations	Sperm removal	Sperm selection	Biased reproductive investment	Chemical signaling	Delayed receptivity or copulation	Cryptic female choice?	References
Penaeoidea (Decapoda)								
Penaeus vannamei (Pacific white shrimp)	O				S		Low frequency of successful spermatophore transfer (20 % in 10 matings). Unpublished data suggest that female sex pheromones induce male searching and chasing behavior	Yano et al. (1988)
Sicyonia dorsalis (Lesser rock shrimp)	O						Incomplete or failed inseminations following copulation.	Bauer (1992)
Amphipoda								
Gammarus palustris (Amphipod)	O					S	Females might have delayed their molts in the absence of a male. Delay of copulation reduces fecundity	Borowski (1988)

(continued)

Table 8.2 (continued)

Species	Failed copulations	Sperm removal	Sperm selection	Biased reproductive investment	Chemical signaling	Delayed receptivity or copulation	Cryptic female choice?	References
Eogammarus oclairi (Amphipod)						O	Females able to delay molting and ovulation when no males are present (same in isopod *Porcellio dilatatus*, Mocquard et al. 1976, and amphipod *Gammarus pulex*, Ward 1984)	Iribarne et al. (1995)
Isopoda								
Paracerceis sculpta (Marine isopod)						O	Females can delay initiation of reproductive molt if males are absent	Shuster (1989a, b, 1990)

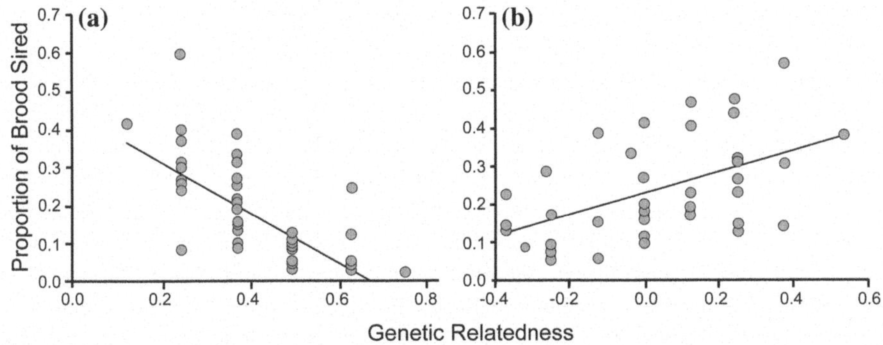

Fig. 8.1 Genetic parentage studies in crustaceans have revealed contrasting patterns. **a** In the shrimp *Caridina ensifera*, paternity is in favor of genetically dissimilar males (genetic similarity index after Nei and Li 1979; Yue and Chang 2010). **b** In contrast, the barnacle *Pollicipes elegans* shows biased paternity in favor of genetically similar males (genetic similarity index after Li et al. 1993; Plough et al. 2014)

to the female achieved higher offspring ratios in the Pacific gooseneck barnacle (*Pollicipes elegans*)—this has been speculated to reflect improved gametic recognition during fertilization, the function and benefit of which will have to be resolved in future studies (Plough et al. 2014) (Fig. 8.1).

More studies that investigate mechanisms and the outcome of postcopulatory sexual selection are clearly needed, although it remains difficult to disentangle the interacting effects of cryptic female choice and sperm competition if female control mechanisms are not known. A straightforward demonstration of cryptic female choice is complicated by the diversity of possible underlying mechanisms, and attempts to separate those into meaningful categories often remain arbitrary or incomplete. For example, Parker (2006) distinguished between ejaculate manipulations (physically block or eject sperm), sperm selection (within the female reproductive tract, or at the ovum surface), and differential reproductive investment (in favor of certain male traits). Such a categorization covers a large fraction of possible cryptic female choice mechanisms, but should not be overemphasized because many other possible mechanisms could be missed. To provide an overview for patterns suggestive of cryptic female choice in crustaceans, we first discuss female exploitation of male behaviors that can facilitate postcopulatory sexual selection through intensifying male competition, and subsequently follow Parker's (2006) categories to explore the potential of cryptic female choice in crustaceans.

8.3.2.1 Exploiting Male Behavior

Mate choice can be facilitated by female behaviors before, during, or after copulation that increase chances of mating with a preferred male. Two such behaviors have been described in crustaceans: (1) release of chemical signals to provoke increased mate searching or male–male contest, and (2) delay of reproductive events such as oviposition, copulation, or receptivity, which may increase encounter probability with favored males. Intensified male–male competition can be provoked through female chemical signaling mediated by sex pheromones (Berry and Breithaupt 2010; Okamura and Goshima 2010), which may also elicit male search and guarding behavior (Yano et al. 1988; Bouchard et al. 1996; Sneddon et al. 2003). The use of sex pheromones for social communication is common and well documented in crustaceans (e.g., Ryan 1966; Atema and Engstrom 1971; Dunham 1978; Gleeson 1980; Imafuku 1986; Díaz and Thiel 2004; Breithaupt and Thiel 2011), and female pheromone signaling to manipulate competitive male behavior could be widespread. Females of blue crabs (*Callinectes sapidus*) apparently release sex pheromones to attract males and may resist male guarding attempts (Gleeson 1980; Jivoff and Hines 1998), but whether or not females employ cryptic mate choice remains to be answered. Attraction of additional, competitive males via chemical signaling might co-occur with a delay of reproduction to increase chances of a successful mating with a preferred male, as it appears to be the case in the hermit crab *Pagurus filholi* (Imafuku 1986; Yamanoi et al. 2006). The ability of females to delay molting and ovulation when no males are present has been reported from a number of crustaceans (e.g., Mocquard et al. 1976; Ward 1984; Shuster 1989b, 1990; Iribarne et al. 1995; Brockerhoff and McLay 2005). Thus, females could use their control over molting to adjust their reproductive timing depending on the social context (presence/absence of preferred males), which might be more realistic if they were able to hide their reproductive status to non-preferred males through suppressed behavioral or chemical signaling as has been suggested for some crabs and lobsters (Christy and Rittschof 2011). For example, females of the American lobster (*H. americanus*) have been reported to delay molting until they mated with a dominant male (Cowan and Atema 1990, but see also Waddy et al. 2013). Similarly, a delay of copulation has been suggested to increase male–male competition in favor of dominant males in marine rock shrimp *Rhynchocinetes typus* (Thiel and Correa 2004).

8.3.2.2 Ejaculate Manipulations

Sperm ejection, physical blockage, or transfer of sperm to a place where it is not used for fertilization might represent commonly overlooked cryptic female choice mechanisms in crustaceans. For example, removal of sperm subsequent to matings has been described for rock shrimps (*R. typus*, Thiel and Hinojosa 2003), and anecdotally also for crayfishes (*Pacifastacus leniusculus*, Berry and Breithaupt 2010) and spiny lobsters (*Panulirus guttatus*, Magallón-Gayón et al. 2011). Selective sperm removal by females remains among the strongest indicators for cryptic female choice because it can be directly observed. Thus, more studies are needed to reveal how commonly it occurs and to experimentally demonstrate its effects on postcopulatory sexual selection in crustaceans. Sperm transfer to unfavorable places could play a role in females of species with multiple sperm storage sites with differential sperm use, or "dead-end" structures where no fertilization takes place. For example, a "bursa" structure located between vagina and spermatheca in female Dungeness crabs (*Metacarcinus magister*) appears to be a dead-end for sperm of secondary males, which cannot overcome the internal sperm plug deposited by the first male (Jensen et al. 1996; Jensen and Bentzen 2012). Therefore, the first male gains paternity despite the presence of a ventral spermatheca that otherwise often leads to last-male precedence (Jensen et al. 1996; Jensen and Bentzen 2012). In other species with a ventral spermatheca, the sperm of the first male is pushed away from the oviduct by the sperm of each subsequent mating, and the separation of the different male's sperm by layers of male-produced "sperm gel" prevents sperm mixing (e.g., Diesel 1988; Urbani et al. 1998; Pardo et al. 2013; Fig. 8.2), which can be viewed as a form of physical blockage of sperm from previous matings that ensures a high fertilization success of the last male. While sperm plugs and sperm gel are usually deposited by males for counteracting sperm competition, Diesel (1989, 1991) suggested that females might induce the formation of mating plugs by forcing seminal fluids into the vagina. This interpretation has not found much support because of the lack of muscular structures that could apply enough force to produce the dense structures of sperm plugs (Jensen et al. 1996). Similarly, the possibility that females might selectively

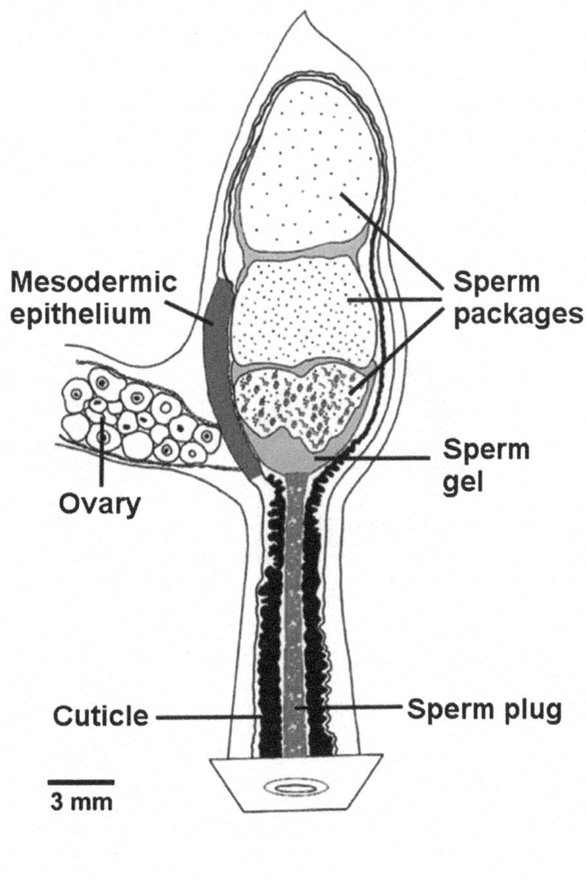

Fig. 8.2 Sperm gel stratification of multiple ejaculates in the spermatheca of *Metacarcinus edwardsii*, preventing sperm mixing and placing the last male's sperm closest to the oviduct (Pardo et al. 2013)

manipulate mating plugs (Diesel 1989) does not seem to apply to most brachyuran crabs in which mating plugs are commonly found (Hartnoll 1969; Christy 1987; Jensen et al. 1996). Although we are not aware of studies showing female-produced mating plugs, some authors have suggested that females might exert cryptic choice by selectively dissolving mating plugs with the help of excretions from the glandular epithelium (Elner et al. 1985; Diesel 1989). However, mating plugs can be very durable in crustaceans (Pardo et al. 2013), and currently, there is no evidence that females have much control over the decay of mating plugs (Bauer and Martin 1991; Oh and Hankin 2004).

8.3.2.3 Selective Sperm Passage ("Sperm Selection")

The possibility of selective sperm passage through the female reproductive tract has rarely been considered in crustaceans. An anatomical feature of the female reproductive tract that could allow for selective sperm passage is a ringlike, muscular "velum" separating dorsal (storage) and ventral (insemination) sections of the spermatheca, which has been found in a few species of the family Majidae (*Inachus phalangium*, *I. communissimus*, *Pisa tetraodon*, *Maja verrucosa*, Diesel 1989; *Libinia spinosa*, Moyano et al. 2010; González-Pisani et al. 2012; Fig. 8.3). The velum has been suggested to selectively regulate sperm use during spawning by reducing the entry of older, stored sperm into the insemination chamber (Diesel 1989; Moyano et al. 2010). However, more studies are needed to resolve the exact function of a velum or similar structures that are found in other crustaceans (e.g., Lee and Yamazaki 1990; Sainte-Marie and Sainte-Marie 1998; Sainte-Marie et al. 2000; Lautenschlager et al. 2010; Moyano et al. 2012). Another example comes from the terrestrial isopod *Armadillidium vulgare*, in which sperm transport from the oviduct to the seminal receptacle has been proposed to be under female control (Ziegler and Suzuki 2011).

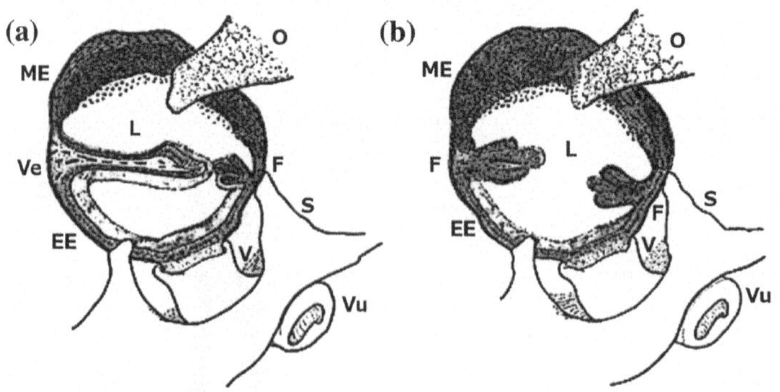

Fig. 8.3 Schematic drawings of spermathecae of two species of spider crabs (Majidae). **a** A velum in *Libinia spinosa* and **b** "folds" in *Leurocyclus tuberculosus* separate the lumen into dorsal (storage) and ventral (insemination) chambers, which might allow for selective use of sperm for insemination. *EE* ectoderm-derived epithelium; *F* folds; *L* lumen; *ME* mesoderm-derived epithelium; *O* ovary; *Ov* oviduct; *S* septum; *V* vagina; *Ve* velum; *Vu* vulva (González-Pisani et al. 2012)

In this species, the immobile sperm appears to be transported with the help of hydrostatic pressure produced by a muscle layer around the main part of the oviduct, which has been suspected to selectively interact with sperm morphology such as tail length (Ziegler and Suzuki 2011). On the other hand, short-term storage of sperm within the oviduct and sperm mixing in case of two subsequent matings may lead to equal fertilization success between the two males (Moreau et al. 2002), which does not support a strong role of cryptic female choice. To some extent, the fate of transferred sperm may also be under female control in species with external fertilization. For example, female hermit crabs might employ cryptic mate choice by selectively breaking up externally deposited spermatophores for spermatozoal release, with the help of mechanical (pereopods, chelipeds) or chemical (glandular system secretions) mechanisms (Contreras-Garduño and Córdoba-Aguilar 2006).

8.3.2.4 Sperm–Egg Interactions

Selective sperm passage may also occur at the ovum surface during sperm–egg interactions, which can play an important role for avoiding genetic incompatibilities or increasing genetic diversity (Zeh and Zeh 1997; Palumbi 1999). The possible consequences for postcopulatory sexual selection have been studied in broadcast spawning invertebrates such as sea urchins (e.g., Evans and Sherman 2013), which could also be relevant for many crustacean species with external fertilization. Relatively few studies on sperm–egg interactions are available for crustaceans (e.g., *Macrobrachium rosenbergii*: Lynn and Clark 1983; *R. typus*: Barros et al. 1986, Dupré et al. 2012; *Sicyonia ingentis*: Clark and Griffin 1988, Chen et al. 1994; *Penaeus monodon*: Ngernsoungnern et al. 2012; ostracods: Matzke-Karasz 2005; Fig. 8.4), offering only limited information on key mechanisms like gamete recognition through glycoproteins attached to sperm and egg surfaces. For example, the presence of egg surface glycoproteins has been suggested to affect fertilization success in *R. typus* (Dupré et al. 2012) and *S. ingentis* (Glas et al. 1996), and activity of proteinases and protein inhibitors on the sperm cells presumably affect sperm–egg

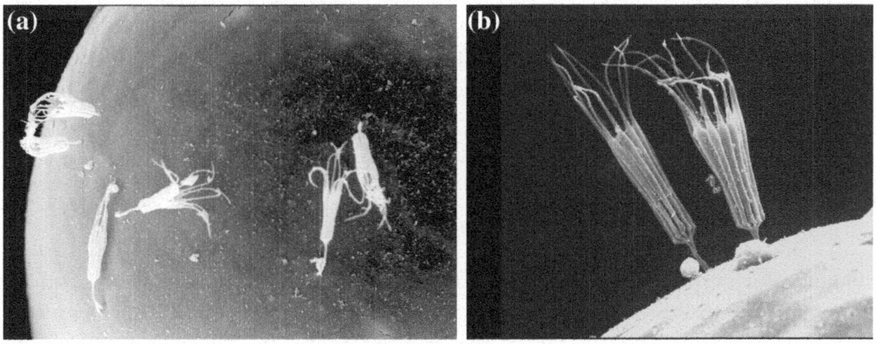

Fig. 8.4 **a** Polyspermic egg of *Rhynchocinetes typus*. **b** Two sperm cells of *R. typus* penetrating an egg (pictures by E. Dupré)

interactions (e.g., giant river prawn *Ma. rosenbergii* (Ma et al. 2010). Such sperm–egg interactions presumably help to avoid unbeneficial polyspermy, sperm limitation or interspecific fertilizations (e.g., Evans and Sherman 2013), but could also provide a cryptic female choice mechanism suited to avoid genetic incompatibilities (Vacquier and Swanson 2011). Hence, the possibility that sperm–egg recognition systems in crustaceans might selectively influence sperm passage through the ovum surface warrants further studies, to better understand its potential significance for cryptic female choice and postcopulatory sexual selection.

8.3.2.5 Differential Reproductive Investment

Cryptic female choice can act through biasing reproductive investment toward the production of offspring from favored males ("differential allocation," Burley 1986; Sheldon 2000; Kindsvater and Alonzo 2014). In crustaceans, differential reproductive investment by females has been suggested for two crayfish species, which are thought to adjust egg size and number depending on male body and chela size (Galeotti et al. 2006; Aquiloni and Gherardi 2008). Adjustment of egg size might be achieved through controlled reabsorption of deutoplasm of the eggs (Aquiloni and Gherardi 2008). Although the exact mechanisms have not yet been investigated, this might be a process that requires time and could be limited to species in which oviposition is delayed for sufficient time after mating. In the freshwater crayfish *Austropotamobius italicus*, females produce larger but fewer eggs with small-sized, large-clawed males, and smaller but more eggs with large-sized, small-clawed males (Galeotti et al. 2006). In contrast, females of the crayfish *Procambarus clarkii* produce larger eggs with larger males, and smaller eggs with smaller males (Aquiloni and Gherardi 2008). To better understand the ecological circumstances and evolutionary significance of such contrasting patterns of differential allocation, further studies are needed to determine how egg size trades off with egg number, and how it affects offspring fitness in each of these species. Because differential allocation in favor of preferred males may interact with compensatory allocation towards offspring from matings with non-preferred males ("reproductive compensation"; e.g., Ratikainen and Kokko 2010; Kindsvater and Alonzo 2014), more studies are needed to investigate targets of mate preferences and fitness consequences of differential reproductive investment.

8.3.3 Selected Case Studies

Evidence for cryptic female choice in crustaceans remains sparse. In the following, we briefly discuss selected species where cryptic female choice has been suggested, or in which a high potential for cryptic female choice exists. Surprisingly, most of these examples come from crustaceans with external fertilization, despite the expectation to find a higher potential for cryptic female choice in species with internal fertilization systems.

8.3.3.1 *Pagurus filholi*

In the hermit crab *P. filholi,* fertilization takes place externally without the help of sperm storage structures. Females cannot escape male precopulatory guarding attempts and ovulate shortly after copulation (Yamanoi et al. 2006). Hence, cryptic female choice opportunities seem limited to female behavior during male guarding and copulation, or selective sperm–egg interactions. Nevertheless, females can delay copulation by withdrawing into their shells and thereby increase chances of re-mating with a more dominant male that displaces the guarding male (Yamanoi et al. 2006). Exploitation of male–male competition relies on high encounter chances with a dominant male, which is promoted by releasing sex pheromones to attract males and provoke male–male combats (Imafuku 1986; Okamura and Goshima 2010). Additional cryptic female choice mechanisms could exist in *P. filholi* or other hermit crabs, including physical manipulation of spermatophores that are deposited externally but within the female's shell (Contreras-Garduño and Córdoba-Aguilar 2006).

8.3.3.2 *Procambarus clarkii*

The red swamp crayfish *P. clarkii* is a North American species mainly known for being a successful invader following commercial introductions all around the world (Gherardi 2006). Both sexes mate multiply, and fertilization takes place at external sperm storage sites ("annulus ventralis") at the posterior end of the female seventh thoracic sternite, where the eggs are released and attached to the pleopods for brooding (Aquiloni and Gherardi 2007). Behavioral experiments revealed precopulatory mate choice by both males and females: Males prefer large, virgin females providing more eggs to fertilize at lower risk of sperm competition, whereas females prefer large males that presumably provide access to male-defended burrows, or indirect genetic benefits of producing large-bodied offspring (Aquiloni and Gherardi 2007). A likely role of postcopulatory selection through cryptic female choice was suggested by Aquiloni and Gherardi (2008), who found that females invest more resources into offspring sired by larger males. While the underlying mechanisms are not fully understood, it could involve reabsorption of yolk substance in the oocyte before spawning when mating with a small male, and producing yolk-rich eggs when sired by large males (Aquiloni and Gherardi 2008). Accordingly, egg size might be adjusted during or after copulation, allowing for the production of more robust larvae after mating with large males. Recent genetic evidence for multiple paternity of broods and skewed offspring ratios is consistent with the assumption of cryptic female choice, although sperm competition and male-driven processes like the deposition of sperm plugs could play an equally important role (Yue et al. 2010), which needs to be addressed in future studies.

8.3.3.3 *Rhynchocinetes typus*

Among the best-supported examples for cryptic female choice in crustaceans is the marine rock shrimp *R. typus* (Fig. 8.5). The mating system has been described as "neighborhood of dominance", in which females can resist but not avoid matings with the abundant subordinate males, and mating success of males largely depends on their ability to win fights over females and guard them against neighboring competitors (Correa and Thiel 2003; Thiel and Correa 2004).

Females prefer to mate with large dominant males, which provide sufficient sperm supply and efficiently guard the females against other males, and are thought to accept subordinate males to minimize costs of harassment ("convenience polyandry," Thiel and Hinojosa 2003). Fertilization is external, with spermatophores being transferred to the ventral region of the female's abdomen (Correa et al. 2000). Multiple matings have been shown repeatedly in behavioral experiments (e.g., Correa et al. 2003; Thiel and Correa 2004; Dennenmoser and Thiel 2008), and a recent genetic study confirmed multiple paternity and unequal offspring ratios (Bailie et al. 2014). Behaviors suggestive of cryptic female choice include active removal of sperm that females received from subordinate but not dominant males (Thiel and Hinojosa 2003). Attraction of males to receptive females by visual cues (Díaz and Thiel 2004) together with a delay of ovulation increases chances of matings with the preferred dominant males (Thiel and Correa 2004). Furthermore, studies on sperm–egg interactions indicate a possible role for selective sperm passage at the ovum surface. The presence of polyspermic eggs

Fig. 8.5 Assembly of different male morphotypes of marine rock shrimp (*Rhynchocinetes typus*) surrounding a female (*arrow*). Note the dominant "robustus" male morphotype in the *upper left corner*, which females prefer over the subordinate males (*below, right, unlabeled*)

with up to 14 sperms attached to the egg surface at the same time (Barros et al. 1986; Fig. 8.4) sets the stage for selective mechanisms that might favor genetically compatible or dissimilar mates. Alternatively, the rare observations of such polyspermic eggs could indicate successful avoidance of polyspermy like in other marine invertebrates (Vacquier and Swanson 2011). Glycoproteins present on the oocyte envelope have been suggested to play a role in sperm recognition and adhesion, with N-acetyl-glucosamine as a potential inhibitor of fertilization (Dupré et al. 2012). Whether sperm–egg recognition mechanisms in rock shrimps or other crustaceans have evolved to avoid polyspermy, or allow for cryptic female choice via selective sperm passage remains to be shown in future studies.

8.4 Crustaceans Compared to Other Invertebrates

Crustaceans inhabit all major habitats including freshwater and terrestrial environments, but the highest diversity is found in the oceans, where all major crustacean taxa have evolved (Brusca and Brusca 2002). Marine crustaceans share the basic conditions of the marine environment with representatives from almost all known invertebrate phyla. Thus, the selective forces driving gamete transfer and fertilization will be similar for all these taxa, offering ample opportunity for a cross-taxonomic comparison of cryptic female choice mechanisms.

Many marine invertebrates shed their gametes freely in the water column, apparently reducing the potential for cryptic female choice to sperm–egg interactions, for which there is indication in several broadcast spawners (ctenophorans—Carré and Sardet 1984; polychaetes—Marshall and Evans 2005; echinoderms—Evans and Marshall 2005). Most research in broadcast spawners has been on sperm–egg interactions after releasing the gametes into the water column (Evans et al. 2013). However, it might also be possible that females delay (or accelerate) gamete release in response to different males. Females might also associate with preferred (compatible) mating partners before spawning their oocytes. If females take an active role in seeking out preferred mates before gamete release, this could be analogous to crustacean females attracting additional (preferred) males before mating.

In some marine invertebrates, only the males shed their gametes, which are then taken up from the water column by the females, a strategy termed "spermcast mating" (Pemberton et al. 2003; Bishop and Pemberton 2006). Females or hermaphrodites acting as females appear to be very efficient in gathering more than sufficient sperm to guarantee fertilization of their oocytes, opening the opportunity for cryptic female choice (Pemberton et al. 2003). Indeed, these females seem to have evolved mechanisms to avoid fertilizations by incompatible sperm (Bishop and Pemberton 2006), although this is considered a relative crude choice mechanism (Pemberton et al. 2004). In crustaceans, spermcast mating has recently been reported for barnacles, which usually employ their penis for sperm transfer (Barazandeh et al. 2013). The high level of multiple paternity in some barnacle species and the fact that genetically related fathers sire more offspring (Plough et al. 2014) are intriguing observations, suggesting that similar choice mechanisms as reported for other spermcasting taxa might be at play.

In species with direct sperm transfer, males usually transfer these in discrete packages, the spermatophores, which may be attached to the external body surface of females, deposited in internalized cuticular spermathecae, or directly injected into their reproductive tract. Externally attached spermatophores are common in cephalopods, and there is good evidence that females can manipulate these spermatophores, by actively detaching or consuming them (Sato et al. 2013; Wegener et al. 2013). Similarly, in crustacean species in which sperm packages are attached externally to the female's body, as for example, in many copepods, shrimps, and lobsters, there is ample opportunity for cryptic female choice (also described above). By forcing males to attach spermatophores to their external body surface (instead of inside their reproductive tract), females might gain easy control over fertilizations.

Transfer of sperm packages directly into the female's reproductive tract might thus be viewed as a male strategy to overcome female control of fertilization. True copulation is common among gastropods. Many gastropods mate multiply, and genetic studies have revealed multiple paternity for diverse species (e.g., Panova et al. 2010; Brante et al. 2011; Xue et al. 2014). Cryptic female choice in gastropods might involve dumping or digestion of sperm, and differential use during fertilization (see Paterson et al. 2001 and citations therein). However, while multiple paternity is common in marine gastropods, little is known about the occurrence of cryptic female choice and the mechanisms involved (Walker et al. 2007; Panova et al. 2010; Xue et al. 2014). This resembles the situation in crustaceans with internal fertilization and highlights the need for future studies.

Males of many soft-bodied invertebrate species such as nudibranchs and flatworms might circumvent the threats (e.g., sperm digestion) that their sperm are facing within the female's reproductive tract by hypodermic insemination in other body parts (Angeloni 2003). This is unlikely to occur in the majority of crustaceans that have a hard exoskeleton, but it could be worthwhile to further explore sperm transfer mechanisms and female choice potential in soft-bodied, parasitic crustaceans such as representatives of the genus *Rhizocephala* in which dwarf males are known to use their antennules for injecting premature spermatogonia cells into much larger female parasites (Høeg 1985; Glenner et al. 2009).

Females may also differentially allocate resources to eggs when mating with different males. This has been recently shown for a tunicate (Hammerschmidt et al. 2011). Also, females of a marine gastropod produced larger eggs after mating with multiple males (Sprenger et al. 2010). Differential allocation to eggs in response to different males has been little studied in crustaceans, but recent studies suggest that this can occur (see above), and it should be investigated more thoroughly in the future. It might be most promising to start with species in which males engage in prolonged precopulatory mate guarding or where fertilization is temporally decoupled from mating, thereby offering females ample time to adjust resource allocation to oocytes.

In summary, many of the mechanisms of cryptic female choice reported from a wide diversity of marine invertebrates are also observed or suggested for crustaceans. The disparity of body forms and reproductive strategies found in crustaceans and the fact that they inhabit many different habitats with variable selective environments makes them an ideal model taxon for future studies on cryptic female choice.

8.5 Conclusions

Despite numerous observations of indicators for cryptic female choice, we barely understand the role of subtle female control mechanisms in most crustacean species. While genetic evidence for multiple paternity and skewed offspring ratios strongly suggest postcopulatory selection, disentangling male- and female-driven processes remains challenging (Anderson and Simmons 2006), and the possible contribution of precopulatory sexual selection through variable timing and number of copulations is poorly understood (Pélissié et al. 2014). More empirical studies on polyandrous crustaceans are needed to reveal how females selectively manipulate sperm use and how they benefit from cryptic female choice. This will ultimately contribute to a better understanding of how pre- and postcopulatory sexual selection interact to generate the diversity of mating systems observed in crustaceans.

Acknowledgments We are very grateful to Enrique Dupré for sharing some of his beautiful figures on sperm–egg interactions with us. We thank Bernard Sainte-Marie, Raymond T. Bauer, and one anonymous reviewer for critical comments that helped to improve the manuscript. Also a big thanks to Annie Mejaes for her help with some of the figures.

References

Almerão M, Bond-Buckup G, de MendonçaM S Jr (2010) Mating behaviour of *Aegla platensis* (Crustacea, Anomura, Aeglidae) under laboratory conditions. J Ethol 28:87–94

Andersson M, Simmons LW (2006) Sexual selection and mate choice. Trends Ecol Evol 21:296–302

Angeloni L (2003) Sexual selection in a simultaneous hermaphrodite with hypodermic insemination: body size, allocation to sexual roles and paternity. Anim Behav 66:417–426

Aquiloni L, Gherardi F (2007) Mutual mate choice in crayfish: large body size is selected by both sexes, virginity by males only. J Zool 274:171–179

Aquiloni L, Gherardi F (2008) Evidence of female cryptic choice in crayfish. Biol Lett 4:163–165

Asakura A (2009) The evolution of mating systems in decapod crustaceans. In: Martin JW, Crandall KA, Felder DL (eds) Decapod crustacean phylogenetics. CRC Press, Taylor and Francis Group, Boca Raton/London/New York, pp 121–182

Atema J, Engstrom DG (1971) Sex pheromone in the lobster, *Homarus americanus*. Nature 232:261–263

Baeza JA, Thiel M (2007) The mating system of symbiotic crustaceans: a conceptual model based on optimality and ecological constraints. In: Duffy JE, Thiel M (eds) Evolutionary ecology of social and sexual systems. Oxford University Press, New York, pp 250–267
Baggio RA, Pil MW, Boeger WA, Patella LA, Ostrensky A, Pie M (2011) Genetic evidence for multiple paternity in the mangrove land crab *Ucides cordateus* (Decapoda: Ocypodidae). Mar Biol Res 7:520–524
Bailie DA, Hynes R, Prodöhl PA (2011) Genetic parentage in the squat lobsters *Munida rugosa* and *M. sarsi* (Crustacea, Anomura, Galatheidae). Mar Ecol Prog Ser 421:173–182
Bailie DA, Fitzpatrick S, Connolly M, Thiel M, Hynes R, Prodöhl PA (2014) Genetic assessment of parentage in the caridean rock shrimp *Rhynchocinetes typus* based on microsatellite markers. J Crust Biol 34:658–662
Baragona MA, Haig-Ladewig LA, Wang SY (2000) Multiple paternity in the grass shrimp *Palaemonetes pugio*. Am Zool 40:935
Barazandeh M, Davis CS, Neufeld CJ, Coltman DW, Palmer AR (2013) Something Darwin didn't know about barnacles: spermcast mating in a common stalked species. Proc R Soc B 280:20122919
Barros C, Dupré E, Viveros L (1986) Sperm-egg interactions in the shrimp *Rhynchocinetes typus*. Gamete Res 14:171–180
Bauer RT, Martin JW (1991) Crustacean sexual biology. Columbia University Press, New York, p 355
Bauer RT (1992) Repetitive copulation and variable success of insemination in the marine shrimp *Sicyonia dorsalis* (Decapoda: Penaeoidea). J Crust Biol 12:153–160
Bauer R (1996) A test of hypotheses on male mating systems and female molting in decapod shrimp, using *Sicyonia dorsalis* (Decapoda: Penaeoidea). J Crust Biol 16:429–436
Becker C, Brandis D, Storch V (2011) Morphology of the female reproductive system of European pea crabs (Crustacea, Decapoda, Brachyura, Pinnotheridae). J Morphol 272:12–26
Berry FC, Breithaupt T (2010) To signal or not to signal? Chemical communication by urine-borne signals mirrors sexual conflict in crayfish. BMC Biol 8:25
Bilodeau AL, Felder DL, Neigel JE (2005) Multiple paternity in the thalassinidean ghost shrimp, *Callichirus islagrande* (Crustacea: Decapoda: Callianassidae). Mar Biol 146:381–385
Birkhead TR, Pizzari T (2002) Postcopulatory sexual selection. Nat Rev Genet 3:262–273
Bishop JDD, Pemberton AJ (2006) The third way: spermcast mating in sessile marine invertebrates. Integr Comp Biol 46:398–406
Borowsky B (1988) Delaying copulation in the amphipod crustacean *Gammarus palustris*: effects on female fecundity and consequences for the frequency of amplexus. Mar Behav Physiol 13:359–368
Bouchard S, Sainte-Marie B, McNeil JN (1996) Indirect evidence indicates female semiochemicals release male precopulatory behaviour in the snow crab, *Chionoecetes opilio* (Brachyura: Majidae). Chemoecology 7:39–44
Brante A, Fernández M, Viard F (2011) Microsatellite evidence for sperm storage and multiple paternity in the marine gastropod *Crepidula coquimbensis*. J Exp Mar Biol Ecol 396:83–88
Breithaupt T, Thiel M (2011) Chemical communication in Crustaceans. Springer, New York
Brockerhoff AM, McLay CL (2005) Factors influencing the onset and duration of receptivity of female purple crabs, *Hemigrapsus sexdentatus* (H. Milne Edwards, 1837) (Brachyura: Grapsidae). J Exp Mar Biol Ecol 314:123–135
Brusca RC, Brusca GJ (2002) Invertebrates. Sinauer Associates, Sunderland
Burfitt AH (1980) Glucose phosphate isomerase inheritance in *Cancer pagurus* L. broods as evidence of multiple paternity (Decapoda, Brachyura). Crustaceana 39:306–310
Burley N (1986) Sexual selection for aesthetic traits in species with biparental care. Am Nat 127:415–445
Carré D, Sardet C (1984) Fertilization and early development in *Beroe ovate*. Dev Biol 105:188–195
Chen TI, Green JD, Clark WH Jr (1994) Sperm penetration of the vitelline envelope of *Sicyonia ingentis* eggs is mediated by a trypsin-like lysine of acrosomal vesicle origin. Dev Growth Differ 36:259–273

Christy JH (1987) Competitive mating, mate choice and mating associations of brachyuran crabs. Bull Mar Sci 4:177–191

Christy JH, Salmon M (1984) Ecology and evolution of mating systems of fiddler crabs (Genus *Uca*). Biol Rev 59:483–509

Christy JH, Rittschof D (2011) Deception in visual and chemical communication in Crustaceans. In: Breithaupt T, Thiel M (eds) Chemical communication in Crustaceans. Springer, New York, pp 313–333

Clark WH Jr, Griffin FJ (1988) The morphology and physiology of the acrosome reaction in the sperm of the decapod, *Sicyonia ingentis*. Dev Growth Differ 30:451–462

Clark R, Caudill CC (2001) Females of the marine amphipod *Jassa marmorata* mate multiple times with the same or different males. Mar Freshw Behav Physiol 34:131–138

Contreras-Garduño J, Córdoba-Aguilar A (2006) Sexual selection in hermit crabs: a review and outlines of future research. J Zool 270:595–605

Contreras-Garduño J, Osorno JL, Córdoba-Aguilar A (2007) Male-male competition and female behavior as determinants of male mating success in the semi-terrestrial hermit crab Coenobita compressus (H. Milne Edwards). J Crusta Biol 27:411–416

Correa C, Baeza JA, Dupré E, Hinojosa IA, Thiel M (2000) Mating behaviour and fertilization success of three ontogenetic stages of male rock shrimp *Rhynchocinetes typus* (Decapoda: Caridea). J Crust Biol 20:628–640

Correa C, Thiel M (2003) Mating systems in caridean shrimp (Decapoda: Caridea) and their evolutionary consequences for sexual dimorphism and reproductive biology. Rev Chil Hist Nat 76:187–203

Correa C, Baeza JA, Hinojosa IA, Thiel M (2003) Male dominance hierarchy and mating tactics in the rock shrimp *Rhynchocinetes typus* (Decapoda: Caridea). J Crust Biol 23:33–45

Cowan DF, Atema J (1990) Moult staggering and serial monogamy in American lobsters, *Homarus americanus*. Anim Behav 39:1199–1206

Dennenmoser S, Thiel M (2008) Effects of social structure on the behaviour and performance of alternative reproductive phenotypes in male rock shrimp, *Rhynchocinetes typus*. Ethology 114:327–339

Díaz ER, Thiel M (2004) Chemical and visual communication during mate searching in rock shrimp. Biol Bull 206:134–143

Diesel R (1988) Discrete storage of multiple-mating sperm in the spider crab *Inachus phalangium*. Naturwissenschaften 75:148–149

Diesel R (1989) Structure and function of the reproductive system of the symbiotic spider crab *Inachus phalangium* (Decapoda: Majidae): observations on sperm transfer, sperm storage, and spawning. J Crust Biol 9:266–277

Diesel R (1990) Sperm competition and reproductive success in the decapod *Inachus phalangium* (Majidae): a male ghost spider crab that seals off rivals' sperm. J Zool 220:213–223

Diesel R (1991) Sperm competition and the evolution of behavior in Brachyura, with special reference to spider crabs (Decapoda, Majidae). In: Bauer RT, Martin JW (eds) Crustacean sexual biology. Columbia University Press, New York, pp 145–163

Duffy JE, Thiel M (2007) Evolutionary ecology of social and sexual systems: Crustaceans as model organisms. Oxford University Press, New York

Dunham PJ (1978) Sex pheromones in Crustacea. Biol Rev 53:555–583

Dupré E, Gómez D, Araya A, Gallardo C (2012) Role of egg surface glycoconjugate in the fertilization of the rock shrimp *Rhynchocinetes typus* (Milne-Edwards, 1837). Lat Am J Aquat Res 40:22–29

Eberhard WG (1994) Evidence for widespread courtship during copulation in 131 species of insects and spiders, and implications for cryptic female choice. Evolution 48:711–733

Eberhard WG (1996) Female control: sexual selection by cryptic female choice. Princeton University Press, Princeton

Eberhard WG (1998) Female roles in sperm competition. In: Birkhead TR, Møller AP (eds) Sperm competition and sexual selection. Academic Press, New York, pp 91–116

Eberhard WG (2000) Criteria for demonstrating postcopulatory female choice. Evolution 54:1047–1050

Elner RW, Gass CA, Campbell A (1985) Mating behaviour of the Jonah crab, *Cancer borealis* Stimpson (Decapoda, Brachyura). Crustaceana 48:34–39

Espinoza-Fuenzalida NL, Acuña E, Hinojosa IA, Thiel M (2012) Reproductive biology of two species of squat lobsters—female receptivity and interbrood intervals. J Crust Biol 32:565–574

Evans JP, Marshall DJ (2005) Male-by-female interactions influence fertilization success and mediate the benefits of polyandry in the sea urchin *Heliocidaris erythrogramma*. Evolution 59:106–112

Evans JP, Sherman CDH (2013) Sexual selection and the evolution of egg-sperm interactions in broadcast-spawning invertebrates. Biol Bull 224:166–183

Franke HD (2000) Alternative male mating behaviours and sperm competition in the marine isopod *Jaera hopeana* (Crustacea). J Reprod Dev 46:47–48

Galeotti P, Rubolini D, Fea G, Ghia D, Nardi PA, Gherardi F, Fasola M (2006) Female freshwater crayfish adjust egg and clutch size in relation to multiple male traits. Proc R Soc B 273:1105–1110

Galeotti P, Rubolini D, Pupin F, Sacchi R, Fasola M (2008) Sperm removal and ejaculate size correlate with chelae asymmetry in a freshwater crayfish species. Behav Ecol Sociobiol 62:1739–1745

Gherardi F (2006) Crayfish invading Europe: the case study of *Procambarus clarkii*. Mar Freshw Behav Physiol 39:175–191

Glas P, Green J, Lynn J (1996) Morphological evidence for a chitin-like glycoprotein in penaeid hatching envelopes. Biol Bull 191:374–384

Gleeson RA (1980) Pheromone communication in the reproductive behaviour of the blue crab, *Callinectes sapidus*. Mar Behav Physiol 7:119–134

Glenner H, Høeg JT, Stenderup J, Rybakov AV (2009) The monophyletic origin of a remarkable sexual system in akentrogonid rhizocephalan parasites: a molecular and larval structural study. Exp Parasitol 125:3–12

González-Gurriarán E, Fernández L, Freire J, Muiño R (1998) Mating and role of seminal receptacles in the reproductive biology of the spider crab *Maja squinado* (Decapoda, Majidae). J Exp Mar Biol Ecol 220:269–285

González-Pisani X, Barón P, López-Greco LS (2012) Functional anatomy of the female reproductive systems of two spider crabs (Decapoda, Majoidea). Invertebr Biol 13:61–74

Gosselin T, Sainte-Marie B, Bernatchez L (2005) Geographic variation of multiple paternity in the American lobster, *Homarus americanus*. Mol Ecol 14:1517–1525

Hammerschmidt K, Pemberton AJ, Michiels NK, Bishop JDD (2011) Differential maternal allocation following mixed insemination contributes to variation in oocyte size in a sea squirt. Mar Ecol Prog Ser 422:123–128

Hartnoll RG (1969) Mating in the Brachyura. Crustaceana 16:161–181

Hartnoll RG (2000) Evolution of brachyuran mating behavior: relation to the female molting pattern. Crustac Issues 12:519–525

Heath DJ, Ratford JR, Riddoch BJ, Childs D (1990) Multiple mating in a natural population of the isopod *Sphaeroma rugicauda*; evidence from distorted ratios in offspring. Heredity 64:81–85

Henmi Y, Murai M (1999) Decalcification of vulvar operculum and mating in the ocypodid crab *Ilyoplax pusilla*. J Zool 247:133–137

Høeg JT (1985) Male Cypris settlement in *Clistosaccus paguri* Lilljeborg (Crustacea, Cirripedia, Rhizocephala). J Exp Mar Biol Ecol 89:221–235

Imafuku M (1986) Sexual discrimination in the hermit crab *Pagurus geminus*. J Ethol 4:39–47

Iribarne O, Fernandez M, Armstrong D (1995) Molting and fertilization of female amphipods *Eogammarus oclairi*: diel molting rhythm and the effect of delayed copulation. Mar Freshw Behav Physiol 27:1–11

Jennions MD, Petrie M (2000) Why do females mate multiply? A review of the genetic benefits. Biol Rev 75:21–64

Jensen PC, Orensanz JM, Armstrong DA (1996) Structure of the female reproductive tract in the Dungeness crab (*Cancer magister*) and implications for the mating system. Biol Bull 190:336–349

Jensen PC, Bentzen P (2012) A molecular dissection of the mating system of the Dungeness crab, *Metacarcinus magister* (Brachyura: Cancridae). J Crust Biol 32:443–456

Jivoff P, Hines AH (1998) Female behavior, sexual competition and mate guarding in the blue crab, *Callinectes sapidus*. Anim Behav 55:589–603

Johnson C (1982) Multiple insemination and sperm storage in the isopod, *Venezillo evergladensis* Schultz 1963. Crustaceana 42:225–232

Jormalainen V, Shuster SM, Wildey H (1999) Reproductive anatomy, precopulatory mate guarding, and paternity in the Socorro isopod, *Thermosphaeroma thermophilum*. Mar Freshw Behav Physiol 32:39–56

Jossart Q, Wattier RA, Kastally C, Aron S, David B, De Ridder C, Rigaud T (2014) Genetic evidence confirms polygamous mating system in a crustacean parasite with multiple hosts. PLoS ONE 93:e90680

Kahrl AF, Laushman RH, Roles A (2014) Evidence for multiple paternity in two species of *Orconectes* crayfish. Can J Zool. 92:985–988

Kelly MW, Grosberg RK, Sanford E (2012) Love the one you're with: proximity determines paternity success in the barnacle *Tetraclita rubescens*. Mol Ecol 21:5088–5097

Kindsvater HK, Alonzo SH (2014) Females allocate differentially to offspring size and number in response to male effects on female and offspring fitness. Proc R Soc B 281:20131981

Krol RM, Hawkins WE, Overstreet RM (1992) Reproductive components. In: Harrison FW, Humes AG (eds) Microscopic anatomy of invertebrates. Wiley-Liss, New York, pp 295–343

Kvarnemo C, Simmons LW (2013) Polyandry as a mediator of sexual selection before and after mating. Phil Trans R Soc B 368:1613

Lautenschlager AD, Brandis D, Storch V (2010) Morphology and function of the reproductive system of representatives of the genus *Uca*. J Morphol 271:1281–1299

Lee TH, Yamazaki F (1990) Structure and function of a special tissue in the female genital ducts of the Chinese freshwater crab *Eriocheir sinensis*. Biol Bull 178:94–100

Li CC, Weeks DE, Chakravarti A (1993) Similarity of DNA fingerprints due to chance and relatedness. Hum Hered 43:45–52

López-Greco LS (2013) Functional anatomy of the reproductive system. In: Watling L, Thiel M (eds) Functional morphology and diversity. Oxford University Press, New York, pp 413–450

Løvlie H, Gillingham MAF, Worley K, Pizzari T, Richardson DS (2013) Cryptic female choice favours sperm from major histocompatibility complex-dissimilar males. Proc R Soc B 280:20131296

Lüpold S, Tomkins JL, Simmons LW, Fitzpatrick JL (2014) Female monopolization mediates the relationship between pre- and postcopulatory sexual traits. Nat Commun 5:3184

Lynn JW, Clark WH Jr (1983) A morphological examination of sperm-egg interaction in the freshwater prawn, *Macrobrachium rosenbergii*. Biol Bull 164:446–458

Ma WM, Qian YQ, Wang MR, Yang F, Yang WJ (2010) A novel terminal ampullae peptide is involved in the proteolytic activity of sperm in the prawn, *Macrobrachium rosenbergii*. Reproduction 140:235–245

MacDiarmid AB, Butler MJ IV (1999) Sperm economy and limitation in spiny lobsters. Behav Ecol Sociobiol 46:14–24

Magallón-Gayón E, Briones-Fourzán P, Lozano-Álvarez E (2011) Does size always matter? Mate choice and sperm allocation in *Panulirus guttatus*, a highly sedentary, habitat-specialist spiny lobster. Behaviour 148:1333–1358

Marshall DJ, Evans JP (2005) The benefits of polyandry in the free-spawning ploychaete *Galeolaria caespitosa*. J Evol Biol 18:735–741

Mathews LM (2007) Evidence for high rates of in-pair paternity in the socially monogamous snapping shrimp *Alpheus angulosus*. Aquat Biol 1:55–62

Matzke-Karasz R (2005) Giant spermatozoon coiled in small egg: fertilization mechanisms and their implications for evolutionary studies on Ostracoda (Crustacea). J Exp Zool 304:129–149
McKeown NJ, Shaw PW (2008) Single paternity within broods of the brown crab *Cancer pagurus*: a highly fecund species with long-term sperm storage. Mar Ecol Prog Ser 368:209–215
Mocquard JP, Besse G, Juchault P, Legrand JJ, Maissiat J, Martin G, Picaud JL (1976) Duree des cycles de mue chez les femelles de l'oniscoide *Porcellio dilatatus* Brandt suivant leur etat sexual et les conditions de'elevage: temperature, photoperiode et groupment. Bull Ecol 7:297–314
Moreau J, Seguin S, Caubet Y, Rigaud T (2002) Female remating and sperm competition patterns in a terrestrial crustacean. Anim Behav 64:569–577
Moyano MPS, Gavio MA, Cuartas EI (2010) Morphology and function of the reproductive tract of the spider crab *Libinia spinosa* (Crustacea, Brachyura, Majoidea): patterns of sperm storage. Helgol Mar Res 64:213–221
Moyano MPS, Luppi T, Gavio MA, Vallina M, McLay C (2012) Receptivity of female *Neohelice granulata* (Brachyura, Varunidae): different strategies to maximize their reproductive success in contrasting habitats. Helgol Mar Res 66:661–674
Moyano MPS, Silva P, Luppi T, Gavio MA (2013) Female mate choice by chemical signals in a semi-terrestrial crab. J Sea Res 85:300–307
Neff BD, Pitcher TE (2005) Genetic quality and sexual selection: an integrated framework for good genes and compatible genes. Mol Ecol 14:19–38
Nei M, Li WH (1979) Mathematical model for studying genetic variation in terms of restriction endonucleases. Proc Natl Acad Sci 76:5269–5273
Nelson K, Hedgecock D (1977) Electrophoretic evidence of multiple paternity in the lobster *Homarus americanus* (Milne-Edwards). Am Nat 111:361–365
Ngernsoungnern P, Ngernsoungnern A, Chaiseha Y, Sretarugsa P (2012) Role of vitelline envelope during fertilization in the black tiger shrimp, *Penaeus monodon*. Acta Histochem 114:659–664
Oh SJ, Hankin DG (2004) The sperm plug is a reliable indicator of mating success in female Dungeness crabs, *Cancer magister*. J Crust Biol 24:314–326
Okamura S, Goshima S (2010) Indirect female choice mediated by sex pheromones in the hermit crab *Pagurus filholi*. J Ethol 28:323–329
Paegelow EAJ (2014) Genetic analyses on the deep-sea shrimp *Acanthephyra pelagica* in the North-West Atlantic. MSc thesis, Dalhousie University, Halifax
Palumbi SR (1999) All males are not created equal: fertility differences depend on gamete recognition polymorphisms in sea urchins. Proc Natl Acad Sci U S A 96:12632–12637
Panova M, Boström J, Hofving T, Areskoug T, Eriksson A, Mehlig B, Mäkinen T, André C, Johannesson K (2010) Extreme female promiscuity in a non-social invertebrate species. PLoS ONE 5:e9640
Pardo LM, Riveros M, Fuentes JP, López-Greco L (2013) Functional morphology of the seminal receptacle in the crab *Metacarcinus edwardsii*. Invertebr Biol 132:386–393
Parker GA (2006) Sexual conflict over mating and fertilization: an overview. Philos Trans R Soc Lond B 361:235–259
Paterson IG, Partridge V, Buckland-Nicks J (2001) Multiple paternity in *Littorina obtusata* (Gastropoda, Littorinidae) revealed by microsatellite analyses. Biol Bull 200:261–267
Pélissié B, Jarne P, Sarda V, David P (2014) Disentangling precopulatory and postcopulatory sexual selection in polyandrous species. Evolution 68:1320–1331
Pemberton AJ, Hughes RN, Manríquez PH, Bishop JDD (2003) Efficient utilization of very dilute aquatic sperm: competition may be more likely than sperm limitation when eggs are retained. Proc R Soc B 270:S223–S226
Pemberton AJ, Sommerfeldt AD, Wood CA, Flint HC, Noble LR, Clarke KR, Bishop JDD (2004) Plant-like mating in an animal: sexual compatibility and allocation trade-offs in a simultaneous hermaphrodite with remote transfer of sperm. J Evol Biol 17:506–518

Plough LV, Moran A, Marko P (2014) Density drives polyandry and relatedness influences paternal success in the Pacific gooseneck barnacle, *Pollicipes elegans*. BMC Evol Biol 14:81

Ra'anan Z, Sagi A (1985) Alternative mating strategies in male morphotypes of the freshwater prawn *Macrobrachium rosenbergii* (De Man). Biol Bull 169:592–601

Ratikainen II, Kokko H (2010) Differential allocation and compensation: who deserves the silver spoon? Behav Ecol 21:195–200

Raviv S, Parnes S, Sagi A (2008) Coordination of reproduction and molt in decapods. In: Mente E (ed) Reproductive biology of Crustaceans enfield. Science Publishers, NH, pp 365–390

Reaney LT, Maurer G, Backwell PRY, Linde CC (2012) Paternity analysis of two male mating tactics in the fiddler crab, *Uca mjoebergi*. Behav Ecol Sociobiol 66:1017–1024

Rojas-Hernandez N, Véliz D, Pardo LM (2014) Use of novel microsatellite markers for population and paternity analysis in the commercially important crab *Metacarcinus edwardsii* (Brachyura: Cancridae). Mar Biol Res 10:839–844

Roy N (2003) Incidence de la polyandrie chez le crabe des neiges, *Chionoecetes opilio* (Brachyura, Majidae). MSc thesis, Université du Québec à Rimouski

Ryan EP (1966) Pheromone: evidence in a decapod Crustacean. Science 151:340–341

Sainte-Marie G, Sainte-Marie B (1998) Morphology of the spermatheca, oviduct, intermediate chamber and vagina of the adult snow crab (*Chionoecetes opilio*). Can J Zool 76:1589–1604

Sainte-Marie G, Sainte-Marie B, Sévigny JM (2000) Ejaculate-storage patterns and the site of fertilization in female snow crabs (*Chionoecetes opilio*; Brachyura, Majidae). Can J Zool 78:1902–1917

Sainte-Marie B, Lovrich GA (1994) Delivery and storage of sperm at first mating of female *Chionoecetes opilio* (Brachyura: Majidae) in relation to size and morphometric maturity of male parent. J Crust Biol 3:508–521

Sainte-Marie B, Gosselin T, Sévigny JM, Urbani N (2008) The snow crab mating system: opportunity for natural and unnatural selection in a changing environment. Bull Mar Sci 83:131–161

Sassaman C (1978) Mating systems in Porcellionid isopods: multiple paternity and sperm mixing in *Porcellio scaber* Latr. Heredity 41:385–397

Sato N, Kasugai T, Munchara H (2013) Sperm transfer or spermatangia removal: postcopulatory behaviour of picking up spermatangium by female Japanese pygmy squid. Mar Biol 160:553–561

Sheldon BC (2000) Differential allocation: tests, mechanisms and implications. Trends Ecol Evol 15:397–402

Shuster SM (1989a) Male alternative reproductive strategies in a marine isopod crustacean (*Paracerceis sculpta*): the use of genetic markers to measure differences in fertilization success among α-, β-, and γ-males. Evolution 43:1683–1698

Shuster SM (1989b) Female sexual receptivity associated with molting and differences in copulatory behaviour among the three male morphs in *Paracerceis sculpta* (Crustacea: Isopoda). Biol Bull 177:331–337

Shuster SM (1990) Courtship and female mate selection in a marine isopod crustacean, *Paracerceis sculpta*. Anim Behav 40:390–399

Shuster SM (2007) The evolution of Crustacean mating systems. In: Duffy JE, Thiel M (eds) Evolutionary ecology of social and sexual systems. Oxford University Press, New York, pp 29–47

Simmons LW (2005) The evolution of polyandry: sperm competition, sperm selection, and offspring viability. Annu Rev Ecol Evol Syst 36:125–146

Simmons LW, Beveridge M, Wedell N, Tregenza T (2006) Postcopulatory inbreeding avoidance by female crickets only revealed by molecular markers. Mol Ecol 15:3817–3824

Slatyer RA, Mautz BS, Backwell PRY, Jennions MD (2011) Estimating genetic benefits of polyandry from experimental studies: a meta-analysis. Biol Rev Camb Philos Soc 87:1–33

Sneddon LU, Huntingford FA, Taylor AC, Clare AS (2003) Female sex pheromone-mediated effects on behaviour and consequences of male competition in the shore crab (*Carcinus maenas*). J Chem Ecol 29:55–70

Sørdalen TK (2012) Multiple paternity assessment and paternity assignment in wild European lobster (*Homarus gammarus*). MSc thesis, Norwegian University of Life Sciences

Sprenger D, Lange R, Michiels NK, Anthes N (2010) Sources of phenotypic variance in egg and larval traits in a marine invertebrate. Evol Ecol 24:185–194

Stebbing PD, Bentley MG, Watson GJ (2003) Mating behaviour and evidence for a female released courtship pheromone in the signal crayfish *Pacifastacus leniusculus*. J Chem Ecol 29:465–475

Streiff R, Mira M, Castro M, Cancela ML (2004) Multiple paternity in Norway Lobster (*Nephrops norvegicus* L.) assessed with microsatellite markers. Mar Biotechnol 6:60–66

Subramoniam T (1993) Spermatophores and sperm transfer in marine crustaceans. In: Blaxter JHS, Southward AJ (eds) Advances in marine biology. Academic Press, New York, pp 129–201

Subramoniam T (2013) Origin and occurrence of sexual and mating systems in Crustacea: a progression towards communal living and eusociality. J Biosci 38:951–969

Tanaka H (2013) The mating behaviour of the seed shrimp *Parapolycope spiralis* (Ostracoda: Cladocopina), with insights into the evolution of mating systems in cryptic interstitial habitats. Biol J Linn Soc Lond 109:791–801

Thiel M, Hinojosa IA (2003) Mating behaviour of female rock shrimp *Rhynchocinetes typus* (Decapoda: Caridea)—indication for convenience polyandry and cryptic female choice. Behav Ecol Sociobiol 55:113–121

Thiel M, Correa C (2004) Female rock shrimp *Rhynchocinetes typus* mate in rapid succession up a male dominance hierarchy. Behav Ecol Sociobiol 57:62–68

Titelman J, Varpe Ø, Eliassen S, Fiksen Ø (2007) Copepod mating: chance or choice? J Plankton Res 29:1023–1030

Todd CD, Stevenson RJ, Reinardy H, Ritchie MG (2005) Polyandry in the ectoparasitic copepod *Lepeophtheirus salmonis* despite complex precopulatory and postcopulatory mate-guarding. Mar Ecol Prog Ser 303:225–234

Toonen RJ (2004) Genetic evidence of multiple paternity of broods in the intertidal crab *Petrolisthes cinctipes*. Mar Ecol Prog Ser 270:259–263

Urbani N, Sainte-Marie B, Sevigny JM, Zadworny D, Kuhnlein U (1998) Sperm competition and paternity assurance during the first breeding period of female snow crab (*Chionoecetes opilio*) (Brachyura: Majidae). Can J Fish Aquat Sci 55:1104–1113

Vacquier VD, Swanson WJ (2011) Selection in the rapid evolution of gamete recognition proteins in marine invertebrates. Cold Spring Harb Perspect Biol 3:a002931

Waddy SL, Hamilton-Gibson N, Aiken DE (2013) Female American lobsters (*Homarus americanus*) do not delay their molt if they cannot find a mate. Invertebr Reprod Dev 57:101–104

Walker D, Porter A, Avise JC (2002) Genetic parentage assessment in the crayfish *Orconectes placidus*, a high-fecundity invertebrate with extended maternal brood care. Mol Ecol 11:2122–2215

Walker D, Power AJ, Sweeney-Reeves M, Avise JC (2007) Multiple paternity and female sperm usage along egg-case strings of the knobbed whelk, *Busycon carica* (Mollusca; Melongenidae). Mar Biol 151:53–61

Ward PI (1984) *Gammarus pulex* control their moult to secure mates. Anim Behav 32:927

Wegener BJ, Stuart-Fox DM, Norman MD, Wong BBM (2013) Strategic male mate choice minimizes ejaculate consumption. Behav Ecol 24:668–671

Xue D, Zhang T, Liu JX (2014) Microsatellite evidence for high frequency of multiple paternity in the marine gastropod *Rapana venosa*. PLoS ONE 9:e86508

Yamanoi T, Yoshino K, Kon K, Goshima S (2006) Delayed copulation as a means of female choice by the hermit crab *Pagurus filholi*. J Ethol 24:213–218

Yano I, Kanna RA, Oyama RN, Wyban JA (1988) Mating behaviour in the penaeid shrimp *Penaeus vannamei*. Mar Biol 97:171–175

Yue GH, Chang A (2010) Molecular evidence for high frequency of multiple paternity in a freshwater shrimp species *Caridina ensifera*. PLoS ONE 5:e12721

Yue GH, Li JL, Wang CM, Xia JH, Wang GL, Feng JB (2010) High prevalence of multiple paternity in the invasive crayfish species, *Procambarus clarkii*. Int J Biol Sci 6:107–115

Zeh J, Zeh DW (1997) The evolution of polyandry II: post-copulatory defences against genetic incompatibility. Proc R Soc B 264:69–75

Ziegler A, Suzuki S (2011) Sperm storage, sperm translocation and genitalia formation in females of the terrestrial isopod *Armadillidium vulgare* (Crustacea, Peracarida, Isopoda). Arthropod Struct Dev 40:64–76

Zimmer M (2001) Why do male terrestrial isopods (Isopoda: Oniscidea) not guard females? Anim Behav 62:815–821

Chapter 9
Female Choice in Damselflies and Dragonflies

Alex Córdoba-Aguilar, Daniel González-Tokman, Ángela Nava-Bolaños, Karina Cuevas-Yáñez, Miguel Rivas and Adriana Nava-Sánchez

Abstract Odonates have been frequently labeled as a taxa where males control female's mating and fertilization decisions. Contrary to this position, in our contribution, we review instances where females can actually show choice of mates. Previous to mating, possible selected male traits are wing pigmentation, ability to defend oviposition sites, body color, and temperature. Females may assess male stimulation during copulation, responding via sperm ejection of previous males' sperm. Benefits females may derive from choosing males that can affect offspring are as follows: an increased ability to withstand pathogen infections (for both male and female offspring) or ability to stimulate, attractiveness, and fighting ability (for male offspring only). Finally, we discuss that even for traits that clearly seem to control female reproductive decisions, i.e., abdominal claspers, there is

A. Córdoba-Aguilar (✉) · D. González-Tokman · Á. Nava-Bolaños · K. Cuevas-Yáñez ·
M. Rivas · A. Nava-Sánchez
Departamento de Ecología Evolutiva, Instituto de Ecología, Universidad Nacional Autónoma de México, Apdo. Postal 70-275 Ciudad Universitaria, 04510 Mexico D.F, Mexico
e-mail: acordoba@iecologia.unam.mx

D. González-Tokman
e-mail: danigt@gmail.com

Á. Nava-Bolaños
e-mail: nb_gasper@hotmail.com

K. Cuevas-Yáñez
e-mail: kcy_999@yahoo.com.mx

M. Rivas
e-mail: miguelrivassoto@gmail.com

A. Nava-Sánchez
e-mail: anavasanchez@gmail.com

D. González-Tokman
Instituto de Ecología, A. C., Antiguo camino a Coatepec 351,
El Haya, 91070 Xalapa, VER, Mexico

A.V. Peretti and A. Aisenberg (eds.), *Cryptic Female Choice in Arthropods*,
DOI 10.1007/978-3-319-17894-3_9

no conclusive evidence that shows that they have evolved and are maintained via male–male competition. Our review thus emphasizes that we are far from admitting that females have little or no reproductive control in this taxa.

9.1 Introduction

Despite the historical rejection/embracement of female choice ideas (reviewed by Cronin 1992), this is a process that explains the evolution and maintenance of secondary sexual traits at least, and presumably, in some evolutionary time during a particular species' evolution (although such evolutionary time and modes have not been investigated in detail yet). Female choice can thus occur along with male–male competition to explain the evolution of secondary sexual traits (Eberhard 2005; Cordero and Eberhard 2003, 2005). What is missing is the relative weight of both processes in the evolution of secondary sexual traits, but this does not imply that female choice is not important.

Odonates have been traditionally assumed to be a taxa whose females have no or reduced control of mating and fertilization decisions (for a thorough discussion of this, see Fincke 1997; Fincke et al. 1997). In this context, odonates are usually thought to be a taxa where the evolution and maintenance of secondary sexual traits is mainly driven via male–male competition (Fincke et al. 1997; for an opposite position, see Cordero Rivera and Córdoba-Aguilar 2010). Contrary to this position, in this chapter, we would like to emphasize an alternative view. Thus, we would like to illustrate instances by which female choice (including cryptic female choice (CFC)) can operate.

9.2 Precopulatory Female Choice

9.2.1 Wing Pigmentation

One of the most conspicuous traits in odonate males is that of wing pigmentation. Males of different families (e.g., Calopterygidae, Megapodagrionidae, Euphaeidae, Diphlebiidae, Pseudolestidae, Platystictidae, Libellulidae) exhibit species-specific patterns of wing pigmentation (reviewed by Corbet 1999; Córdoba-Aguilar and Cordero-Rivera 2005; Svensson and Walter 2013) (Fig. 9.1). Wing pigmentation can take different patterns. It can cover the entire wing area (e.g., some *Calopteryx* species) or be present as a small spot (e.g., most *Hetaerina* species). It can also come in white, red, and black colors. At the intraspecific level, however, pigmented patterns can vary in area (e.g., Álvarez et al. 2013) and color aspect (e.g., UV patterns; Guillermo-Ferreira et al. 2014). Evolutionary explanations for wing pigmentation are related to a male–male competition process or a mixture between this process and female choice. These differences seem species specific. For example, in *Calopteryx* species, males fight with other males for the possession of mating territories, during which males use their wing pigmentation to communicate their energetic condition (i.e., those males in better energetic condition have larger or more uniform pigmented areas; Córdoba-Aguilar 2002). However, acquiring a territory guarantees

Fig. 9.1 A sample of *Hetaerina* males showing their wing pigmentation: *H. vulnerata* (**a**), *H. occisa* (**b**), *H. americana* (**c**), and *H. titia* (**d**)

just a higher rate of encounters with females, as males still need to court females via exhibiting their pigmented wing patterns when females visit them (e.g., Siva-Jothy 1999). Interestingly, wing pigmentation in *Calopteryx* species may also be used after copulation. After mating, males use their pigmented wings to face females whenever these start flying (Waage 1973). It is unclear whether such male post-copulatory interaction is related to indicate females where to lay eggs or "convince" them to still remain within the male's defended territory and use his sperm (Córdoba-Aguilar and Cordero-Rivera 2005). In any case, wing pigmentation in these species is used for both male–male competition and female choice. Conversely, in species such as *Hetaerina*, males use their wing pigmentation solely for territory disputes as there is no overt precopulatory or post-copulatory courtship as in *Calopteryx* (Grether 1996). In any case, although our information regarding female choice for wing pigmentation comes mainly from a handful of species: a few calopterygids (reviewed by Córdoba-Aguilar and Cordero-Rivera 2005) and one libellulid (Moore 1990), there is a lot to learn. For example, we lack evidence of whether there is a genetic basis for pigmentation and/or why females also show wing pigmentation patterns (Córdoba-Aguilar et al. 2003).

9.2.2 Defending and Showing Oviposition Sites

After copulation, males of some species take females to aquatic places to oviposit. This can take place while the couple remains in tandem (the male holds the female mesostigmal plates (in Zygoptera) or head (in Anisoptera) with his anal appendages; Corbet 1999). The male then lands on the surface of an aquatic plant or floating debris and so female starts ovipositing (Corbet 1999). During the course of taking the female, males still have to repel males that try to dislodge him from the female (Córdoba-Aguilar et al. 2009a, b). We propose that female choice can take place during such post-copulatory instances by a) checking whether the male is robust enough to withstand other males' dislodging attempts and b) assessing whether males have taken females to a good oviposition site. There is no evidence that supports the first possibility. In *Hetaerina*, it is known that bigger males are capable to withstand such dislodging attempts more successfully compared to smaller males, but this is no indication that females favor these males (Córdoba-Aguilar et al. 2009a, b; however, see our last section of this chapter for the possible role of claspers in female choice). In any case, the best supporting result would be to see whether females, whose male mates have been successful, favor such males by laying more eggs. For the second possibility, it is known that there is variation in oviposition site quality in terms of egg hatching (Siva-Jothy et al. 1995). However, again, it is unclear whether females use male choice of oviposition site to influence reproductive success. We know that females may clearly separate wing pigmentation pattern ("attractiveness") from oviposition site, meaning that females can assess both variables (Córdoba-Aguilar 2002). Interestingly, while in tandem, males may take the female to other oviposition places. In this case, it is hard to visualize how males would be able to assess oviposition quality and one way males could do that is by looking whether other couples are also present. Such copying of oviposition decisions is actually used by females when these look for an oviposition site that is safe (e.g., McMillan 2000a) or of good quality (e.g., Byers and Eason 2009). The copying could be a proxy, for example, of predation risk (e.g., Rehfeld 1992). So far, nevertheless, copying has not been interpreted as a way by which males can try to convince females to increase a mating male's fertilization success.

9.2.3 Male Body Color

If anything, one of the most conspicuous aspects of odonate morphology is their body color. Male body color seems far more diverse and complex than wing pigmentation. Similar to wing pigmentation, body color also varies intraspecifically in the same sex. Such variation is in terms of color type and aspect, stripe presence/absence, and color type combinations in the thorax, abdomen, and/or legs. Despite such variation, little is known in terms of body color evolutionary explanations. We know that both males and females are capable of assessing color differences (Huang et al. 2014). Males can actually "tell" the difference when exposed to contrasting female color and behavior such as female morphs that differ in these two aspects (Cordero et al. 1998; Takahami and Watanabe 2011; Van Gossum et al. 2011; Huang et al. 2014; however, see Iserbyt and Van Gossum 2011). However, there has been little effort to understand variation in male body color in terms of female choice. Research in *Calopteryx maculata* has shown that territorial status (being a territory owner or not) can be predicted by males with blue-colored thorax that are more likely being territory owners (Fitzstephens and Getty 2000). Additionally, diet can impact thoracic color as males can remain blue for longer (Fitzstephens and Getty 2000). A closest manipulation to investigate whether male color is related to female choice was done in *Platycypha caligata*, a chlorocyphid, whose males show their white anterior surface of the six legs to females prior to mating (Jennions 1998). Manipulation of the white area (either covering 25 % or the entire surface) showed no changes in mating or courting rate which suggested no role for these white spots in precopulatory female choice. Other than these references, there is little evidence for body color in terms of sexual selection.

9.2.4 Male Body Temperature

Odonate males frequently expose themselves to sun rays. One obvious benefit is that by increasing their temperature, males can engage in sexual activities (Corbet 1999). In *Mnais costalis*, increases in temperature allow males to court more intensively and be more preferred by females (Tsubaki et al. 2010). This ability is acquired more quickly by territorial males than by nonterritorial males, given that the former are more exposed to sunny spots than the latter (Sarejima and Tsubaki Samejima and Tsubaki 2009). These different pieces of information have been interpreted as female choice for hotter males in which females assess the quality of a male (being better able to defend her against other males by being more active) or an oviposition site (the warmer the site, the faster the egg development) (Tsubaki et al. 2010). A sunny place can also have other indirect benefits such as being more able to detect and avoid a predator which is a very common hazard in odonates, a hypothesis that deserves further examination. Conversely, a sunny spot may bear some inconveniences too. Although odonate males have ways to cool down (e.g., by exposing less body area), some environments may render animals to thermal shocks. For example, in some cases, *Calopteryx haemorrhoidalis* fighting males that accidentally fall into the water die very quickly presumably because of a temperature shock as the river temperature is far too low compared to that of the terrestrial environment (Cordero Rivera and Andrés 2002). On the other hand, being in an illuminated place would make a male more likely to be seen by predators, a hypothesis that, to our knowledge, has not been clarified in odonates. The evolution of male body temperature in the context of sexual selection in odonates clearly requires more research.

9.3 Post-copulatory Female Choice

9.3.1 Copulation Duration

Odonates vary tremendously in terms of copulation duration with disparate mating times of 5–7 h (even precopulatory tandem has been observed to last one day in extraordinary circumstances; e.g., Cordero et al. 1992) and as short as a few seconds (reviewed by Córdoba-Aguilar et al. 2009a, b). The control of such duration has been assumed not to be influenced by females but only by males in the following adaptive terms: (a) sperm displacement. According to this, duration of copulation will be dependent on how long it takes males to displace the sperm females have stored from previous matings (Córdoba-Aguilar and Cordero-Rivera 2008); (b) territory defense and mating opportunities. A recent comparative analysis uncovered that mating in males of nonterritorial species takes longer than that in males of territorial species (Córdoba-Aguilar et al. 2009a, b). This is a prediction that goes well with the fact that by spending too long in copulation, the chances of losing a territory, and thereby mating chances, increase, a prediction that at the intraspecific level actually applies well (e.g., Siva-Jothy and Tsubaki 1989a, b); and (c) time of the day. Several studies even in distantly related odonate species have shown that the earlier the mating starts in the day, the longer the copulation takes (reviewed by Córdoba-Aguilar and Cordero-Rivera 2008). The explanation is that the mating male performs an in-copula guarding to avoid other males to remate with the mating male's mate (Cordero 1990). However, if copulation starts late in the day, when there are not so many male competitors, copulation takes less long (Cordero 1990). In fact, this competitor-based reason applies well to explain why copulation duration is also influenced by temperature (as early in the morning is colder than midday) and male disturbance, as these two variables correlate with the number of competitor males (reviewed by Córdoba-Aguilar and Cordero-Rivera 2008). However, females may also influence copulation duration in terms of CFC (Córdoba-Aguilar and Cordero-Rivera 2008). For example, a classical and elegant paper by Andrés and Cordero-Rivera (2000) showed that not only copulation duration was far longer than needed to displace female's stored sperm but also explained male fertilization success. According to Andrés and Cordero-Rivera (2000), such long copulation duration by males was the means to court females.

9.3.2 Sperm Ejection

It has been well documented that after copulation, females of a number of animals spell sperm (e.g., spiders, Calbacho-Rosa et al. 2013; birds, Dean et al. 2011; primates, Reeder 2003). Such ejection has been until recently examined in light of sexual selection. In odonates, sperm ejection has been documented in a few species (Córdoba-Aguilar 1999, 2006; González-Soriano and Córdoba-Aguilar 2003; Tajima and Watanabe 2013); however, it seems widespread (Córdoba-Aguilar 2006). The one piece of evidence that indicates such widespread nature is that females entering copulation or after having finished oviposition have a much smaller sperm volume in their sperm storage organs compared to recently mated females (see Table 1 of Córdoba-Aguilar 2006). Such reduction is sometimes as much as 50 % (e.g., *Argia moesta*, Waage 1986) which cannot be understood in terms of sperm use for fertilization. In fact, from the species reviewed by Córdoba-Aguilar (2006), we have seen such a reduction in *Argia tezpi* (33 % reduction) and *Protoneura cara* (47 %) (all authors' unpublished data based on records of sperm masses via measuring remaining volumes of sperm from both sperm storage organs). In at least three odonate species, it has been documented that females may use such behavior as a CFC mechanism. In *Pantala flavescens*, *Enallagma praevarum*, and *Ischnura denticollis*, different females consistently eject similar volumes of sperm and lay similar numbers of eggs when mated with the same males (Córdoba-Aguilar 2006). This suggests that in case female choice is operating, females may use the same criteria (whatever this can be) to bias their oviposition decisions. One criterion that may be operating is choosing males on the basis of their territorial status. Support for this comes from *Paraphlebia quinta*, a tropical species with two male morphs that differ in aspect and territorial status (territorial and nonterritorial) (González-Soriano and Córdoba-Aguilar 2003). In this species, females tend to eject more sperm after having copulated with nonterritorial males than with territorial males (González-Soriano and Córdoba-Aguilar 2003). Nevertheless, the possibility that the sperm ejected after mating is the sperm removed by the last male (Lindeboom 1998) needs further study with molecular methods.

9.3.3 Oviposition Duration

After copulation, females are usually accompanied either in tandem or at a distance by the mating male, to lay their eggs (Corbet 1999). There is variation in the duration of oviposition, and one cause may be that females are exerting cryptic choice as suggested by Eberhard (1996). For example, females may prefer a certain male by performing a longer oviposition, which means laying a larger egg batch (Eberhard 1996) if egg laying rate is constant. A mating male, in fact, flies in front of the female or chases her when she changes her original landing site. In

Fig. 9.2 A *Calopteryx splendens* male showing the ventral area of his last abdominal segments during female oviposition. Photograph courtesy by A. Cordero Rivera

Calopteryx species, males show their pigmented wings while letting themselves be taken by the river surface for a few seconds while in front of the female (e.g., Córdoba-Aguilar 2000). During such behavior, males expose the ventral area of the last abdominal segments, an area that is conspicuous in color and that is only used during such situation (Córdoba-Aguilar 2000; Fig. 9.2). Such oviposition duration may also have other causes, however: (a) presence/absence of mating male. It is known that males may help the female to resist mating attempts by other males which may help female to save energy (e.g., McMillan 2000b) or reduce the female's risk to be captured by predators (e.g., Zeiss et al. 1999) or drown (Fincke 1986); (b) quality of oviposition site. Females usually prove how convenient an oviposition substrate can be (e.g., if it is not too hard to impede egg hatching or too dry to desiccate the egg) using mechanoreceptors and gustatory sensilla in the cutting valves (e.g., Rebora et al. 2013); (c) abiotic factors. Eggs can be laid in places where there is not enough oxygen or that are too cold or hot; (d) reduced energetic costs by harassing males. In the same fashion that females may accept to mate even when they are not ready to do so (e.g., even when they do not have mature eggs to lay), females may accept to continue in oviposition postures even when they do not have eggs to lay. The reason for this is that females may try to reduce the energetic costs of males chasing them.

Although such situation has not been tested, a female may save more energy by remaining motionless, pretending to be ovipositing, so the male may "assume" he is increasing his fertilization success as he is not "aware" that the female is not actually laying any eggs. Supporting information for this is that females actually remain perched on an oviposition substrate without performing oviposition.

9.4 What Do Odonate Females Choose?

Our evidence of what females choose seems restricted to (a) pathogen-resistant males. Studies in Calopterygidae indicate that females choose more pigmented males as these are more resistant to natural and non-natural pathogens such as mites, protozoan gregarines, and bacteria (reviewed by Forbes and Robb 2008). Based on sexual selection theory, such choice supposes that there is a genetic basis for resistance which can be passed on to offspring (reviewed by Puurtinen et al. 2009). Whether such inheritance is actually the case for odonates, it is not known. Related to this, one case leading to intraspecific variation in parasitism is that of seasonal variation as varying climatic conditions may affect parasite survival (e.g., Forbes et al. 2012); (b) sexy sons. Sperm ejection patterns by females during or after copulation have been associated with aedeagal stimulation carried out by the mating male (Córdoba-Aguilar 1999, 2006; Tajima and Watanabe 2013). Miller (1987) was the first to show this via direct manipulation of *Ischnura elegans* females. This was corroborated in at least two calopterygids (Córdoba-Aguilar 1999, 2006) and another ischnurin (Tajima and Watanabe 2013): Males use their aedeagus to stimulate a number of mechanoreceptor sensilla embedded in the female vagina. These sensilla have evolved initially to be stimulated by a passing egg so that the sensilla promote contractions of the spermatheca inducing sperm release and fertilization (Miller 1987). Since the aedeagus uses the same site where eggs pass, the aedeagus produces a super stimulus that elicits an extraordinarily large amount of ejected sperm (Córdoba-Aguilar 1999, 2009a, b; Tajima and Watanabe 2013). Aedeagal width has been positively selected as the wider the aedeagus, the more the ejected sperm (Córdoba-Aguilar 1999). It remains to be seen whether females actually benefit by having sexy sons (with similarly good stimulatory abilities as their male parents); and (c) male condition. Although condition can be reflected at the level of being pathogen resistant, as indicated above, aspects of male vigor can be extended to situations such as having enough energetic reserves and muscular mass to defeat male competitors that would try to dislodge the mating male prior and after copulation (previous and during oviposition). Tests should be designed to see how these three variables can be investigated in terms of female choice.

9.5 Female Choice Despite Male Coercion?

Odonate males are frequently "equipped" with traits that seem little to do with a female choice scenario. One example of such traits is the abdominal appendages ("claspers") that males use to grasp females during mating and tandem. The interspecific variation in this trait is amazing for both dragonflies (Fig. 9.3) and damselflies (Fig. 9.4). However, there has been literally no research on whether such claspers are selected via male–male competition, female choice or both, and/or

Fig. 9.3 Males of a sample of anisopteran species illustrating (insert) their claspers. **a** *Scalmogomphus guizhousensis*; **b** *Davidius trox*; **c** *Heliogomphus scorpio*; and **d** *Aeshna mixta*. Photograph courtesy by Tom Kompier for **a–c**, and by Harold Gough for **d**

natural selection. In relation to male–male competition, we know that such claspers promote a good grasping which may serve to avoid being dislodged by other males. One indication of this is that (a) females may frequently be damaged by several dragonflies (Dunkle 1991) and (b) there is positive selection acting during the time of grasping in nonterritorial males of *Hetaerina americana* (Grether 1996). With respect to female choice, it is interesting that females of some *Enallagma* species have mechanoreceptor sensilla embedded in their mesostigmal plates (Robertson and Paterson 1982), the sites where the claspers make contact (Fig. 9.5a, b). One would expect that such sensilla should not be present if claspers serve only for grasping, as also stated by Eberhard (1985, 1996). However, notice that sensilla presence could be also expected if these structures allow rejecting interspecific matings. Another source of information may also be taken as an indication of a female choice process for clasper evolution. Several studies have examined the static allometry (i.e., the proportional size of a particular structure with respect to body size) of clasper length and found that there is no positive allometry (i.e., large animals with proportionally larger structures than small animals) (Outomuro and Cordero Rivera 2012). According to Eberhard and collaborators (1998), positive allometry would be expected for structures whose mode of selection is via physical endurance like grasping the female in the case of odonates. The fact that such pattern is not consistent in odonates reveals that other processes may be involved leaving room for a female choice mechanism. There is also a natural selection explanation for the morphology of both claspers and mesostigmal plates. Using a group of *Argia* species, McPeek and collaborators have found that

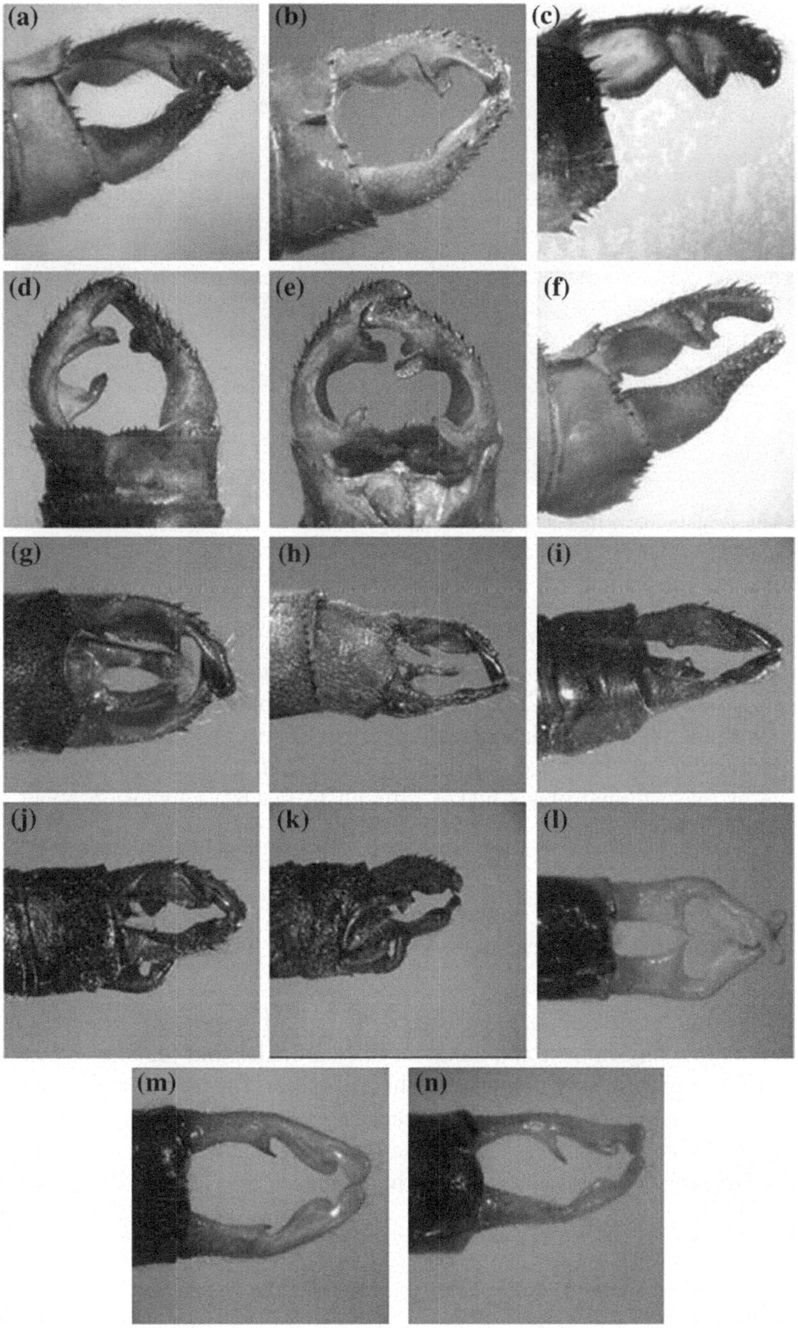

Fig. 9.4 Abdominal appendages for males of *Heteragrion brianmayi* (**a**), *H. freddiemercury* (**b**), *H. johndeaconi* (**c**), *H. muryense* (**d**, **e**), *H. rogertaylori* (**f**), *Lestes minutus* (**g**), *L. paulistus* (**h**), *Mnesarete guttifera* (**i**), *M. rhopalon* (**j**), *M. rhopalon* (**k**), *Perissolestes cornutus* (**l**), *P. paprzyckii* (**m**, **n**). Views are as follows: lateral (**a–c**, **f**), dorsal (**g–l**, **m**), and ventral (**e**, **k**, **n**). Photograph courtesy by F. Lencioni

(a) (b)

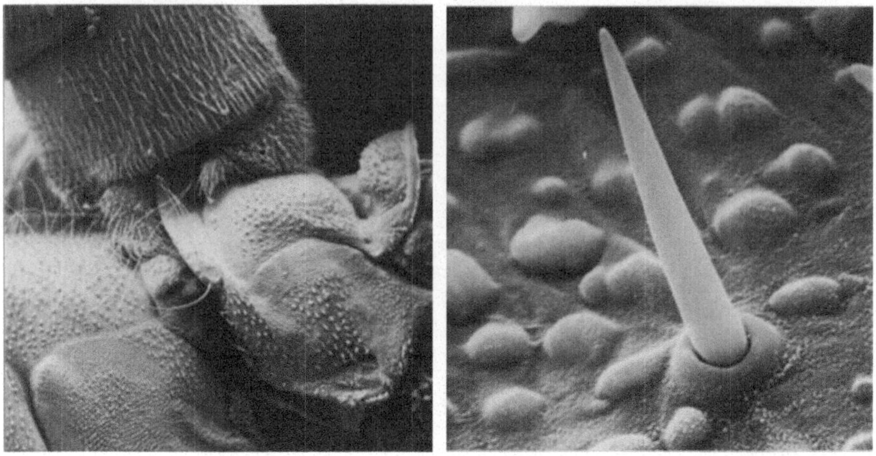

Fig. 9.5 **a** Female mesostigmal plates in *Enallagma dubium* showing hairylike emerging sensilla. Photograph courtesy from M. May and M. Mc Peek

the most likely explanation for the intra- and interspecific variation is that of species recognition, given that they found a rate of punctuated evolution for male and female traits (McPeek et al. 2008, 2009, 2011). This punctuated evolution is compatible with the idea of speciation with no effect of sexual selection. The fact that no experimental tests have been carried out manipulating the morphology of either claspers or mesostigmal plates makes difficult to reach a conclusion about their evolutionary maintenance.

9.6 Concluding Remarks

Far from being a male-dominated group, we have outlined several examples based on morphology, physiology, and behavior in both males and females, which may serve as instances used for females to choose males. These instances, in most if not all cases, seem speculative partly because no experimental tests have been carried out. Future researchers should use their imagination to see the extent that females drive the coevolutionary process in odonates.

Acknowledgments Thanks to R. Guillermo-Ferreira and, our long-lasting colleague, A. Cordero Rivera for their constructive comments to a previous version. Our research in odonates has been generously supported by PAPIIT grants 204610, 211506, 212506, 216808, and 222312. Thanks to R. I. Martínez Becerril for their logistic help.

References

Álvarez H, Serrano-Meneses M, Reyes-Márquez I, Jiménez-Cortés J, Córdoba-Aguilar A (2013) Allometry of a sexual trait in relation to diet experience and alternative mating tactics in two rubyspot damselflies (Calopterygidae: *Hetaerina*). Biol J Linn Soc 108(3):521–533

Andrés J, Cordero Rivera A (2000) Copulation duration and fertilization success in a damselfly: an example of cryptic female choice? Anim Behav 59(4):695–703

Byers CJ, Eason PK (2009) Conspecifics and their posture influence site choice and oviposition in the damselfly *Argia moesta*. Ethology 115(8):721–730

Calbacho-Rosa L, Galicia-Mendoza I, Dutto MS, Córdoba-Aguilar A, Peretti AV (2013) Copulatory behavior in a pholcid spider: males use specialized genitalic movements for sperm removal and copulatory courtship. Naturwissenschaften 100(5):407–416

Corbet P (1999) Dragonflies: behavior and ecology of odonata. Comstock Publishing Associates Cornell University Press, New York

Cordero A (1990) The adaptive significance of the prolonged copulations of the damselfly, *Ischnura graellsii* (Odonata: Coenagrionidae). Anim Behav 40(1):43–48

Cordero Rivera A, Andrés JA (2002) Male coercion and convenience polyandry in a Calopterygid damselfly (Odonata). J Insect Sci 2:14. www.insectscience.org/2.14

Cordero A, Santolamazza-Carbone S, Utzeri C (1992) A twenty-four-hours-lasting tandem in *Coenagrion scitulum* (Ramb.) in the laboratory (Zygoptera: Coenagrionidae). Not Odonatol 3(10):166–167

Cordero A, Santolamazza Carbone S, Utzeri C (1998) Mating opportunities and mating costs are reduced in androchrome female damselflies, *Ischnura elegans* (Odonata). Anim Behav 55:185–197

Cordero C, Eberhard WG (2003) Female choice of sexually antagonistic male adaptations: a critical review of some current research. J Evol Biol 16(1):1–6

Cordero C, Eberhard WG (2005) Interaction between sexually antagonistic selection and mate choice in the evolution of female responses to male traits. Evol Ecol 19(2):111–122

Córdoba-Aguilar A (1999) Male copulatory sensory stimulation induces female ejection of rival sperm in a damselfly. Proc Biol Sci 266(1421):779–784

Córdoba-Aguilar A (2000) Reproductive behaviour of the territorial damselfly *Calopteryx haemorrhoidalis asturica* Ocharan (Zygoptera: Calopterygidae). Odonatologica 29(4):295–305

Córdoba-Aguilar A (2002) Wing pigmentation in territorial male damselflies, *Calopteryx haemorrhoidalis*: a possible relation to sexual selection. Anim Behav 63:759–766

Córdoba-Aguilar A (2006) Sperm ejection as a possible cryptic female choice mechanism in Odonata (Insecta). Physiol Entomol 31(2):146–153

Córdoba-Aguilar A (2009a) Seasonal variation in genital and body size, sperm displacement ability, female mating rate, and male harassment in two calopterygid damselflies (Odonata: Calopterygidae). Biol J Linn Soc 96(4):815–829

Córdoba-Aguilar A (2009b) A female evolutionary response when survival is at risk: male harassment mediates early re-allocation of resources to increase egg number and size. Behav Ecol Sociobiol 63(5): 751–763

Córdoba-Aguilar A, Cordero-Rivera A (2005) Evolution and ecology of Calopterygidae (Zygoptera: Odonata): status of knowledge and research perspectives. Neotrop Entomol 34(6):861–879

Córdoba-Aguilar A, Cordero-Rivera A (2008) Cryptic female choice and sexual conflict. In: Córdoba-Aguilar A (ed) Dragonflies damselflies: model organisms for ecological and evolutionary research. Oxford University Press, New York, pp 189–202

Cordero Rivera A, Córdoba-Aguilar A (2010) Selective forces propelling genitalic evolution in Odonata. In: Leonard L, Córdoba-Aguilar A (eds) The evolution of primary sexual characters in animals. Oxford University Press, Oxford, pp 332–352

Córdoba-Aguilar A, Salamanca-Ocaña JC, Lopezaraiza M (2003) Female reproductive decisions and parasite burden in a calopterygid damselfly (Insecta: Odonata). Anim Behav 66:81–87

Córdoba-Aguilar A, Serrano-Meneses MA, Cordero-Rivera A (2009a) Copulation duration in nonterritorial odonate species lasts longer than in territorial species. Ann Entomol Soc Am 102(4):694–701

Córdoba-Aguilar A, Serrano-Meneses MA, Contreras-Garduño J (2009b) The lek mating system of *Hetaerina* damselflies (Insecta: Calopterygidae). Behaviour 146(2):189–207

Cronin H (1992) The ant and the peacock. Altruism and sexual selection from Darwin to today. Cambridge University Press, Cambridge

Dean R, Nakagawa S, Pizzari T (2011) The risk and intensity of sperm ejection in female birds. Am Nat 178(3):343–354

Dunkle SW (1991) Head damage from mating attempts in dragonflies (Odonata: Anisoptera). Entomol News 102(1):37–41

Eberhard WG (1985) Sexual selection and animal genitalia. Harvard University Press, New York

Eberhard WG (1996) Female control: sexual selection by cryptic female choice. Princeton University Press, Princeton

Eberhard WG (2005) Evolutionary conflicts of interest: are female sexual decisions different? Am Nat 165(5):S19–S25

Eberhard WG, Huber BA (1998) One size fits all? Relationships between the size and degree of variation in genitalia and other body parts in twenty species of insects and spiders. Evolution 52(2):415–431

Fincke OM (1986) Underwater oviposition in a damselfly (Odonata: Coenagrionidae) favors male vigilance, and multiple mating by females. Behav Ecol Sociobiol 18(6):405–412

Fincke OM (1997) Conflict resolution in the Odonata: implications for understanding female mating patterns and female choice. Biol J Linn Soc 60(2):201–220

Fincke OM, Waage JK, Koenig WD, Choe JC, Crespi BJ (1997) Natural and sexual selection components of odonate mating patterns. In: Choe JC, Crespi BJ (eds) The evolution of mating systems in insects and arachnids. Cambridge University Press, New York, pp 58–74

Fitzstephens D, Getty T (2000) Colour, fat and social status in male damselflies, *Calopteryx maculata*. Anim Behav 60(6):851–855

Forbes MR, Mlynarek JJ, Allison J, Hecker KR (2012) Seasonality of gregarine parasitism in the damselfly, *Nehalennia irene*: understanding unimodal patterns. Parasitol Res 110(1):245–250

Forbes MR, Robb T (2008) Testing hypotheses about parasite-mediated selection using odonate hosts. In: Córdoba-Aguilar A (ed) Dragonflies damselflies: model organisms for ecological and evolutionary research. Oxford University Press, New York, pp 175–188

González-Soriano E, Córdoba-Aguilar A (2003) Sexual behaviour in *Paraphlebia quinta* Calvert: male dimorphism and a possible example of female control (Zygoptera: Megapodagrionidae). Odonatologica 32(4):345–353

Grether GF (1996) Intrasexual competition alone favors a sexually dimorphic ornament in the rubyspot damselfly *Hetaerina americana*. Evolution 50(5):1949–1957

Guillermo-Ferreira R, Therézio EM, Gehlen MH, Bispo PC, Marletta A (2014) The role of wing pigmentation, UV and fluorescence as signals in a neotropical damselfly. J Insect Behav 27(1):67–80

Huang S, Chiou T, Marshall J, Reinhard J (2014) Spectral sensitivities and color signals in a polymorphic damselfly. PloS One 9. doi:10.1371/journal.pone.0087972

Iserbyt A, Van Gossum H (2011) Show your true colour: cues for male mate preference in an intra-specific mimicry system. Ecol Entomol 36(5):544–548

Jennions MD (1998) Tibial coloration, fluctuating asymmetry and female choice behaviour in the damselfly *Platycypha caligata*. Anim Behav 55:1517–1528

Lindeboom M (1998) Post-copulatory behaviour in *Calopteryx* females (Insecta, Odonata, Calopterygidae). Int J Odonatol 1:175–184

McMillan VE (2000a) Aggregating behavior during oviposition in the dragonfly *Sympetrum vicinum* (Hagen) (Odonata: Libellulidae). Am Midl Nat 144(1):11–18

Mcmillan VE (2000b) Postcopulatory behavior in Libellula pulchella Drury (Odonata : Libellulidae) and female tactics for avoiding male interference with oviposition. J Insect Behav 13(4):573–583

McPeek MA, Shen L, Farid H (2009) The correlated evolution of three-dimensional reproductive structures between male and female damselflies. Evolution 63(1):73–83

McPeek MA, Shen L, Torrey JZ, Farid H (2008) The tempo and mode of three-dimensional morphological evolution in male reproductive structures. Am Nat 171(5):E158–E178

McPeek MA, Symes LB, Zong DM, McPeek CL (2011) Species recognition and patterns of population variation in the reproductive structures of a damselfly genus. Evolution 65(2):419–428

Miller PL (1987) An examination of the prolonged copulations of *Ischnura elegans* (Vander Linden) (Zygoptera: Coenagrionidae). Odonatologica 16(1):37–56

Moore AJ (1990) The evolution of sexual dimorphism by sexual selection: the separate effects of intrasexual selection and intersexual selection. Evolution 44(2):315–331

Outomuro D, Cordero-Rivera A (2012) Allometry of secondary, primary, and nonsexual traits in the beautiful demoiselle (*Calopteryx virgo meridionalis*). Can J Zool 90(9):1094–1101

Puurtinen M, Ketola T, Kotiaho JS (2009) The good-genes and compatible-genes benefits of mate choice. Am Nat 174(5):741–752

Rebora M, Piersanti S, Dell'Otto A, Gaino E (2013) The gustatory sensilla on the endophytic ovipositor of Odonata. Arthropod Struct Dev 42(2):127–134

Reeder DM (2003) The potential for cryptic female choice in primates: behavioral, anatomical and physiological considerations. In: Jones CB (ed) Sexual selection and reproductive competition in primates: new perspectives and directions. American Society of Primatologists, Norman, pp 255–303

Rehfeldt GE (1992) Aggregation during oviposition and predation risk in *Sympetrum vulgatum* L. (Odonata: Libellulidae). Behav Ecol Sociobiol 30:317–322

Robertson HM, Paterson HE (1982) Mate recognition and mechanical isolation in *Enallagma* damselflies (Odonata: Coenagrionidae). Evolution 36(2):243–250

Samejima Y, Tsubaki Y (2009) Body temperature and body size affect flight performance in a damselfly. Behav Ecol Sociobiol 64(4):685–692

Siva-Jothy MT (1999) Male wing pigmentation may affect reproductive success via female choice in a calopterygid damselfly (Zygoptera). Behaviour 136(10):1365–1377

Siva-Jothy MT, Gibbons DW, Pain D (1995) Female oviposition-site preference and egg hatching success in the damselfly *Calopteryx splendens xanthostoma*. Behav Ecol Sociobiol 37(1):39–44

Siva-Jothy MT, Tsubaki Y (1989a) Variation in copulation duration in *Mnais pruinosa pruinosa* Selys (Odonata: Calopterygidae) 1. Alternative mate securing tactics and sperm precedence. Behav Ecol Sociobiol 24:39–45

Siva-Jothy MT, Tsubaki Y (1989b) Variation in copulation duration in *Mnais pruinosa pruinosa* Selys (Odonata: Calopterygidae) 2. Causal factors. Behav Ecol Sociobiol 25:261–267

Svensson EI, Waller JT (2013) Ecology and sexual selection: evolution of wing pigmentation in calopterygid damselflies in relation to latitude, sexual dimorphism, and speciation. Am Nat 182(5):E174–E195

Tajima Y, Watanabe M (2013) Male secondary genitalia mimic the female egg during oviposition for sperm displacement in the non-territorial damselfly *Ischnura asiatica* (Brauer) (Zygoptera: Coenagrionidae). Odonatologica 3(42):229–236

Takahashi Y, Watanabe M (2011) Male mate choice based on ontogenetic colour changes of females in the damselfly *Ischnura senegalensis*. J Ethol 29(2):293–299

Tsubaki Y, Samejima Y, Siva-Jothy MT (2010) Damselfly females prefer hot males: higher courtship success in males in sunspots. Behav Ecol Sociobiol 64(10):1547–1554

Van Gossum H, Bots J, Heusden J, Hammers M, Huyghe K, Morehouse NI (2011) Reflectance spectra and mating patterns support intraspecific mimicry in the colour polymorphic damselfly *Ischnura elegans*. Evol Ecol 25(1):139–154

Waage JK (1973) Reproductive behavior and its relation to territoriality in *Calopteryx maculata* (Beauvois) (Odonata: Calopterygidae). Behaviour 47:240–256

Waage JK (1986) Evidence for widespread sperm displacement ability among Zygoptera (Odonata) and the means for predicting its presence. Biol J Linn Soc 28(3):285–300

Zeiss C, Martens A, Rolff J (1999) Male mate guarding increases females' predation risk? A case study on tandem oviposition in the damselfly *Coenagrion puella* (Insecta: Odonata). Can J Zool 77(6):1013–1016

Chapter 10
What Is Indirect Cryptic Female Choice? Theoretical Considerations and an Example from a Promiscuous Earwig

Yoshitaka Kamimura

Abstract Cryptic female choice (CFC) does not necessarily involve discriminative responses in the female nervous system to sperm from multiple mates. Even without any active sperm-choice mechanisms, polyandrous females can gain genetic benefits by having an arena in which genetically superior sperm are "automatically" sorted. In this chapter, possible mechanisms in this CFC category, termed "indirect CFC," are reviewed. A simple theoretical model is developed to examine the hypothesis that females obtain genetic benefits by allowing only partial displacement of stored sperm by subsequent mates. The model predicts that such restricted sperm displacement automatically grants genetic benefits when genetically superior males copulate more times per encounter with the female than less fit males. The promiscuous earwig species, *Euborellia plebeja*, provides an empirical example of this type of indirect CFC. The elongated female sperm-storage organs allow only partial removal and displacement of stored sperm by shorter male genitalia, resulting in a 20 % gain in paternity per mating. In staged mating trials, large males dominated male–male competition for burrows housing females, resulting in a significant increase in paternity by repeated matings with the same female. A numerical simulation based on this mating pattern showed that restricted sperm displacement (~20 % per mating) is optimal for females to accumulate sperm from larger males. Given that male body size is heritable, females were estimated to gain a 1.4 % increase in their sons' mating success as a genetic benefit. Advantages and disadvantages of indirect CFC are discussed and compared with precopulatory mate choice and direct CFC.

Y. Kamimura (✉)
Department of Biology, Keio University, Hiyoshi, Yokohama, Japan
e-mail: kamimura@fbc.keio.ac.jp

A.V. Peretti and A. Aisenberg (eds.), *Cryptic Female Choice in Arthropods*,
DOI 10.1007/978-3-319-17894-3_10

10.1 Introduction: Direct and Indirect Mechanisms for Collecting Good/Sexy Sperm

Darwin (1871) recognized female choice as one of the two major categories of sexual selection that operate in precopulatory phases, along with male–male competition for mates. Even without any direct benefits (i.e., increase in the number of offspring raised via enjoyment of nuptial gifts, suitable habitats, and/or paternal care of offspring from males) arising from choosing mates, theories predict that female preference for a certain male trait can evolve if the genetic benefits—that is, the increase in the genetic quality of offspring—outweigh the associated costs (see Chap. 1). The term "female choice" may give the impression that females *directly* compare and *actively* choose some males over others through differential responses of the female nervous system to varying male attributes (direct female choice). However, Wiley and Poston (1996) pointed out another category of female choice. Even without any direct and active choice behavior, females can gain genetic benefits by promoting male–male competition for mates. For example, reproductively active males sometimes gather in certain places, which females visit only for mating (Höglund and Alatalo 1995). Males usually show exaggerated mating displays based on which females actively choose mating partners (Höglund and Alatalo 1995). However, this mating system, termed lekking, also promotes more severe male–male competition within the arena. As a result, even without active female choice of males, females in a lek likely mate with males of higher genetic quality than females that do not show lekking behavior (Wiley and Poston 1996). Wiley and Poston (1996) extensively reviewed this "indirect female choice" and defined it as female characteristics that restrict the mate pool, but which lack behavioral/morphological changes in response to male attributes. By contrast, in direct female choice, females directly discriminate male attributes through central or peripheral neural filtering (Wiley and Poston 1996). Wiley and Poston (1996) also included another mechanism, female morphology that allows only males with matching morphology to mate, as a type of direct female choice (Wiley and Poston 1996). They proposed that instead of neural filtering, female morphology directly selects males based on their attributes (e.g., male genital morphologies matching the female genitalia). However, as discussed below (Sect. 10.2.3), female morphologies that complicate genital coupling can also promote harsher male–male competition for mates, imposing selection pressures similar to those of lekking behavior. Thus, I adopt a single criterion to distinguish direct and indirect mechanisms of female choice: the presence or absence of neural filtering of male attributes by females.

Are indirect mechanisms also feasible for cryptic female choice (CFC), which can be seen as the postcopulatory version of sexual selection via female choice? As discussed in every chapter of this book, females usually have many opportunities for control over the processes that occur after genital coupling and insemination. When a heritable variation in male abilities to override such female control exists, polyandrous females "automatically" gain genetic benefits by having sons that inherit the superior abilities of their fathers (sexy-son effects; Keller and Reeve 1995). In addition, when such abilities are positively and genetically correlated with other male or female fitness measures, further benefits arise (good-gene effects; Yasui 1997). In the following two sections of this chapter, I examine and review the possible mechanisms of indirect CFC. A detailed theoretical consideration is given for the possibility that incomplete sperm displacement, which is imposed by the morphology of female sperm-storage organs, works as an efficient mechanism of indirect CFC. In the latter half of this chapter, I review our previous studies of an earwig species as an empirical example of the incomplete-sperm-displacement model.

10.2 Possible Mechanisms of Indirect CFC

Based on an extensive literature review, Eberhard (1996) listed nine principal mechanisms of CFC that generally apply to animals with multiple supporting examples (Table 10.1). To identify examples of CFC, Eberhard (1996) adopted the following five criteria: (1) Female responses to some conspecific males differ from those to others; (2) such discrimination occurs under natural conditions; (3) the discrimination results in differences in reproductive success for the males involved; (4) female biases are associated with particular male characteristics; (5) variation among males in characters used by females to discriminate is associated with genetic differences. However, as Eberhard (1996) himself pointed out by reviewing many possible examples, the first criterion, differential responses by females, is not indispensable in the sense that the female nervous system is involved in the response. For example, in eggs of the marine ctenophore *Beroe ovata*, the female pronucleus "visits" multiple male pronuclei (resulting from polyspermy) before it "chooses" one as the counterpart of fusion (Carré and Sardet 1984). Although the significance of this phenomenon is currently unknown, allowing both it and polyspermy in a single cell (ovum) can work as a mechanism of CFC without any involvement of the nervous system. As reviewed below, we can envisage much simpler mechanisms of indirect CFC that lack alterations in female morphological or physiological characteristics. In this and the next sections, I discuss possible mechanisms of indirect CFC in relation to the nine principal categories proposed by Eberhard (1996) (Table 10.1, but in a different order from that of Eberhard 1996).

10.2.1 Remating with Another Male

One of the most prevalent and well-supported mechanisms of CFC is female remating with another male when the quality of previous males is lower (category I of Table 10.1). For example, in several socially monogamous bird species, females actively seek extra-pair matings with high-quality males (e.g., Smith 1988; Kempenaers et al. 1996). No doubt exists that neural mechanisms are involved

Table 10.1 Nine principal categories of CFC identified by Eberhard (1996) and possible corresponding indirect mechanisms.

Direct mechanisms	Female property as the target of selection	Possible indirect mechanisms (male traits selected; section in which discussed)
I. Sometimes remate with another male	Probability of remating	Indiscriminate remating with males (larger ejaculate size; Sect. 10.2.1)
II. Sometimes discard sperm of current male	Amount of sperm ejected	Dumping of constant amount of sperm (larger ejaculate size; Sect. 10.2.2)
III. Sometimes discard sperm of previous males	Amount of sperm ejected (or allowed to be displaced)	Dumping of constant amount of sperm (larger ejaculate size; Sect. 10.2.2) Allowance of partial sperm displacement by having exaggerated storage organs (larger genitalia, matching in male–female genital traits, larger ejaculate size, and/or greater ability to secure mates to allow repeat matings; Sects. 10.3–10.5)
IV. Sometimes prevent complete intromission and ejaculation	Strength of resistance	Having obstacle morphologies that make intromission and/or ejaculation difficult (more vigorous mating attempts and/or matching in male–female genital traits; Sect. 10.2.3)
V. Sometimes forcefully terminate mating before sperm are transferred	Strength of resistance	Promoting takeover of mates by making rapid ejaculation difficult (more vigorous males that are effective at mating takeovers; Sect. 10.2.3)
VI. Sometimes fail to transport sperm to storage organs or fertilization sites	Probability of transportation	Digestion or killing of a constant amount of sperm before transportation (higher ejaculate size, higher swimming speed, and/or higher viability of sperm; Sect. 10.2.2) Chemical attraction of sperm to the storage or fertilization site (sperm swimming speed; Sect. 10.2.2); Production of hostile environments at the storage or fertilization sites (sperm viability or compatibility; Sect. 10.2.2)
VII–IX. Sometimes fail to ovulate, fail to mature eggs, or reduce rate or number of offspring produced	Probability of ovulation/probability of maturing eggs/rate or number of offspring produced	Ovulating eggs/maturing eggs/producing offspring proportional to the amount of nutrition provided by the focal males (larger nuptial gifts; Sect. 10.2.4)

In direct CFC, the female nervous system is involved in modulating female properties (middle column) based on male attributes. On the other hand, female properties do not vary (or vary only passively) with male attributes in the possible indirect CFC mechanisms (the right column) in the decision making (i.e., direct CFC). Let us consider an indirect version of this mechanism in which females repeatedly accept mating attempts regardless of the attributes of the current and the previous males. Can this mechanism—that is, indiscriminate multiple matings by females (indiscriminate FMM)—work as an indirect CFC? Males usually provide females with sperm enough to fertilize all eggs produced in the reproductive bout (e.g., Bateman 1948), and such increases in the number of mates likely result in enhanced risk and intensity of sperm competition. When the number of sperm in the ejaculate has a genetic basis (sexy sperm) or is positively correlated with male quality (good sperm), and when sperm from multiple males are used for fertilization based on numerical representation in the female storage organ (fair raffle sensu Parker 1990), indiscriminate FMM automatically brings genetic benefits to promiscuous females as an increase in the number of offspring inheriting high-quality genes relative to monandrous females (Fig. 10.1a vs. 10.1b). Similar logic holds in cases where sperm quality (e.g., swimming speed from a storage organ to the site of fertilization) varies with male quality. Because of indiscriminate acceptance of courting males, the sperm store of a polyandrous female is inevitably diluted by the sperm of lower quality males, making indiscriminate FMM less effective than direct CFC (Fig. 10.1b vs. 10.1c). Thus, this type of FMM is likely to evolve only when (1) "directly" assessing male quality in pre- or postcopulatory phases is impossible or highly costly, and (2) the cost of additional matings does not outweigh the genetic benefits (Fig. 10.1).

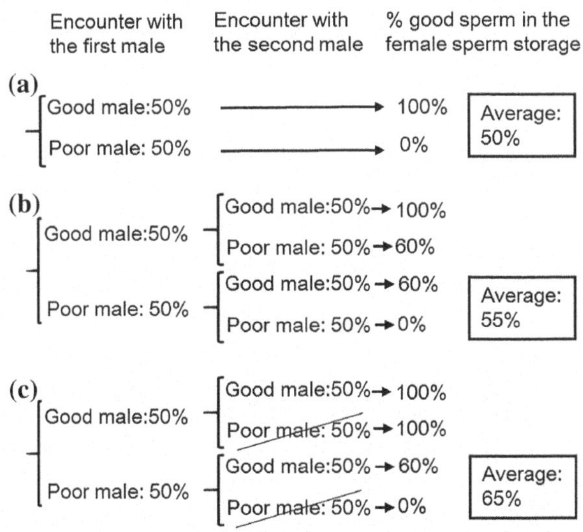

Fig. 10.1 Genetic benefits of the "indiscriminate polyandry" model (**b**) compared to monandry (**a**), in which females accept only the first mating attempt, and direct CFC (**c**), in which females reject poor males as second mates as indicated by the strike-through (i.e., only seek remating with a good male). The probabilities of encountering a good or poor male are assumed to be even. Good males provide sperm at 1.5-fold the rate of poor males: If a female mates with one good and one poor male, the expected percentage of offspring sired by the good male is 60 %. See main text for details

To my knowledge, no empirical studies have quantified the genetic benefits raised by this mechanism of indirect CFC. Female chimpanzees (*Pan troglodytes*) and some Old World monkeys living in multi-male and multi-female groups show marked sexual skin swelling at the peak of estrous. Although the most plausible function of this sexual signal is to induce male–male competition and mate guarding by high-ranking males (precopulatory indirect female choice), promiscuous matings and multiple paternity are also common in these primates (Dixson and Anderson 2004). Relatively large testis size in these animals indicates strong sperm competition working on ejaculate size, and the sexual skin swellings can also function as a sexual signal toward unspecified males to induce severe sperm competition (Harvey and May 1989).

10.2.2 Discarding of Sperm and Failure to Transport Sperm to Storage Organs or Fertilization Sites

As the second principal CFC mechanism, females sometimes eject sperm from the current male (category II of Table 10.1). A well-described example comes from the feral fowl *Gallus gallus domesticus*, in which coercive mating by subordinate, unpreferred males is common. When subsequent mating by a dominant male is unavailable, females frequently expel the semen of the subordinate male from the cloaca (Pizzari and Birkhead 2000). This mechanism obviously is effective at preferentially biasing the resultant paternity toward high-ranking males. This raises the question whether an indirect version of sperm dumping also works: Can females acquire genetic benefits by sperm dumping even when their nervous system cannot discriminate male quality? Although no empirical study has investigated such indiscriminate sperm dumping, let us consider an imaginary case in which females always discard a constant amount of ejaculate, D, after each mating regardless of mate quality. The volume of ejaculate transferred in a single mating, V, is positively correlated with male quality. At each mating, females eject D before transferring sperm of the remaining ejaculate ($V - D$) to a storage organ and use them after several matings in a fair raffle pattern (thorough mixing of $V - D$ from multiple males in the storage organ). As shown in Fig. 10.2, unless D is too large to maintain fertility, this simple mechanism can increase the proportion of offspring sired by males of higher quality. In other words, a small variation in male quality can be amplified in the resultant paternity success by constant, indiscriminate sperm dumping (Fig. 10.2). Instead of dumping sperm from the current mate, a similar effect can be obtained by dumping a constant amount of previous sperm before each mating. For indirect CFC related to the use of sperm from previous males (category III in Table 10.1), further mechanisms are discussed in detail later (Sects. 10.3–10.5).

Females of some animals possess an organ (or cells) that digests or kills part of the sperm transferred by males (e.g., female bedbugs; Carayon 1966). The environment of the female reproductive tract of mammals, including humans, is generally so hostile (or at least not conducive) to sperm that most die before arriving at fertilization sites (see pp. 336–337 in Eberhard 1996 for a review), indicating the importance

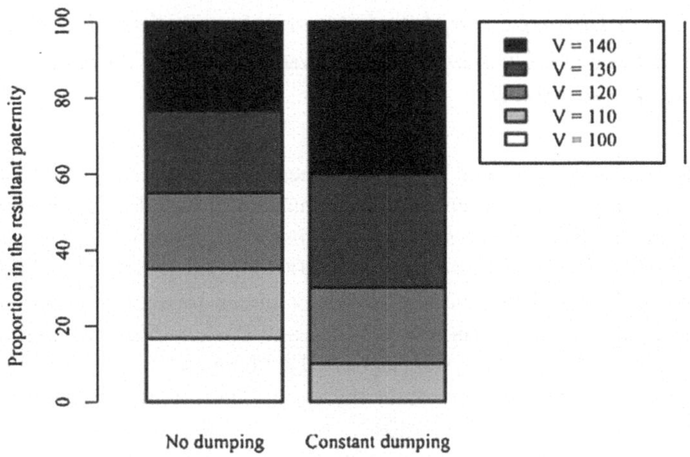

Fig. 10.2 The "constant-indiscriminant sperm dumping" model. The graph shows a case of a female mating with five males of varying ejaculate size (100, 110, 120, 130, and 140), which is positively correlated with male genetic quality. Note that the female is expected to have more superior sons if she discharges a constant amount (100) of each ejaculate (*the right column*) instead of accepting all ejaculates without dumping (*the left column*). The dumping increases the variation in relative paternity success of males (range: 16.7–23.3 % without dumping vs. 0–40 % with dumping)

of a huge number of sperm to overcome high mortality in the female genital tract. Thus, the severe environment of the female genital tract evidently imposes a selection pressure on the amount of sperm in an ejaculate, from which genetic benefits inevitably arise if the amount of sperm has a genetic basis (Harvey and May 1989). Such digestion or killing of sperm can occur at various stages, including before the transfer of sperm to storage organs or fertilization sites (category VI in Table 10.1). Like the constant sperm-dumping model, digestion or killing of a constant amount of sperm from each ejaculate can function as an indirect CFC mechanism. One key assumption is a positive correlation between male quality and the size (or quality) of ejaculates, which was shown for at least some animals (e.g., Chargé et al. 2013). Several studies have shown that sperm-attractant substances are released from fertilization sites such as eggs or ovaries (Eisenbach and Giojalas 2006; Yanagimachi et al. 2013). Given the hostile conditions of the corridor to the fertilization site, attraction mechanisms likely also select for faster swimming sperm.

Finally, fertilization itself is likely a difficult task for sperm, at least in some animals. Eggs or other female components (e.g., ovarian fluids) of many animals have been reported to possess chemical barriers reducing the possibility of fertilization by sperm of distantly related conspecific males, as well as heterospecific sperm (e.g., Evans and Sherman 2013; Yeates et al. 2013). While such gamete incompatibility apparently satisfies the definition of indirect CFC and may be effective at avoiding outbreeding depression, there is ongoing debate on whether this mechanism is also effective as an indirect CFC for sexy or good sperm, as a precursor for the rapid divergence of gamete recognition systems (Evans and Sherman 2013).

10.2.3 *Prevention of Complete Intromission and Ejaculation or Forceful Termination of Mating Before Sperm are Transferred*

Success in mounting a female does not necessarily mean success in insemination. Females sometimes prevent complete intromission and ejaculation or forcefully terminate mating before sperm are transferred (Eberhard 1996; categories IV and V in Table 10.1; see also Simmons 2001). Such female-borne barriers do not necessarily involve resistance by females and can be accomplished simply by having obstructive structures (note that Wiley and Poston 1996 treated this mechanism as a case of direct CFC; see Sect. 10.1). Females of several water striders bear a pair of elaborate projections at the caudal end of the abdomen (Arnqvist and Rowe 1995; Perry and Rowe 2012). By artificially shortening or extending the projections, Arnqvist and Rowe (1995) clearly showed that this structure functions to thwart harassing males and to reduce mating frequency. This result suggests that only superior males (more vigorous and/or with genitalia matching the female's morphology) can mate with females with longer projections. In the seed bug *Lygaeus simulans*, the length of male genitalia determines ejaculation success (Tadler 1999). Although no manipulative experiment has been conducted, the long, convoluted seminal duct in the females, which makes complete insertion of the male genitalia difficult (Gschwentner and Tadler 2000; Micholitsch et al. 2000), is a candidate of the postcopulatory version of this indirect female choice mechanism. When females behaviorally resist courting males "with unmodulated vigor" regardless of the male phenotype, this should also work as an indirect CFC mechanism akin to morphological obstacles. In such cases, showing the constancy and indifference of resistance would be very difficult.

In many arthropods, insemination occurs long after the onset of genital coupling (Eberhard 1996). If females determine the length of this preejaculation period (but without any modulation in response to the male's attributes) and if it promotes competition and female takeover of by males, this would certainly represent an indirect CFC. Again, although such takeovers are widely reported for many animal taxa including arthropods, to my knowledge, the genetic benefits have not been empirically estimated.

10.2.4 *Reduction in the Rate or Number of Offspring Produced, Failure to Ovulate, or Failure of Eggs to Mature*

Females sometimes control the rate and/or numbers of offspring production through failures of ovulation or egg maturation, or by other mechanisms (Eberhard 1996; categories VII–IX in Table 10.1). Because modulation of these processes usually occurs at a time/site separated both chronologically and spatially from the time/site of insemination, it is difficult to rule out, either theoretically or empirically, differential responses of the female nervous system to varying male or ejaculate attributes. However, even for these categories of CFC, we can envision potential indirect mechanisms. Let us consider an animal in which males donate a nuptial gift during mating and where the volume of the gift is positively correlated with male quality. If females produce mature eggs after each mating and the nutrients from the gift limit the number of eggs produced in each oviposition bout without any differential responses in the female nervous system, such a case can be categorized as indirect CFC. In crickets *Allonemobius socius*, larger males provide larger nuptial gifts (hemolymph), which promote female egg production (Fedorka and Mousseau 2002a, b). Male body size is highly heritable in this species (Fedorka et al. 2007), indicating that this species is a promising candidate of this CFC mechanism.

10.3 Why Females Do not Allow Complete Sperm Displacement: A CFC Hypothesis

Removal of rival sperm from female sperm-storage organs by male genitalia is widely known (e.g., Simmons 2001). Although this tactic is usually discussed from the perspectives of males for avoiding or mitigating sperm competition, females potentially affect the outcome of sperm removal as well. Mounting evidence shows that females sometimes have exaggerated morphologies in their sperm receptacle/storage organs that allow only partial manipulation of sperm from previous males by a current male. For example, males of several damselflies were reported to make repeated abdominal movements to remove as much rival sperm as possible from the female sperm-storage organ (Siva-Jothy and Tsubaki 1989; Cordero et al. 1995). Similar sperm-removal behaviors were also reported for a cuttlefish (Wada et al. 2005). The morphology of female sperm-storage organs sometimes does not permit male genitalia to access stored sperm, rendering complete sperm manipulation physically impossible (reviewed by Córdoba-Aguilar et al. 2003 for cases in Odonata). Thus, the shape of a sperm-storage organ is an important determinant of postcopulatory sexual selection (see also Requena and Alonzo 2014). These examples suggest that, by possessing exaggerated sperm-storage organs, females can passively select males that are better at removing sperm and/or males that transfer a larger ejaculate that is difficult for subsequent males to completely remove. As introduced and discussed in the following sections, the functional significance of an elongated sperm-storage organ (a spermatheca) has been studied for a promiscuous earwig species, *Euborellia plebeja* Dohrn (Kamimura 2013). In this species, the elongated spermatheca is thought to function as part of another indirect CFC mechanism. To explore how a long (or large) sperm-storage organ, which allows only incomplete sperm manipulation by males, might function in indirect CFC in promiscuous female earwigs, I develop a simple theoretical model below. Most of the model's specific assumptions relate to the mating ecology of *E. plebeja*.

10.3.1 Model

For simplicity, let us consider a population consisting of two discrete types of males. One shows high mate-guarding ability, reflecting high genetic quality. I refer to males of this type as good males. The other type, poor males, is of lower genetic quality and is less efficient at mate guarding. Although many male and female traits can be positively correlated with genetic quality (the good sperm hypothesis), for convenience, it is sufficient to assume here that male genetic quality positively correlates with only a single trait, the efficiency of mate guarding (the sexy sperm hypothesis). Good males mate *Tg* times on average with every female encountered, while poor males mate *Tp* times ($Tp < Tg$) because of their lower mate-guarding ability. We can assume that good males can more efficiently take over females guarded by a poor male and that this causes the difference in the average number of matings per encounter with a female. When the encounter rates with good and poor males are *Eg* and *Ep*, respectively ($Eg + Ep = 1$), polyandrous females that encounter *N* males mate $N\ (Eg\ Tg + Ep\ Tp)$ times. The encounter rates are not necessarily equal to the numerical representation of the two types of males in the population (for example, in cases where good males more efficiently access females). Since we assume that the male's ability to secure a mate for a prolonged duration has a genetic basis, when repeated matings by males result in an increase in paternity, females gain genetic benefits in the form of production of "good sons" sired by good fathers. From a female perspective, let us consider the optimal value of the last male paternity for gathering good sperm from good males. Here, I define the last male paternity (P_{last}) as the paternity gain from a single mating with a sperm-saturated female ($0 \leq P_{\text{last}} \leq 1$; Kamimura 2005): This means that a proportion, P_{last}, of the stored sperm is displaced by males at each mating. The paternity gain from the previous matings—that is, the paternity of the previous males and that gained from the previous matings by the focal male—can be defined as $r = 1 - P_{\text{last}}$. I assume that female sperm storage is full by the first mating and that P_{last} does not vary with male quality. Let us consider a female that has mated with $N - 1$ males, with the proportion of good sperm S_{N-1} in sperm storage. Sperm from previous males are completely mixed. Now, she encounters another good male (*N*th male) with a probability of *Eg*.

After one mating with the male, the proportion of good sperm changes to:

$$S_{N(1)} = rS_{N-1} + P_{\text{last}}.$$

After one more mating with the same male, and assuming thorough mixing of sperm between the two consecutive matings, the proportion of good sperm changes to:

$$S_{N(2)} = rS_{N(1)} + P_{\text{last}} = r^2 S_{N-1} + \left(1 - r^2\right).$$

Repeating this procedure for *Tg* times, we obtain the expected proportion of good sperm after the *N*th encounter with a good male as follows:

$$r^{Tg}S_{N-1} + \left(1 - r^{Tg}\right).$$

However, when the female encounters a poor male as her Nth mate, the proportion of good sperm is reduced to $r^{Tp}\, S_{N-1}$ due to the male mating Tp. Considering the encounter rates, the expected proportion of good sperm after the Nth encounter is as follows:

$$S_N = Eg\left[r^{Tg}S_{N-1} + \left(1 - r^{Tg}\right)\right] + Epr^{Tp}S_{N-1}. \tag{10.1}$$

The expected proportion of good sperm in a female after the first encounter (S_1) is simply expressed as the encounter rate with good males (Eg). Using this value, we can solve the recurrence formula (10.1) as follows:

$$S_N = \left(r^{Tg}Eg + r^{Tp}Ep\right)^{N-1}\left[Eg - \frac{Eg\left(1 - r^{Tg}\right)}{1 - r^{Tg}Eg - r^{Tg}Ep}\right] + \frac{Eg\left(1 - r^{Tg}\right)}{1 - r^{Tg}Eg - r^{Tg}Ep}. \tag{10.2}$$

Figure 10.3 shows a number of example solutions. When females sample only one male (monandry), the last male paternity (P_{last}) has no effect on the expected proportion of good sperm (S_N) because it is simply determined by the encounter rate (Eg). When females mate with multiple males, S_N shows a unimodal curve with a peak at $P_{\text{last}} < 0.5$; the smaller the P_{last}, the higher the number of males that must be sampled to maximize S_N. However, regardless of the number of males sampled (N), no increase in S_N from Eg is observed when $P_{\text{last}} = 1$ ($r = 0$) or $P_{\text{last}} = 0$ ($r = 1$), where only sperm from the last or from the first male are stored, respectively. The observed increase in S_N for moderately low P_{last} is higher when good males mate many more times per encounter than poor males (compare Fig. 10.3a, c, and e with Fig. 10.3b, d, and f, respectively). Changes in the encounter rates (Eg, Ep) directly affect the minimum values of S_N, but they have little effect on the qualitative pattern of the graphs (Fig. 10.3a, b vs. Fig. 10.3c, d vs. Fig. 10.3e, f), unless Eg or Ep are extremely small (i.e., in cases where almost no genetic variation is available for females, data not shown).

10.3.2 Interpretation of Model Predictions

Why is moderately low sperm displacement effective in acquiring sperm from superior males? To illustrate the underlying logic, let us consider an example in which three males successively visit a female. Both the first and last male are males of poor quality that can mate only once with the female ($Tp = 1$). After a single mating with the female, the first male is displaced by the second male. The second male is superior and can guard the female for a longer, during which he mates with her twice ($Tg = 2$). However, the second male is finally displaced by a third male of lower quality. As discussed below, such takeovers by males of lower

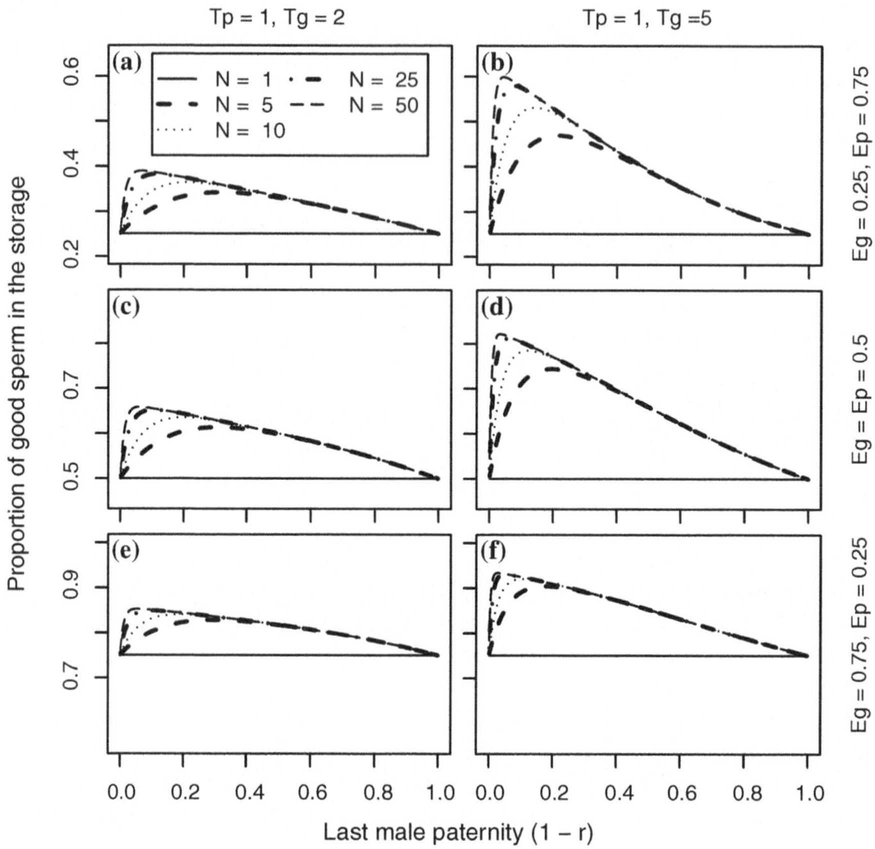

Fig. 10.3 Expected proportion of sperm from good males in female sperm storage calculated based on Eq. 10.2. *N* denotes the number of mates

quality do occur in nature. Given this mating sequence (poor–good–poor), how can the female maximize the quality of the progeny? If the female allows only a very small amount of sperm to be displaced per mating (e.g., $P_{last} = 0.01$), the majority of offspring will be sired by the first poor mate irrespective of the subsequent repeated matings by the superior second male. Instead, if the female allows almost complete displacement of sperm by males (e.g., $P_{last} = 0.8$), most of the sperm from the first male will be displaced by sperm from the second good male during the two subsequent matings. However, this effect will also be canceled by the final mating with the third, poor mate. This illustrates that to properly reflect differences in male quality (expressed as the difference in mating frequency per encounter) in the resultant paternity share, an optimal value exists between values that are too low ($P_{last} \approx 0$) or too high ($P_{last} \approx 1$), which maximizes the average quality of sons. If females can sample many (e.g., 30) males before oviposition, they have many chances to gather good sperm a little at a time from multiple good males. In this case, one would expect that a lower P_{last} value would be favored, as it reduces the risk of "cancelation" by the subsequent arrival of poor males. This effect is shown in Fig. 10.3 as a gradual decrease in the optimal P_{last} values with increases in the number of males sampled.

10.4 A Case Study of a Promiscuous Earwig

A fundamental question is whether any empirical examples exist for the indirect CFC mechanism modeled in the previous section. In this section, our previous work on the earwig species *E. plebeja* (Anisolabididae) is introduced in detail as the empirical case for which we quantitatively estimate the genetic benefits (Kamimura and Iwase 2010; Kamimura 2013).

Dermaptera (earwigs) is a polyneopteran insect order with ~2200 described species from mainly tropical and warm temperate regions (Popham 2000; Haas et al. 2012). Although several species are considered pests in gardens and agriculture or pest control agents, most earwig species have no direct relationship with human activities. Accordingly, mating biology has been studied in only a few common species and published information on the ecology of most is extremely scarce. Earwigs have diverse genital structures that differ in the size, number, and laterality of both male and female genital components, making them valuable study models of genital evolution (reviewed in Kamimura 2014; Fig. 10.4). Nevertheless, few studies on postcopulatory sexual selection have been published on this group of insects. As introduced in detail below (Sects. 10.4 and 10.5), all relevant studies examined the functional aspects of elongated male and female genitalia in several earwig species in the family Anisolabididae (but see also Kamimura 2004, 2014 for earwigs with multiple sperm-storage organs that can potentially function as sperm-choice devices).

10.4.1 Study Species and Problems

E. plebeja is a small flightless earwig with vestigial wing pads on the thorax (Fig. 10.5a–c) inhabiting open habitats. Both males and females are highly promiscuous (Baijal and Srivastava 1974; Kamimura, 2003b, 2005). Under laboratory conditions, females mate several to dozens of times in a 15-h period (Kamimura 2005). Several (usually less than three) matings are enough to saturate the single, elongated, and fine-tubed spermatheca (Fig. 10.5f). Even after saturation, females continue to have frequent repeated matings with males, which cannot be explained merely as a sperm-supply function (Kamimura 2005). These behaviors suggest severe sperm competition in this species.

Earwig males have a genital (phallic) organ on the midline of the body in the reflected membrane above the penultimate sternite (subgenital plate or

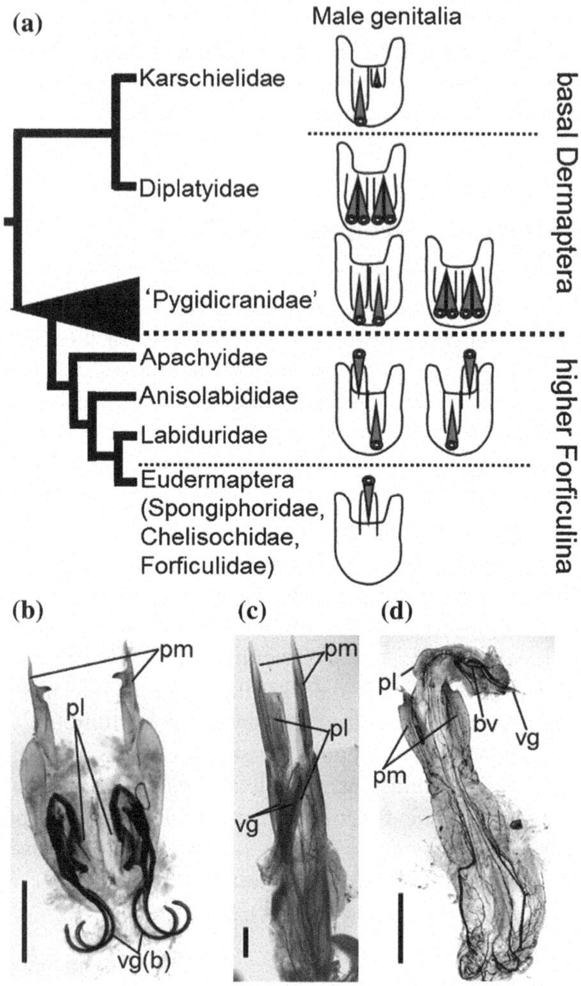

Fig. 10.4 The most elaborate phylogenetic hypothesis of the Forficulina (modified from Haas and Klass 2003) with schematic drawings of male genitalia of each group (**a**). Although Haas and Kukalovà-Peck (2001) originally proposed that Karschiellidae is the first offshoot of the present Dermaptera, while paraphyletic Diplatyidae is besal-most to the other extent Forficulina, subsequent detailed examination of female genitalia by Klass (2003) supported the monophyly of two taxa, Diplatyidae (*Haplodiplatys* + *Diplatys*) and Karschiellidae + Diplatyidae (Haas and Klass 2003). Examples of male genitalia (**b**–**d**): **b**, *Diplatys annandalei* Burr (Diplatyidae); **c**, *Allostethus indicum* (Burmeister) (Labiduridae); **d**, *Proreus coalescens* (Borelli) (Chelisochidae). Abbreviations: bv, basal vesicle; pl, penis lobe; pm, paramere; vg, virga; vg(b), virga (bifurcated). Scale bars, 500 μm. [Reproduced from Kamimura (2014) with permission from the Entomological Society of Japan.]

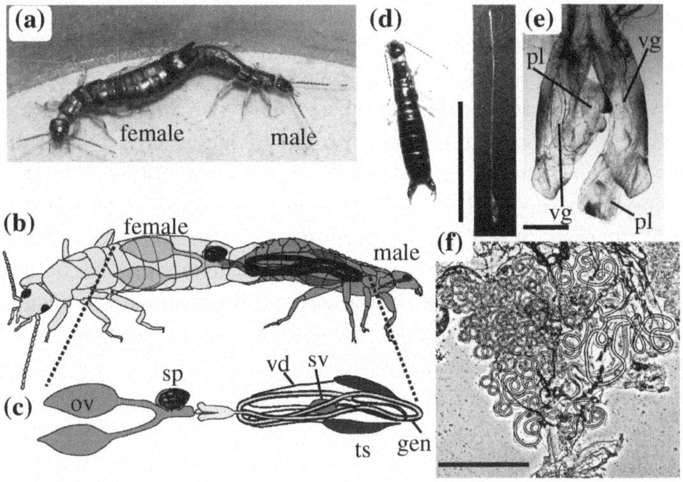

Fig. 10.5 Mating and genital structures of *Euborellia plebeja*. A picture (**a**) and schematic drawings (**b**, **c**) of a mating pair. Male and female internal reproductive organs and genitalia (**c**) are overlaid on **b** as silhouettes. (**d**) A male (*left*) and the genitalia dissected out of the male (*right*), with the distal portion magnified in **e**. (**f**) Spermatheca. Paired virgae (vir), one of which is inserted into the spermatheca for sperm transfer, are encased in the elongated male genitalia (gen). Abbreviations: gen, male genitalia (sensu stricto); ov, ovary; pl, penis lobe; sp, spermatheca; sv, seminal vesicle; ts, testis; vd, vas deferens; vg, virga; scale bar in **d** = 10 mm and in **e** and **f** = 500 μm. [Reproduced with partial modification from Kamimura (2013) with permission from Elsevier.]

coxosternum of abdominal segment IX). In *E. plebeja* and related anisolabidids, male genitalia are highly elongated (Fig. 10.5b–d; Kamimura 2014). In this group, the genitalia include a pair of penes (Fig. 10.5e), each of which bears an elongated, heavily sclerotized process called a virga. The virga contains the terminal part of the ejaculatory duct and has the gonopore on its tip. Flash-fixation experiments on mating pairs revealed that male *E. plebeja* use one of their paired virgae, which are as long as their bodies (on average 15.8 mm; Fig. 10.5d), to remove rival sperm from the spermatheca (Kamimura 2000, 2003a). The mechanism of sperm removal is as follows. First, a male inserts the virga deeply into the spermatheca without ejaculating. He then extracts the virga while ejaculating semen from its tip and simultaneously removing rival sperm using a fringe-like projection on the virgal tip (Kamimura 2000). However, the spermatheca in this species is twice the length of the female body (mean ± SD = 33.6 ± 5.7 mm; Kamimura 2000, 2005; Fig. 10.5f). As predicted from the considerable difference in their lengths, males can remove only a portion of the stored sperm (Kamimura 2000) and a paternity analysis experiment revealed that P_{last} is only ~20 % (mean ± SD = 0.193 ± 0.182; range, 0–0.650; median 0.154) for a single mating with a female whose sperm-storage organ has been saturated by a rival male (Kamimura 2005). Importantly, the estimated values do not vary with the duration of the last mating (Kamimura 2005), male body size ($r = 0.014$, $P = 0.96$), or male genital length ($r = 0.68$, $P = 0.12$; reanalysis of the data of Kamimura 2005), satisfying the assumption of the theoretical model (Sect. 10.3.1). Another experiment in which the mating order of two males was not controlled also showed that the best fit between predicted and realized paternity success is obtained when assuming $P_{last} = 0.22$ (Kamimura 2005). Females undergo repeated oviposition (up to five clutches under laboratory conditions; ~ 40 eggs per clutch) at intervals of ~27 days (Kamimura 2003b) and thus are likely to collect sperm from several to dozens of males before depositing each clutch. Field-caught females lay egg batches with low genetic relatedness as a result of mixed paternity, clearly demonstrating that female promiscuity and incomplete sperm displacement are not laboratory artifacts (Kamimura 2003b).

Thus, if the key assumption of the model, that mating frequency per encounter is positively correlated with male quality, is satisfied, females of this species likely gain genetic benefits from low P_{last} values. In a previous study (Kamimura 2013), a staged mating experiment and allozyme-based paternity analysis were conducted to test whether this condition holds true for *E. plebeja*. Based on the results of the staged mating experiment and a previous genetic study of male body size, a simulation study was conducted to examine the relationships among the number of mates for females, P_{last} values, and expected offspring quality (body size of offspring) to test the hypothesis that promiscuity and restricted sperm displacement work cooperatively as an indirect CFC mechanism in this species (Kamimura 2013).

10.4.2 Staged Mating Experiment

The staged mating experiment was designed to clarify the relationship between male mating success and paternity. Mating trials involving two males and one female ($N = 53$) were conducted under controlled laboratory conditions using laboratory-raised virgin adults (see Kamimura 2013 for details). One of the two paired males in each triplet was randomly chosen and marked with a small spot of yellow paint on the prothorax; post hoc analyses detected no significant effects from marking on any male fitness traits examined (Kamimura 2013). A mating arena (87 mm in diameter) contained six small burrows (20 mm in diameter) that were covered with a transparent plastic board, simulating the natural habitat of *E. plebeja* with many harborages. Although the burrow lid was transparent, earwigs spent most of their time in burrows, possibly because of their positive thigmotaxic nature (93.0 % of matings occurred when at least one partner was hiding in a burrow). Mating, which was defined as a single bout of genital coupling, was considered to be repeated when a pair repeated genital coupling without relocating to a different burrow. Mating that occurred outside the burrows ($N = 53$; 7 %) was never repeated by a pair in the same place within the arena. I refer to the coexistence of a male and a female in a given location, together with the series of repeated matings between them, as "cohabitation." Mating that occurred outside the burrows is considered a single mating per cohabitation. Thus, the cohabitation rates of males correspond to the encounter rates (*Eg* and *Ep*) of the model introduced previously. When male–male combat (defined when at least one male attacked an opponent with his forceps) for burrows (with or without a female) occurred, I deemed the male that occupied the burrow after combat to be the winner. After individuals were euthanized by freezing, pronotum width (PW) was measured to the nearest 0.026 mm for all males; PW is a good index of body size as log-transformed PW is highly correlated with log-transformed body weight ($r^2 = 0.77 - 0.80$; Y. Kamimura, unpublished data). To analyze paternity success, some mating trials ($N = 23$) were conducted using individuals of known allozyme genotypes (Kamimura 2003b, 2005).

Females mated 14.3 ± 9.2 times (median, 12; range, 2–42) during the 15-h observation period, for a total mating duration of 205.7 ± 179.7 min (median, 144.6; range, 8.4–769.9 min). Among the 565 male–female encounters observed, 512 occurred in burrows. Males visited burrows occupied by females in 441 cases (86.1 %), and matings were repeated between the same pair (3.33 ± 2.68 times) in 82 cases (14.5 %). Overall, larger males had a higher mating frequency (binomial test, $P = 0.0089$) and total mating duration ($P = 0.018$) than smaller males.

The mating frequency of males increased with male body size (Fig. 10.6a) due to repeated matings by large males (Fig. 10.6c) rather than increased cohabitation rate (Fig. 10.6b). Among the 756 observed matings, 54 (7.1 %) were terminated because they were disrupted by another male that entered the burrow occupied by the mating pair. In 22 trials in which both males mated at least once and mating disruption occurred, the probability that a mating would be terminated by a disruption was lower for larger males ($P = 0.0015$).

With the exception of cases in which a non-mating male visited the burrow of a mating pair, two males never occupied the same burrow. When a male entered a burrow that housed another male, combat inevitably occurred and resulted in either the takeover or defense of the burrow. In all cases, the winner of the combat was easily identified by his consequent occupation of the burrow. Large males had a higher probability of winning in combat ($P = 0.0091$), indicating they had higher competitive abilities for resources (burrows and females). A logistic regression analysis revealed that paternity success was significantly and positively related to male mating success ($F_{1,21} = 5.08$, $P = 0.035$). As expected from the low P_{last} (~20 %), mating order had little effect on paternity success: The resultant paternity of males that mated first or last did not significantly differ from parity on average (Kamimura 2013). In contrast, males that mated more frequently than their opponent gained a greater share of paternity than was expected from a random determination of paternity: 0.67 (95 %CL, 0.54–0.78; Kamimura 2013), indicating that mating frequency is a more important determinant of the resultant paternity than mating order (see Sect. 10.3.2).

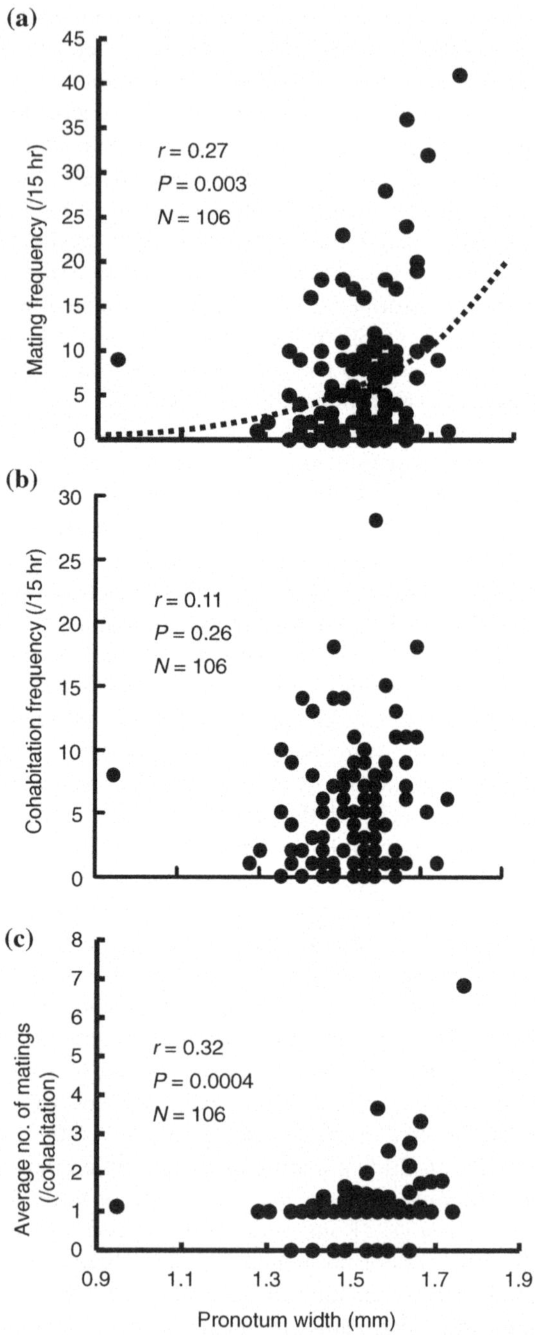

Fig. 10.6 Total mating frequency (**a**), cohabitation (mate acquisition) success (**b**), and mating frequency per cohabitation (**c**) for male *Euborellia plebeja* in relation to body size (pronotum width). Correlations were tested by bootstrapping. The *dotted line* in **a** was fitted by a Poisson regression [$y = \exp(3.255x - 3.063)$; $F_{1104} = 10.4$, $P = 0.0017$]. [Reproduced with partial modification from Kamimura (2013) with permission from Elsevier.]

10.4.3 Simulation Study

The staged mating experiment revealed that males compete for burrows that house females and that male body size is a reliable index of resource-holding potential (RHP). Moreover, the observed pattern of mating in which males with a high RHP tend to repeatedly mate with a female satisfies the condition of the indirect CFC model discussed previously (Sect. 10.3.1). By applying a half-sib mating design, a previous study showed that male body size is heritable in this species (narrow-sense heritability $h^2 = 0.41$; Kamimura and Iwase 2010). These findings led us to further questions: Do females of this species realize the optimal P_{last} to accumulate sperm from males with high RHP? How large is the genetic benefit? To answer these questions, I conducted a simulation study using real data sets of mating patterns, including the relationships obtained between male body size and cohabitation frequency (Fig. 10.6b) and between male body size and mating frequency per cohabitation (Fig. 10.6c). In the simulation (Kamimura 2013), males were randomly resampled based on their cohabitation frequency (Fig. 10.6b, as a proxy for *Ep* and *Eg* of the theoretical model). For each cohabitation, mating frequency was randomly determined by resampling from the real data set of mating frequency per cohabitation for the chosen male. By repeating this random resampling, for a given number of males that were sampled (1, 5, 10, 15, or 30 males), mating sequences for 10,000 females were determined. Female sperm storage was assumed to be saturated by the first single mating. Despite the fact that several matings (usually less than three) are required to saturate the sperm-storage organs in *E. plebeja* (Kamimura 2005), this effect is negligible when females mate an average of 10–20 times, which typically occurs during 15-h pairings in this species (Fig. 10.6a; Kamimura 2005). For a given value of P_{last} (from 0 to 1, in increments of 0.05) and a given mating sequence, one can calculate the resultant paternity share of each male in the manner discussed in Sect. 10.3.1. Given the paternity share values and the heritability of male body size, one can calculate the expected average body size of sons for each female. First, for convenience and simplicity, the body sizes of sons and their sires were assumed to be identical. That is, the body size of male offspring that were sired by a male of body size x was assumed to be x. The actual narrow-sense heritability of male body size ($h^2 = 0.41$; Kamimura and Iwase 2010) was then taken into account to consider the outcome between the benefits and costs of mating.

Figure 10.7a summarizes the results of the simulation based on actual male mating pattern data (each dot represents the average of 10,000 iterations). Like the theoretical model (Fig. 10.3), it shows that by sampling several to dozens of males, and with P_{last} values ~0.2, females can produce larger sons. Although the exact mating frequency in the wild is currently unknown, females likely mate with several dozen males during a 27-day oviposition interval (Kamimura 2003b). Thus, the simulation revealed that the actual measured value of P_{last}, which was determined by a controlled mating experiment (shown in Fig. 10.7a with its 95 % confidence limits; Kamimura 2005), is nearly optimal for accumulating

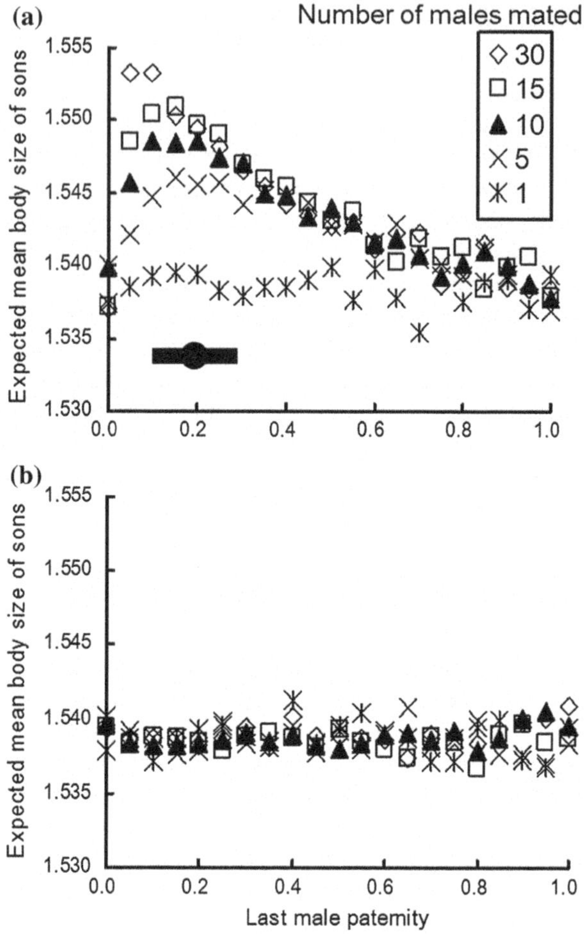

Fig. 10.7 Results of the simulation of the relationship between last male paternity (P_{last}) and expected offspring size. Results based on real data sets and on virtual data sets assuming all males mate only once per cohabitation are shown in **a** and **b**, respectively. In **a**, the *closed circle* and *associated bars* represent the empirical estimate of last male paternity and its 95 % confidence limits, respectively. To take into account the actual narrow-sense heritability of male body size ($h^2 = 0.41$; Kamimura and Iwase 2010), scale of the vertical axes should be multiplied by 0.41. See text for details. [Reproduced with partial modification from Kamimura (2013) with permission from Elsevier.]

sperm from large males. For the sake of evaluating the effect of size-dependent remating success, I conducted a similar simulation using a virtual data set in which all males mated only once per cohabitation regardless of their body size. Figure 10.7b shows the results of this "null" model. In this case, P_{last} had little effect on the accumulation of sperm from larger sires. Therefore, the positive correlation between male size and the frequency of repeated matings per cohabitation is essential for the selective accumulation of sperm from high RHP males. By assuming females copulate with 15 males before oviposition, polyandry results in an increase in the body size of sons (Fig. 10.7a; 1.54–1.55), which leads to a corresponding 3.6 % increase in the mating success of sons (expected from a Poisson regression; Fig. 10.6a). Even including a reduction due to the heritability of male body size ($h^2 = 0.41$; corresponding to a change of 1.540–1.544 in the body size of sons), the expected genetic benefit is a 1.4 % increase in the mating success of sons.

10.5 Discussion

10.5.1 Interrelationships Between CFC and FMM

As reviewed in this chapter, indirect CFC is potentially ubiquitous. Furthermore, the list of indirect CFC mechanisms (Table 10.1) must be far from complete. However, well-established empirical examples are extremely rare for indirect CFC. Compared with direct mechanisms, many possible indirect CFC mechanisms are inconspicuous, and their scarcity in empirical studies could be due to the difficulty of negative proof, such as the absence of discriminative responses by the female nervous system. To this end, female arthropod morphological traits, of which the dimensions of many exoskeletal parts cannot be altered in response to male attributes, likely include many candidate models for future studies of indirect CFC.

Indirect CFC mechanisms can bring genetic benefits to potentially polyandrous females by (1) providing an arena for more intense male–male or sperm competition or (2) by passively discriminate males or sperm without involvement of any neural filtering (Table 10.1). As in direct CFC (Eberhard 1996), no indirect CFC mechanism reviewed here (Table 10.1) can work unless females at least potentially mate with multiple males. However, FMM alone does not necessarily bring any genetic benefits to females without proper direct or indirect CFC mechanisms. For example, when sperm from the last mate are exclusively used for fertilization ($P_2 = 1$) regardless of male attributes, indiscriminate FMM does not bring any genetic benefits at all, unless genetically superior males tend to be the last mate (see Fig. 10.1).

Furthermore, although FMM is a prerequisite for CFC, it does not necessarily evolve for the sake of CFC. Several studies demonstrated genetic variation in the propensity of female rematings, indicating that FMM is an evolvable trait (e.g., Singh and Singh 2001; Harano and Miyatake 2009). However, females cannot always realize their desired number of matings due to environmental constraints, such as sperm shortage (e.g., Charlat et al. 2007). On the other hand, males can coercively impose FMM. Thus, it is important to distinguish and separately discuss the evolution of FMM for the sake of CFC and the evolution of CFC mechanisms coupled with FMM. For example, when the direct costs of additional matings outweigh the associated genetic benefits, FMM is unlikely to have evolved "for" CFC. Although this condition strongly suggests sexual conflict over mating, it does not preclude the possibility that female reproductive characteristics (such as a voluminous sperm-storage organ that can accommodate sperm from multiple ejaculates) evolved as devices for CFC. Even when FMM is imposed by males (coercive matings), female morphology or physiology can evolve as a CFC mechanism to collect genetic benefits that compensate for the imposed mating costs. Thus, two conditions must be satisfied for CFC to be a major explanatory factor in the evolutionary maintenance of FMM: (1) genetic benefits arising from additional matings outweigh the associated costs and (2) no other reasons for

FMM, such as direct benefits or coercion by males, exist or are negligible compared to the genetic benefits (see also Chap. 7 for synergistic interactions between direct benefits and CFC for the evolution of FMM).

10.5.2 Alternative Hypotheses for the Function of Elongated Spermatheca in E. plebeja

The example of the earwig *E. plebeja* is unique for having quantitative estimates of the genetic benefits due to indirect CFC. The study showed that a moderately low value of P_{last} ($\approx$0.2), which arises from restricted sperm displacement from an elongated spermatheca, is optimal for producing large sons. Females of earwigs from various families have elongated spermathecae irrespective of their promiscuity and/or whether the males have short and simple virgae, which are likely ineffective at removing sperm (e.g., see Klass 2003; Schneider and Klass 2013; Kamimura and Lee 2014a, b). In these species, the lengthy spermathecae likely possess as-yet-unknown functions other than the indirect CFC mechanism discussed above.

It is usually unclear why females develop exaggerated morphologies in sperm-storage organs that make sperm manipulation difficult for males. Polyandry with restricted sperm displacement may be an adaptive strategy for obtaining fresh young sperm (e.g., Lodesani et al. 2004; Tsuchiya and Hayashi 2010), while minimizing the risk of an infertile mating or mating with insufficient sperm supply (Wedell et al. 2000; García-González 2004; South and Lewis 2011). For female *E. plebeja*, infertile matings seldom occur (Kamimura 2003b). Since females with clutches that have low hatch rates also tend to have depleted sperm stocks in their spermathecae, intermittent matings are considered important for sperm replenishment (Kamimura 2003b). However, the retention of 80 % of stored sperm and the observed frequent matings are redundant with respect to these functions (Kamimura 2005).

As a result of incomplete sperm displacement, female *E. plebeja* lay egg batches of mixed paternity (Kamimura 2003b). For species with parental care by males, mixed or uncertain paternity may be directly advantageous for females by inducing protection of offspring by males (e.g., Hartley et al. 1995; Slatyer et al. 2012; cf. Sheldon 2002; Griffin et al. 2013 for an opposing view). However, as in other species of earwigs, male *E. plebeja* do not provide paternal care to offspring (Baijal and Srivastava 1974; Kamimura 2000, 2003b, 2013). Alternatively, production of genetically diverse offspring may be advantageous in unpredictable environments. However, such genetic benefits (or indirect benefits) may only arise under restricted conditions: very small populations or when genetically diverse offspring cooperate to increase their average performance (reviewed by Yasui 1998). Thus, the indirect CFC mechanism likely represents the major function of the elongated spermatheca of *E. plebeja*.

10.5.3 Other Possible Indirect CFC in E. plebeja and Related Species

Although the sperm-removal function has been shown only in *E. plebeja* (Kamimura 2000, 2005), the elongated virgae of this species and related anisolabidid earwigs (Fig. 10.5d) can be considered adaptations for sperm competition (Kamimura and Matsuo 2001; van Lieshout and Elgar 2011a). Thus, in *E. plebeja*, females seem to counteract males by having longer spermathecae that permit for only incomplete sperm manipulation. Further elongation of the virgae may be limited by a counter-selection pressure in the form of increased fragility during mating: The virgae of anisolabidid earwigs sometimes break during mating (Kamimura and Matsuo 2001; Kamimura 2003a).

Since several CFC mechanisms can operate simultaneously in a female (Eberhard 1996), examining other possible direct and indirect CFCs in these earwigs would be worthwhile. If heritable variation exists in the male's ability to remove and displace sperm from the elongated spermatheca, females likely gain further genetic benefits by FMM (see Sect. 10.3). In *E. plebeja*, although phenotypic variation in male genitalia is restricted compared to that of body size (the allometric slope $b = 0.653$), these two morphological traits showed similar levels of narrow-sense heritability (Kamimura and Iwase 2010). van Lieshout and Elgar (2011a) examined the benefits of longer genitalia, almost as long as the body length, in male *Mongolabis brunneri* (Dohrn) (van Lieshout and Elgar (2011a) placed this species in the genus *Euborellia*; cf. Srivastava 1999). This species belongs to the same subfamily, Anisolabidinae, and exhibits male and female genital morphologies similar to those of *E. plebeja*. Although they did not measure and control the mating frequency of each male, the mating experiments revealed that a second male with short virgae earned a reduced P_2. Therefore, in addition to repeated matings for sperm displacement, male *M. brunneri* likely gain further benefits in paternity success by having longer genitalia. Although the paternity gain per mating (P_{last}) did not vary with male attributes in *E. plebeja* (body size or genital length; Sect. 10.4.1), the results for *M. brunneri* warrant further studies in the related anisolabidids to examine whether such a mechanism is more widespread. In *M. brunneri*, males with longer genitalia also tend to mate for longer durations, although the significance of this trend is still unclear (van Lieshout 2011).

10.5.4 Alternative Hypotheses for FMM in E. plebeja

As discussed above (Sect. 10.2.1), CFC processes can be considered a major cause of the propensity for FMM only when associated mating costs are outweighed by the genetic benefits of polyandry and when direct benefits cannot explain FMM. The frequent matings of *E. plebeja* are redundant with respect to sperm replenishment (see Sect. 10.5.2). Males do not donate detectable nutrients during mating or provide paternal care to offspring (Baijal and Srivastava 1974; Kamimura 2000). Because the forceps are located at the caudal end and male and female genitalia are located on the ventral side of the abdomen, male earwigs usually rotate their abdomen nearly 180° around the anteroposterior axis and walk backward to establish an end-to-end mating posture (Fig. 10.5a, b). Female quiescence is necessary to establish genital coupling; thus, male earwigs usually cannot coercively mate with an unwilling female (cf. Briceño and Eberhard 1995 for an exceptional case in the genus *Pseudomarava*). Because of this mating posture, male earwigs may be constrained in terms of evolving adaptations that relate to sexual conflict (Kamimura 2014). No harassment behavior to unreceptive females has been reported for *E. plebeja* (Baijal and Srivastava 1974; Kamimura 2000, 2003b, 2005). Therefore, females are unlikely to mate frequently for material benefits or to avoid the costs of sexual harassment. The thin virgae of anisolabidid earwigs sometimes break during mating, and broken pieces remain in the spermatheca (Kamimura and Matsuo 2001). However, this very rarely occurs in nature (<0.3 % of wild-caught females; Kamimura and Matsuo 2001) and when it does, it usually does not affect subsequent matings or oviposition by females (Kamimura 2003a). The evidence listed above indicates that the costs of mating for female *E. plebeja* are very low. By eliminating other possible explanations, frequent matings in *E. plebeja*, which seem redundant for sperm replenishment alone, are best explained by the indirect CFC hypothesis.

Thus, both the propensity for multiple mating and the elongated spermathecae of female *E. plebeja* can be viewed as adaptations for CFC. However, as discussed above (Sects. 10.2 and 10.3), without an appropriate value for last male paternity, polyandry itself brings no genetic benefits (Figs. 10.3, 10.7). The opposite is also true: Without multiple mating, moderately low P_{last} is not functional (Figs. 10.3, 10.7). This argument raises the question of which female trait evolved first. As introduced above, females of various earwigs, some of which are less promiscuous than *E. plebeja*, also have elongated spermathecae, supporting the view that the evolution of elongated spermathecae preceded that of frequent FMM in *E. plebeja* and related anisolabidids. To confirm this hypothesis, further studies of the behavior and phylogenetic positions of many earwig species are warranted. Nevertheless, the low realized P_{last}, which is nearly optimum with respect to genetic benefits, indicates subsequent evolutionary modulation of the spermathecal length in relation to FMM and male genital traits.

10.5.5 Ecological Conditions for the Evolution of Indirect CFC

Eberhard (1996; p. 96) pointed out ecological conditions that make precopulatory female choice difficult or impossible and favor the evolution of CFC: (1) When males are able to force females to copulate (convenience polyandry); or (2) when males donate physical resources (material benefit polyandry), females are expected to have reduced chances or incitements for precopulatory choice. Eberhard (1996) also argued that females can exercise both pre- and postcopulatory choices and that these conditions can therefore be viewed as "weak signals." However, when females can effectively and precisely assess the quality of potential mates before copulation (e.g., when male courtship functions as an honest signal that reveals male quality), their resulting choices should reduce the variation in genetic quality among mates. This likely makes postcopulatory choice a more difficult and "unprofitable" task. We can thus expect a negative correlation (trade-offs) between the intensity of pre- and postcopulatory female choices.

Although females of *E. plebeja* rarely show aggressive behavior to courting males (no obvious precopulatory mate choice), females are unlikely to mate frequently for material benefit or to avoid the costs of sexual harassment (see Sects. 10.5.2, 10.5.3). For terrestrial earwigs like *E. plebeja*, there is another ecological condition making the precopulatory choice of mates by females difficult and unreliable. In contrast to animals that mate in open spaces, many terrestrial earwig species mate in very narrow and dark habitats, where precopulatory mate choice based on visual or tactile stimuli is limited. Moreover, van Lieshout and Elgar (2011b), who examined the matings of *M. brunneri*, pointed out a unique feature in earwig mating. Because of their end-to-end mating posture (e.g., Fig. 10.5a), males cannot use forceps for male–male combat during mating. Due to this owner-specific positional disadvantage, smaller or weaker *M. brunneri* males can sometimes break up the mating of a male with a higher RHP (van Lieshout and Elgar 2011b). Similar patterns were observed for *E. plebeja*: Of 54 matings that were terminated by another male, smaller males won in 20 cases (reanalyzed from the data of Kamimura 2013). As revealed here, larger *E. plebeja* males do not necessarily visit females frequently (Fig. 10.6b) and judging the quality of a male at the time of its arrival and before male–male combat takes place is difficult for females. Restricted sperm displacement may be a robust strategy for collecting superior sperm under such conditions when executing precopulatory mate choice is difficult. Recently, Lüpold et al. (2014) showed that even with precopulatory selection, postcopulatory sexual selection remains an important determinant of male fitness when males with high RHP cannot monopolize mates. The disadvantage intrinsic to the mating posture may be seen as a special case of this trend.

10.6 Interrelationships Between Pre- and Postcopulatory Sexual Selection

The preceding discussion does not mean that precopulatory sexual selection is unimportant in *E. plebeja* and related earwigs. In fact, male–male competition, which dictates the number of matings per encounter, is an essential factor in indirect CFC in *E. plebeja* (compare Fig. 10.7a, b). At least two indirect CFC mechanisms discussed here confer genetic benefits by promoting more intense male–male competition for mates rather than sperm competition: (1) retarding ejaculation after mounting that may promote takeovers by genetically superior males and (2) the mechanism discussed above for *E. plebeja* (Table 10.1). In both cases, disturbance, either during mating or between repeated matings during cohabitation, by intruder males is a key factor that relates the pre- and postcopulatory processes. The owner-specific positional disadvantage due to mating posture likely facilitates mating takeovers in various groups of earwigs (van Lieshout and Elgar 2011b; Kamimura 2014). However, such positional disadvantages are not unique to earwigs. Many other male animals likely experience disadvantages during mating, providing opportunities for similar mechanisms of indirect CFC.

10.7 Conclusions

Although we can envision possible CFC mechanisms that lack any involvement of the female nervous system (indirect CFC) for every principal category identified by Eberhard (1996), well-substantiated examples are scarce. In females of the earwig *E. plebeja*, extensive multiple matings and elongated sperm-storage organs, which allow only partial displacement of stored sperm by males, are thought to work cooperatively as an indirect CFC mechanism. Given that (1) larger males can secure a mate for a prolonged duration during which they can repeatedly mate and that (2) this resource-holding power has an additive genetic basis, this mechanism likely brings genetic benefits for promiscuous females in the form of a 1.4 % increase (on average) in the mating success of sons. As a component of precopulatory sexual selection, male–male competition dictates the number of matings per encounter. Such interrelationships between pre- and postcopulatory sexual selection should be addressed in future studies on CFC in earwigs and other organisms.

Acknowledgments I thank A. Aisenberg and A. Peretti for the invitation to write this chapter, and G.S. Requena and an anonymous reviewer for useful comments on a previous version of this manuscript. A theoretical model of this chapter (Sect. 10.3) was developed as part of my unpublished doctoral thesis. I am also grateful to T. Suzuki, T. Kusano, and F. Hayashi for their supervision during the study. Writing of this manuscript was supported in part by a Keio Gijuku Academic Development Fund grant from Keio University.

References

Arnqvist G, Rowe L (1995) Sexual conflict and arms races between the sexes—a morphological adaptation for control of mating in a female insect. Proc R Soc Lond B 261:123–127

Baijal HN, Srivastava GK (1974) Notes on the biology of *Euborellia plebeja* (Dohrn). Indian J Entomol 36:23–27

Bateman AJ (1948) Intra-sexual selection in Drosophila. Heredity 2:349–368

Briceño RD, Eberhard WG (1995) The functional morphology of male cerci and associated characters in 13 species of tropical earwigs (Dermaptera: Forficulidae, Labiidae, Carcinophoridae, Pygidicranidae). Smithsonian Contr Zool 555:1–63

Carayon J (1966) Paragenital system. In: Usinger RL (ed) Monograph of Cimicidae (Hemiptera–Heteroptera). Thomas Say Foundation, Entomological Society of America, College Park, Maryland, pp 81–167
Carré D, Sardet C (1984) Fertilization and early development in *Beroe ovata*. Dev Biol 105:188–195
Chargé R, Teplitsky C, Hingrat Y et al (2013) Quantitative genetics of sexual display, ejaculate quality and size in a lekking species. J Anim Ecol 82:399–407
Charlat S, Reuter M, Dyson EA et al (2007) Male-killing bacteria trigger a cycle of increasing male fatigue and female promiscuity. Curr Biol 17:273–277
Cordero A, Santolamazza-Carbone S, Utzeri C (1995) Male disturbance, repeated insemination and sperm competition in the damselfly *Coenagrion scitulum* (Zygoptera: Coenagrionidae). Anim Behav 47:437–449
Córdoba-Aguilar A, Uhia E, Cordero Rivera A (2003) Sperm competition in Odonata (Insecta): the evolution of female sperm storage and rivals' sperm displacement. J Zool 261:381–398
Darwin C (1871) The descent of man, and selection in relation to sex. Murray, London
Dixson AF, Anderson MJ (2004) Effects of sexual selection upon sperm morphology and sexual skin morphology in primates. Int J Primatol 25:1159–1171
Eberhard WG (1996) Female control: sexual selection by cryptic female choice. Priceton University Press, Princeton
Eisenbach M, Giojalas LC (2006) Sperm guidance in mammals—an unpaved road to the egg. Nat Rev Mol Cell Biol 7:276–285
Evans JP, Sherman CDH (2013) Sexual selection and the evolution of egg-sperm interactions in broadcast-spawning invertebrates. Biol Bull 224:166–183
Fedorka KM, Mousseau TA (2002a) Tibial spur feeding in ground crickets: larger males contribute larger gifts (Orthoptera: Gryllidae). Florida Entomol 85:317–323
Fedorka KM, Mousseau TA (2002b) Material and genetic benefits of female multiple mating and polyandry. Anim Behav 64:361–367
Fedorka KM, Winterhalter WE, Mousseau TA (2007) The evolutionary genetics of sexual size dimorphism in the cricket *Allonemobius socius*. Heredity 99:218–223
García-González F (2004) Infertile matings and sperm competition: the effect of "nonsperm representations" on intraspecific variation in sperm precedence patterns. Am Nat 164:457–472
Griffin AS, Alonzo SH, Cornwallis CK (2013) Why do cuckolded males provide paternal care? PLoS Biol 11:e1001520
Gschwentner R, Tadler A (2000) Functional anatomy of the spermatheca and its duct in the seed bug *Lygaeus simulans* (Heteroptera : Lygaeidae). Eur J Entomol 97:305–312
Haas F, Kukalovà-Peck J (2001) Dermaptera hindwing structure and folding: new evidence for familial, ordinal and superordinal relationships within Neoptera (Insecta). Eur J Entomol 98:445–509
Haas F, Klass KD (2003) The basal phylogenetic relationships in the Dermaptera. In: Klass KD (ed) Proceedings of the first dresden meeting on insect phylogeny: phylogenetic relationships within the insect orders, dresden, september 19–21, 2003. Entomol Abhand 61:138–142
Haas F, Hwen JTC, Tang HB (2012) New evidence on the mechanics of wing unfolding in Dermaptera. Arthropod Syst Phylogeny 70:95–105
Harano T, Miyatake T (2009) Bidirectional selection for female propensity to remate in the bean beetle, *Callosobruchus chinensis*. Popul Ecol 51:89–98
Hartley IR, Davies NB, Hatchwell BJ et al (1995) The polygynandrous mating system of the alpine accentor, *Prunella collaris*. II: multiple paternity and parental effort. Anim Behav 49:789–803
Harvey PH, May RM (1989) Out for the sperm count. Nature 337:508–509
Höglund J, Alatalo RV (1995) Leks. Priceton University Press, Princeton
Kamimura Y (2000) Possible removal of rival sperm by the elongated genitalia of the earwig, *Euborellia plebeja*. Zool Sci 17:667–672
Kamimura Y (2003a) Effects of broken male intromittent organs on the sperm storage capacity of female earwigs, *Euborellia plebeja*. J Ethol 21:29–35

Kamimura Y (2003b) Effects of repeated mating and polyandry on the fecundity, fertility and maternal behaviour of female earwigs, *Euborellia plebeja*. Anim Behav 65:205–214
Kamimura Y (2004) Mating behaviour and insemination in *Diplatys flavicollis*, an earwig with double-barreled penises and a variable number of female sperm-storage organs. J Zool 262:37–46
Kamimura Y (2005) Last male paternity of *Euborellia plebeja*, an earwig with elongated genitalia and sperm removal behavior. J Ethol 23:35–41
Kamimura Y (2013) Promiscuity and elongated sperm storage organs work cooperatively as a cryptic female choice mechanism in an earwig. Anim Behav 85:377–383
Kamimura Y (2014) Pre- and postcopulatory sexual selection and the evolution of sexually dimorphic traits in earwigs (Dermaptera). Entomol Sci 17:139–166
Kamimura Y, Matsuo Y (2001) A "spare" compensates for the risk of destruction of the elongated penis of earwigs (Insecta: Dermaptera). Naturwissenschaften 88:468–471
Kamimura Y, Iwase R (2010) Evolutionary genetics of genital size and lateral asymmetry in the earwig *Euborellia plebeja* (Dermaptera: Anisolabididae). Biol J Linn Soc 101:103–112
Kamimura Y, Lee C-Y (2014a) Mating and genital coupling in the primitive earwig species *Echinosoma denticulatum* (Pygidicranidae): implications for genital evolution in dermapteran phylogeny. Arthropod Syst Phylogeny 72:11–21
Kamimura Y, Lee C-Y (2014b) Genital morphology and mating behaviour of *Allostethus* (Dermaptera), an earwig genus of enigmatic phylogenetic position. Arthropod Syst Phylogeny 72:331–343
Keller L, Reeve HK (1995) Why do females mate with multiple males? The sexually selected sperm hypothesis. Adv Stud Behav 24:291–315
Kempenaers B, Verheyen GR, Van den Broeck M et al (1996) Extra-pair paternity results from female preference for high quality males in the blue tit. Nature 357:494–496
Klass K-D (2003) The female genitalic region in basal earwigs (Insecta: Dermaptera: Pygidicranidae s.l.). Entomol Abhand 61:173–225
van Lieshout E (2011) Male genital length and mating status differently affect mating behaviour in an earwig. Behav Ecol Sociobiol 65:149–156
van Lieshout E, Elgar MA (2011a) Longer exaggerated male genitalia confer defensive sperm-competitive benefits in an earwig. Evol Ecol 25:351–362
van Lieshout E, Elgar MA (2011b) Owner positional disadvantage in contests over mating prevents monopolization of females. Anim Behav 82:753–758
Lodesani M, Balduzzi D, Galli A (2004) A study on spermatozoa viability over time in honey bee (*Apis mellifera ligustica*) queen spermathecae. J Apicul Res 43:27–28
Lüpold S, Tomkins JL, Simmons LW et al (2014) Female monopolization mediates the relationship between pre- and postcopulatory sexual traits. Nat Commun 5:3184
Micholitsch T, Krugel P, Pass G (2000) Insemination and fertilization in the seed bug *Lygaeus simulans* (Heteroptera : Lygaeidae). Eur J Entomol 97:13–18
Parker GA (1990) Sperm competition games: raffles and roles. Proc R Soc B Biol Sci 242:120–126
Perry JC, Rowe L (2012) Sexual conflict and antagonistic coevolution across water strider populations. Evolution 66:544–557
Pizzari T, Birkhead TR (2000) Female feral fowl eject sperm of subdominant males. Nature 405:787–789
Popham EJ (2000) The geographical distribution of the Dermaptera (Insecta) with reference to continental drift. J Nat Hist 34:2007–2027
Requena GS, Alonzo SH (2014) Female sperm use and storage between fertilization events drive sperm competition and male ejaculate allocation. Evolution 68:3433–3444
Schneider K, Klass K-D (2013) The female genitalic region in Eudermaptera (Insecta: Dermaptera). Zool Anzeiger 252:183–203
Sheldon BC (2002) Relating paternity to paternal care. Phil Trans R Soc Lond Ser B 357:341–350

Simmons LW (2001) Sperm competition and its evolutionary consequences in the insects. Princeton University Press, Princeton

Singh SR, Singh BN (2001) Female remating in *Drosophila ananassae*: bidirectional selection for remating speed. Behav Genet 31:361–370

Siva-Jothy MT, Tsubaki Y (1989) Variation in copulation duration in *Mnais pruinosa pruinosa* Selys (Odonata: Calopterygidae) 1: alternative mate-securing tactics and sperm precedence. Behav Ecol Sociobiol 24:39–45

Slatyer RA, Jennions MD, Backwell PRY (2012) Polyandry occurs because females initially trade sex for protection. Anim Behav 83:1203–1206

Smith SM (1988) Extra-pair copulations in the black-capped chickadee: the role of the female. Behaviour 107:15–23

South A, Lewis SM (2011) The influence of male ejaculate quantity on female fitness: a meta-analysis. Biol Rev 86:299–309

Srivastava GK (1999) On the higher classification of Anisolabididae (Insecta: Dermaptera) with check-list of genera. Rec Zool Surv India 97:73–100

Tadler A (1999) Selection of a conspicuous male genitalic trait in the seedbug *Lygaeus simulans*. Proc R Soc B Biol Sci 266:1773–1777

Tsuchiya K, Hayashi F (2010) Factors affecting sperm quality before and after mating of calopterygid damselflies. PLoS ONE 5:e9904

Wada T, Takegaki T, Mori T et al (2005) Sperm displacement behavior of the cuttlefish *Sepia esculenta* (Cephalopoda: Sepiidae). J Ethol 23:85–92

Wedell N, Gage MJG, Parker GA (2000) Sperm competition, male prudence and sperm-limited females. Trends Ecol Evol 17:313–320

Wiley RH, Poston J (1996) Indirect mate choice, competition for mates, and coevolution of the sexes. Evolution 50:1371–1381

Yanagimachi R, Cherr G, Matsubara T et al (2013) Sperm attractant in the micropyle region of fish and insect eggs. Biol Rep 88:1–11

Yasui Y (1997) A 'good-sperm' model can explain the evolution of costly multiple mating by females. Am Nat 149:573–584

Yasui Y (1998) The 'genetic benefits' of female multiple mating reconsidered. Trends Ecol Evol 13:246–250

Yeates SE, Diamond SE, Einum S et al (2013) Cryptic choice of conspecific sperm controlled by the impact of ovarian fluid on sperm swimming behavior. Evolution 67:3523–3536

Chapter 11
Cryptic Female Choice in Crickets and Relatives (Orthoptera: Ensifera)

Karim Vahed

Abstract In his pioneering work on cryptic female choice, Eberhard identified a wide range of mechanisms that potentially allow multiply-mated females to bias paternity in favour of certain types of male following the start of copulation. The aim of this chapter is to review critically the empirical evidence for a range of these mechanisms of cryptic female choice in crickets and relatives (Orthoptera: Ensifera), while taking into account coevolutionary interactions between the sexes. There is compelling evidence that female crickets control the duration of spermatophore attachment and/or the uptake of sperm to the sperm storage organ to bias paternity in favour of males expressing a variety of favourable traits, or in favour of non-kin males. There is also some evidence that females can bias paternity to favour males with certain traits by choosing to remain with them for repeated mating. For other potential mechanisms of cryptic female choice, such as differential allocation of resources to the production of eggs, there is currently insufficient evidence to distinguish male-induced effects from cryptic female choice (if, indeed, such a distinction can be made). The evidence that mechanisms of cryptic female choice have resulted in coevolutionary adaptations in males is strong: males have evolved a wide range of behaviours to facilitate ejaculate transfer by deterring the female from removing the ampulla of the spermatophore prematurely, for example. How such adaptations affect the form and intensity of cryptic female choice and whether or not they result in ongoing sexually antagonistic coevolution deserve further investigation.

11.1 Introduction

For the last three decades, crickets and their relatives (order Orthoptera; suborder Ensifera) have proved to be excellent model organisms with which to examine cryptic female choice. Thornhill (1983) defined cryptic female choice as

K. Vahed (✉)
University of Derby, Derby, UK
e-mail: K.Vahed@Derby.ac.uk

A.V. Peretti and A. Aisenberg (eds.), *Cryptic Female Choice in Arthropods*,
DOI 10.1007/978-3-319-17894-3_11

"Female-influenced processes occurring during and/or after copulation that bias offspring production more towards one male than other mates (or potential mates)" (see Chap. 1). Arnqvist (2014) rephrased this definition as "A pattern of non-random post-copulatory fertilisation success among male phenotypes that is caused by a trait in females".

In his pioneering work on this subject, Eberhard (1996) identified at least 20 mechanisms of cryptic female choice, drawing his examples from a wide range of animal taxa. Cryptic female choice is just one aspect of post-copulatory sexual selection, however, and must be considered together with sperm competition (Parker 1970; Simmons 2001) and sexually antagonistic coevolution (Holland and Rice 1998; Arnqvist and Rowe 2005; see Chap. 1). The latter two processes may be linked: if an adaptation that allows the male to bias the paternity of the female's eggs in his favour results in a decrease in the fitness of the female, then females will be selected to resist manipulation by males. Males, in turn, will be selected to overcome such resistance, and cycles of sexually antagonistic coevolution may result (e.g. Arnqvist and Nilsson 2000). The aim of this chapter is to review critically the empirical evidence for a range of potential mechanisms of cryptic female choice in the Ensifera and to consider how such processes may shape, or be shaped by, the behaviour, morphology or physiology of the male.

The suborder Ensifera can be divided into six main extant superfamilies (Eades et al. 2014; Table 11.1). The vast majority of research on cryptic female choice in the Ensifera has been conducted on a small number of species of field crickets (Grylloidea: Gryllidae: Gryllinae) including the members of the genera *Gryllus*, *Acheta*, *Teleogryllus* (Tribe: Gryllini) and *Gryllodes* (Tribe: Modicogryllini). This reflects, in part, the relative ease with which these crickets can be reared in captivity, together with various aspects of their reproductive biology (see below). While the majority of examples in this review will therefore relate primarily to these grylline crickets, the aim of this chapter is to consider cryptic female choice within the Ensifera in general, where possible.

Table 11.1 Key extant superfamilies and families within the Ensifera (following Eades et al. 2014)

Superfamily	Family	Common English names
Grylloidea	Gryllidae	Field crickets and Tree crickets
	Gryllotalpidae	Mole crickets
	Mogoplistidae	Scaly crickets
	Myrmecophilidae	Ant-loving crickets
Rhaphidophoroidea	Rhaphidophoridae	Cave crickets
Schyzodactyloidea	Schizodactylidae	Dune crickets/Splay-footed crickets
Hagloidea	Haglidae	Hump-winged crickets
Stenopelmatoidea	Anostostomatidae	Wetas and King crickets
	Gryllacrididae	Leaf-rolling crickets
	Stenopelmatidae	Jerusalem crickets
Tettigonioidea	Tettigoniidae	Bushcrickets/Katydids

11.1.1 Distinguishing Cryptic Female Choice from Other Processes

One aim of this chapter is to attempt to consider the possible influence of the male in determining the outcome of cryptic female choice. It is not, however, my intention to suggest that a clear distinction can necessarily be made between the influence of the male and that of the female on post-copulatory sexual selection. Attempting to make such a distinction is not at all straightforward, for a variety of reasons (see Chap. 1; Arnqvist 2014). Eberhard (1996), for example, pointed out that in organisms with internal fertilisation, mechanisms of sperm competition must rely, to a large extent, upon the morphology, physiology and behaviour of the female. Furthermore, the definition of cryptic female choice does not require that females actively assess males: the term encompasses any female trait that biases fertilisation success among male phenotypes, including apparently "passive" processes (e.g. those related to the morphology of the female's sperm storage organ and its ducts, Chap. 1; Arnqvist 2014). As a consequence, virtually all mechanisms of sperm competition in internally fertilising organisms potentially fall within the definition of cryptic female choice (Arnqvist 2014).

It is similarly conceivable that mechanisms of cryptic female choice and adaptations in the male to bias sperm use in his favour are merely two sides of the same coin: the male and female aspects of the same process (Vahed 1998, 2007a). If, for example, the extent of development of a male adaptation to increase ejaculate transfer (such as a nuptial gift, see Sect. 11.4.2) is an honest indicator of an aspect of the male's genetic quality, then the female may gain indirect, genetic benefits by receiving more ejaculate from that male. On the other hand, it is also conceivable that such adaptations in males to bias sperm use in their favour could represent manipulation by males (see, e.g. Córdoba-Aguilar 2002; reviewed in Arnqvist and Rowe 2005; Vahed 2007a). If the resulting fertilisation bias and/or the mechanism by which it is caused is counter to the female's net fitness or is selectively neutral, then the term "female choice" would not be appropriate. Considering the overall effect on female fitness might therefore potentially allow a distinction to be made between sexually antagonistic coevolution and cryptic female choice (Chap. 1). A caveat, however, is that a history of counteradaptation by females may mean that male-imposed costs to the female are not currently evident (Arnqvist and Rowe 2005).

Even apparent manipulation by males does not, of course, rule out benefits to the female: if the "manipulative" trait is heritable, then the female may benefit through having sons that are better at manipulating females and in obtaining fertilisation (Eberhard 1996; Chap. 1). It is difficult either to confirm or rule out the existence of such indirect benefits empirically, or (more crucially) to quantify their magnitude (Chap. 1; Arnqvist 2014). Thus, the question of whether any reduction in the fitness of the female due to manipulation by the male is greater or less than the reproductive pay-off she receives from indirect benefits can rarely be answered (Chap. 1).

While it may not be possible to rule out cryptic female choice when considering cases of sperm competition and sexually antagonistic coevolution, the converse is not necessarily true: there are likely to be cases of cryptic female choice in which the role of the male can be ruled out. An elegant example of this was recently provided by Tuni et al. (2013), who demonstrated that biases in the uptake of sperm favouring unrelated males in *Teleogryllus oceanicus* were under female, rather than male, control (see Sect. 11.2.2.1).

11.1.2 Reproduction in Crickets and Relatives

There are several aspects of the reproductive biology of ensiferans that make them ideal to examine cryptic female choice. Importantly, it appears that females of the majority of species examined so far routinely mate with more than one male: while there has been long-standing evidence of polyandry in many ensiferan species (reviewed in Brown and Gwynne 1997), recent studies have quantified the degree of polyandry in a range of species (principally gryllids and tettigoniids) in the field (Bretman and Tregenza 2005; Vahed 2006; Simmons et al. 2007; Simmons and Beveridge 2010). These studies have also confirmed that females tend to store the sperm from more than one male within their spermatheca (the sperm storage organ).

Pair formation in the Grylloidea, Tettigonioidea and Hagloidea generally occurs via phonotaxis of females to the male's calling song (produced by tegminal stridulation), although vibratory signals (e.g. tremulation and drumming) and pheromonal communication can also be involved in this process, especially once the pair have made contact (reviewed in Loher and Dambach 1989; Brown and Gwynne 1997; Zuk and Simmons 1997; Gwynne 2001). In the other ensiferan taxa, tegminal stridulation does not occur, so males are limited to the latter methods only (Brown and Gwynne 1997; Field and Jarman 2001; Weissman 2001). At close range, males of some taxa (e.g. many gryllids) switch to courtship song, which is distinct from the calling song (Loher and Dambach 1989; Brown and Gwynne 1997). The pair typically antennate one another when they meet and this may allow mutual assessment based on cuticular hydrocarbons (Tregenza and Wedell 1997). A receptive female generally mounts the male, who often assists by backing underneath her. Glandular secretions from "alluring glands" on the male's dorsal tergites can encourage the female to mount in some taxa (e.g. in most tettigoniids, Engelhardt 1915), while in other taxa, such as the tree crickets (Gryllidae; Oecanthinae), females feed on secretions from the male's metanotal glands before, during and after copulation (reviewed in Brown and Gwynne 1997).

The way in which copulation is achieved differs between the various ensiferan taxa: in the Tettigoniidae, for example, the male generally uses teeth on the inner side of his cerci to link with pits situated either on the base of the female's ovipositor or on, or adjacent to, her subgenital plate (Rentz 1972; Vahed et al. 2014). In the Gryllidae, the cerci have a sensory function and are not used to grasp the female (Alexander and Otte 1967). Instead, in many species, the hooklike epiphallus is inserted under the female's subgenital plate (Alexander and Otte 1967;

Loher and Rence 1978). In the Stenopelmatidae and a few other ensiferan families, hooks on the male's supra-anal plate are used to secure the female during copulation (Weissman 2001). A structure known as the "gin trap" that is formed by projections from the male's tenth abdominal tergite and projections on the margin of the eighth abdominal tergite is used to grasp the female in the Hagloidea and some anostostomatids (Brown and Gwynne 1997; Field and Jarman 2001). In a few species, pair formation and copulation appear to be coercive (e.g. Vahed 2002; Mello 2007; Vahed and Carron 2008; Haley and Gray 2012). In Alpine bushcrickets (Tettigoniidae; *Anonconotus* spp.), for example, the male silently "stalks" the female and then leaps on her, using sharply pointed, pincerlike cerci to maintain hold of the female. The tips of the cerci grip the cuticle of the sides of the female's abdomen, causing visible wounds (Vahed 2002; Vahed and Carron 2008).

An important aspect of reproduction in the Ensifera in the context of cryptic female choice is that sperm are generally transferred in a spermatophore (Boldyrev 1915; Alexander and Otte 1967; Gwynne 1997). A sperm tube from the ejaculate-containing ampulla is inserted into the female's genital chamber, while the majority of the ampulla generally remains external to the female. In gryllids, this tube is relatively long and reaches up into part of the spermathecal duct (Khalifa 1949), while in tettigoniids and other ensiferans, the tube is much shorter and only reaches into the genital chamber (Boldyrev 1915; Viscuso et al. 2002). In most tettigoniids, and some species of a variety of ensiferan families, attached to the ampulla is a gelatinous mass known as the spermatophylax (reviewed in Gwynne1997, 2001; Vahed 1998). This is consumed by the female following the end of copulation. In most species, copulation ends following spermatophore transfer, but in some species, the male remains firmly attached to the female, while sperm is transferred from the ampulla (Boldyrev 1915; Vahed 1996, 1997, Vahed et al. 2014). The amount of sperm transferred per spermatophore (Vahed and Gilbert 1996) and duration of sperm transfer vary across species, especially within the Tettigoniidae (reviewed in Gwynne 2001). As a consequence of the external spermatophore and gradual sperm transfer, there is considerable potential for female ensiferans to manipulate the amount of sperm received from each male by removing spermatophores prior to complete sperm transfer (Sakaluk 1984; Simmons 1986; Sect. 11.2.1).

The spermatheca in ensiferans is generally spherical or ovoid, with flexible walls which expand to hold numerous ejaculates (e.g. Simmons 1986; Vahed 2003a; Sturm 2008; Brundo et al. 2011; de Carvalho and Shaw 2010). Simmons (1986), for example, reported that the spermatheca of *Gryllus bimaculatus* can accommodate sperm from at least 38 spermatophores, while Vahed (2006) found that the spermatheca of *Anonconotus baracunensis* (Tettigoniidae) contained up to 44 ejaculates. The capacity of the spermatheca is likely to vary across species, however (as does ejaculate size; Vahed 2006). The spermatheca has only one opening and is linked to the genital chamber by the spermathecal duct, which can be long and convoluted in some species (e.g. Sturm 2008). Eggs are matured in paired ovaries and in most species continue to be developed and laid throughout the female's adult life (reviewed in Loher and Dambach 1989). As in most insects, eggs are only fertilised as they pass the exit of the spermathecal duct on their passage through the genital chamber, while they are being laid (Sugawara and Loher 1986; Sugawara 1993).

11.2 Mechanisms of Cryptic Female Choice in Crickets

Of the various potential mechanisms of cryptic female choice detailed by Eberhard (1996), those for which most empirical evidence is available within the Ensifera include the following: manipulation of the duration of spermatophore attachment; control of the uptake of sperm to the spermatheca; remaining to remate with the same male, as opposed to rapidly seeking a mating with a different male; adjusting the rate of oviposition and/or the differential allocation of resources to eggs; and resistance during copulation.

11.2.1 Manipulation of the Duration of Spermatophore Attachment

One of the most intensively studied potential mechanisms of cryptic female choice in crickets involves control by females of the duration of ejaculate transfer by manipulating externally attached spermatophores. That female ensiferans often remove and eat or kick off spermatophores prior to complete sperm transfer has been well documented in a wide range of ensiferans (Gerhardt 1914; Boldyrev 1915; Alexander and Otte 1967; Brown and Gwynne 1997). In addition to functioning as a mechanism of female choice, premature spermatophore removal could reflect sexual conflict over the optimal volume of ejaculate transfer per mating (Vahed 2007a) or could result from females treating spermatophores as a food source (e.g. Simmons 1988).

11.2.1.1 Experimental Evidence that Females Use Spermatophore Removal as a Form of Mate Choice

There is a considerable body of evidence from laboratory studies of field crickets of the tribe Gryllini (Gryllidae: Gryllinae) that females appear to use the timing of spermatophore removal to bias paternity to favour certain males over others (see Table 11.2).

In many gryllids (e.g. *Gryllus*, *Teleogryllus* and *Gryllodes*), full sperm transfer from the ampulla to the spermatheca takes about 40–60 min (Sakaluk 1984; Simmons 1986; Bussière et al. 2006; Fig. 11.1). By counting the number of sperm in the spermatheca at set intervals following the spermatophore transfer, Simmons (1986, in *G. bimaculatus*) and Bussière et al. (2006, in *Teleogryllus commodus*)

Table 11.2 Studies of grylline crickets that have examined cryptic female choice, in terms of whether or not females manipulate the duration of attachment of the ampulla of the spermatophore, in relation to specific traits in their mates

Species	Male "trait" potentially under selection	Effect on ampulla attachment statistically significant?	Guarding male present or absent?	Author
Gryllus bimaculatus	Body mass	Yes (favouring heavier males, but mate-guarding duration could have been a confounding factor)	Present	Simmons (1986)
G. bimaculatus	Body mass	No	Present and absent	Wynn and Vahed (2004)
G. bimaculatus	Body mass relative to female body mass	No	Absent	Bateman et al. (2001)
G. bimaculatus	Body mass relative to that of the female's previous mate (in the females second and third mating)	Mixed (favouring males heavier than the female's previous mate—see text)	Absent	Bateman et al. (2001)
G. bimaculatus	Degree of relatedness to the female	Yes (favouring less related males)	Present	Simmons (1991)
G. vocalis	Novel male (as opposed to a previous mate)	Yes (favouring the novel male)	Absent	Gershman (2009)
Acheta domesticus	Body mass	No	Absent	Mautz and Sakaluk (2008a)
Teleogryllus commodus	Body mass	No	Absent	Drayton et al. (2013)
T. commodus	Body mass	No	Present and absent	Bussière et al. (2006)
T. commodus	Body size (pronotum width)	No	Present and absent	Bussière et al. (2006)
T. commodus	Precopulatory "attractiveness" (based on how quickly males were mounted in precopulatory trials)	Yes (favouring "attractive" males)	Present and absent	Bussière et al. (2006)
T. commodus	Precopulatory "attractiveness" relative to that of the female's previous mates	No	Present and absent	Bussière et al. (2006)

(continued)

Table 11.2 (continued)

Species	Male "trait" potentially under selection	Effect on ampulla attachment statistically significant?	Guarding male present or absent?	Author
T. commodus	Courtship call properties (e.g. trill number, trill length and chirp interpulse duration) and male body size (based on mass and pronotum width)	Yes (but multivariate selection was nonlinear and the form of selection depended upon whether or not males were allowed to guard)	Present and absent	Hall et al. (2008)
T. commodus	Cuticular hydrocarbon profiles (relative abundances of 22 individual compounds)	Yes (multivariate selection was nonlinear and suggestive of stabilising selection)	Absent	Thomas and Simmons (2009a, b)
T. commodus	Precopulatory "attractiveness" (based on how quickly males were mounted in precopulatory trials)	No	Absent	Drayton et al. (2013)
T. commodus	Male age	No	Absent	Drayton et al. (2013)
T. commodus	Immune challenged males (injected with bacterial lipopolysaccharides) versus control males	No	Absent	Drayton et al. (2013)
T. oceanicus	"Attractiveness" of courtship song (using playback of artificially constructed songs that were determined to be "preferred" or "non-preferred" in previous precopulatory trials)	Yes (favouring more "attractive" song)	Absent	Rebar et al. (2011)
T. oceanicus	"Attractiveness" (see above) of calling song of the current mate relative to that of the previous mate	Yes (favouring more "attractive" song)	Absent	Rebar et al. (2011)
Platygryllus primiformis	Male body length relative to that of the female (smaller or larger than the female)	Yes (favouring males larger than themselves)	Absent	Bateman (1998)

(continued)

Table 11.2 (continued)

Species	Male "trait" potentially under selection	Effect on ampulla attachment statistically significant?	Guarding male present or absent?	Author
P. primiformis	Male fighting success (determined in precopulatory trials, in the absence of a female), which covaried with male body mass	Yes (favouring greater fighting success)	Present and absent	Parker (2009)
Gryllodes sigillatus	Familiar versus novel males	Yes (favouring the novel male, but only when both males were inbred rather than outbred)	Absent	Gershman and Sakaluk (2010)
G. sigillatus	Degree of relatedness to the female (siblings vs. unrelated males)	No	Present	Stockley (1999)
G. sigillatus	"Attractiveness" (based on how quickly males were mounted in precopulatory trials)	Yes (favouring "attractive" males)	Present (although the effects of mate guarding were taken into account by analysing the time until the female first attempted to remove the ampulla rather than actual ampulla removal)	Ivy and Sakaluk (2007)
G. sigillatus	Current male "attractiveness" relative to that of the previous mate	No	Present (but see above)	Ivy and Sakaluk (2007)
G. sigillatus	Hind wing length	Yes (favouring shorter wings, but note that this was driven by differences in nuptial gift size)	Present	Sakaluk (1997)
G. sigillatus	Male body mass	Yes (favouring larger males, but note that this was driven by differences in nuptial gift size)	Present	Sakaluk and Smith (1988), Sakaluk (1985)

(continued)

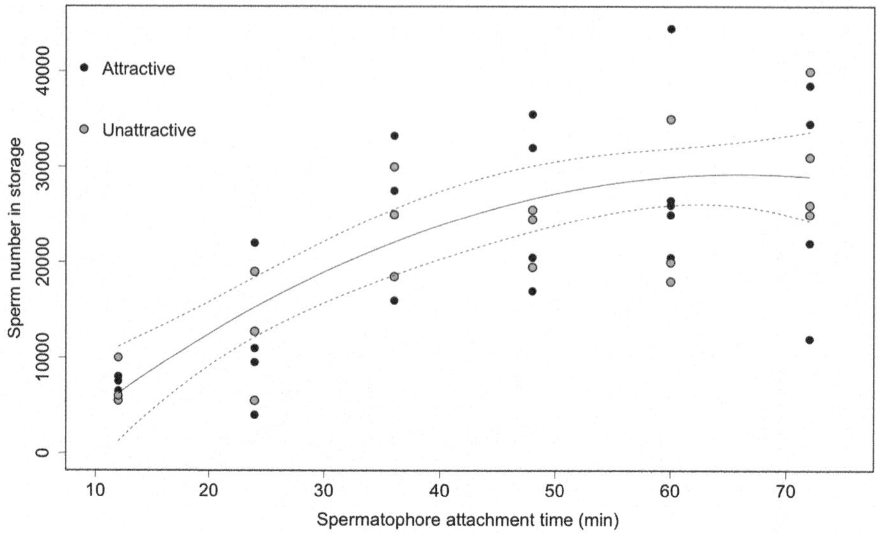

Fig. 11.1 Sperm transfer to the spermatheca as a function of spermatophore attachment duration in *Teleogryllus commodus*. There was no difference in sperm transfer between attractive males (*closed symbols*) and unattractive males (*open symbols*). Adapted from Bussière et al. (2006)

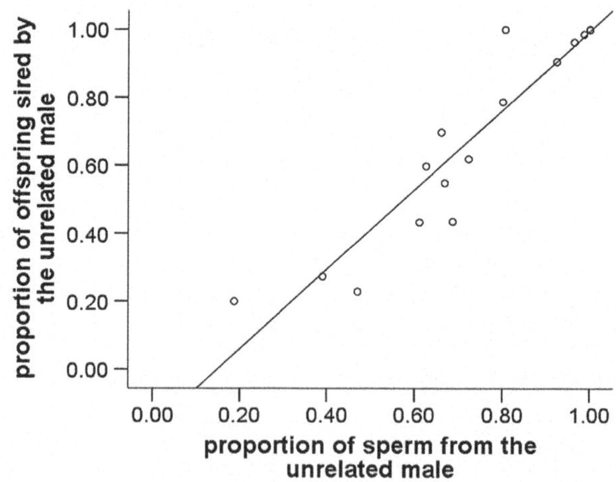

Fig. 11.2 The relationship between the proportion of sperm in the spermatheca from a focal male and the proportion of offspring sired by that male in *Gryllus bimaculatus* (females were mated both to a sibling and to an unrelated male, in either order). The proportion of a male's sperm in the spermatheca almost perfectly predicts paternity. Adapted from Bretman et al. (2009)

found that the rate of sperm transfer was greatest towards the beginning of this period, in females on their first mating. Using "competitive PCR" (based on amplifying microsatellite markers) to quantify the relative contribution of sperm from two different males to the spermatheca in non-virgin females, however, Hall et al. (2010) found that the proportion of sperm in storage for the second male to mate increased linearly with the duration of spermatophore attachment for that male in *T. commodus*. There is strong evidence that in *G. bimaculatus* (Fig. 11.2), *Gryllus vocalis* and *Gryllodes sigillatus*, when a female mates with more than one male, the proportion of sperm from a given male within the spermatheca determines the proportion of offspring subsequently sired by that male (Simmons 1987a; Sakaluk and Eggert 1996; Gershman 2009; Bretman et al. 2009: but see Simmons et al. 2003 who found no such effect in *T. oceanicus*). Therefore, by manipulating the duration of spermatophore attachment, females can potentially control the proportion of her eggs fertilised by the male.

The studies listed in Table 11.2 have examined the duration of spermatophore attachment in relation to a variety of traits or aspects of the male that produced the spermatophore. There is evidence that females appear to alter spermatophore attachment in relation to absolute male body size and/or mass (Simmons 1986; Hall et al. 2008; but see Wynn and Vahed 2004; Bussière et al. 2006; Mautz and Sakaluk 2008a; Drayton et al. 2013 who found no effect), male body size relative to the female's body size (Bateman 1998; but see Bateman et al. 2001), male fighting success (which was related to male body mass, Parker 2009), the attractiveness or properties of the male's courtship song (Hall et al. 2008; Rebar et al. 2011), the male's cuticular hydrocarbon profiles (Thomas and Simmons 2009a) and the overall "attractiveness" of the male, as determined in precopulatory trials (Fig. 11.3, Bussière et al. 2006; Ivy and Sakaluk 2007; but see Drayton et al. 2013).

There is also evidence that the characteristics of the female's previous mate can influence subsequent cryptic mate choice in terms of the relative body mass of the males (Bateman et al. 2001), the relative attractiveness of the courtship song (Rebar et al. 2011), but not the relative "attractiveness" of the males (Bussière et al. 2006; Ivy and Sakaluk 2007). Females also appear to alter spermatophore

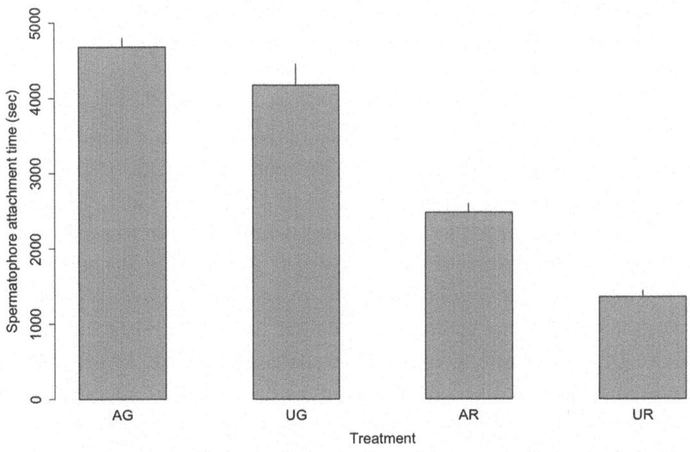

Fig. 11.3 Spermatophore attachment duration (mean ± SE) as a function of the male's attractiveness (latency to copulation in precopulatory trails) and the presence or absence of post-copulatory mate guarding in *Teleogryllus commodus*. AG, male attractive and permitted to guard; AR, male attractive and removed after spermatophore transfer; UG, male unattractive and permitted to guard; UR, male unattractive and removed after spermatophore transfer. Spermatophore attachment is longer for attractive males. Mate guarding prolongs spermatophore attachment for both types of male. Adapted from Bussière et al. (2006)

attachment to favour novel males (i.e. with which they have not previously mated; Gershman 2009; Gershman and Sakaluk 2010; see also Ivy et al. 2005) and in favour of males that are not close relatives (Simmons 1991; but see Stockley 1999).

Because females are unlikely to base mate choice decisions on single traits and because different traits could possibly trade-off against one another (reviewed in Candolin 2003), studies that have measured multiple traits are likely to give a more realistic picture of selection. One approach has been to take into account a wide range of traits in males using multivariate selection analysis (Hall et al. 2008; Thomas and Simmons 2009a). This appears to be a very powerful approach, although the results are more challenging to interpret than analyses based on single traits. Furthermore, as with analyses based on single traits, we can never be certain that the traits measured are those upon which females base their choices. Another approach has been to examine post-copulatory responses of females to overall male precopulatory "attractiveness" (e.g. Bussière et al. 2006; Ivy and Sakaluk 2007; Drayton et al. 2013). The measure of attractiveness commonly used is the latency to mounting or copulation, i.e. the time from initial introduction of a pair to the start of mounting (Shackleton et al. 2005). One potential problem with this approach, however, is that the latency to mounting could also be influenced by male mate choice (reviewed in Bonduriansky 2001) and/or the degree of courtship persistence of the male. Although males cannot force copulation in most grylline crickets that have been studied so far, some males might be better at persuading the female to mount by backing under her more persistently.

11.2.1.2 The Role of the Male in Influencing the Duration of Spermatophore Attachment in These Studies

In the Gryllinae, males typically engage in "mate-guarding" or "mate harassment" behaviour following copulation (Fig. 11.4; reviewed in Sakaluk 1991; Zuk and Simmons 1997; Wynn and Vahed 2004; Bussière et al. 2006; Hall et al. 2008; Parker and Vahed 2010; Rodríguez-Muñoz et al. 2011; see Sect. 11.4.1 for further discussion): they remain close to the female, usually within antennal contact, and perform a series of behaviours directed towards the female if she moves, including body rocking, aggressive chirping and even biting. One hypothesis for this behaviour is that it functions to deter the female from removing the ampulla of the spermatophore before complete ejaculate transfer (Gerhardt 1914; Sakaluk 1991). There is some support for this hypothesis in several grylline species: in *T. commodus*, for example, spermatophore attachment is almost twice as long if the male is allowed to guard than if he is removed following mating (Hall et al. 2008).

In some of the studies listed in Table 11.2, the male was allowed to remain with the female following copulation. This potentially introduces a confounding factor. The positive relationship between the duration of spermatophore attachment and male body size in *G. bimaculatus* demonstrated by Simmons (1986), for example, could have been caused by male mate-guarding behaviour rather than female control: the duration of spermatophore attachment also correlated positively with the

Fig. 11.4 A pair of *Gryllus bimaculatus* shortly after spermatophore transfer. The male (on the left) "guards" the female following copulation. This has been found to prolong the duration of attachment of the spermatophore (*S*) in some crickets

duration of mate guarding, and larger males may have been more efficient guards (Simmons 1991; Zuk and Simmons 1997). Females might, however, use the ability of the male to guard to inform their post-copulatory choices (see Sect. 11.4.1).

Bussière et al. (2006) and Hall et al. (2008) examined the influence of mate guarding on post-copulatory female choice in *T. commodus* by measuring the relationship between the duration of spermatophore attachment and male traits, both in the presence and in the absence of a mate-guarding male. Bussière et al. (2006) found that females took significantly longer to remove spermatophores of "attractive" rather than "unattractive" males (as determined in precopulatory trials) both when the male was absent and when he was allowed to guard (Fig. 11.3). Guarding, however, significantly increased the duration of spermatophore attachment for both attractive and unattractive males. Interestingly, unattractive males guarded the female more vigorously than attractive males, and the extent to which the presence of the mate-guarding male increased spermatophore attachment duration was greatest for unattractive males. Bussière et al. (2006) concluded that guarding by males actively restricts the efficiency of cryptic female choice. Hall et al. (2008) similarly found that post-copulatory mate guarding by males alters the form and intensity of cryptic female choice for courtship call characteristics and male body size: the presence of a guarding male significantly increased the duration of spermatophore attachment and reduced the overall variation in this measure of cryptic female choice. The opportunity for sexual selection was therefore seven times smaller when males were allowed to "harass" females after mating than when females were in sole control of the duration of spermatophore attachment (Hall et al. 2008). Irrespective of the presence of the male following copulation, spermatophore attachment duration still generated strong selection on male courtship call characteristics and male body size. The form of selection on these characters, however, depended upon whether or not the guarding male was present.

A further potentially confounding factor in some of the studies in Table 11.2 is the presence of a spermatophylax in *Gryllodes* spp. (Sakaluk 1985). There is much evidence that the spermatophylax in *Gryllodes* functions to prolong ampulla attachment by occupying the female's jaws and thereby delaying her from eating the ampulla: spermatophylax size determines the duration of feeding by the female and ampulla attachment time (e.g. Sakaluk 1984, 1985; see Sect. 11.4.2 for further discussion). Sakaluk (1997) found that in *G. sigillatus*, ampulla attachment duration was longer for males lacking hind wings than for males with fully developed hind wings. Males with fully developed hind wings (which are capable of flight) produce smaller spermatophylaces, presumably as a result of a trade-off between resources allocated to flight muscles and those allocated to nuptial gift production (Sakaluk 1997). The shorter duration of ampulla attachment in long-winged males was therefore driven by differences in spermatophylax mass (females consumed smaller spermatophylaces more quickly and ate the ampulla shortly afterwards). Nevertheless, Sakaluk (1997) argued that the resulting bias in sperm uptake (and therefore in fertilisation) against long-winged males constituted cryptic female choice. It could be argued, however, that the difference in ampulla attachment was caused by an adaptation in males (the spermatophylax) rather than a female trait and therefore does not constitute cryptic female choice. On the other hand, the cryptic female choice "trait" here could be the tendency of the female to eat the ampulla only after consuming the spermatophylax (Eberhard 1996; see Sect. 11.4.2 for further discussion). In some ecological circumstances (i.e. those favouring dispersal), however, the fertilisation bias against long-winged males could be counter to the female's fitness.

11.2.1.3 Other Factors Influencing the Duration of Spermatophore Attachment

In addition to the "traits" in Table 11.2, the mating status of the female can have a substantial effect on the duration of ampulla attachment. This has been demonstrated in several studies of *Gryllus* spp. (Bateman et al. 2001; Wynn and Vahed 2004; Gershman 2009) and in *T. commodus* (Drayton et al. 2013). In *G. bimaculatus*, for example, the mean duration of ampulla attachment for females on their first, second and third mating was as follows: 162, 82 and 50 min, respectively (Bateman et al. 2001). These patterns fit with the theoretical expectation that females should be less selective on their first mating to avoid the costs of remaining unmated (Kokko and Mappes 2005).

The duration for which females leave spermatophores attached was found to be highly repeatable and to have a high level of heritability in *Acheta domesticus* (Fleischman and Sakaluk 2004a; Mautz and Sakaluk 2008b). In *T. commodus*, Hall et al. (2013) similarly found that 25 % of genetic variance in spermatophore removal time was explained by the female, rather than the male, additive genetic component (see Simmons et al. 2014 for similar results in *T. oceanicus*). This suggests that the timing of spermatophore removal is largely determined by the female's genotype, irrespective of the quality of her mate (Mautz and Sakaluk 2008b). At the same time, these results also provide evidence that the extent to which females use ampulla removal as a mechanism of cryptic female choice can itself respond to selection.

11.2.2 Control of Sperm Uptake/Storage in the Spermatheca (Beyond Control of Spermatophore Attachment)

11.2.2.1 Experimental Evidence of Biases in Sperm Uptake

There is a growing body of evidence that female crickets use mechanisms beyond the control of ampulla attachment duration to determine the relative amount of sperm stored in the spermatheca from a given male and the proportion of eggs fertilised by that male. In a study designed to examine whether polyandrous mating allows females to avoid costs of inbreeding in *G. bimaculatus*, Tregenza and Wedell (2002) found evidence that females bias sperm use in favour of unrelated males. Females that had mated with two siblings had lower egg hatching success than those that had mated with two non-siblings. When females had mated with both a sibling and a non-sibling, however, they had the same hatching success as females mated with two unrelated males (Tregenza and Wedell 2002; but see Jennions et al. 2004, who, in a very similar experiment, found no evidence that females preferentially use sperm from non-relatives in *T. commodus*).

More direct evidence for such a bias in sperm use was provided by studies of the paternity of offspring of females mated to both a related and an unrelated male (Stockley 1999; Bretman et al. 2004; Simmons et al. 2006). Using microsatellite markers to assign paternity, both Bretman et al. (2004) and Simmons et al. (2006) found that the unrelated male sired a greater proportion of each female's offspring than the related (sibling) male in *G. bimaculatus* and *T. oceanicus*, respectively, especially when they were first in the mating order. In *Gryllodes supplicans*, the trend was also for the share of paternity to be higher for unrelated males, but the mean proportion of eggs fertilised by the first male to mate did not differ significantly between treatments (Stockley 1999). While these studies indicate that females preferentially use sperm from unrelated males, they do not provide clues of the mechanisms involved, i.e. whether the fertilisation bias resulted from a bias in sperm storage or from the selective use of stored sperm.

Bretman et al. (2009) used a novel molecular technique ("competitive microsatellite PCR" or "competitive PCR", see also Bussière et al. 2006) to quantify the relative amount of sperm from related and unrelated males stored in the female's spermatheca and their relative contribution to the paternity of offspring in *G. bimaculatus*. As in previous studies (Stockley 1999; Bretman et al. 2004; Simmons et al. 2006), Bretman et al. (2009) mated females to both a sibling and an unrelated male (in either order) and found that unrelated males contributed significantly more than half of the sperm to the spermatheca (Fig. 11.5a) and, consequently, sired significantly more than half of the females' offspring (Fig. 11.5b). Furthermore, there was a positive correlation between the proportion of sperm in the spermatheca from the unrelated male and the proportion of offspring sired by the unrelated male (Fig. 11.2).

Hall et al. (2010) similarly quantified the relative amount of sperm in the spermatheca contributed by two males differing in the level of "attractiveness" (latency to copulation in precopulatory trials) in *T. commodus*. For a given duration of spermatophore attachment, Hall et al. (2010) found that "attractive" males contributed more sperm to the spermatheca than "unattractive" males. This result appears to contrast with Bussière et al. (2006), who found that there was no difference between attractive and unattractive males in the rate of sperm transfer to the spermatheca in the same species (Fig. 11.1). In Bussière et al.'s (2006) study, however, the females were on their first mating, which could have made them less selective (see Kokko and Mappes 2005).

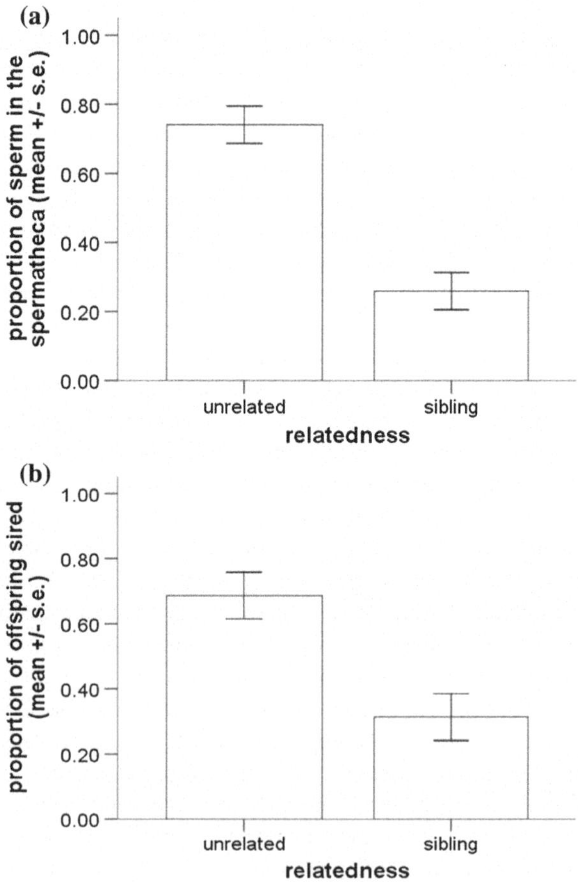

Fig. 11.5 The relative contribution to stored sperm (**a**) and to paternity (**b**) of a sibling male versus an unrelated male (females were mated to both in either order) in *Gryllus bimaculatus*. The unrelated male contributed significantly more to (**a**) the proportion of sperm in the spermatheca (mean ± SE) and (**b**) the proportion of offspring sired (mean ± SE). Adapted from Bretman et al. (2009)

Together, the studies above strongly suggest that the fertilisation biases in favour of unrelated males demonstrated in previous studies (Tregenza and Wedell 2002; Bretman et al. 2004; Simmons et al. 2006) were caused by biases in the degree of uptake of sperm into the spermatheca (rather than, for example, the selective use in fertilisation of sperm from different males). Although previous studies have demonstrated that females remove spermatophores of closely related males (Simmons 1991) or "unattractive" males (Bussière et al. 2006) sooner (see section Table 11.2), the bias in sperm uptake does not appear to have been caused by differences in spermatophore attachment duration because this was standardised for both of the female's mates (Tregenza and Wedell 2002; Bretman et al. 2004; Simmons et al. 2006; Bretman et al. 2009; Hall et al. 2010; Tuni et al. 2013): in most of these studies, males were allowed to guard the female for a set period of time following spermatophore transfer, which was apparently successful in ensuring that the spermatophore remained attached. The duration of spermatophore attachment could, however, have been controlled more precisely in most of these studies.

Another possible cause of biases in sperm storage is ejaculate adjustment by the male in relation to the quality of the female (see, e.g. Thomas and Simmons 2008). It seems unlikely that this factor accounted for differences in sperm storage between sibling and unrelated males in Bretman et al's (2009) study, however, because males were already carrying fully formed spermatophores prior to the introduction of the female. Furthermore, in *T. oceanicus* at least, males do not appear to reduce ejaculate expenditure (in terms of ejaculate substances that affect sperm viability) when mating with siblings (Simmons and Thomas 2008). Tuni et al. (2013) demonstrated conclusively that ejaculate adjustment by the male did not account for the bias in sperm storage favouring unrelated males in *T. oceanicus* (see Simmons et al. 2006): like Bretman et al. (2009), Tuni et al. (2013) used competitive PCR to assess the relative contribution to the female's spermatheca of sperm from a sibling and a non-sibling male. The identity of the guarding male in the second mating, however, was manipulated: immediately after they had transferred their spermatophores, unrelated males were swapped with sibling males (i.e. who were siblings of the female) and vice versa. Tuni et al. (2013) found that when they were guarded by a sibling male, females stored less sperm from their actual mate, even if their actual mate was a non-sibling. This strongly suggests that the differential uptake of sperm was under female control and that mate assessment by females (possibly based on the males' cuticular hydrocarbons) occurred during the guarding period (Tuni et al. 2013).

It is conceivable that the difference in sperm uptake between "attractive" and "unattractive" males demonstrated by Hall et al. (2010) was driven by differences between males in ejaculate characteristics: larger males, for example, tend to produce larger spermatophores (Rodríguez-Muñoz et al. 2008). However, Hall et al. (2010) found that male size did not correlate with their measure of male attractiveness. Overall, the evidence is very strong that female crickets are able to selectively regulate the amount of sperm transferred from the spermatophore to the spermatheca, as predicted by Eberhard (1996).

11.2.2.2 Physiological Mechanisms Accounting for Bias in Sperm Uptake

The physiological mechanisms that allow female crickets to control the uptake of sperm to the spermatheca remain to be determined. The wall of the spermathecal duct in grylline crickets such as *G. bimaculatus* and *T. commodus* includes a muscular layer (Essler et al. 1992; Sturm 2008). It has been proposed that this could be used to facilitate or inhibit sperm transfer (Eberhard 1996; Tuni et al. 2013). Transport of sperm along the 25-mm-long, convoluted spermathecal duct to the spermatheca by peristaltic contractions of these muscle fibres is known to occur in *T. commodus* (Sturm 2005). "Zipperlike" projections of the wall of the spermathecal duct could also block the entry of sperm if the female were to narrow the lumen of the duct (Sturm 2008). Further research in this area could examine how exposure to siblings versus non-siblings affects the rate of peristaltic contractions of the spermathecal duct muscles and/or the diameter of the lumen of the duct. A further possibility is that females selectively digest sperm within the spermathecal tube: spermolytic activity has been documented within the spermathecal tube in some tettigoniids (Viscuso et al. 1996; Brundo et al. 2011).

11.2.2.3 Interspecific Fertilisation Biases

Post-copulatory mechanisms that allow females to bias sperm uptake and storage in favour of members of the correct species or subspecies could be seen as part of the continuum of mate choice for genetic compatibility. Tyler et al. (2013) allowed female *G. bimaculatus* to mate with both a conspecific male and a male *Gryllus campestris* and (in either order) then used competitive PCR to quantify the relative number of sperm in the spermatheca from each male and the proportion of nymphs that each of them sired. Tyler et al. (2013) found that females stored more sperm from the conspecific male. The proportion of sperm from the male of each species in the spermatheca, however, did not equate with the male's fertilisation success (in contrast to females mated to two conspecific males); almost all nymphs were sired by the conspecific male. This suggests the existence of two forms of post-copulatory choice (or at least barriers to fertilisation) favouring the conspecific male: a bias in sperm uptake and a bias in sperm use in fertilisation (Tyler et al. 2013). There are, however, several possible mechanisms to account for these results, not all of which involve cryptic female choice (see Tyler et al. 2013). A greater success of conspecific over heterospecific sperm when in competition for fertilisation has also been demonstrated in *Allonemobius* (Gryllidae: Nemobiinae) species (Gregory and Howard 1994; Marshall 2007). This could be caused, in part, by differences between species in the length of the spermathecal duct (Marshall 2007).

11.2.3 Repeated Mating with the Same Male/Choosing to Mate Subsequently with a Different Male

Females could exercise choice by remaining with a favoured male for repeated copulations or, conversely, by mating more rapidly with a novel or better quality male following copulation with a poor-quality male (Eberhard 1996). Whether these processes and the paternity bias resulting from the relative number of copulations with two different males constitute cryptic, as opposed to "overt" female choice, however, is debatable (Eberhard 1996). Repeated mating within pairs occurs in numerous ensiferan taxa (reviewed in Alexander and Otte 1967; Brown and Gwynne 1997; Vahed 1998; see also Sect. 11.4.1). Undisturbed pairs of *G. bimaculatus* left for 7 h in the laboratory, for example, remated a mean of 5.3 times (Simmons 1988). The extent of repeated mating can be much higher than this, however: in *Ornebius aperta* (Grylloidea: Mogoplistidae), the female may receive up to 58 successive spermatophores from her mate over 3 h (Andrade and Mason 2000). For males, such extreme repeated mating appears to be an adaptation to compensate for the tendency of females to remove spermatophores before complete sperm transfer (Boldyrev 1915; Laird et al. 2004). In *O. aperta*, for example, the female typically removes and eats each spermatophore within just a few seconds following the end of copulation (Laird et al. 2004).

In both *G. bimaculatus* and *G. campestris*, the proportion of the female's eggs fertilised by a given male was found to increase with the number of times that male mated relative to rival males (Simmons 1987a; Rodríguez-Muňoz et al. 2011). This confirms that by remaining to remate with a male, the female can bias paternity in his favour. There is some evidence that female *G. bimaculatus* use this as a mechanism of mate choice (Simmons 1986, 1989). Simmons (1986) found that in large enclosures in the laboratory, females spent significantly longer with larger males (but only when mating occurred in shelters, as opposed to in the open arena). It is possible that this could have been driven by a positive relationship between mate-guarding ability and male mass. Simmons (1986) and Rodríguez-Muňoz et al. (2011), however, observed that in *Gryllus*, guarding males seemed to be unable to prevent the female from leaving after copulation if she chose to do so. Simmons (1989) found that female *G. bimaculatus* made more attempts to leave when guarded by a sibling than when guarded by an unrelated male.

It may be in the female's interests to remain with a male for repeated mating if the rate of spermatophore production is condition dependent. In support of this, Zuk (1987) found that in *G. veletis*, the number of spermatophores a male was able to produce in 24 h correlated negatively with the male's level of gregarine parasite infection. In *O. aperta*, females were more likely to remain for repeated mating with males that were in good condition (which had been raised on a high-quality diet) than with males in poor condition (Andrade and Mason 2000). This did not seem to be due to an inability of such males to maintain spermatophore production, however, because females often left poor condition males even when they still had another spermatophore ready (Andrade and Mason 2000).

Studies of ensiferans have provided some evidence that females seek additional matings sooner after having mated with a poor-quality male (Brown 1997; Chadwick-Johnson et al. 1999). In the tree cricket *Oecanthus nigricornis* (Gryllidae), Brown (1997) found that females disfavoured smaller males by subsequently mating more quickly with another male. However, contrary to expectation, Brown (1997) also found that females disfavoured males in better condition in this manner. In the humpback cricket *Cyphoderris strepitans* (Haglidae), the female feeds on hemolymph from the male's fleshy hind wings during mounting (Dodson et al. 1983). Chadwick-Johnson et al. (1999) found that females paired with dewinged males remated more quickly with a replacement male than females that had received a hemolymph meal on her previous mating. This did not appear to be due to substances in the hemolymph that inhibit female receptivity (Weddle and Sakaluk 2003). It is unclear how a female would benefit from biasing paternity against a male that had provided a smaller gift, however, unless the ability of males to feed females is an indicator of male genetic quality (see Sect. 11.4.2).

The selective pressures associated with sperm competition will select for males to delay or prevent the female from mating with another male (reviewed in Simmons 2001). There is some evidence that substances in the ejaculate can inhibit remating in the Ensifera, and in a few species, males produce mating plugs (see Sect. 11.4.6).

11.2.4 Resistance During Copulation

Resistance by the female during copulation (consisting of biting and kicking the male and rapid locomotion during copulation) is found in ensiferan species in which males use grasping devices to force or prolong copulation (reviewed in Vahed and Carron 2008; Vahed et al. 2014). In some species, resistance by females can result in injury to the male (Rentz 1963; Kuridwa and Kasuya 2009). One hypothesis for the function of this behaviour is that it is a form of mate screening (Eberhard 1996; Baena and Eberhard 2007). An alternative hypothesis is that resistance by the female reflects sexual conflict over the occurrence or duration of copulation (Arnqvist and Rowe 2005). In the case of the former hypothesis, resistance should enable the female to gain indirect genetic benefits (e.g. by screening for male condition and/or the form of the male's mate-clasping devices), while in the case of the latter hypothesis, resistance should enable the female to avoid costs associated with superfluous copulations or with receiving more ejaculate. In either case, resistance behaviour could potentially be seen as a form of communication (Baena and Eberhard 2007; Blyth and Gilburn 2011; Peretti and Aisenberg 2011).

Vahed et al (2014) examined the relationship between resistance by females, the way males grasp females during copulation and copulation duration in 44 species of tettigoniid. Resistance by females during copulation was associated with species in which males used modified grasping or piercing cerci (as opposed to the normal "lock-and-key"-based mechanism, Rentz 1972) to prolong copulation following spermatophore transfer (see Sect. 11.4.3). The duration of copulation prior to spermatophore transfer, on the other hand, was not associated with resistance by females. This suggests that resistance by females results from intersexual conflict over the duration of ejaculate transfer in these tettigoniids (Vahed et al 2014), although it does not rule out the possibility that resistance might act as a form of mate screening (Eberhard 1996). If this was the case, however, it is not clear why resistance generally occurred only following spermatophore transfer (with some exceptions, see Vahed 2002) and did not generally occur during prolonged copulation prior to spermatophore transfer.

11.2.5 The Rate of Oviposition Following Mating and/or Differential Allocation of Resources to Offspring

A female could potentially exercise post-copulatory mate choice by increasing the rate of oviposition following mating with a given male (Thornhill 1983; Eberhard 1996). This could involve allocating more resources to ovaries to allow the production of more or larger eggs ("differential allocation", reviewed in Sheldon 2000) and/or laying existing eggs at a faster rate. There is some support for this mechanism of cryptic female choice in crickets. Simmons (1987b) found that female *G. bimaculatus* that had been allowed to choose their own mates laid a greater proportion of their eggs than females allocated either a large or a small male. The degree of polyandry differed between these treatments and therefore could have been a confounding factor, however. Head et al. (2005, 2006) examined the number of eggs laid by female *A. domesticus* in relation to the level of attractiveness of their mates (latency to copulation in precopulatory trials). Head et al. (2005) found that females that had mated to a series of "attractive" males laid eggs of a greater width in the first week than females mated to a series of "unattractive" males, but there were no significant differences in the number of eggs laid. Head et al. (2006), however, did find that females mated to "attractive" males laid significantly more eggs (although this was only the case for daughters of "attractive" males and not for daughters of "unattractive" males).

Bretman et al. (2006) and Bertram and Rook (2012) examined the relationship between the rate of oviposition and the level of dominance of the mating male in *G. bimaculatus* and *Gryllus assimilis*, respectively. Both studies found that females that mated with the more dominant males laid more eggs over the experimental period. It is not clear, however, whether the increased rate of oviposition might have been caused by the properties of the male's ejaculate (see Sect. 11.4.6) rather than reflecting cryptic female choice (Bretman et al. 2006): Thomas and Simmons (2009b), for example, found that in *T. oceanicus*, the proportion of live sperm in the ejaculate was greater in dominant males than in subordinate males. Furthermore, there is evidence that males transfer substances in the ejaculate that influence the rate of oviposition following mating (Sect. 11.4.6).

11.3 Evidence for Ultimate Benefits of Cryptic Female Choice to the Female

The main focus of this chapter is to evaluate the empirical evidence for different proximate mechanisms of cryptic female choice in the Ensifera. Providing empirical evidence for the ultimate, evolutionary benefits of cryptic female choice to the female is more challenging. Benefits of female choice in general include the following: direct benefits (e.g. the acquisition of good resources), direct genetic benefits (e.g. obtaining genetically compatible sperm, leading to higher hatching success) and indirect genetic benefits. The latter encompass the acquisition of genes to enhance the survival of offspring ("good genes") and "Fisherian" benefits: genes to enhance the ability of offspring (particularly male offspring) to acquire mates (reviewed in Kokko et al. 2003, 2006; Hunt et al. 2004; Bussière et al. 2008). The finer details of these benefits of mate choice, how they are interrelated and their relative strength have been the subject of much debate. While a detailed examination of the benefits of mate choice in general is outside the scope of this chapter, studies of crickets have provided some insights into the ultimate benefits of post-copulatory female choice.

Good resources provided by males in the Ensifera include various forms of nuptial gift (see Sect. 11.4.2), sperm itself (reviewed in Wedell et al. 2002) and, in some cases, burrows or other shelters (e.g. Rodríguez-Muñoz et al. 2011). Unlike precopulatory mate choice for good resources, however, cryptic female choice for good resources per se does not seem logical (see Sect. 11.4.2). Having mated with a male, females would not benefit from biasing the fertilisation of eggs in favour of males that provided good resources, unless the resources acted as an indicator of good genes. While such post-copulatory mate choice might impose selection that could favour the future evolution of enhanced male-provided resources, it would not give an individual female the ability to gain extra resources from her current mate (Simmons and Parker 1989). One way in which an individual female could potentially use post-copulatory choice to obtain better quality direct benefits, however, would be if she were to avoid remating with males that had provided poor-quality direct benefits in previous copulations (see Wagner et al. 2007).

Studies of field crickets (Gryllidae) have provided evidence that females stand to gain direct genetic benefits by preferentially using sperm from genetically compatible males (see Sects. 11.2.1 and 11.2.2). There is evidence that female field crickets can bias spermatophore attachment and/or the uptake or storage of sperm to favour males that are less closely related to themselves (and therefore presumably more genetically compatible). The benefit of mating with unrelated males has been demonstrated in some cricket species: the eggs of females mated with non-kin have a higher hatching success than those of females mated with close kin (e.g. in *G. bimaculatus*, Tregenza and Wedell 2002, and *T. commodus*, Jennions et al. 2004).

A post-copulatory preference for novel males over previous mating partners has been demonstrated in crickets (in *G. vocalis*, Gershman 2009, and in *G. sigillatus*, Gershman and Sakaluk 2010; see Sect. 11.2.1.1). Such a preference will favour the storage and use of sperm from more than one male, which in turn may lead to genetic benefits to the female (e.g. through enhanced genetic diversity in offspring and/or an increase in the probability of receiving genetically compatible or competitively superior sperm, reviewed in Jennions and Petrie 2000; McNamara et al. 2014). Polyandrous mating has been shown to enhance offspring hatching success and/or survival in field crickets (e.g. Tregenza and Wedell 1998; Ivy and Sakaluk 2005; Gershman 2007). Fleischman and Sakaluk (2004a), however, found no measurable benefits of cryptic female choice in *A. domesticus*. Females were allowed to mate with five different males. In one group, females were prevented from exercising cryptic female choice by confining them to a narrow tube for 60 min after copulation, to prevent them from removing the spermatophore. In the other group, females were allowed to determine the duration of spermatophore attachment. There were no significant differences in, for example, nymphal production, nymphal survival or nymphal development time between the two groups, leading Fleischman and Sakaluk (2004a) to conclude that there was therefore no evidence of genetic benefits of cryptic female choice in terms of allowing the female to obtain genetically compatible sperm or genes for higher offspring quality.

The studies reviewed in this chapter suggest that females may use post-copulatory mechanisms to select males based on a range of traits that could confer indirect genetic benefits, including male body size (Simmons 1986; Bateman 1998; Bateman et al. 2001; Hall et al. 2008), condition (Bertram and Rook 2012), courtship call characteristics (Hall et al. 2008; Rebar et al. 2011), cuticular hydrocarbon profiles (Thomas and Simmons 2009a), male fighting success or dominance (Bretman et al 2006; Parker 2009) and overall attractiveness (as measured by the latency to copulation in precopulatory trials; Bussière et al. 2006; Ivy and Sakaluk 2007). Rodríguez-Muñoz et al. (2008), however, found that additive genetic effects explained relatively little phenotypic variance for a range of reproductive and non-reproductive traits in male *G. bimaculatus* (using laboratory-bred individuals that were recently derived from a single field population). These traits included male size (thorax width), spermatophore mass, male calling duration, male fighting ability, latency to copulation and male development time and lifespan. The authors concluded that in the population sampled, females cannot use either pre- or post-copulatory choice to acquire "good genes" for their offspring. Other studies of the heritability of sexually selected traits in crickets have had mixed results. For example, unlike Rodríguez-Muñoz et al. (2008), Simmons (1987c) did find that male body size was heritable in *G. bimaculatus*, while in captive populations of *A. domesticus*, some studies have found that body size shows significant heritability (e.g. Ryder and Siva-Jothy 2001), while others have not (e.g. Mautz and Sakaluk 2008b). It is possible that the low or non-existent additive genetic variance in such a wide range of both male and female traits in Rodríguez-Muñoz et al.'s (2008) study reflects the history of the particular population sampled. Their field population might, for example, have experienced a genetic bottleneck in its recent history (Rodríguez-Muñoz et al. 2008). It is also worth noting that while a lack of heritability in a trait would suggest that a female is unlikely to gain indirect genetic benefits by selecting males with that trait, it could also indicate that there has been strong selection on that trait in the past (Eberhard 1996).

Despite the lack of additive genetic variance of individual sexually selected traits found by Rodríguez-Muñoz et al. (2008), evidence was provided in support of one aspect of "good genes" selection: that sexually selected traits may act as condition-dependent indicators of genes for overall "genetic quality", possibly reflecting the ability of the male to acquire resources (reviewed in Hunt et al. 2004; Bussière et al. 2008). As predicted by this hypothesis, many of the sexually selected traits that were quantified in male *G. bimaculatus* by Rodríguez-Muňoz et al. (2008) were positively correlated with one another. In *G. assimilis*, Bertram and Rook (2012) similarly found positive correlations between male body size, male condition (higher residual mass, a trait possibly selected for via cryptic female choice in this species) and various aspects of the male's calling song.

There is also some evidence for the hypothesis that benefits to the female of cryptic choice in crickets are related to "Fisherian" benefits: the production of attractive sons (Head et al. 2005, 2006). One prediction of this hypothesis is that genes for the preferred trait(s) in the male and genes for the degree of preference for that trait in females should show genetic covariance (Fisher 1930; Lande 1981). In support of this prediction, Head et al. (2006) found that in *A. domesticus*, the extent of post-copulatory preference for male size depended upon whether the female was the daughter of an "attractive" or an "unattractive" male: daughters of "attractive" males exhibited a stronger cryptic preference for larger males (expressed in terms of laying more eggs in the week following mating). Further support for "Fisherian" benefits in *A. domesticus* was provided by Head et al. (2005): females that had mated with a series of "attractive" males tended to have sons who were more "attractive" (see also Wedell and Tregenza 1999).

Assessing sexual selection by considering cryptic female choice alone could be misleading (Hunt et al. 2009). Precopulatory female choice can, for example, sometimes counteract post-copulatory choice (e.g. Danielsson 2001). This does not seem to be the case in crickets, however. Several studies have compared pre- and post-copulatory choice for certain male traits or combinations of traits in crickets. These studies have generally found that pre- and post-copulatory selection for the traits in question is re-enforcing rather than in opposition (in *T. commodus*, Bussière et al. 2006; in *T. oceanicus*, Rebar et al. 2011; Thomas and Simmons 2009a; in *Platygryllus primiformis*; Parker 2009). Despite there being similarities in episodes of pre- and post-copulatory selection, Thomas and Simmons (2009a) demonstrated that there were differences in the form and intensity of selection: cryptic female choice generated weaker selection than did precopulatory choice. The effects of competition between males for fertilisation on selection for the traits in question should also be taken into account (Simmons 2001; Hall et al. 2008; Hunt et al. 2009; Sects. 11.2.1.2 and 11.4).

11.4 Male–Female Interactions During and After Copulation: From Courtship to Coercion

The existence of cryptic female choice will select for adaptations in males that function to induce the female to respond in a way that favours the male's reproductive success (Eberhard 1996). Such adaptations could range from coercion (e.g. Vahed et al. 2014) or manipulation via "sensory traps" (reviewed in Arnqvist and Rowe 2005) to more subtle forms of persuasion, which Eberhard (1991, 1996) referred to as "copulatory courtship" (see also Peretti and Aisenberg 2011, in which evidence is presented that copulatory courtship can also occur alongside coercive strategies). Copulatory courtship encompasses attempts by males to elicit favourable responses from females both during and immediately after copulation and can entail both behavioural and morphological adaptations, including stimulation via the male's genitalia (Eberhard 1985, 1991, 1996, 2010). Selectively responding to the form or intensity of copulatory courtship signals could benefit the female by providing her with indirect genetic benefits (see Sect. 11.3, above) or perhaps by ensuring that the female accepts compatible sperm from the correct species or subspecies (Eberhard 1996).

The existence of a range of adaptations in male crickets and relatives to deter the female from eating the spermatophore prematurely was recognised around a century ago: Gerhardt (1914) and Boldyrev (1915) provided examples of a range of such "countermeasures" in different ensiferan species which include feeding the female with glandular secretions following spermatophore transfer, feeding the female with a spermatophylax, remaining attached to the female during ejaculate transfer, rapidly repeated multiple mating with the female and post-copulatory mate guarding.

The influence of these behaviours on cryptic female choice will be considered in this section, together with other adaptations used by male ensiferans to increase fertilisation success in the face of both sperm competition and cryptic female choice.

11.4.1 "Courtship" that Occurs Between Repeated Copulations

In many cricket species, especially those in which pairs routinely engage in repeated mating, males direct a variety of post-copulatory behaviours towards the female (reviewed in Alexander and Otte 1967, Brown and Gwynne 1997; Vahed 1998 see also Preston-Mafham 2000; Andrade and Mason 2000; Sadowski et al. 2002; Zefa et al. 2008; de Carvalho and Shaw 2010; Prokop and Maxwell 2008; 2011), which could be interpreted either as a form of copulatory courtship (Eberhard 1991, 1996) or as mate "harassment" (Hall et al. 2008). The former implies intersexual cooperation, while the latter implies intersexual conflict.

Bussière et al (2006) and Hall et al. (2008) interpreted post-copulatory mate guarding in field crickets as a form of mate harassment. They demonstrated that post-copulatory mate guarding can prolong the duration of spermatophore attachment and thereby influence the form and intensity of cryptic female choice (see Sect. 11.2.1.2). However, if mate-guarding ability or intensity is an honest signal of male genetic quality, this may not result in sexual conflict (Simmons 1990). Evidence for this does not seem to be strong: Bussière et al. (2006) found that in *T. commodus*, males classed as "unattractive" (i.e. that the females were less willing to mount in precopulatory trials) actually guarded more intensely than "attractive" males, while in *G. bimaculatus*, Simmons (1990) similarly found that guarding was more intense in males that were more heavily infected with protozoan gut parasites. There is also little evidence to support the hypothesis that females might benefit from biasing paternity in favour of better guards through having sons who are better at guarding (i.e. gain indirect, "Fisherian", benefits, see Sect. 11.3). Hall et al. (2013) found no evidence for a heritable component to mate-guarding intensity in *T. commodus*.

Despite evidence that mate guarding in grylline crickets is "harassment" by males, guarding behaviour does contain some behavioural elements (such as antennal rotation and body rocking, see Simmons 1990; Parker and Vahed 2010) that do not appear to affect the ability of females to remove spermatophores directly, so it might be interpreted instead as courtship signals. Similarly, in *Nisitrus* sp. (Gryllidae: Eneopterinae), the guarding male repeatedly raises one or more hind legs as an apparent visual signal that he has produced the next spermatophore: the female responds to this signal by beginning to consume the previous spermatophore in readiness to receive the next (Preston-Mafham 2000). Male *O. aperta* (Mogoplistidae) produce vibratory signals between the transfer of successive spermatophores to their mate, which might act as an honest signal of male body mass (Andrade and Mason 2000). Such vibratory signals are also found in *Leptogryllus ookala* (Gryllidae: Oecanthinae). In this species, post-copulatory movement by mounted females increases the rate of tremulation and cercal tapping by males, which induces the female to remain still (Brown 1999).

In the Hawaiian swordtail cricket, *Laupala cerasina* (Gryllidae: Triginidiinae), courtship involves a series of repeated copulations over several hours, in which the male transfers a series of spermless "microspermatophores" (on average transferring seven) in advance of a larger, sperm-containing "microspermatophore" (de Carvalho and Shaw 2010). By replacing some females after the male had transferred all of his microspermatophores, de Carvalho and Shaw (2010) were able to examine the effect of the protracted courtship on sperm transfer. Interestingly, the extent of sperm transfer from the spermatophore to the spermatheca was greater for females that had received protracted courtship and microspermatophores than it was for females that had received only the macrospermatophore. This could suggests that females selectively respond to the transfer of microspermatophores by facilitating ejaculate transfer, although it does not rule out other processes such as the transfer of substances in the microspermatophore that enhance sperm transfer (de Carvalho and Shaw 2010).

11.4.2 Nuptial Gifts and Cryptic Female Choice

Nuptial gifts occur in several ensiferan clades (reviewed in Brown and Gwynne 1997; Vahed 1998). Gifts that are consumed during copulation and/or ejaculate transfer include metanotal gland secretions in tree crickets (Gryllidae; Oecanthinae) and other grylloids, hemolymph from tibial spurs in ground crickets

(Gryllidae; Nemobiinae) and the spermatophylax. The latter occurs in most tettigoniids and in various other ensiferans including some rhaphidophorids, haglids, stenopelmatids, anostostomatids, gryllacridids and a few gryllids (reviewed in Gwynne 2001).

There is much evidence for Boldyrev's (1915) hypothesis that nuptial gifts delay the female from eating the ampulla of the spermatophore. Studies of nuptial gifts in these ensiferans have demonstrated that the presence or size of the nuptial gift increases the duration of ampulla attachment (e.g. Sakaluk 1984; Wedell and Arak 1989; Reinhold and Heller 1993; Brown 1997; Fedorka and Mousseau 2002) and results in a greater share of paternity of the female's eggs for the donating male (Wedell 1991; Sakaluk and Eggert 1996). Furthermore, comparative studies have shown that the production of a larger spermatophylax is associated with the production of a larger ampulla and more sperm, controlling for male body weight and phylogeny (Wedell 1993; Vahed and Gilbert 1996).

It had been proposed that nuptial gifts represent male rather than female control of the paternity of the female's offspring: by using the gift to exploit the female's gustatory responses, the male is able to distract the female, allowing him to transfer more ejaculate than may be in the female's reproductive interests and potentially to counter cryptic female choice (Sakaluk 2000; Vahed 2007a; Warwick et al. 2009; Vahed et al. 2014). With reference to the spermatophylax in *G. sigillatus*, however, Eberhard (1996) stated that the view that the nuptial gift represented male control of the duration of insemination was an "illusion": females were setting the "rule of the game" by following the rule of thumb of "terminate sperm transfer after consuming the nuptial gift". Eberhard (1996) pointed out that females could choose to discard the nuptial gift and terminate insemination prematurely. This does indeed sometimes occur (Sakaluk 1984, 1985, 1987), although the spermatophylax in *G. sigillatus* appears to contain phagostimulants, in the form of free amino acids, which encourage the female to maintain feeding (Warwick et al. 2009). Interestingly, Gershman et al. (2012) found significant differences in the amino acid profiles of discarded and accepted spermatophylaces in this species. In many tettigoniids with large spermatophylaces, however, the spermatophylax is firmly cemented to the ampulla and is not detachable (as it is in *Gryllodes*) (personal observation). The neck of the ampulla is also very firmly attached to the female's genital opening, so rubbing off or discarding the spermatophylax or ampulla without eating through the spermatophylax is not an option. It is notable that male *G. sigillatus* employ a "belt and braces" approach to prolonging ampulla attachment: in addition to producing a spermatophylax, they also guard females, which has been found to be effective in delaying females from eating the ampulla (Ivy and Sakaluk 2007).

Positive relationships between gift size and the number of sperm transferred by a male have been interpreted as demonstrating cryptic female choice for good resources (Thornhill 1976). However, cryptic choice for good resources does not seem logical: while permitting a greater degree of insemination for males that have provided larger gifts will impose selection on males that could ultimately lead to the evolution of larger gifts, such selection does not provide a means by which an individual female can extract a greater level of direct material benefits from her mate (Simmons and Parker 1989). Cryptic female choice based on gift characteristics does, however, potentially make sense in terms of indirect genetic benefits, e.g. if the gift is a condition-dependent signal of male genetic quality (Eberhard 1996; Vahed 2007b). In tettigoniids, comparative evidence suggests that large spermatophylaces are costly to produce: there is a positive relationship between the duration of the male's sexual refractory period and spermatophylax size across species (Vahed 2007b). Furthermore, a range of studies involving nuptial gifts in the Ensifera have provided evidence that nuptial gift size is correlated positively with aspects of male fitness such as male body mass, the degree to which males are infected by parasites and male foraging ability, reflected by the effects of diet quality on gift size (reviewed in Vahed 2007a).

A problem with the concept that females could use gift size as an indicator of male genetic quality, however, is plasticity in gift size (reviewed in Vahed 2007a). Male bushcrickets (Tettigoniidae) and tree crickets (Gryllidae: Oecanthinae), for example, can tailor gift size in relation to the size of ejaculate they are able to produce at the time of mating, the time elapsed since the previous mating, the number of previous mates, the risk of sperm competition, the female encounter rate and the characteristics of the female (reviewed in Vahed 2007a). Such variation will obviously reduce the extent to which gift size will act as an honest indicator of male genetic quality. It has even been proposed that attractive, high-quality males should provide *smaller* gifts due to their higher expected mating frequency (allocating limited gift resources over a greater number of matings) (Bussière 2002).

11.4.3 Prolonged Copulation During Ejaculate Transfer: Copulatory Structures and Coercion?

In some ensiferan species, males remain attached to the female during ejaculate transfer, effectively blocking the female from removing the spermatophore before sperm transfer is complete (Boldyrev 1915; Vahed 1996; Wedell 1998; reviewed in Vahed 1997; Vahed et al. 2014). Alexander and Otte (1967) proposed that in species with prolonged copulation, the male's genitalia should be adapted to prevent the disengagement of the female. This prediction was supported by Vahed et al. (2014): in tettigoniid species with brief copulation following spermatophore transfer, teeth on the male's cerci engage with pits or grooves on the female's subgenital plate or the base of the ovipositor (a "lock-and-key" system, Rentz 1972). In contrast, prolonged copulation following spermatophore transfer was associated with three different types of "modified" cerci: a) those that contact the female in multiple places, b) cerci that encircle the female's abdomen, or c) cerci that pierce the unsclerotised abdominal cuticle (leaving melanised scars in some cases). In addition, resistance by the female during copulation only occurred in species with modified cerci (in which copulation was generally prolonged) (see Sect. 11.2.4).

11.4.4 "Titillators": Copulatory Structures and Courtship?

Titillators, which occur in certain tettigoniid subfamilies, are very strong candidates for genital structures in males that have evolved through cryptic female choice and play a role in copulatory courtship (reviewed in Vahed et al. 2011). These are sclerotised, paired genital structures (Fig. 11.6) that are inserted into the female's genital chamber and are moved rhythmically back and forth prior to the transfer of the spermatophore (Boldyrev 1928; Duijm et al. 1983; Hartley and Warne 1984; Vahed et al. 2011). Titillators show considerable variation in morphology between species and are widely used as important taxonomic characters (e.g. Harz 1969). In the Tettigoniidae, titillators tend to be present in taxa in which the duration of copulation prior to spermatophore transfer is relatively long (Vahed et al. 2011) and, therefore, in which there is, arguably, more time for copulatory courtship.

Boldyrev (1928) noted that audible scratching sounds accompanied movements of the titillators within the female's genital chamber in the bushcricket *Bradyporus multituberculatus* (Tettigoniidae: Bradyporinae). He proposed that titillators function to "irritate by scratching the walls of the genital chamber and, perhaps to enlarge that chamber before the spermatophore is introduced to it". If the word "irritate" is replaced by "stimulate", this description fits closely with Eberhard's (1985; 1996; 2010) concept of genitalia as copulatory courtship devices. Indeed, the name "titillator" itself implies a stimulatory function. Experimental evidence appears to support this hypothesis: when the tips of the titillators were ablated in the tettigoniids *Ephippiger ephippiger* (Bradyporinae; Duijm et al. 1983) and in *Metrioptera roeselii* (Tettigoniidae; Wulf et al., manuscript), pairs remained linked in copulation for the normal duration, but the tube of the spermatophore was not inserted correctly into the female's genital chamber and consequently spermatophores failed to remain attached (in 50 % of cases in *E. ephippiger* and in 29 % in *M. roesellii*). These results, however, could also be consistent with the hypothesis that males use

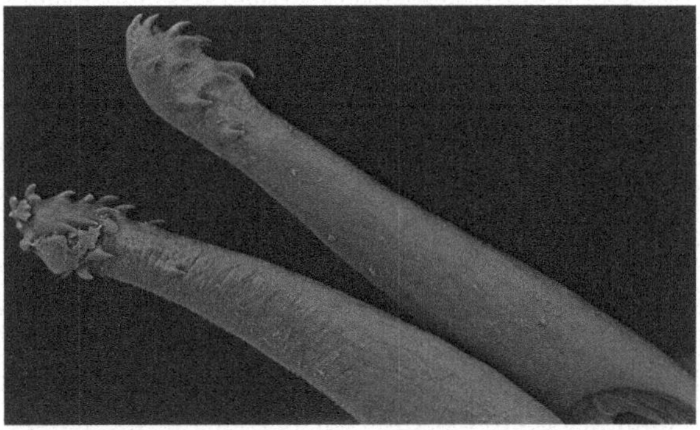

Fig. 11.6 Electron micrograph of a pair of titillators in the bushcricket *Metrioptera saussuriana*. The tips of the titillators are shown, which are moved rhythmically back and forth within the female's genital chamber prior to spermatophore transfer and could function in copulatory courtship courtship

titillators to mechanically open up the entrance to the female's genital chamber or act as a guiding rod for the spermatophore tube, as proposed by Hartley and Warne (1984). Stimulation by titillators could also represent sensory exploitation by the male, mimicking stimulation associated with the passage of an egg through the tract (see von Helversen and von Helversen 1991; Córdoba-Aguilar 2002). Furthermore, studies involving the ablation of parts of genitalia have also been criticised for telling us little about the effects of natural variation in genital morphology on male reproductive success (Simmons 2014). Distinguishing between such hypotheses for titillator function (e.g. by selective numbing of the portions of the female contacted by the titillators, see Eberhard 2010) is a challenge for further research on this topic.

11.4.5 Ejection/Removal of Sperm from the Spermatheca

In a few ensiferan species, males remove, or stimulate the female to release, previous males' sperm from the spermatheca prior to transferring their own (Ono et al. 1989; von Helversen and von Helversen 1991). While such behaviour is clearly adaptive for the male (reviewed in Simmons 2001), Eberhard (1996) argued that the role of the female in such processes should not be overlooked. The tree cricket *Truljalia hibinonis* (Gryllidae: Podoscirtinae) is apparently unique amongst the Ensifera in that sperm appeared to be transferred directly into the spermatheca via the male's genitalia, without the use of a spermatophore (Ono et al. 1989). Sperm are injected in such a way that they volumetrically displace the sperm of the previous male, resulting in the last male's sperm occupying 88 % of the spermatheca (Ono et al. 1989). Eberhard (1996) proposed that the morphology of the female's reproductive tract (such as the very short spermathecal tube) reflects selection on females to assist this process.

In the bushcricket *Metaplastes ornatus* (Tettigoniidae; Phaneropterinae), females appear to play a more active role in the ejection of stored sperm (von Helversen and von Helversen 1991). Mating in this species consists of two stages: in the first stage, which can be repeated 2–19 times, the male inserts his highly modified, keeled, subgenital plate into the females' genital chamber and moves it back and forth repeatedly prior to withdrawing it; in the second stage, the male transfers his spermatophore in the normal way. By using interrupted mating, von Helversen and von Helversen (1991) demonstrated that the first stage of mating results in the ejection of a substantial proportion of the sperm of rival males. The mechanism of sperm ejection appears to be based on the movements of the keel of the subgenital plate mimicking the passage of an egg past the exit of the spermathecal duct (von Helversen and von Helversen 1991). This suggests that sperm ejection works by exploiting a pre-existing sensory bias in the female (see Arnqvist and Rowe 2005) and is not necessarily adaptive for the female. On the other hand, the possibility exists that females might selectively cooperate with males depending upon their phenotype and that such selective sperm ejection might act as a mechanism of cryptic female choice (Eberhard 1996).

11.4.6 Allohormones in the Ejaculate and Mating Plugs

In many insects, males transfer accessory gland proteins in their ejaculate that can act as allohormones, affecting the oviposition and remating behaviour of their mates (reviewed in Gillott 2003; Avila et al. 2011). In gryllids, there is some evidence for a dose-dependent effect of ejaculate substances on the rate of oviposition (e.g. Stanley-Samuelson et al. 1986; Marshall et al. 2009) and on the receptivity of the female (Loher et al. 1993; but see Fleischman and Sakaluk 2004b). The ejaculate in gryllids is known to contain a complex range of proteins and peptides, some of which are linked with the induction of oviposition (Marshall et al. 2009) and some with other functions such as maintaining sperm viability (Simmons and Beveridge 2011). In *T. commodus*, prostaglandins appear to be involved in the control of oviposition behaviour in the female and males are known to transfer prostaglandin precursors in their ejaculate (Stanley-Samuelson et al. 1986, 1987; reviewed in Eberhard 1996; Arnqvist and Rowe 2005).

In the Tettigoniidae, experiments in which the duration of attachment of the spermatophore has been varied have demonstrated dose-dependent effects of substances in the ejaculate on both the rate of oviposition following mating and the time taken for the female to subsequently mate with another male (Gwynne 1986; Wedell and Arak 1989; Simmons and Gwynne 1991; Vahed 2003b). Furthermore, comparative studies have found that larger spermatophore ampullae (and thus presumably larger ejaculates) are associated with longer sexual refractory periods in females (Wedell 1993; Vahed 2007b) and a lower lifetime degree of polyandry (Vahed 2006).

The extent to which these substances reflect intersexual conflict rather than cooperation is not clear (see Eberhard 1996). A female might benefit from responding to allohormones in the ejaculate. For example, females could gain indirect genetic benefits from biasing paternity in favour of males with larger ejaculates if ejaculate size is condition dependent and/or if the ability of the male to "manipulate" his mate is heritable (Eberhard and Cordero 1995). In addition, it may be in the female's interests to lay eggs at a faster rate immediately following mating and to refrain from seeking additional copulations while doing so (Eberhard 1996). That females have the power to selectively cooperate with males in this context was demonstrated by Simmons and Gwynne (1991): in the tettigoniid *Kawanaphila nartee*, females appear to be able to stop responding to ejaculate substances that induce a delay in remating when they are exposed to conditions that favour polyandry.

In a few ensiferan species, the male appears to be successful in preventing the female from remating for the rest of her life by using a mating plug (Mello 2007). In *Adenophallusia* spp. (Gryllidae: Eneopterinae) and *Aracamby* spp. (Gryllidae: Phalangopsinae), the mating plug is secreted from the male's phallic glands, while in *Eimanacris* spp. (Phalangopsinae), it is formed from the neck of the spermatophore (Mello 2007). Although Mello (2007) observed that plugged females would attempt to mount other males, they were apparently unable to receive further spermatophores. By imposing a lifetime of monandry, these adaptations would appear to prevent the female from exercising cryptic female choice. The extent to which female cooperation is needed in the formation of the mating plug and whether females are able to remove freshly deposited mating plugs deserves further investigation in these species.

11.5 Conclusions

Support for some of the potential mechanisms of cryptic female choice proposed by Eberhard (1996) is very strong within the Ensifera. There is compelling evidence that female crickets use control of the duration of spermatophore attachment to bias paternity in favour of males in relation to a variety of traits (Sect. 11.2.1). There is also a growing body of evidence that females can control the uptake of sperm to the spermatheca for a given duration of spermatophore attachment, as predicted by Eberhard (1996) (Sect. 11.2.2). For other potential mechanisms of cryptic female choice, such as differential allocation of resources to the production of eggs, there is currently insufficient evidence to distinguish between manipulation by males and cryptic female choice.

An obvious avenue for further research is the physiological mechanisms that underlie female control of the uptake of sperm to the spermatheca. Approaches that have been used to examine how females control the movement of sperm from the spermatheca during fertilisation (e.g. Sugawara 1993) could be applied to examine how females might control the travel of sperm in the opposite direction. The rate or extent of contractions of muscles surrounding the spermathecal duct (or the rate of firing of the neurones that control them) during sperm uptake, for example, could be compared for females exposed to either sibling or non-sibling males.

The evidence that mechanisms of cryptic female choice, such as manipulation of spermatophore attachment by the female, have resulted in coevolutionary adaptations in males is also very strong (Sect. 11.4). As documented a century ago (Gerhardt 1914; Boldyrev 1915), males have evolved a wide range of behaviours to facilitate ejaculate transfer by deterring the female from removing the ampulla of the spermatophore prematurely. How such adaptations affect the form and intensity of cryptic female choice (see Hall et al. 2008) deserves further investigation. Whether these adaptations potentially decrease female fitness and result in ongoing sexually antagonistic coevolution, or whether they instead act as honest signals of male genetic quality and therefore do not adversely affect female fitness, is currently unclear. In some cases, elements of behaviour directed towards the female during or after copulation could be viewed as "copulatory courtship" (Eberhard 1991, 1996). Strong candidates for this category include movements of genital "titillators" in bushcrickets and visual and vibratory signals produced by males in between the transfer of successive spermatophores to the female (Sects. 11.4.1 and 11.4.4). Both of these phenomena are potentially fruitful subjects for further research.

Arnqvist (2014) called for further studies of the selective pressures responsible for the maintenance of cryptic female choice traits. He pointed out that data on the consequences to female fitness of phenotypic variation in cryptic female choice traits are virtually non-existent. There is considerable potential to use cricket species to address this issue. The tendency of females to remove spermatophores prematurely, for example, has been found to be highly repeatable within individuals and to show a high level of heritability in *A. domesticus* (Fleischman and Sakaluk 2004a; Mautz and Sakaluk 2008b; see also Hall et al. 2013; Simmons et al. 2014). There is scope for further studies of the fitness consequences of such variation in this cryptic female choice trait (see Fleischman and Sakaluk 2004a).

The copulatory and post-copulatory behaviour of the ensiferan species that have been examined so far is incredibly diverse (reviewed in Gerhardt 1913; Boldyrev 1915; Alexander and Otte 1967; Brown and Gwynne1997; Field and Jarman 2001; Vahed et al. 2014). Yet there remain very many ensiferan subfamilies, families (e.g. the Gryllacrididae) and even superfamilies (e.g. the Schyzodactyloidea) about which very little is known in this respect. One of the greatest resources for the study of evolutionary biology is the very same biological diversity that evolutionary theory seeks to explain. Further detailed observations (see Peretti and Córdoba-Aguilar 2007) of copulatory and post-copulatory behaviour in a wide range of species within focal ensiferan taxa will undoubtedly provide many novel insights into the nature of post-copulatory sexual selection, including the mechanisms of cryptic female choice, forms of copulatory courtship or coercion and the resulting coevolutionary dynamics between the sexes.

Acknowledgements I thank A. Peretti and A. Aisenberg for inviting me to write this chapter, D. T. Gwynne and L. Bussière for their insightful comments on a previous version of the manuscript, and T. Tregenza, A. Bretman and L. Bussière for permission to use and for supplying copies of figures. I am also grateful to W.G Eberhard, G. Lehmann, S. Mowles, C. Tuni, T. Tregenza and A. Bretman for discussion and to K. Bellis and G. Vahed for their support.

References

Alexander RD, Otte D (1967) The evolution of genitalia and mating behaviour in crickets (Gryllidae) and other Orthoptera. Mis Publ Mus Zool Univ Mich 133:1–62

Andrade MCB, Mason AC (2000) Male condition, female choice, and extreme variation in repeated mating in a scaly cricket, *Ornebius aperta* (Orthoptera: Gryllidae: Mogoplistinae). J Insect Behav 13:483–497

Arnqvist G (2014) Cryptic female choice. In: Shuker DM, Simmons LW (eds) The evolution of insect mating systems. Oxford University Press, Oxford, pp 204–220

Arnqvist G, Nilsson T (2000) The evolution of polyandry: multiple mating and female fitness in insects. Anim Behav 60:145–164

Arnqvist G, Rowe L (2005) Sexual conflict. Princeton University Press, Princeton, NJ

Avila FW, Sirot LK, LaFlamme BA, Rubinstein CD, Wolfner MF (2011) Insect seminal fluid proteins: identification and function. Annu Rev Entomol 56:21–40

Baena ML, Eberhard WG (2007) Appearances deceive: female "resistance" behaviour in a sepsid fly is not a test of male ability to hold on. Ethol Ecol Evol 19:27–50

Bateman PW (1998) Assortative mating by both sexes in the cricket *Platygryllus primiformis* (Orthoptera: Gryllidae; Gryllinae). Trans Am Ent Soc 124:63–68

Bateman PW, Gilson LN, Ferguson JWH (2001) Male size and sequential mate preference in the cricket *Gryllus bimaculatus*. Anim Behav 61:631–637

Bertram SM, Rook V (2012) Relationship between condition, aggression, signalling, courtship, and egg laying in the field cricket, *Gryllus assimilis*. Ethology 118:360–372
Blyth JE, Gilburn AS (2011) The function of female behaviours adopted during premating struggles in the seaweed fly, *Coelopa frigida*. Anim Behav 81:77–82
Boldyrev BT (1915) Contributions à l'étude de la structure des spermatophores et des particularités de la copulation chez Locustodea et Gryllodea. Horae Soc Ent Rossicae 6:1–245
Boldyrev BT (1928) Biological studies on *Bradyporus multituberculatus* F.W. (Orth., Tettig). Eos 4:13–56
Bonduriansky R (2001) The evolution of male mate choice in insects: a synthesis of ideas and evidence. Biol Rev 76:305–339
Bretman A, Newcombe D, Tregenza T (2009) Promiscuous females avoid inbreeding by controlling sperm storage. Mol Ecol 18:3340–3345
Bretman A, Rodríguez-Muñoz R, Tregenza T (2006) Male dominance determines female egg laying rate in crickets. Biol Lett 2:409–411
Bretman A, Tregenza T (2005) Measuring polyandry in wild populations: a case study using promiscuous crickets. Mol Ecol 14:2169–2179
Bretman A, Wedell N, Tregenza T (2004) Molecular evidence of post-copulatory inbreeding avoidance in the field cricket *Gryllus bimaculatus*. Proc R Soc Lond B 271:159–164
Brown WD (1997) Female remating and the intensity of female choice in black-horned tree crickets, *Oecanthus nigricornis*. Behav Ecol 8:66–74
Brown WD (1999) Mate choice in tree crickets and their kin. Annu Rev Entomol 44:397–428
Brown WD, Gwynne DT (1997) Evolution of mating in crickets, katydids and wetas (Ensifera). In: Gangwere SK, Muralirangen MC, Muralirangen M (eds) The bionomics of grasshoppers katydids and their kin. CAB International, Wallingford, pp 281–318
Brundo MV, Longo G, Sottile L, Trovato M, Vitale D, Viscuso R (2011) Morphological and ultrastructural organisation of the spermatheca of some Tettigoniidae (Insecta, Orthoptera). Ital J Zool 78:53–62
Bussière LF (2002) A model of the interaction between 'good genes' and direct benefits in courtship feeding animals: when do males of high genetic quality invest less? Phil Trans R Soc Lond 357:309–317
Bussière LF, Hunt J, Jennions MD, Brooks R (2006) Sexual conflict and cryptic female choice in the black field cricket, *Teleogryllus commodus*. Evolution 60:792–800
Bussière LF, Hunt J, Stölting Jennions MD, Brooks R (2008) Mate choice for genetic quality when environments vary: suggestions for empirical progress. Genetica 134:69–78
Candolin U (2003) The use of multiple cues in mate choice. Biol Rev 78:575–595
Chadwick Johnson J, Ivy TM, Sakaluk SK (1999) Female remating propensity contingent on sexual cannibalism in sagebrush crickets, *Cyphoderris strepitans*: a mechanism of cryptic female choice. Behav Ecol 10:227–233
Córdoba-Aguilar A (2002) Sensory trap as the mechanism of sexual selection in a damselfly genitalic trait (Insecta: Calopterygidae). Am Nat 160:594–601
Danielsson I (2001) Antagonistic pre- and post-copulatory sexual selection on male body size in a water strider (*Gerris lacustris*). Proc R Soc Lond B 268:77–81
de Carvalho TN, Shaw KL (2010) Elaborate courtship enhances sperm transfer in the Hawaiian swordtail cricket, *Laupala cerasina*. Anim Behav 79:819–826
Dodson G, Morris GK, Gwynne DT (1983) Mating behaviour in the primitive orthopteran genus *Cyphoderris* (Haglidae). In: Gwynne DT, Morris GK (eds) Orthopteran mating systems: sexual competition in a diverse group of insects. Westview Press, Boulder, Colorado, pp 305–318
Drayton JM, Boeke JEK, Jennions MD (2013) Immune challenge and pre- and post-copulatory female choice in the cricket *Teleogryllus commodus*. J Insect Behav 26:176–190
Duijm M, Oudman L, Veldstra BG (1983) Copulation in *Ephippiger* (Orthoptera: Tettigoniidae). Tijd V Entomol 126:91–96
Eades DC, Otte D, Cigliano MM, Braun H (2014) Orthoptera species file online. Version 2.0/4.1. http://Orthoptera.SpeciesFile.org

Eberhard WG (1985) Sexual selection and animal genitalia. Harvard University Press, Cambridge, MA
Eberhard WG (1991) Copulatory courtship in insects. Biol Rev 66:1–31
Eberhard WG (1996) Female control: sexual selection by cryptic female choice. Princeton University Press, Princeton, NJ
Eberhard WG (2010) Evolution of genitalia: theories, evidence, and new directions. Genetica 138:5–18
Eberhard WG, Cordero C (1995) Sexual selection by cryptic female choice on male seminal products—a new bridge between sexual selection and reproductive physiology. Trends Ecol Evol 10:493–496
Engelhardt V (1915) On the structure of the alluring gland in *Isophya acuminata* Br-W. Bull Soc Entomol Moscow 1:58–63
Essler H, Herzog EH, Musiol IM, Pohlhammer K (1992) Morphology of the receptacular complex in the cricket *Teleogryllus commodus* (Saltatoria: Ensifera: Gryllidae). Entomol Gener 17:219–232
Fedorka KM, Mousseau TA (2002) Tibial spur feeding in ground crickets: larger males contribute larger gifts (Orthoptera: Gryllidae). Fla Entomol 85:317–323
Fisher RA (1930) The genetical theory of natural selection. Clarendon Press, Oxford
Field LH, Jarman TH (2001) Mating behaviour. In: Field LH (ed) The biology of wetas, king crickets and their allies. CAB International, Wallingford, pp 317–332
Fleischman RR, Sakaluk SK (2004a) No direct or indirect benefits to cryptic female choice in house crickets (*Acheta domesticus*). Behav Ecol 15:793–798
Fleischman RR, Sakaluk SK (2004b) Sexual conflict over remating in house crickets: no evidence for an anti-aphrodisiac in males' ejaculates. Behaviour 141:633–646
Gerhardt U (1914) Copulation und spermatophoren von grylliden und locustiden. II. Zool Jahrb Syst 37:1–64
Gershman SN (2007) Female *Gryllus vocalis* field crickets gain diminishing returns from increasing number of matings. Ethology 113:1099–1106
Gershman SN (2009) Postcopulatory female choice increases the fertilisation success of novel males in the field cricket, *Gryllus vocalis*. Evolution 63:67–72
Gershman SN, Mitchell C, Sakaluk SK, Hunt J (2012) Biting off more than you can chew: sexual selection on the free amino acid composition of the spermatophylax in decorated crickets. Proc R Soc Lond B 279:2531–2538
Gershman SN, Sakaluk SK (2010) Mate quality and novelty influence post-copulatory female choice in decorated crickets, *Gryllodes sigillatus*. Ethology 116:1113–1117
Gillott C (2003) Male accessory gland secretions: modulators of female reproductive physiology and behavior. Ann Rev Entomol 48:163–184
Gregory PG, Howard DJ (1994) A postinsemination barrier to fertilisation isolates two closely related ground crickets. Evolution 48:705–710
Gwynne DT (1986) Courtship feeding in katydids: investment in offspring or in obtaining fertilisations? Am Nat 128:342–352
Gwynne DT (1997) The evolution of edible "sperm sacs" and other forms of courtship feeding in crickets, katydids and their kin (Orthoptera: Ensifera). In: Choe J, Crespi B (eds) The evolution of mating systems in insects and arachnids. Cambridge University Press, Cambridge, pp 110–129
Gwynne DT (2001) Katydids and bush-crickets: reproductive behaviour and the evolution of the Tettigoniidae. Cornell University Press, Ithaca
Haley EL, Gray DA (2012) Mating behavior and dual-purpose armaments in a camel cricket. Ethology 118:49–56
Hall MD, Bussière LF, Demont M, Ward P, Brooks RC (2010) Competitive PCR reveals the complexity of poscopulatory sexual selection in *Teleogryllus commodus*. Mol Ecol 19:610–619
Hall MD, Bussière LF, Hunt J, Brooks R (2008) Experimental evidence that sexual conflict influences the opportunity, form and intensity of sexual selection. Evolution 62:2305–2315

Hall MD, Lailvaux SP, Brooks RC (2013) Sex-specific evolutionary potential of pre- and postcopulatory reproductive interactions in the field cricket, *Teleogryllus commodus*. Evolution 67:1831–1837

Hartley JC, Warne AC (1984) Taxonomy of the *Ephippiger ephippiger* complex (*ephippiger*, *cruciger* and *cunii*) with special reference to the mechanics of copulation (Orthoptera: Tettigoniidae). Eos 60:43–54

Harz K (1969) Die Orthopteren Europas/The Orthoptera of Europe I. Dr W Junk, Gravenhage

Head ML, Hunt J, Brooks R (2006) Genetic association between male attractiveness and female differential allocation. Biol Lett 2:341–344

Head ML, Hunt J, Jennions MD, Brooks R (2005) The indirect benefits of mating with attractive males outweigh the direct costs. PLoS Biol 3(2):e33

Holland B, Rice WR (1998) Chase-away sexual selection: antagonistic seduction versus resistance. Evolution 52:1–7

Hunt J, Brueker CJ, Sadowski JA, Moore AJ (2009) Male-male competition, female mate choice and their interaction: determining total sexual selection. J Evol Biol 22:13–26

Hunt J, Bussière LF, Jennions MD, Brooks R (2004) What is genetic quality? Trends Ecol Evol 19:329–333

Ivy TM, Sakaluk SK (2005) Polyandry promotes enhanced offspring survival in decorated crickets. Evolution 59:152–159

Ivy TM, Sakaluk SK (2007) Sequential mate choice in decorated crickets: females use a fixed internal threshold in pre- and postcopulatory choice. Anim Behav 74:1065–1072

Ivy TM, Weddle CB, Sakaluk SK (2005) Females use self-referent cues to avoid mating with previous mates. Proc R Soc Lond B 272:2475–2478

Jennions MD, Hunt J, Graham R, Brooks R (2004) No evidence for inbreeding avoidance through postcopulatory mechanisms in the black field cricket, *Teleogryllus commodus*. Evolution 58:2472–2477

Jennions MD, Petrie M (2000) Why do females mate multiply? A review of genetic benefits. Biol Rev 75:21–64

Khalifa A (1949) The mechanism of insemination and the mode of action of the spermatophore in *Gryllus domesticus*. Quart J Micro Sci 90:281–292

Kokko H, Brooks R, Jennions MD, Morley J (2003) The evolution of mate choice and mating biases. Proc R Soc Lond B 270:653–654

Kokko H, Jennions MD, Brooks R (2006) Unifying and testing models of sexual selection. Annu Rev Ecol Evol Syst 37:43–66

Kokko H, Mappes J (2005) Sexual selection when fertilisation is not guaranteed. Evolution 59:1876–1885

Kuridwa T, Kasuya E (2009) Longer copulation duration increases the risk of injury during copulation in the male bell cricket *Meloimorpha japonica*. Entomol Sci 12:141–146

Laird G, Gwynne DT, Andrade MCB (2004) Extreme repeated mating as a counter-adaptation to sexual conflict? Proc R Soc Lond B 271:S402–S404

Lande R (1981) Models of speciation by sexual selection on polygenic traits. Proc Nat Acad Sci USA 78:3721–3725

Loher W, Dambach M (1989) Reproductive behavior. In: Huber F, Moore TE, Loher W (eds) Cricket behavior and neurobiology. Cornell University Press, Ithaca, New York, pp 43–82

Loher W, Rence B (1978) The mating behaviour of *Teleogryllus commodus* (Walker) and its central and peripheral control. Z Tierpsychol 46:225–259

Loher W, Weber T, Huber F (1993) The effect of mating on phonotactic behaviour in *Gryllus bimaculatus* (De Geer). Physiol Ent 18:57–66

Marshall JL (2007) Rapid evolution of spermathecal duct length in the *Allonemobius socius* complex of crickets: species, population and *Wolbachia* effects. PLoS ONE 2:e720

Marshall JL, Huestis D, Hiromasa Y, Wheeler A, Oppert C, Marshall SA, Tomich JM, Oppert B (2009) Identification, RNAi knockdown, and functional analysis of an ejaculate protein that mediates a postmating, prezygotic phenotype in a cricket. PLoS ONE 4(10):e7537

Mautz BS, Sakaluk SK (2008a) The effects of age and previous mating experience on pre- and post-copulatory mate choice in female house crickets (*Acheta domesticus* L.). J Insect behav 21:203–212

Mautz BS, Sakaluk SK (2008b) Heritable variation in the timing of spermatophore removal, a mechanism of post-copulatory female choice in crickets. J Evol Biol 21:1366–1370

McNamara KB, Lieshout EV, Simmons LW (2014) A test of the sexy-sperm and good-sperm hypothesis for the evolution of polyandry. Behav Ecol (advance access). doi:10.1093/behe co/aru067

Mello FAG (2007) Female monopolisation and paternity assurance in South American crickets (Orthoptera, Grylloidea): mating plugs, extra claspers and forced copulation. Papéis Avulsos de Zoologia Museu de Zoologia da Universidade de São Paulo 47:245–257

Ono T, Siva-Jothy MT, Kato A (1989) Removal and subsequent ingestion of rival's semen during copulation in a tree cricket. Physiol Ent 14:195–202

Parker DJ (2009) Pre- and post-copulatory mate choice in *Platygryllus primiformis*: cryptic female choice and sexual conflict. BioscienceHorizons 2:164–171

Parker DJ, Vahed K (2010) The intensity of pre- and post-copulatory mate guarding in relation to spermatophore transfer in the cricket *Gryllus bimaculatus*. J Ethol 28:245–249

Parker GA (1970) Sperm competition and its evolutionary consequences in the insects. Biol Rev 45:525–567

Peretti AV, Aisenberg A (2011) Communication under sexual selection hypotheses: challenging prospects for future studies under extreme sexual conflict. Acta Ethol 14:109–116

Peretti AV, Córdoba-Aguilar A (2007) On the value of fine-scaled behavioural observations for studies of sexual coercion. Ethol Ecol Evol 19:77–86

Preston-Mafham K (2000) Diurnal mating behaviour of a *Nisitrus* sp. Cricket (Orthoptera: Gryllidae) from Sumatra. J Nat Hist 34:2241–2250

Prokop P, Maxwell MR (2008) Interactions between multiple forms of nuptial feeding in the wood cricket *Nemobius sylvestris* (Bosc): duel spermatophores and male forewings. Ethology 114:1173–1182

Prokop P, Maxwell MR (2011) Sexual conflict over spermatophore attachment in a nuptially feeding cricket. Ethology 117:520–528

Rebar D, Zuk M, Bailey NW (2011) Mating experience in field crickets modifies pre- and post-copulatory female choice in parallel. Behav Ecol 22:303–309

Reinhold K, Heller KG (1993) The ultimate function of nuptial feeding in the bushcricket *Poecilimon veluchianus* (Orthoptera: Tettigoniidae: Phaneropterinae). Behav Ecol Sociobiol 32:55–60

Rentz DCF (1963) Biological observations on the genus *Decticita* (Orthoptera: Tettigoniidae). Wasmann J Biol 21:91–94

Rentz DCF (1972) The *lock and key* as an isolating mechanism in katydids. Am Sci 60:750–755

Rodríguez-Muñoz R, Bretman A, Hadfield JD, Tregenza T (2008) Sexual selection in the cricket *Gryllus bimaculatus*: no good genes? Genetica 132:287–294

Rodríguez-Muñoz R, Bretman A, Tregenza T (2011) Guarding males protect females from predation in a wild insect. Curr Biol 21:1716–1719

Ryder JJ, Siva-Jothy MT (2001) Quantitative genetics of immune function and body size in the house cricket, *Acheta domesticus*. J Evol Biol 14:646–653

Sadowski JA, Grace JL, Moore AJ (2002) Complex courtship behaviour in the striped ground cricket, *Allonemobius socius* (Orthoptera: Gryllidae): does social environment affect male and female behaviour? J Insect Behav 15:69–84

Sakaluk SK (1984) Male crickets feed females to ensure complete sperm transfer. Science 223:609–610

Sakaluk SK (1985) Spermatophore size and its role in the reproductive behaviour of the cricket, *Gryllodes supplicans* (Orthoptera: Gryllidae). Can J Zool 63:1652–1656

Sakaluk SK (1987) Reproductive behaviour of the decorated cricket, *Gryllodes supplicans* (Orthoptera: Gryllidae): calling schedulaes, spatial distribution, and mating. Behaviour 100:202–225

Sakaluk SK (1991) Post-copulatory mate guarding in decorated crickets. Anim Behav 41:207–216
Sakaluk SK (1997) Cryptic female choice predicted on wing dimorphism in decorated crickets. Behav Ecol 8:326–331
Sakaluk SK (2000) Sensory exploitation as an evolutionary origin to nuptial food gifts in insects. Proc R Soc Lond B 267:339–343
Sakaluk SK, Eggert AK (1996) Female control of sperm transfer and intraspecific variation in sperm precedence: antecedents to the evolution of a courtship food gift. Evolution 50:694–703
Sakaluk SK, Smith R (1988) Inheritance of male parental investment in an insect. Am Nat 132:594–601
Shackleton MA, Jennions MD, Hunt J (2005) Fighting success and attractiveness as predictors of male mating success in the black field cricket, *Teleogryllus commodus*: the effectiveness of no-choice tests. Behav Ecol Sociobiol 58:1–8
Sheldon BC (2000) Differential allocation: tests, mechanisms and implications. Trends Ecol Evol 15:379–402
Simmons LW (1986) Female choice in the field cricket *Gryllus bimaculatus* (De Geer). Anim Behav 34:1463–1470
Simmons LW (1989) Kin recognition and its influence on mating preferences of the field cricket, *Gryllus bimaculatus* (de Geer). Anim Behav 38:68–77
Simmons LW (1987a) Sperm competition as a mechanism of female choice in the field cricket, *Gryllus bimaculatus*. Behav Ecol Sociobiol 21:197–202
Simmons LW (1987b) Female choice contributes to offspring fitness in the field cricket, *Gryllus bimaculatus* (De Geer). Behav Ecol Sociobiol 21:313–321
Simmons LW (1987c) Heritability of a male character chosen by females of the field cricket, *Gryllus bimaculatus*. Behav Ecol Sociobiol 21:129–133
Simmons LW (1988) The contribution of multiple mating and spermatophore consumption to the lifetime reproductive success of female field crickets (*Gryllus bimaculatus*). Ecol Entomol 13:57–69
Simmons LW (1990) Post-copulatory guarding, female choice and the levels of gregarine infections in the field cricket, *Gryllus bimaculatus*. Behav Ecol Sociobiol 26:403–407
Simmons LW (1991) Female choice and the relatedness of mates in the field cricket, *Gryllus bimaculatus*. Anim Behav 41:493–501
Simmons LW (2001) Sperm competition and its evolutionary consequences in the insects. Princeton University Press, Princeton, NJ
Simmons LW (2014) Sexual selection and genital evolution. Austral Entomol 53:1–17
Simmons LW, Beveridge M (2010) The strength of postcopulatory sexual selection within natural populations of field crickets. Behav Ecol 21:1179–1185
Simmons LW, Beveridge M (2011) Seminal fluid affects sperm viability in a cricket. PLoS ONE 6(3):e17975
Simmons LW, Beveridge M, Kennington WJ (2007) Polyandry in the wild: temporal changes in female mating frequency and sperm competition intensity in natural populations of the tettigoniid *Requena verticalis*. Mol Ecol 16:4613–4623
Simmons LW, Beveridge M, Wedell N, Tregenza T (2006) Postcopulatory inbreeding avoidance by female crickets only revealed by molecular markers. Mol Ecol 15:3817–3824
Simmons LW, Gwynne DT (1991) The refractory period of female katydids (Orthoptera: Tettigoniidae): sexual conflict over the re-mating interval? Behav Ecol 2:276–282
Simmons LW, Lovegrove M, Almbro M (2014) Female effects, but no intrinsic male effects on paternity outcome in crickets. J Evol Biol 27:1644–1649
Simmons LW, Parker GA (1989) Nuptial feeding in insects: mating effort versus paternal investment. Ethology 81:332–343
Simmons LW, Thomas ML (2008) No postcopulatory response to inbreeding by male crickets. Biol Lett 4:183–185

Simmons LW, Wernham J, Garcia-Gonzalez F, Kamien D (2003) Variation in paternity in the field cricket *Teleogryllus oceanicus*: no detectable influence of sperm numbers of sperm length. Behav Ecol 14:539–545

Stanley-Samuelson DW, Jurenka RA, Blomquist GJ, Loher W (1987) Sexual transfer of prostaglandin precursor in the field cricket, *Teleogryllus commodus*. Physiol Entomol 12:347–354

Stanley-Samuelson DW, Peloquin JJ, Loher W (1986) Egg-laying in response to prostaglandin injections in the Australian field cricket, *Teleogryllus commodus*. Physiol Entomol 11:213–219

Stockley P (1999) Sperm selection and genetic incompatibility: does relatedness of mates affect male success in sperm competition? Proc R Soc Lond B 266:1663–1669

Sturm R (2005) Motoric activity of the receptacular complex in the cricket *Teleogryllus commodus* (Insecta: Orthoptera: Gryllidae). Entomolgische Abhandlungen 62:2185–2192

Sturm R (2008) Morphology and histology of the *ductus receptaculi* and accessory glands in the reproductive tract of the female cricket, *Teleogryllus commodus*. J Insect Sci 8(35). doi:10.1673/031.008.3501

Sugawara T (1993) Oviposition behaviour in the cricket *Teleogryllus commodus* : mechanosensory cells in the genital chamber and their role in the switch-over of steps. J Insect Physiol 39:335–346

Sugawara T, Loher W (1986) Oviposition behaviour of the cricket *Teleogryllus commodus*: observation of external and internal events. J Insect Physiol 32:179–188

Thomas ML, Simmons LW (2008) Male-derived cuticular hydrocarbons signal sperm competition intensity and affect ejaculate expenditure in crickets. Proc R Soc Lond B 276:383–388

Thomas ML, Simmons LW (2009a) Sexual selection on cuticular hydrocarbons in the Australian field cricket, *Teleogryllus oceanicus*. BMC Evol Biol 9:162

Thomas ML, Simmons LW (2009b) Male dominance influences pheromone expression, ejaculate quality, and fertilisation success in the Australian field cricket, *Teleogryllus oceanicus*. Behav Ecol 20:1118–1124

Thornhill R (1976) Sexual selection and nuptial feeding behaviour in *Bittacus apicalis* (Insecta: Mecoptera). Am Nat 110:529–548

Thornhill R (1983) Cryptic female choice and its implications in the scorpionfly *Harpobittacus nigriceps*. Am Nat 122:765–788

Tuni C, Beveridge M, Simmons LW (2013) Female crickets assess relatedness during mate guarding and bias storage of sperm towards unrelated males. J Evol Biol 26:1261–1268

Tregenza T, Wedell N (1997) Definitive evidence of cuticular hydrocarbons in a cricket. Anim Behav 54:979–984

Tregenza T, Wedell N (1998) Benefits of multiple mates in the cricket *Gryllus bimaculatus*. Evolution 52:1726–1730

Tregenza T, Wedell N (2002) Polyandrous females avoid costs of inbreeding. Nature 415:71–73

Tyler F, Harrison XA, Bretman A, Veen T, Rodrígue-Muňoz R, Tregenza T (2013) Multiple postmating barriers to hybridisation in field crickets. Mol Ecol 22:1640–1649

Vahed K (1996) Prolonged copulation in oak bushcrickets (Tettigoniidae; Meconematinae; *Meconema thalassinum* and *M. meridionale*). J Orthop Res 5:199–204

Vahed K (1997) Copulation and spermatophores in the Ephippigerinae (Orthoptera: Tettigoniidae): prolonged copulation is associated with a smaller nuptial gift in *Uromenus rugosicollis* Serville. J Orthop Res 6:83–89

Vahed K (1998) The function of nuptial feeding in insects: a review of empirical studies. Biol Rev 73:43–78

Vahed K (2002) Coercive copulation in the Alpine Bushcricket *Anonconotus alpinus* Yersin (Tettigoniidae: Tettigoniidae: Platycleidini). Ethology 108:1065–1075

Vahed K (2003a) The structure of spermatodoses in shield-back bushcrickets (Orthoptera: Tettigoniidae: Tettigoniinae). J Morphol 257:45–52

Vahed K (2003b) Increases in egg production in multiply mated female bushcrickets *Leptophyes punctatissima* are not due to substances in the nuptial gift. Ecol Entomol 28:124–128

Vahed K (2006) Larger ejaculate volumes are associated with a lower degree of polyandry across bushcricket taxa. Proc R Soc B 273:2387–2394

Vahed K (2007a) All that glisters is not gold: sensory bias, sexual conflict and nuptial feeding in insects and spiders. Ethology 113:105–127

Vahed K (2007b) Comparative evidence for a cost to males of "manipulating" females in bushcrickets. Behav Ecol 18:507–512

Vahed K, Carron G (2008) Comparison of forced mating behaviour in four taxa of *Anonconotus*, the Alpine bushcricket. J Zool 276:313–321

Vahed K, Gilbert FS (1996) Differences across taxa in nuptial gift size correlate with differences in sperm number and ejaculate volume in bushcrickets (Orthoptera: Tettigoniidae). Proc R Soc Lond B 263:1257–1265

Vahed K, Gilbert JDJ, Weissman DB, Barrientos-Lozano L (2014) Functional equivalence of grasping cerci and nuptial food gifts in promoting ejaculate transfer in katydids. Evolution 68:2052–2065

Vahed K, Lehmann AW, Gilbert JDJ, Lehmann GUC (2011) Increased copulation duration before ejaculate transfer is associated with larger spermatophores, and male genital titillators, across bushcricket taxa. J Evol Biol 9:1960–1968

Viscuso R, Brundo MV, Sottile L (2002) Mode of transfer of spermatozoa in Orthoptera Tettigoniidae. Tiss Cell 34:337–348

Viscuso R, Barone N, Sottile L, Narcisi L (1996) Spermiolytic activity of the epithelium of the of Rhacocleis annulata fieber (Orthoptera: Tettigoniidae). Int J Insect Morphol Embryol 25:135–144

von Helversen D, von Helversen O (1991) Pre-mating sperm removal in the bushcricket *Metaplastes ornatus* Ramme 1931 (Orthoptera, Tettigoniidae, Phaneropteridae). Behav Ecol Sociobiol 28:391–396

Wagner WE Jr, Smith AR, Basolo AL (2007) False promises: females spurn cheating males in a field cricket. Biol Lett 3:379–381

Warwick S, Vahed K, Raubenheimer D, Simpson S (2009) Free amino acids as phagostimulants in cricket nuptial gifts: support for the "Candymaker" hypothesis. Biol Lett 5:194–196

Weddle CB, Sakaluk SK (2003) Ingestion of male haemolymph and mating propensity of female sagebrush crickets: no evidence of a male-derived antiaphrodisiac. Anim Behav 65:83–88

Wedell N (1991) Sperm competition selects for nuptial feeding in a bushcricket. Evolution 45:1975–1978

Wedell N (1993) Spermatophore size in bushcrickets: comparative evidence for nuptial gifts as a sperm protection device. Evolution 47:1203–1212

Wedell N (1998) Sperm protection and mate assessment in the bushcricket *Coptaspis* sp. 2. Anim Behav 56:357–363

Wedell N, Arak A (1989) The wartbiter spermatophore and its effect on female reproductive output (Orthoptera: Tettigoniidae, *Decticus verrucivorus*). Behav Ecol Sociobiol 24:117–125

Wedell N, Gage MJG, Parker GA (2002) Sperm competition, male prudence and sperm-limited females. Trends Ecol Evol 17:313–320

Wedell N, Tregenza T (1999) Successful fathers sire successful sons. Evolution 53:620–625

Weissman DB (2001) Communication and reproductive behaviour in North American Jerusalem Crickets (*Stenopelmatus*) (Orthoptera: Stenopelmatidae). In: Field LH (ed) The biology of wetas, king crickets and their allies. CAB International, Wallingford, pp 351–373

Wynn H, Vahed K (2004) Male *Gryllus bimaculatus* guard females to delay them from mating with rivals and to obtain repeated copulations. J Insect Behav 17:53–66

Zefa E, Martins LP, Szinwelski N (2008) Complex mating behaviour in *Adelosgryllus rubricephalus* (Orthoptera, Phalangopsidae, Grylloidea). Iheringia Sér Zool Porto Alegre 98:325–328

Zuk M (1987) The effects of gregarine parasites, body size and time of day on spermatophore production and sexual selection in crickets. Behav Ecol Sociobiol 21:65–72

Zuk M, Simmons LW (1997) Reproductive strategies of the crickets (Orthoptera:). In: Choe J, Crespi B (eds) The Evolution of Mating Systems in Insects and Arachnids. Cambridge University Press, Cambridge, pp 89–109

Chapter 12
Sexual Selection Within the Female Genitalia in Lepidoptera

Carlos Cordero and Joaquín Baixeras

Abstract The genitalia of male and female Lepidoptera are complex organs, composed of several structures that exhibit great diversity of shapes, sizes, and positions, suggesting that they have evolved in a relatively rapid and divergent way. In this chapter, we explore the selective pressures responsible for the evolution of genital morphology in the Lepidoptera, emphasizing the possible role of post-copulatory intersexual selection (PCIS) mechanisms (cryptic female choice and sexual conflict). Our exploration is in great extent speculative because knowledge on the functional morphology of genitalia in this group is limited. We start by describing the complexity and diversity of genitalia in Lepidoptera, discussing the available information about the role of the different structures in copulation. Then, we discuss possible ways in which PCIS and other types of selective pressures could have influenced the evolution of genitalia by developing illustrative hypothesis for several structures whose function is not well understood. Finally, we describe in detail recent experimental and comparative studies aimed at understanding the function and selective pressures responsible for the evolution of a female genital trait known as signum.

12.1 Introduction

In his seminal book "Sexual Selection and Animal Genitalia," William Eberhard put forward the idea that the bewildering variety and complexity of male genital morphology observed in most animal groups is mainly a product of sexual

C. Cordero (✉)
Departamento de Ecología Evolutiva, Instituto de Ecología,
Universidad Nacional Autónoma de México, D.F., Mexico City, Mexico
e-mail: cordero@ecologia.unam.mx

J. Baixeras
Instituto Cavanilles de Biodiversidad y Biología Evolutiva,
Universidad de Valencia, Paterna, Spain
e-mail: joaquin.baixeras@uv.es

A.V. Peretti and A. Aisenberg (eds.), *Cryptic Female Choice in Arthropods*,
DOI 10.1007/978-3-319-17894-3_12

selection by cryptic female choice (CFC) (Eberhard 1985). The elegant argumentation, as well as the broad range, and even-handed use of the evidence reviewed resulted in a masterpiece that paved the way for the modern study of genital evolution, a research area that has attracted increasing attention from scientists since the publication of Eberhard's book (Ah-King et al. 2014; Schilthuizen 2014). Eberhard's main focus was explaining the enormous variety and complexity of male genitalia, resulting from a pattern of rapid and divergent evolution (for critical discussions of this pattern, see Huber 2003; Song and Bucheli 2010), and his answer was female choice of males with the most stimulating genitalia. He called this type of female choice "cryptic" because it occurs within the female genitalia, out of view of the observer. Therefore, Eberhard's theory implies the coevolution between male genitalia and the ability of females to discriminate between males based on their genital morphology and behavior, thus the idea that the function of copulation for males is not only transferring ejaculates, but practicing internal courtship with their genitalia (Eberhard 1985). Currently, it is widely accepted that sexual selection is the main cause of male genital divergence and complexity, although there is some disagreement about the relative importance of intrasexual selection (sperm competition), CFC, and sexual conflict (Schilthuizen 2014; Simmons 2014). However, the role of other selective forces, such as natural selection via parasites or sperm aging (Reinhardt 2010) or selection against hybridization (the lock-and-key mechanism; Masly 2012; Simmons 2014), should not be discarded without detailed consideration.

In his discussion of the "lock-and-key" hypothesis (the idea that genital evolution is driven by selection against hybridization), Eberhard (1985) mentions that female genital morphology is usually not species-specific but relatively uniform, which implies that female's mechanisms of CFC generally involve only the nervous and endocrine systems. However, Simmons (2014) claims that the absence of rapid and divergent evolution in female genital morphology probably is not real, but results of the inadequate study of female genitalia by taxonomists (see also Ah-King et al. 2014). Although Eberhard (1985) discarded this explanation by mentioning that in several groups in which female genitalia have been considered by taxonomists, they tend to be uniform, there is an increasing number of studies showing that male genital morphology and female genital morphology have coevolved in several groups of animals producing complex and diverse structures that play a broad gamut of functions (reviews in Sánchez et al. 2011; Simmons 2014).

The genitalia of male Lepidoptera are used as examples in several contexts in Eberhard's book, but he explicitly mentions this order as one of the animal groups in which "female genitalia are relatively uniform while male genitalia are diverse and species-specific" (Eberhard 1985, p. 30). However, there is a large amount of taxonomic papers and books that clearly show that the female genitalia of many (probably most) Lepidoptera are complex organs (Figs. 12.1b and 12.2), composed of several structures that exhibit great diversity of shapes, sizes, positions, and, possibly, functions, suggesting that they have evolved in a relatively rapid and divergent way (Sánchez et al. 2011; Lincango et al. 2013). As explained in

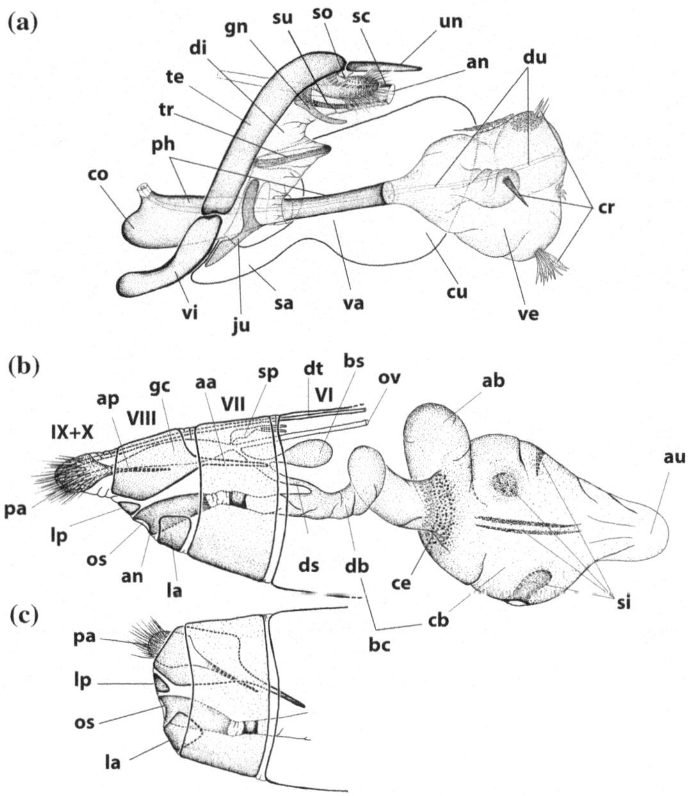

Fig. 12.1 Generalized Lepidoptera male and female genitalia structure. **a** Male genitalia, lateral view left valva removed. **b** Female genitalia, lateral view. **c** Female genitalia retracted. 1a abbreviations (male): *an* anal cone; *co* coecum penis; *cr* cornuti; *cu* cucullus; *di* diaphragma; *du* ductus ejaculatorius; *gn* gnathos; *ju* juxta; *ph* phallus; *sa* sacculus; *sc* scaphium; *so* socii; *su* subscaphyum; *te* tegumen; *tr* transtilla; *un* uncus; *va* valva; *ve* vesica; *vi* vinculum. 1b abbreviations (female): *aa* apophysis anterior; *ab* accessory bursae; *an* antrum; *ap* apophysis posterior; *au* appendix bursae; *bc* bursa copulatrix; *bs* bulla seminalis; *cb* corpus bursae; *ce* cervix; *db* ductus bursae; *ds* ductus seminalis; *dt* digestive tube; *gc* genital chamber; *la* lamella antevaginalis; *lp* lamella postvaginalis; *os* ostium; *ov* oviduct; *pa* anal papillae; *si* signum; *sp* spermatheca

the following section, although there are hundreds of detailed descriptions of the genitalia of female Lepidoptera, knowledge about the functions of its different structures is rather poor. Such knowledge is fundamental, not only to explain the copulatory mechanisms (which also appears to be rather variable; Bayard 1944; Callahan 1958, 1960; Ferro and Akre 1975; Naumann 1987; Miller 1988; Fänger and Naumann 1998; Justus and Mitchell 1999; Sihvonen 2007; Mikkola 2008), but to understand the selective pressures responsible for the evolutionary origin and diversification of female genital structures and of the male genital traits interacting (and possibly coevolving) with them.

Fig. 12.2 Female genitalia diversity. **a** *Bombyx mori* (Bombycidae). **b** *Ethmia bipunctella* (Ethmiidae). **c** *Etiella zinckenella* (Pyralidae). **d** *Pachycnemia hippocastanaria* (Geometridae). **e** *Pararge aegeria* (Nymphalidae). **f** *Noctua pronuba* (Noctuidae). *Scale bars* **a** and **b** = 1.0 mm; **c**–**e** = 500 μm; **f** = 2.0 mm. Arrows indicate the point where the ductus seminalis meets the ductus bursae. Some bursa copulatrix may be completely membranous (as **a**) or include deeply spiny areas or plates (as **d**) among other sclerotized structures. See the ductus bursae, short (as in **a**), spiralized (**b** or **c**), or stout (**d** or **f**). The bulla seminalis is sometimes sclerotized enough to be prepared (**f**)

Our aim in this work was to explore the main selective pressures responsible for the evolution of male genital morphology and female genital morphology in the Lepidoptera, emphasizing the possible role of post-copulatory intersexual selection (PCIS) mechanisms. Our exploration is in great extent speculative because knowledge on the functional morphology of genitalia in this diverse group is limited. We began with an overview of the complexity and diversity of genital morphology in Lepidoptera, including the available information about the role of the different structures in copulation. Then, we discuss possible ways in which PCIS and other types of selective pressures could have influenced the evolution of genitalia, first, by developing illustrative hypothesis for some structures whose function is not well understood and then by describing studies on the function and evolution of a female genital trait known as signum (plural signa).

12.2 Complexity and Diversity of Male and Female Genitalia: Form and Function

The insect reproductive system is frequently divided into external and internal genitalia. This distinction is just purely anatomical as functionally internal and external parts must interact continuously. In fact, the whole reproductive system works in conjunction to successfully resolve a complex pairing, but the information we have on the whole process is quite often fragmentary. In the Lepidoptera, male genitalia (Figs. 12.1a and 12.3) have been generally considered more informative in taxonomical terms than female genitalia (Figs. 12.1b and 12.2), what could be interpreted as an indicator of more diverse male genitalia. But this does not mean that female genitalia are not diverse, and most lepidopterists would agree

Fig. 12.3 Examples of male genitalia diversity in the Lepidoptera. **a** and **d**, *Bombyx mori* (Bombycidae). **b** and **e**, *Ethmia bipunctella* (Ethmiidae). **c** and **f**, *Etiella zinckenella* (Pyralidae). **g** and **l**, *Pachycnemia hippocastanaria* (Geometridae). **h** and **m**, *Pararge aegeria* (Nymphalidae). **i** and **n**, *Noctua pronuba* (Noctuidae). *Scale bars* **a** and **i** = 1.0 mm; **b**, **c**, **g**, and **h** = 500 μm. The shape and characteristics of the valvae (the clasping organs) vary considerably among species from digitiform (**a**) to really elaborate structures (**b**, **i**). Other structures are equally extravagant; for instance, generalized uncus are fingerlike (**g**, **h**), but many shapes are possible. The juxta (supporting the phallus) varies from a plate (**a**) to forklike structures (**c**) sometimes projected in spiny processes (**g**). Note that the phallus is shown separately below every genitalia at same scale (*posterior on the right, anterior on the left*). See Fig. 12.4 to see the vesica everted

in considering female genitalia as a powerful source of anatomical information. Skeletal structures have been studied in much detail as they are key diagnostic characters in taxonomy, and the information available in this sense is enormous. However, the difficulties in homologizing many of these structures are a constant problem. A complete description of these structures is beyond the scope of this text. A succinct description follows for those not familiar with Lepidoptera morphology. Kristensen (2003a) provides a general detailed comparative analysis of the morphology, and Klots (1970) details the terminology commonly used both in male and female genitalia. Both authors' ideas are thoroughly followed in these sections.

12.2.1 Male Genitalia Structure

The external male genitalia includes basically all the organs associated to the abdominal segments IX and X (Figs. 12.1a and 12.3), anatomically considered the genital segments. The abdominal segment IX of the males forms a modified ring to which the rest of structures attach. The dorsal part of this ring is called tegumen and the ventral part vinculum. A membrane (the diaphragma) closes the posterior end of the segment. The digestive tube opens dorsally in the diaphragma through the anal cone, and the genital duct opens midventrally through the phallus. Different areas of the diaphragma may develop sclerotized plates. The most conspicuous of them is normally the juxta, a usually furcate plate that bears the phallus from below. A prominent well-sclerotized structure articulates dorsally with the tegumen, the uncus, a remnant of segment X. Two paired structures attach to the posterior part of the tegumen: the gnathos—a pair of fingerlike sclerotized process and the socii—a pair of setose lobes. Paired sclerotized bands may accompany the anal cone dorsally (scaphium) and/or ventrally (subscaphyum). The phallus is the intromittent organ, normally a robust subcylindrical structure extremely variable in shape (Figs. 12.1a and 12.3d–f, j–l) that may exhibit hooks, spines, and other extravagant sculpturing. The phallus internally includes the endophallus (called vesica), an eversible membranous bag frequently bearing teeth, hooks, and other sclerotized structures called cornuti (Fig. 12.4b–f). The phallus may be so strong that is able to produce traumatic copulation (Bieman and Witter 1982; Brower et al. 2007). Unlike other genital structures, usually symmetric, both the phallus and the vesica are essentially asymmetric structures (see also Sect. 12.2.3).

Male genital structural diversity is especially notable on the valvae, two paired lateral structures derived from genital appendages (the "gonopods" morphologically) of the abdominal segment IX. The development of ectopical structures (appendages) in connection with genital functions is seen as a source of diversity from the evolutionary development point of view (Minelli 2002). The valvae fit perfectly into this idea. Their complexity and enormous variation is intriguing. In Fig. 12.3, a reduced but illustrative enough range of shapes of male genitalia

Fig. 12.4 Phallus with vesica everted (the ductus ejaculatorius has been also everted). **a** *Bombyx mori* (Bombycidae). **b** *Ethmia bipunctella* (Ethmiidae). **c** *Etiella zinckenella* (Pyralidae). **d** *Pachycnemia hippocastanaria* (Geometridae). **e** *Pararge aegeria* (Nymphalidae). **f** *Noctua pronuba* (Noctuidae). *Scale bars* **a**, **c**, **d** and **e** = 500 μm, **b** = 250 μm, **f** = 1.0 mm. Three-dimensional disposition may be distorted by preparation. The vesica may be a simple and relatively small bag (as in **a** or **e**) or complex and considerably ornate by cornuti (**c**)

belonging to selected representatives of some of the most specious families of Lepidoptera is shown. Valvae may be ornate by a considerable diversity of lobes, spines, hairs, especially on the internal side where are supposed to interact with the female abdominal external surface. No comparable organ is present in the female system whose organization remains mainly internal to the body. However, the functional contribution of every concrete structure of the valvae remains obscure. The differences found among the cases studied do not justify the extreme diversity in shape these organs express. Most authors emphasize the mechanical and sensorial function, but other functions are possible. For example, valvae may also include pleats with androconia (as in some Notodontidae and Geometridae, among other families) and thus may be involved in pheromone dissemination (Miller 1988), which also means a potential interaction with other males. Valvae are nearly always symmetric although there are well-known cases of asymmetry in the Lepidoptera (Huber et al. 2007; Huber 2010). Functional asymmetry (independent of anatomical asymmetry) has been also observed (Scott 1978).

The musculature of male genitalia has been explored also for comparative morphological purposes. Forbes (1939) provided a pioneering study of male genitalia muscles. The impressive series of studies commanded by Kuznetsov and Stekolnikov in the Russian school are summarized in Kuznetsov and Stekolnikov (2001). Unfortunately, the internal organization of the male reproductive system has no comparable level of knowledge, and what we know is based on relatively scattered studies on model Lepidoptera as *Bombyx mori* (Bombycidae) by Osanai et al. (1988), *Calpodes* (Hesperiidae) by Lai-Fook (1982a, b), *Cydia pomonella* (Tortricidae) by Ferro and Akre (1975), *Dioryctria abietella* (Pyralidae) by Fatzinger (1970), *Manduca sexta* (Sphingidae) by Reinecke et al. (1983) or some noctuids (Callahan 1958, Callahan and Chapin 1960, Callahan and Cascio 1963, Buntin and Pedigo 1983) among others (see Kristensen 2003b for a complete revision). The general impression is that internal reproductive structures tend to be more conservative, less diverse, than external structures (Eberhard 1985). Interestingly, works by Mitter (1988) and Simonsen (2006a) demonstrate that there may be taxon-dependent differences in the male internal reproductive system even among related species. The potential correlation between the differences in the internal and external structures and their contribution to the reproductive system remains unexplored. We know insects in which the differences in the internal morphology (soft tissues) are more pronounced than the differences in external genitalia (hard tissues) (Masly 2012). Our understanding of functional morphology and evolutionary trends in genitalia would change dramatically if this scenario were confirmed for the Lepidoptera.

The pregenital abdominal segments I to VII may be involved in sexual behavior including copulation. Pockets with sex scales, eversible structures (termed in general coremata), hair pencils, and glandular areas are common among males. The contribution of these secondary sexual characters to the mating behavior remains unknown in most cases. Better known are the skeletal modifications of the immediate pregenital segment VIII. It is the case of the geometrid genera *Scopula* and *Eupithecia*. *Scopula* males bear two ventrolateral appendices (cerata) on sternum VIII that interact mechanically with the female lamella antevaginalis (see Sect. 12.2.2) during mating (Hausmann 1999) or even are involved in female stimulation (Sihvonen 2007). The interaction is so strong that Mikkola (1994) suggested for similar structures in *Eupithecia* that the recognition function of the valvae could well have been replaced by these structures.

12.2.2 Female Genitalia Structure

The generalized condition in animals is that male genitalia support a single set of functions all of them related to copulation (in morphological terms, assisting the male gonopore); meanwhile, female genitalia support both copulation and oviposition functions (gonopore- and ovipore-assisting functions). However, in most Lepidoptera, the female abdomen has two independent orifices, one for mating and another for oviposition (ditrysian condition). Only some primitive groups retain a single pore for both functions (a cloaca, common with the anus). Analogous configurations are reported in Cicadidae (Homoptera), Anthocoridae (Heteroptera) and some few Coleoptera (Matsuda 1976). Some chelicerates and flatworms complete the restrictive list of animals with two separate genital orifices (Eberhard 1985). Figure 12.1b illustrates the basic structure of the ditrysian genital configuration. The ovipore opens on the IX + X segment laterally flanked by the anal papillae. The bursa copulatrix is a bottlelike structure that receives the spermatophore and other substances from the male during mating and opens through the ostium in sternum VIII. The anterior and posterior areas to the ostium (lamella antevaginalis and postvaginalis, respectively; ventral and dorsal during mating; see Fig. 12.1c) are often sclerotized configuring what is called the sterigma. Sternum VII may be also modified and strongly influenced by the highly variable configuration of the sterigma. The bursa copulatrix includes in its more typical configuration a saclike structure (the corpus bursae) separated from the ostium by a tubular conduct (the ductus bursae). The area where the ductus meets the corpus bursae is called cervix and tends to be sclerotized including the presence of plates, teeth, etc. The ductus seminalis connects the bursa copulatrix with the genital chamber and may include a vesicle-like expansion known as the bulla seminalis. The exact position where the ductus seminalis meets the bursa is of taxonomic importance and a highly variable character. Particularly interesting is the presence of sclerotized areas (collectively called signum) on the bursa copulatrix as well as microsculpturing (Lincango et al. 2013). As in the case of the male phallus and vesica, the structure of the bursa copulatrix is usually asymmetrical (see Sect. 12.2.3). The structure and adaptive significance of the signa in connection with the structure of the spermatophore has been recently reexamined and will be discussed in Sect. 12.4.

If the ditrysian model is behind the enormous diversification of the group or is simply an incidental morphological trait are conjectures that the morphologists have carefully avoided (e.g., Dugdale 1974). Matsuda (1976) suggested that the development of a second orifice may have relaxed the concentration of functions on a single opening. No doubt the diversity we find in both the mating structures as well as those related to oviposition has been facilitated by the specialization in the orifices avoiding conflicts between sexual selection (around the gonopore) and natural selection (around the ovipore). However, the true morphological novelty of the ditrysian model could be not necessarily the separation of the genital pores but the presence of the ductus seminalis (Fig. 12.1b), a direct internal connection between the bursa copulatrix and the genital chamber, the primitive vagina (Kristensen 2003a; Grimaldi and Engel 2005). This configuration would have increased the female control on the insemination isolating the sperm, eggs, and oviduct from male manipulation during mating (Eberhard 1985).

12.2.3 The Function: Mating

The information available allows some generalizations on mating position, the way of coupling, the intromission of the phallus into the bursa copulatrix, and the insemination mechanism. Further steps until fertilization (e.g., spermatophore break or sperm migration) do not require male physical involvement. It is, however, impossible to even suggest an approximate function for every anatomical structure in such a complex scenario. In this respect, the experiments on partial mutilation of structures of the genitalia (Sengün 1944 and Lorkovic 1952 on males, and Tschudi-Rein and Benz 1990 on females) as well as of artificial pairing (e.g., Clarke and Sheppard 1956) should be mentioned. Even if we agree that their contribution to the understanding of the functional morphology of the genitalia should be considered with reserves (see Shapiro and Porter 1989, for criticism), they certainly suggest that some of the organs involved in the mating may have secondary roles or that their real adaptive significance should be looked for out of the mating behavior.

12.2.3.1 The Mating Position

Insects adopt variable positions during mating. According to Alexander (1964), a position in which the female is placed above the male, with both male and female genitalia interacting dorsoventrally (inverse genital correlation) and symmetrically, should be considered as the plesiomorphic condition. Deviations from this ground plan are common including male-above, belly-to-belly, and end-to-end positions. But these modalities refer to the position of the body, not the genitalia. In most of the cases, the plesiomorphic relative inverse position of the genitalia is retained. In order to accommodate the position, the male abdomen generally must be flexed or even twisted (terminal inversion); quite often the genitalia become asymmetric. In the Lepidoptera mating position is rather variable, from side-by-side to end-to-end, also depending on the support, but surprisingly male abdomen and female abdomen interact with no terminal inversion. The respective dorsal and ventral structures of both male and female invariably correlate during mating. Interestingly, Huber et al. (2007) have suggested that the asymmetries observed in the phallus–vesica in the male, and the bursa copulatrix in the female could well be the expression of internal twisting of the genitalia. Thus, the generalized Lepidoptera mating position would fit the terminal inversion rule but only internally, an attractive hypothesis that certainly deserves attention.

12.2.3.2 The Coupling and Clasping

The potential functional implication of the genitalia in mating was early intuitively understood (Gosse 1882; Jordan, 1896) but Norris (1932), Bayard (1944), and Hannemann (1954a, b) described the basics of the mechanism based on empirical data. The valvae of the male genitalia ventrolaterally and the uncus—and associated structures—dorsally, grip the terminal female abdomen, the phallus must be then introduced in the bursa copulatrix. This means that basically, the valvae must interact with female abdominal sclerites, the uncus with the anal papilla and the lamella postvaginalis, and the area around the phallus (especially the juxta) must interact with the sterigma. The available evidence indicates that the exact point where these interactions take place is variable. When mating, the female musculature retracts and raises the IX-X segments that do not interfere with phallus intromission (Fig. 12.1c). The same movement exposes the ostium bursae to the posterior end of the abdomen. Norris (1932) observed in *Ephestia* and *Plodia* (Pyralidae) how the uncus pushed the anal papillae down into the female abdomen. Arnold and Fischer (1977) described the skeletomuscular mechanism involved in the copulation of *Speyeria* (Nymphalidae). They agreed with previous observations in that the valvae are responsible for holding the abdomen (VII segment), but they observed at the same time some part of the valva (concretely the claspers—harpe—in the sense by Sibatani et al. 1954) grasping laterally the anal papillae. The anal papillae rested on two pouches on the male diaphragm. Whether this means some simple mechanical positioning of the papillae to avoid interference or a definitive participation in mating is unknown. Observations by Stekolnikov (1965) and Miller (1988) in selected noctuoids coincide describing how the uncus slides through a membranous invagination between the anal papillae to reach the caudal part of the lamella postvaginalis. The socii would collaborate to push downward the papillae. In *Carcharodus* (Hesperiidae), De Jong (1978) found that the uncus is bent down to hook the distal part of the lamella postvaginalis (postvaginal plate in the article) and the valvae grasp the ventrolateral part of the female intersegmental membranes VII–VIII. This intersegmental area is in *Carcharodus,* a wide membranous area that strongly inflates (copulative pouch) during mating. This copulative pouch is grasped by the male valvae. The pressure the male operates is internally transmitted to the dorsal volume of the segment pushing more tightly the female genital plate to the male, thus facilitating penetration. The effect pushes the sternite VIII from horizontal to vertical position, and consequently, the ostium (and the genital plate) passes also to be vertical and exposed to the male; the anal papillae are pushed to a dorsal position. De Jong's idea suggests the involvement of hydrostatic forces combined with muscles what is slightly different from standard interpretations that base the movements on muscle activity.

The anal cone (ventral to the uncus) must be retracted during mating and work by Simonsen (2006b) supports this idea. Direct evidences are, however, necessary to clarify the exact role—if any—of the anal cone and associated structures (e.g., scaphium and subscaphium) during copulation. In *Scopula*, Sihvonen (2007) observed that the socii were inserted in a membranous area between the anal papilla and the ostium. Ferro and Akre (1975) described the mating position of the genitalia for *Cydia pomonella* (Tortricidae). In the genus *Cydia,* the dorsal elements of the genitalia (uncus, scaphium, socii) are drastically reduced if present and thus, there is no way for close mechanical interaction between the anal papilla and any of these structures. Basically, the arrangement of the organs during mating is the same (valvae hold segment VII) except that the anal papillae are kept dorsally with no grasping by the male. Certainly, the flat disposition of the anal papillae in tortricids compared to the more protruding in nymphalids, geometrids, and pyraloids may facilitate the coupling. Anyway, this suggests that the dorsal interactions during mating are not so relevant for the copulation but, depending on the structure of the ovipositor, the male genitalia must have developed some strategy to avoid any interference with the ovipositor. In this sense, the dorsal male structure would reflect the counterbalance of divergent interests (oviposition and mating).

12.2.3.3 The Intromission of the Phallus

Little is known about the direct interaction between the juxta (bearing the phallus) and the corresponding structure, the sterigma (bearing the ostium), in the female. Both are seen as reinforced structures for support and muscle attachment of their respective pores. However, it makes sense to think about some contact between these areas. Again in the geometrid *Scopula,* Sihvonen (2007) found the dorsal arm of the valvae involved in opening the lamella antevaginalis of the female genitalia by pushing it down mechanically to expose the ostium. The two arms of the juxta are then responsible for physical coupling with the female genitalia. Similar processes of the juxta are observable in other geometrids (see Fig. 12.3g). Okagaki et al. (1955) pointed out the importance of the articulation and muscles in the juxta–valvae area with special reference to the Geometridae. Covell (1970) describes extreme cases of the North American geometrid fauna in which the juxta protrudes as an elongated tube that couples with the ductus bursae of the female genitalia. It is obvious that the evolutionary pathway in some geometrid moths replacing valval functions is unusual but demonstrates how flexible the organs involved are. In Zygaenidae, studies by Naumann (1987) and Fänger and Naumann (1998) beautifully described the interaction between the female and male sclerites around phallus and ostium, respectively, and they concluded that the male sclerites (called laminae dorsalis and ventralis in Zygaenidae) not only play a mechanical role but also participate in the stimulation of the female. Although most authors suggest that the hemolymph pressure must be responsible for the vesica eversion certainly, Naumann (1987) makes a relevant objection. Not being any connection between the phallus-vesical space and the body cavity is certainly difficult to justify this possibility. The injection of the spermatophore secretions via ductus ejaculatorius seems a more adequate way to evaginate the vesica. Studies by Osanai et al. (1988) showed several powerfully muscled regions in the male genital tract that may justify this conclusion.

Much more attention has received the interaction between the phallus (aedeagus or penis in many papers) and the bursa copulatrix. In the most generalized case, the male introduces the phallus through the ostium opened on the sterigma. The complex articulation of the phallus with the need of a sclerotized area of support (juxta) and correlation between the length and protrusion of the phallus to reach the ostium, as well as inverted cases in which the ostium is protruded from the female to reach the phallus, were subject or early studies by Chapman (1916) and Bayard (1944) (see Jolivet 2008 for a revision of the concept of inverted copulation). The phallus must interact during penetration with the ductus bursae at least in its more external part. The ductus bursae may be extremely complex, full of spines, coiled, etc. It is difficult to imagine a simple interaction, and the details of this are unknown. Generally, the ductus seminalis is present at some point of the ductus and the phallus or the vesica will block the pore during mating. Sihvonen (2007) made a detailed account of corresponding anatomical structures between the phallus and ductus bursae in *Scopula* (Geometridae), but no generalization is possible. The distal tip of the phallus must reach the cervix and must anchor there to ensure firm attachment. As mentioned above (see Sect. 12.2.2), this area tends to be sclerotized and plates and spiny areas are common. Ferro and Akre (1975) observed interlocking between the cornuti and the sclerotized plate of the cervix in *Cydia pomonella*, and Mikkola (1993) considered this cervical interaction as supporting the "lock-and-key hypothesis." Lincango et al. (2013) in their study of the bursa copulatrix wall of Tortricidae suggested that this cervical sclerotization could also be correlated with the strengthening of the wall muscles of the corpus bursae attached to the signum. Obviously, both hypotheses are compatible. However, many male Lepidoptera have a short phallus that is firmly attached to the juxta making impossible penetration of the phallus. Miller (1988) studied a case easily generalizable in the Dioptinae *Cyanotricha*. In this species, only a small ventral tooth on the aedeagus inserts into the ostium, and the vesica is everted all through the ductus. The cornuti firmly attach to the spinulous area of the bursa copulatrix. Thus, the phallus may be in contact directly with the internal side of the ductus bursae, but if it is firmly attached to the juxta, then it is the vesica that is in contact with the ductus bursae. Intermediate possibilities are endless. Mechanical interactions of phallus–bursa copulatrix in noctuids inspired the "lock-and-key" hypothesis (Callaghan and Chapin 1960), and this interaction has been well documented (Mikkola 1992, 2008; Mutanen et al. 2006). Technical difficulties for vesica eversion in small moths have been a limiting factor to extend these studies to other groups of Lepidoptera, but technical progress in dissection promises some improvement in this area (Dang 1993; Matthews 1998; Zlatkov 2011). The cornuti not only operate as attachment devices. Cordero (2010) reviewed all the potential functions of cornuti including caltrop cornuti (Cordero and Miller 2012) (see next section). They are diverse structures even within a single family (e.g., see recent analysis of cornuti in Tortricinae by Anzaldo et al. 2014). Finally, it is interesting to consider the work by Justus and Mitchell (1999) that found interpopulation variation affecting phallus length and bursal sclerotization connected with larval diet, something that would extend the problem of bursa-phallus coupling to the field of insect–plant relationship.

12.2.3.4 The Insemination

The corpus bursa is the scenario of insemination. The male (through the phallus and vesica) must transfer seminal products as well as other substances in order to build a spermatophore. The sequence of introduction of the different secretions is known in detail in some few cases (Khalifa 1950; Lai-Fook 1984; Fänger and Naumann 1998; Blanco et al. 2009). The contribution of the female to the production of the spermatophore is unclear. In addition, the male must manipulate the spermatophore inside the bursa copulatrix leaving the spermatophore aperture close to the point where the ductus seminalis meets the bursa copulatrix. In cases in which the female pierces the spermatophore by the signa, the exit for the spermatozoa could be bypassed through the spermatophore wall; however, this possibility remains unclear (Tschudi-Rein and Benz 1990). The literature available on the composition of the spermatophore is extensive. However, although it is well known that the shape of the spermatophore is taxon dependent (Petersen 1907; Williams 1941; Callahan 1960), the information on its morphological diversity is rather poor. The corpus bursae behaves as a digestive chamber. Male provides nutrients to the female (Boggs and Gilbert 1977; Lai-Fook 1991) with the spermatophore as well as reproductive stimulants (Park et al. 1998). The digestive character—with absorption—of the bursa copulatrix has been demonstrated not only on nutrients and spermatophore supplied by the male but also on oosortion of oocytes (Koshio et al. 2002), a particular function that shares with the bulla seminalis (Lum 1984; Lum and Baker 1989). Aging females would increase survival by recovering nutrients from oocytes through digestion in the bursa and bulla. However, the remains of this digestion would obstruct the passage of sperm reducing in the end the female reproductive success (Lum 1982). The wall of the bursa copulatrix and its associated muscle layer has been examined in few cases (Rogers and Wells 1984; Lai-Fook 1986; Lincango et al. 2013). The description of the epithelium and its potential role (secretory or absorptive) were not coincident. However, the structure and function of the epithelium could be affected by the physiological moment or position in the bursa. The question deserves more extensive research.

12.3 Selection on Genitalia in Lepidoptera: Some Illustrative Hypotheses

In this section, we explain the different types of PCIS pressures that could be responsible for the evolutionary origin and elaboration of male and female genitalia in Lepidoptera. As an illustration of the way in which sexual selection could act on genital traits, we develop hypotheses for some of the structures mentioned in the previous section. Our treatment is far from comprehensive; we rather chose a few structures and propose hypotheses about the selective pressures acting on them. We also discuss how genital adaptations evolved in one sex result in selective pressures on the genital structures of the opposite sex with which they interact. In almost all cases treated in this section, there is no evidence to test the hypotheses.

There are two main mechanisms of PCIS that could act on male genital morphology and female genital morphology and result in coevolution between males and females: CFC and sexual conflict (Eberhard 2010; Schilthuizen 2014; Simmons 2014). In the first case, females could bias the paternity of their offspring in favor of males whose genital configuration and behavior provide the best stimulation during copulatory courtship; this preference would be favored because of the positive effect on female fitness of producing male offspring that are good stimulators (Eberhard 1985, 1996, 2010). If female stimulation is affected by morphological traits of her genitalia, these could coevolve with the genital traits of males providing the stimulus. One potential example is valval flexion. Scott (1978) and Platt (1978) observed males, from species of *Erynnis* skippers and *Limenitis* butterflies, rhythmically flexing both or just one of their valvae during copulation, which resulted in the "squeezing" of the female distal abdomen, or in the "scraping over" or "raking" of the females' sternae with the valval "ornamentations" (as mentioned in the previous section, the ornamentations present on the inner side of each valva are frequently very elaborated). Platt (1978) and Eberhard (1985) suggest that these valval flexions have a stimulatory function that is compatible with the CFC explanation, and Scott (1978) observed that, during copulation, some of the valval ornaments are pressing on, or hooked in, specific portions of the female genital tract, portions that could have evolved specific configurations of mechanical receptors on them.

Another potential example of structures subject to CFC is that of cornuti (the sclerotized structures located on the endophallus described in the previous section; Fig. 12.4b–d, f), which could provide internal stimulation to the female during copulation. If this is the case, CFC could favor males whose cornuti provide the best stimulation because they are located in the proper positions and have the appropriate size and shape (Cordero 2010). Female stimulation could be influenced by properties of the female genitalia, such as the degree of sclerotization and the distribution of sensory structures on different portions of the bursa copulatrix (Rogers and Wells 1984; Lincango et al. 2013), which in turn could evolve to improve the ability of females to choose the males that provide the best stimulation (i.e., the best copulatory courtship). Particularly interesting are the deciduous cornuti that are shed from the endophallus and remain in the bursa copulatrix after copulation. These structures could continue stimulating the female after copulation, for example, every time the female contracts her abdomen near the area surrounding the bursa copulatrix.

On the other hand, sexual conflict occurs if the reproductive interests of males and females do not coincide and members of one sex use genital structures to manipulate the behavior or physiology of the opposite sex, thus increasing their fitness; in response, the manipulated sex could evolve counteracting genital traits that prevent or reduce such manipulation (Arnqvist and Rowe 2002). For example, the star-shaped caltrop cornuti, shed by males during copulation in many Notodontidae species (Miller 1991, 2009; Miller et al. 1997), could pierce the bursa copulatrix if the female receives further spermatophores and thus prevent female remating

(Cordero and Miller 2012). Since monandry could be detrimental for females (for example, by reducing the amount of spermatophore-derived nutrients), sclerotized shieldlike structures in areas of contact with cornuti could evolve in the female's bursae copulatrix (Cordero 2010; Cordero and Miller 2012). Female structures that damage the male genitalia could also evolve. These structures could be used by females, for example, to terminate the copulation before sperm transfer, but after the male already transferred at least part of the nutritious secretions contained by the spermatophore. Potential examples are the signa spines of the tortricid *Tylopeza zelotypa*, where Anzaldo et al. (2014) found signa spines inserted in the endophallus, or the menacing strong spines present in the ductus bursae of *Issikiopteryx* (Lecithoceridae) species (Fan and Li 2008; Liu and Wang 2013).

As mentioned above, although nowadays most researchers accept that sexual selection has played a fundamental role in genital evolution, still remain disagreements on the relative importance of the mechanisms described above (Simmons 2014). Although the controversy about the relative importance of female choice and sexual conflict has attracted the most attention (Hosken and Stockley 2004; Eberhard 2010; Arnqvist and Rowe 2002; Schilthuizen 2014; Simmons 2014), determining the relative importance of the different mechanisms is an empirical problem that needs to take into account the fact that more than one mechanism could influence the evolution of a particular genital structure and that the different mechanisms could interact in complex ways (Cordero 2005, 2010; Cordero and Eberhard 2005; Simmons 2014). We illustrate these ideas with some speculations about the evolution of cornuti (Fig. 12.4b–d, f) and the portions of the female genitalia that interact with cornuti during copulation (Figs. 12.1b and 12.2). Let us begin with the origin of cornuti. Consider a species without cornuti in which sometimes the intromittent organ of a copulating male is forcefully dislodged from the female genitalia before ejaculation is complete. A mutant male, whose endophallus bears, for example, a couple of small sclerotized projections that make forceful dislodgement less likely will have increased fitness, compared to normal males without the incipient cornuti, because he will transfer full ejaculates to a larger number of females. The type of selection responsible for the origin of cornuti depends on the causes of the forceful dislodgement of the intromittent organ from the female genitalia. If the cause is harassment from, or struggling with, competing males, the cause would be intrasexual selection. If the dislodgement is caused by sudden and energetic movements used by the female to test the strength of the male, it would be female choice. If the dislodgement is caused by sudden and energetic movements used by mated females to prevent multiple mating (i.e., resistance behavior), it would be sexual conflict. If the reasons are the squalls that commonly blow in the mating habitat, it would be natural selection. Now consider that genital dislodgement could have been caused by more than one of the previous causes and you will have an idea of the complexities of assigning relative importance to the different selective pressures.

If we decide not to trouble ourselves with the origin question, and are only interested in understanding the selective pressures responsible for the maintenance of cornuti in a particular species, we should consider further possible functions. For example, the cornuti could provide internal stimulation to the female (Cordero

2010) and cryptic female choice (of a different nature than the one considered above) would be responsible for the maintenance and further elaboration of cornuti. Other possibility is that males used their cornuti to pierce the female genital tract to delay female remating, until the female is able to heal the damage provoked by cornuti, in which case sexual conflict (also of a different nature than the one considered above) would be responsible for the maintenance and elaboration of cornuti. Different selective pressures could act simultaneously (for example, they could prevent dislodgement due to sudden violent winds and to violent competitor males attempting to displace them from the female, depending on the circumstances) or at different times during the life of the male (for example, in a species in which monandry is adaptive for females, males could use cornuti for internal stimulation of virgin females, and against resistance behavior of already mated females that they manage to copulate with). It is clear that, deciding the relative importance of each selective pressure for the origin and maintenance of cornuti will be, in most cases, very difficult (and very interesting, of course). Identification of the selective pressures responsible for the evolution of female genital coadaptations to cornuti would be an equally complex (and fascinating) process.

Finally, one mechanism that has been invoked to explain the complexity and diversity of genitalia in Lepidoptera is species isolation by lock and key. Although the term lock and key may have been used sometimes by morphologists and taxonomists as the simple correspondence of anatomical parts that couple as positive and negative molds without presuming any special selective value (Kristensen 2003a), the hypothesis behind this idea is that the morphology of male and female genitalia coevolved to allow that only individuals belonging to the same species mate successfully, preventing costly interspecific matings (Mikkola 2008 and references therein). Eberhard (2010, p. 6) considers that the lock-and-key mechanism as a general explanation of the evolutionary patterns shown by genitalia "is probably in the process of being discarded definitively," while other authors propose that the lock-and-key mechanism implies female choice and should be considered one end of a continuum that includes typical CFC at the opposite end (Ryan and Rand 1993; Castellano and Cermelli 2006; Simmons 2014). One of the theoretical problems in this controversy is that species isolation also benefits from sexual selection mechanisms; thus, there will be always space for those claiming for an anti-hybrid mechanism (Shapiro and Porter 1989). Eberhard (1985, 2010) also acknowledges that the lock-and-key mechanism could explain genital evolution in particular cases, although he did not mention any specific example.

12.4 Sexual Coevolution of Signa and Spermatophore Envelopes

In the previous section we developed hypotheses on the way in which selection could act on some parts of the genitalia of Lepidoptera (Figs. 12.1 and 12.2). Determining the relative importance (if any) of the selective pressures considered by these hypotheses requires detailed knowledge about the function (before, during, and after copulation, and also on the effects of shape, size, and position on performance), genetics, and phylogenetic history of the structures. Unfortunately, the evidence for most genital structures of Lepidoptera is limited and does not permit testing the hypotheses. In this section, we briefly review studies on signa, the sclerotized genital structures located in the internal wall of the corpus bursae of many Lepidoptera species (Figs. 12.1b and 12.2), probably the only genital structure of Lepidoptera for which part of the above-mentioned studies have been made (if we exclude the extraordinary genital photoreceptors—four cells—found in the genitalia of several butterfly species—as far as we know, they have not been observed in moths—studied in great detail by Dr. Kentaro Arikawa and collaborators and summarized, for example, in Arikawa 2001).

Signa are present in numerous species belonging to most families of Lepidoptera. Their shape, size, texture, position, and number vary widely (e.g., Sánchez et al. 2011; Lincango et al. 2013), and Figs. 12.1b and 12.2b–f only provide a glimpse of this variation. Such variation suggests that signa evolved rapidly and divergently (Sánchez et al. 2011). The phylogenetic reconstruction of the Lepidoptera has received much attention (Regier et al. 2013), and the available information on the ground plan of the order, including primitive subgroups and relation with sister orders, is excellent (Kristensen 1999). However, the origin of signa from the phylogenetic point of view is unclear. Some sclerotized structures are found in the basal family Micropterigidae. True signa appear at least in Nepticulidae, and some spinulose areas may be found in other non-Ditrysia groups. Signa are also found among basal Ditrysia as Tineoidea and Gracillarioidea. As Kristensen (1984) has stated, there must have been considerable parallelism in the early evolution of the Lepidoptera. Thus, what we know is that there is a considerable plastic potentiality for producing signa and other sclerotizations on the bursa copulatrix and that they have been lost and gained several times along the history of Lepidoptera (Sánchez et al. 2011 and references therein). In some taxa, signa are considered phylogenetically informative traits (e.g., Penz 1999; Rubinoff and Powell 1999). Interestingly, in several species, belonging to different taxa, signa are absent (e.g., Fig. 12.2a). Thus, the pattern of evolution of signa is similar to the pattern observed in male genitalia that led Eberhard (1985) to propose his famous sexual selection hypothesis (see Introduction). Eberhard built his theory by reconsidering the implications of the intimate contact between male and female genitalia during copulation, suggesting that an understanding of the role of signa in sexual interactions is necessary for the comprehension of its evolution.

The British entomologist Howard E. Hinton considered four hypotheses on the function of signa and, apparently based on his own observations, concluded that these structures assist in tearing open the spermatophore (Hinton 1964). Galicia et al. (2008) added three new functional hypotheses and made an experimental study in four butterfly species that led them to a similar conclusion. The two spine-shaped signa of *Callophrys xami* (Lycaenidae) and the small spines covering the, also paired, "L-" or "V"-shaped signa of two species of *Eueides* (Nymphalidae) and of *Heliconius ismenius* (Nymphalidae) were observed piercing or sectioning the spermatophore envelope. Releasing sperm from the spermatophore is not the reason why signa tear open the spermatophore envelope, since sperm is transferred to the spermatheca before the spermatophore envelope is broken (this has also been observed in other lepidopterans; Drummond 1984). Thus, signa allow the female to gain access to the nutritious and hormonal resources contained in the spermatophore (Galicia et al. 2008). Previous studies also support the idea that the main function of signa is to break the spermatophore envelope (Rogers and Wells 1984; Tschudi-Rein and Benz 1990).[1] These results led us to pose two questions: (1) why do males transfer spermatophores whose envelope requires a sharp or acute device to be broken? and (2) why in some species females lack signa?

In most Lepidoptera species studied, at least a fraction of the females mates with more than one male (i.e., they are polyandrous; Drummond 1984; Torres-Vila et al. 2004) and expose males to sperm competition. However, female sexual receptivity usually decreases after mating and is gradually recovered with time (Drummond 1984; Wedell 2005). The large size of the spermatophore (relative to the storage capacity of the corpus bursae) and the substantial amount of time invested in copulation, probably selected for females that delay remating until the spermatophore of the last copulation has been partially digested. The length of the female postmating period of non-receptivity (the refractory period) is determined in a more or less complex way by several factors (Wedell 2005), among which the degree of corpus bursae distention provoked by the spermatophore is very important (Sugawara 1979; Drummond 1984; Wedell 2005). These observations led Drummond (1984) to propose that in polyandrous species sperm competition favored males producing spermatophores with an increased content of chitin that decreases the rate at which they are digested within the corpus bursae and, therefore, increases the length of the refractory period of the female. This idea was extended to include the evolution of thicker spermatophore envelopes that are more difficult to break (Cordero 2005). Since spermatophores contain valuable nutritious resources for females (Boggs and Gilbert 1977; Boggs 1995; Torres-Vila et al. 2004), an increased refractory period could decrease female fitness, thus generating a sexually antagonist selective pressure on females to recover control of their rate of spermatophore digestion and of the duration of their refractory period (Cordero 2005). From this argument, the hypothesis that females evolved signa as devices that allow a faster breaking of the spermatophore envelope, thus moving the duration of their refractory period back to the female optimum, was derived (Cordero 2005). This hypothesis (that we called the sexually antagonistic coevolution hypothesis or SAC; Sánchez and Cordero 2014) is supported by the above-mentioned evidence that one of the main functions of signa is breaking the spermatophore envelope.

[1]Galicia et al. (2008) also mention that there is evidence suggesting that in some species, signa could play different or additional functions. For example, in some species, the signa is shaped like one or more relatively large plates that could protect the corpus bursae from the damage that cornuti could inflict (see Ferris 2004 for a possible example). These alternatives need to be studied.

Another prediction of the SAC hypothesis is that signa will evolve in polyandrous species and will tend to be absent in monandrous taxa. Sánchez et al. (2011) tested this prediction by means of a comparative study of 37 taxa, finding statistical support for the prediction. They found that: (1) polyandry and presence of signa are plesiomorphic for the taxa analyzed; (2) as expected from the SAC hypothesis, 93 % of polyandrous species had signa, in comparison with only 33.3 % in monandrous species; (3) the phylogenetically controlled analysis also found support for the predicted association between female mating pattern and presence of signa; (4) monandry evolved eight times and in five of them, its evolution was associated with the loss of signa; and (5) signa were lost seven times, five of them in taxa that evolved monadry. Summarizing, Sánchez et al. (2011) found statistical support for the predicted association between female mating pattern and presence of signa, but there were several exceptions that merit further study.

The SAC hypothesis also predicts that in species in which monandry is adaptive for females, the spermatophore envelopes will be thinner than in polyandrous species (Cordero 2005). Evidence supporting this prediction was found in a study comparing the thickness of the spermatophore envelopes of two *Heliconius* butterfly species belonging to the monandrous clade lacking signa (H. *hortense* and *H. charithonia*), with those of one species belonging to the polyandrous clade with signa (*H. ismenius*) (Sánchez and Cordero 2014). Further evidence supporting this prediction can be extracted from a paper by Matsumoto and Suzuki (1995) on mating plugs and mating patterns in six genera of Japanese Papilionidae. The two virtually monogamous genera (*Luehdorfia* and *Parnassius*) lack signa, and their spermatophore envelopes are thin membranes (in the words of Matsumoto and Suzuki, they lack a "capsule"), whereas the two slightly polyandrous genera (*Atrophaneura* and *Pachliopta*) have a "small signum" and a "relatively thick" spermatophore envelope ("capsule"), and, finally, the two polyandrous genera (*Papilio* and *Graphium*) possess a "signum" and the spermatophores envelopes are "thick" (quantitative details of the reanalysis of Matsumoto and Suzuki's data are in Sánchez et al. 2011). In contrast, Lincango et al. (2013) found well-developed signa and strong associated muscles in some selected Tortricidae in which a single large and thick spermatophore was found. This extreme case could represent an example of male coercion in which the male has imposed monandry to the female via an internal plug. We must, however, consider that these males must invest a large amount of energy to produce those large spermatophores together with the fact that their cornuti are deciduous. Their capacity for remating is questionable, and so females could have imposed monogyny to their partners.

Data from an investigation looking for additive genetic variance in the morphology of the spine-shaped signa of the butterfly *C. xami* (Jiménez et al. 2011) found significant additive genetic variance for signa length and width in one of the two populations studied, and also (and more interesting from the point of view of the SAC hypothesis) significant extranuclear maternal effects on some components of size and shape of signa, which could facilitate the evolution of optimal signa size and shape, even if these female adaptations decrease male fitness.

The SAC hypothesis (Cordero 2005) provides a relatively well-supported explanation for the evolutionary origin of signa (Sánchez et al. 2011; Sánchez and Cordero 2014) and possibly for part of its diversity. However, the great diversity of signa shapes, sizes, positions, and numbers observed in different taxa seems hard to explain simply as a response to the evolution of spermatophores that are harder to break open, as a consequence of selective pressures derived from the development of thicker spermatophore envelopes or of increased amounts of chitin in the spermatophore. We think that comprehension of the evolutionary diversification of signa will be obtained by considering functions additional to breaking the spermatophore envelope. We think that the selective pressures responsible for these functions will help us understand the diversity observed in this fascinating genital structure. Two hypotheses seem to us particularly promising: the idea that signa protect the corpus bursae from damage inflicted by structures of the male genitalia (e.g., cornuti), and the possibility that signa act as courtship devices stimulating the male to transfer more and better ejaculates (i.e., a "role-reversed" version of Eberhard's hypothesis on the evolution of male genitalia). Both hypotheses invoke sexual selection and require detailed investigations of the sexual interactions taking place within the female genitalia of Lepidoptera.

12.5 Conclusions

In a recent review, Simmons (2014) concludes that although sexual selection is nowadays considered the main driver of genital evolution, the relative importance of the different types of sexual selection (Hosken and Stockley 2004; Eberhard 2010; Simmons 2014) and the possible role of the "lock-and-key" hypothesis (Eberhard 1985; Mikkola 2008) are still open questions whose solutions will require cost-benefit studies of mating interactions in a variety of conditions and species, as well as estimates of the tempo and mode of genital evolution in different taxa. These studies need a detailed understanding of the functional morphology of female genital traits, as well as of the patterns of variation and covariation with male genitalia. Furthermore, the role of natural selection (Reinhardt 2010) should not be discarded without a detailed examination that sometimes will require experimental studies. The same organ may express slightly different roles depending on how it interacts with other biological and functional aspects. We cannot isolate the structure of the bursa copulatrix and their signa from the structure of the spermatophore, the phallus, vesica and cornuti, and so a rather long list of variables. Networking (Proulx et al. 2005), instead of singular approaches, could be used to explore the complexity and diversity of genitalia observed in the Lepidoptera, an ideal group to undertake this research program.

Acknowledgments We thank Dr. Laura Sirot for very useful comments on a previous version of the manuscript. CC thanks Jaime Camacho and Raúl Martínez for technical help and PAPIIT/DGAPA (IN208413) for financial support. JB contribution was supported by project BFU2012-39816-CO2-01 of the Spanish Ministerio de Economía y Competitividad.

References

Ah-King M, Barron AB, Heberstein ME (2014) Genital evolution: why are females still understudied? PLoS Biol 12(5):e1001851

Alexander RD (1964) The evolution of mating behavior in arthropods. In: Royal Entomological Society London, Symposia, vol 2, pp 78–94

Anzaldo SS, Dombroskie J, Brown JW (2014) Morphological variation, taxonomic distribution, and phylogenetic significance of cornuti in tortricinae (Lepidoptera: Tortricidae). Proc Entomol Soc Wash 116:1–31

Arikawa K (2001) Hindsight of butterflies. Bioscience 51:219–225

Arnold RA, Fisher RL (1977) Operational mechanism of copulation and oviposition in *Speyeria* (Lepidoptera: Nymphalidae). Ann Entomol Soc Amer 70:455–468

Arnqvist G, Rowe L (2002) Comparative analysis unveils antagonistic coevolution between the sexes in a group of insects. Nature 415:787–789

Bayard A (1944) Observations sur l'accouplement de *Dryas paphia* (L.) (Lep. Nymphalidae). Bull Soc Entomol France 49:92–95

Bieman DN, Witter JA (1982) Mating wounds in *Malacosoma*: an insight into bed bug mating behavior. Fla Entomol 65:377–378

Blanco CA, Guadalupe Rojas M, Groot AT, Morales-Ramos J, Abel CA (2009) Size and chemical composition of *Heliothis virescens* (Lepidoptera: Noctuidae) spermatophores. Ann Entomol Soc Amer 102:629–637

Boggs CL (1995) Male nuptial gifts: Phenotypic consequences and evolutionary implications. In: Leather SR, Hardie J (eds) Insect reproduction. CRC Press, Cleveland, pp 215–242

Boggs CL, Gilbert LE (1977) Male contribution to egg production in butterflies: evidence for transfer of nutrients at mating. Science 206:83–84

Brower LP, Oberhauser KS, Boppré M et al (2007) Monarch sex: ancient rites or recent wrongs. Antenna 31:12–18

Buntin GD, Pedigo LP (1983) Morphology of the Male and female reproductive systems of *Plathypena scabra* (F.) (Lepidoptera: Noctuidae). J Kansas Entomol Soc 56:377–386

Callahan PS (1958) Serial morphology as a technique for determination of reproductive patterns in the Corn Earworm, *Heliothis zea* (Boddie). Ann Entomol Soc Amer 51:413–428

Callaghan PS (1960) A morphological study of spermatophore placement and mating in the subfamily Plusiinae (Noctuidae, Lepidoptera). Proc XI Inter Congr Entomol 1:339–345

Callahan PS, Chapin JB (1960) Morphology of the reproductive systems and mating in two representative members of the family Noctuidae, *Pseudaletia unipuncta* and *Peridroma margaritosa*, with comparison to *Heliothis zea*. Ann Entomol Soc Amer 53:763–782

Callahan PS, Cascio T (1963) Histology of the reproductive tracts and transmission of sperm in the Corn Earworm, *Heliothis zea*. Ann Entomol Soc Amer 56:535–556

Castellano S, Cermelli P (2006) Reconciling sexual selection to species recognition: a process-based model of mating decision. J Theor Biol 242:529–538

Chapman TA (1916) On the pairing of the plebeiid blue butterflies (Lycaeninae, tribe Plebeiini). Trans Entomol Soc Lond 1916:156–180

Clarke CA, Sheppard PM (1956) Hand-pairing of butterflies. Lep News 10:47–53

Cordero C (2005) The evolution of signa in female Lepidoptera: natural and sexual selection hypotheses. J Theor Biol 232:443–449

Cordero C (2010) On the function of cornuti, sclerotized structures of the endophallus of Lepidoptera. Genetica 138:27–35

Cordero C, Eberhard WG (2005) Interaction between sexually antagonistic selection and mate choice in the evolution of female responses to male traits. Evol Ecol 19:11–122

Cordero C, Miller JS (2012) On the evolution and function of caltrop cornuti in Lepidoptera—potentially damaging male genital structures transferred to females during copulation. J Nat Hist 46:701–715

Covell CV (1970) A revision of the North American species of the genus *Scopula* (Lepidoptera: Geometridae). Trans Amer Entomol Soc 96:101–221

Dang PT (1993) Vesicas of selected tortricid and small lepidopterous species with descriptions of new techniques of vesica eversion (Lepidoptera: Tortricidae, Oecophoridae, Gelechiidae, and Nepticulidae). Can Entomol 125:785–799

De Jong R (1978) Functional morphology of the genitalia of *Carcharodus boeticus* Stauderi Rev. (Lepidoptera, Hesperiidae). Nether J Zool 28:206–212

Drummond BA III (1984) Multiple mating and sperm competition in the Lepidoptera. In: Smith RL (ed) Sperm competition and the evolution of animal mating systems. Academic Press, Orlando, pp 291–371

Dugdale JS (1974) Female genital configuration in the classification of Lepidoptera. NZJ Zool 1:127–146

Eberhard WG (1985) Sexual selection and animal genitalia. Harvard University Press, Cambridge

Eberhard WG (1996) Female control: sexual selection by cryptic female choice. Princeton University Press, Princeton

Eberhard WG (2010) Evolution of genitalia: theories, evidence, and new directions. Genetica 138:5–18

Fan X, Li H (2008) The genus *Issikiopteryx* (Lepidoptera: Lecithoceridae): checklist and descriptions of new species. Zootaxa 1725:53–60

Fänger H, Naumann CM (1998) Genital morphology and copulatory mechanism in *Zygaena trifolii* (Esper, 1783) (Insecta, Lepidoptera, Zygaenidae). Acta Zool 79:9–24

Fatzinger CW (1970) Morphology of the reproductive organs of *Dioryctria abietella* (Lepidoptera: Pyralidae (Phycitinae)). Ann Entomol Soc Am 63(5):1256–1261

Ferro DN, Akre RD (1975) Reproductive morphology and mechanics of mating of the Codling Moth, *Laspeyresia pomonella*. Ann Entomol Soc Amer 68:417–424

Forbes WTM (1939) The muscles of the lepidopterous male genitalia. Ann Entomol Soc Amer 32:1–10

Galicia I, Sánchez V, Cordero C (2008) On the function of signa, a genital trait of female Lepidoptera. Ann Entomol Soc Amer 101:786–793

Gosse PH (1882) On the clasping-organs ancillary to generation in certain groups of the Lepidoptera. Trans Linn Soc Lond S 2(2):265–345

Grimaldi D, Engel MS (2005) Evolution of the insects. Cambridge University Press, New York

Hannemann HJ (1954a) Zur funktionellen anatomie des männlichen kopulationsapparates von *Argynis paphia* (L.). Zool Anz 152:266–274

Hannemann HJ (1954b) Zur muskelfunktion der weiblichen genitalsegmente von *Argynnis paphia* (L.) (Lep.). Zool Anz 153:149–154

Hausmann A (1999) Falsification of an entomological rule: polymorphic genitalia in geometrid moths. Spixiana 22:83–90

Hinton HE (1964) Sperm transfer in insects and the evolution of haemocoelic insemination. In: Highnam KC (ed) Insect reproduction. In: Symposium on Royal Entomology Society London, pp 95–107

Hosken DJ, Stockley P (2004) Sexual selection and genital evolution. Trends Ecol Evol 19:87–93

Huber BA (2003) Rapid evolution and species specificity of arthropod genitalia: fact or artifact? Organ Div Evol 3:63–71

Huber BA (2010) Mating positions and the evolution of asymmetric insect genitalia. Genetica 138:19–25

Huber BA, Sinclair BJ, Schmitt M (2007) The evolution of asymmetric genitalia in spiders and insects. Biol Rev 82:647–698

Jiménez-Ambriz G, Mota D, Cordero C (2011) Genetic variation in a female genital trait evolved by sexual coevolution. Genetica 139:1241–1249

Jolivet P (2008) Inverted copulation. In: Capinera JL (ed) Encyclopedia of entomology. Springer, Berlin, pp 2041–2044

Jordan K (1896) On mechanical selection and other problems. Nov Zool 3:426–525

Justus KA, Mitchell BK (1999) Reproductive morphology, copulation, and inter-populational variation in the diamondback moth, *Plutella xylostella* (L.) (Lepidoptera: Plutellidae). Int Insect Morphol Embriol 28:233–246

Khalifa A (1950) Spermatophore production in *Galleria melonella* L. (Lepidoptera). Proc Roy Entomol Soc Lond A 25:33–42

Klots AB (1970) Lepidoptera. In: Tuxen SL (ed) Taxonomist's glossary of genitalia in insects, 2nd edn. Munksgard, Copenhagen, pp 115–130

Koshio C, Tanaka Y, Kudo S-I (2002) The collapsed eggs found in the bursa copulatrix of a plum moth, *Illiberis rotundata* Jordan (Zygaenidae: Procridinae): an unusual egg resorption system? J Lep Soc 56:62–65

Kristensen NP (1984) Studies on the morphology and systematics of primitive Lepidoptera (Insecta). Steenstrupia 10:141–191

Kristensen NP (1999) Handbook of zoology 4: Arthropoda Insecta. Lepidoptera, moths and butterflies, 1: evolution, systematics, and biogeography. De Gruyter, Berlin, pp 1–491

Kristensen NP (2003a) Skeleton and muscles: adults. In: Kristensen NP (ed) Handbook of zoology 4: Arthropoda: Insecta, part 36, Lepidoptera, moths and butterflies, 2: morphology, physiology, and development. De Gruyter, Berlin, pp 39–131

Kristensen NP (2003b) Reproductive organs. In: Kristensen NP (ed) Handbook of zoology 4: Arthropoda: Insecta, part 36, Lepidoptera, moths and butterflies, 2: morphology, physiology, and development. De Gruyter, Berlin, pp 427–447

Kuznetsov VI, Stekolnikov AA (2001) New approaches to the system of Lepidoptera of world fauna on the base of the functional morphology of abdomen. Nauka, St. Petersburg (In Russian)

Lai-Fook J (1982a) Structure of the noncutiular simplex of the internal male reproductive tract of *Calpodes ethlius* (Hesperiidae, Lepidoptera). Can J Zool 60:1184–1201

Lai-Fook J (1982b) Structure, function, and possible evolutionary significance of the constrictions in the male reproductive system of *Calpodes ethlius* (Hesperiidae, Lepidoptera). Can J Zool 60:1828–1836

Lai-Fook J (1984) The spermatophore of the skipper, *Calpodes ethlius* (Hesperiidae, Lepidoptera): the serum sac. Can J Zool 62:1135–1142

Lai-Fook J (1986) The virgin bursa copulatrix of the butterfly *Calpodes ethlius*. Tissue Cell 18:545–558

Lai-Fook J (1991) Absorption of phosphorus from the spermatophore through the cuticle of the bursa copulatrix of the butterfly, *Calpodes ethlius*. Tissue Cell 23:247–259

Lincango P, Fernández G, Baixeras J (2013) Microstructure and diversity of the bursa copulatrix wall in Tortricidae (Lepidoptera). Arthrop Struc Dev 42:247–256

Liu S, Wang S (2013) Three new species of the genus *Issikiopteryx* Moriuti, 1973 (Lepidoptera: Lecithoceridae: Lecithocerini) from China. Zootaxa 3669:037–042

Lorkovic Z (1952) L'accouplement artificiel chez les Lépidoptères et son application dans les recherches de la fonction de l'appareil génital des insectes. Physiol Compar Oecol 3:313–319

Lum PTM (1982) Degenerating oocytes: a physical barrier to sperm translocation in female *Plodia interpunctella* (Hübner) and *Ephestia cautella* (Walker). J Georgia Entomol Soc 17:421–427

Lum PTM (1984) Ultrastructure of the bulla seminalis in *Cadra cautella* Walker (Pyralidae: Lepidoptera). Int J Insect Morphol Embryol 13:303–309

Lum PTM, Baker JE (1989) Role of bursa copulatrix in oosortion in *Cadra cautella* (Lepidoptera: Pyralidae). J Entomol Sci 24:373–377

Masly JP (2012) 170 years of "lock-and-key": genital morphology and reproductive isolation. Int J Evol Biol 2012. doi:10.1155/2012/247352

Matsuda R (1976) Morphology and evolution of the insect abdomen. International series in pure and applied biology, zoology division, vol 56. Pergamon Press, Oxford

Matsumoto K, Suzuki N (1995) The nature of mating plugs and the probability of reinsemination in Japanese Papilionidae. In: Scriber JM, Tsubaki Y, Lederhouse RC (eds) Swallowtail butterflies: their ecology and evolutionary biology. Scientific Publishers, Gainesville, pp 145–154

Matthews M (1998) The CSIRO vesica everter: a new apparatus to inflate and harden eversible and other weakly sclerotized structures in insect genitalia. J Nat Hist 32:317–327
Mikkola K (1992) Evidence for lock-and-key mechanism in the internal genitalia of the *Apamea* moths (Lepidoptera, Noctuidae). Syst Entomol 17:145–153
Mikkola K (1993) The lock and key mechanism in the internal genitalia of the noctuid and geometrid moths (Lepidoptera) in relation to the speciation concepts. Fol Baer 6:149–157
Mikkola K (1994) Inferences about the function of genitalia in the genus *Eupithecia*, with description of a new organ (Lepidoptera, Geometridae). Nota Lepidop (Suppl 5):73–78
Mikkola K (2008) The lock-and-key mechanisms of the internal genitalia of the Noctuidae (Lepidoptera): how are they selected for? Eur J Entomol 105:13–25
Miller JS (1988) External genitalic morphology and copulatory mechanism of *Cyanotricha necyria* (Felder) (Dioptidae). J Lepid Soc 42:103–115
Miller JS (1991) Cladistics and classification of the Notodontidae (Lepidoptera: Noctuoidea) based on larval and adult morphology. Bull Amer Mus Nat Hist 204:1–230
Miller JS (2009) Generic revision of the Dioptinae (Lepidoptera: Noctuoidea: Notodontidae). Part 1: Dioptini. Part 2: Josiini. Bull Amer Mus Nat Hist 321:1–1022
Miller JS, Janzen DH, Franclemont JG (1997) New species of *Euhapigiodes*, new genus, and *Hapigiodes* in Hapigiini, new tribe, from Costa Rica, with notes on their life history and immatures (Lepidoptera: Notodontidae). Trop Lepid 8:81–99
Minelli A (2002) Homology, limbs and genitalia. Evol Devel 4:127–132
Mitter C (1988) Taxonomic potential of some internal reproductive structures in *Catocala* (Lepidoptera: Noctuidae) and Related Genera. Ann Entomol Soc Amer 81:10–18
Mutanen M, Kaitala A, Mönkkönen M (2006) Genital variation within and between three closely related *Euxoa* moth species: testing the lock-an-key hypothesis. J Zool 268:109–119
Naumann CM (1987) Functional morphology of the external male and female genitalia in *Zygaena* Fabricius, 1775 (Lepidoptera: Zygaenidae). Entomol Scand 18:213–219
Norris MJ (1932) Contributions towards the study of insect fertility. I. The structure and operation of the reproductive organs of the genera *Ephestia* and *Plodia* (Lepidoptera, Phycitidae). Proc Zool Soc Lond 3:595–611
Okagaki H, Sibatani A, Ogata M, Okada Y (1955) Male genitalia of Lepidoptera: morphology and nomenclature II. Morphological significance of saculus and furca. Ann Entomol Soc Amer 48:438–442
Osanai M, Kasuga H, Aigaki T (1988) Functional morphology of the glandula prostatica, ejaculatory valve, and ductus ejaculatorius of the Silkworm *Bombyx mori*. J Morphol 198:231–241
Park YI, Ramaswamy SB, Srinivasan A (1998) Spermatophore formation and regulation of egg maturation and oviposition in female *Heliothis virescens* by the male. J Insect Physiol 44:903–908
Penz CM (1999) Higher level phylogeny for the passion vine butterflies (Nymphalidae, Heliconiinae) based on early stage and adult morphology. Zool J Linn Soc 127:278–344
Petersen W (1907) Über die Spermatophoren der Schmetterlinge. Z Wissens Zool 88:117–130
Platt AP (1978) Editor's note. J Lepid Soc 32:305
Proulx SR, Promislow DEL, Phillips PC (2005) Network thinking in ecology and evolution. Trends Ecol Evol 20:345–353
Regier JC, Mitter C, Zwick A et al (2013) A large-scale, higher level, molecular phylogenetic study of the insect Order Lepidoptera (moths and butterflies). PLoS ONE 8(3):e58568
Reinecke LH, Reinecke JP, Adams TS (1983) Morphology of the male reproductive tract of mature larval, pupal and adult tobacco hornworms (Lepidoptera: Sphingidae), *Manduca sexta*. Ann Entomol Soc Am 76:365–375
Reinhardt K (2010) Natural selection and genital variation: a role for the environment, parasites and sperm ageing? Genetica 138:119–127
Rogers SH, Wells H (1984) The structure and function of the bursa copulatrix of the monarch butterfly (*Danaus plexippus*). J Morphol 180:213–221

Rubinoff D, Powell JA (1999) Description of *Diedra*, new genus (Lepidoptera: Tortricidae: Tortricinae: Archipini), and three new species, based on phylogenetic analysis. Ann Entomol Soc Amer 92:473–478

Ryan MJ, Rand AS (1993) Species recognition and sexual selection ass a unitary problem in animal communication. Evolution 47:647–657

Sánchez V, Cordero C (2014) Sexual coevolution of spermatophore envelopes and female genital traits in butterflies: evidence of male coercion? PeerJ 2:e247

Sánchez V, Hernández-Baños BE, Cordero C (2011) The evolution of a female genital trait widely distributed in Lepidoptera: comparative evidence for an effect of sexual selection. PLoS ONE 6(8):e22642

Scott JA (1978) Mid-valval flexion in the left valva of asymmetric genitalia of *Erynnis*. J Lepid Soc 32:304–305

Schilthuizen M (2014) Nature's nether regions. Viking, New York

Sengün A (1944) Experimente zur sexuell-mechanischen isolation. Rev Fac Sci Univ Istanbul 9:239–253

Shapiro AM, Porter AH (1989) The lock-and-key hypothesis: evolutionary and biosystematic interpretation of insect genitalia. Ann Rev Entomol 34:231–245

Sibatani A, Ogata M, Okada Y, Okagaki H (1954) Male genitalia of Lepidoptera: morphology and nomenclature: I, Divisions of the valvae in Rhopalocera, Phalaenidae (= Noctuidae) and Geometridae. Ann Entomol Soc Amer 47:93–106

Sihvonen P (2007) Mating behavior and copulation mechanisms in the genus *Scopula* (Geometridae: Sterrhinae). Nota Lepid 30:299–313

Simmons LW (2014) Sexual selection and genital evolution. Austral Entomol 53:1–17

Simonsen TJ (2006a) Glands, muscles and genitalia. Morphological and phylogenetic implications of histological characters in the male genitalia of Fritillary butterflies (Lepidoptera: Nymphalidae: Argynnini). Zool Scr 35:231–241

Simonsen TJ (2006b) The male genitalia segments in fritillary butterflies: comparative morphology with special reference to the "rectal plate" in *Issoria* (Lepidoptera: Nymphalidae). Eur J Entomol 103:425–432

Song H, Bucheli SR (2010) Comparison of phylogenetic signal between male genitalia and non-genital characters in insect systematics. Cladistics 26:23–35

Stekolnikov AA (1965) Functional morphology of the copulatory apparatus in some Lepidoptera. Entomol Rev 44:143–149

Sugawara T (1979) Stretch reception in the bursa copulatrix of the butterfly *Pieris rapae crucivora*, and its role in behavior. J Ins Physiol A 130:191–199

Torres-Vila LM, Rodríguez-Molina MC, Jennions MD (2004) Polyandry and fecundity in the Lepidoptera: can methodological and conceptual approaches bias outcomes? Behav Ecol Sociobiol 55:315–324

Tschudi-Rein K, Benz G (1990) Mechanisms of sperm transfer in female *Pieris brassicae* (Lepidoptera: Pieridae). Ann Entomol Soc Amer 83:1158–1164

Wedell N (2005) Female receptivity in butterflies and moths. J Exp Biol 208:3433–3440

Williams JL (1941) The relations of the spermatophore to the female reproductive ducts in Lepidoptera. Entomol News 52:61–65

Zlatkov B (2011) A preliminary study of everted vesicae of several leafrollers (Tortricidae). Nota Lepid 33:285–300

Chapter 13
Who's Zooming Who? Seminal Fluids and Cryptic Female Choice in Diptera

Laura K. Sirot and Mariana F. Wolfner

You walked in on the sly
Scopin' for love.
You thought I'd be naïve and tame
But I beat you at your own game.
Take another look and tell me, baby
Who's zoomin' who?
Aretha Franklin, Preston Glass, Narada Michael Walden 1985

Abstract Dipteran females have many opportunities to influence the reproductive success of their mates. After each mating, females may influence their mates' post-copulatory reproductive success by choosing whether and where to store sperm, whether and when to remate and lay eggs, and how much to invest in eggs fertilized by different males. Female neural, endocrine, and muscular mechanisms are necessary for these processes to occur. We review physiological experiments that have borne this out. Further evidence from many Diptera shows that seminal fluid proteins (Sfps) also influence female post-copulatory processes including ones that affect sperm use. The most comprehensive evidence comes from *Drosophila melanogaster*, whose seminal proteome is well characterized. In this species, studies of sequence variation, including in natural populations, and of gene-specific knockdown in the laboratory, have identified male and female genes whose actions influence and/or correlate with post-copulatory processes in the female. Furthermore, co-evolution between Sfps and female reproductive proteins suggests their involvement in common functional pathways. We review the evidence for the interaction of Sfp-mediated effects and cryptic female choice (CFC), with a focus on *D. melanogaster* and evidence from other Diptera as available.

L.K. Sirot (✉)
Department of Biology, College of Wooster, Wooster, OH 44691, USA
e-mail: Lsirot@wooster.edu

M.F. Wolfner
Department of Molecular Biology and Genetics, Cornell University, Ithaca, NY 14853, USA
e-mail: mariana.wolfner@cornell.edu

A.V. Peretti and A. Aisenberg (eds.), *Cryptic Female Choice in Arthropods*,
DOI 10.1007/978-3-319-17894-3_13

Finally, we conclude by assessing what is known and as yet unknown about the interface between CFC and Sfps and by suggesting avenues for further research in this fascinating area.

13.1 Introduction

Just as there can be tremendous variation in males' abilities to attract or compete for a mate, there also can be tremendous variation between males in their post-copulatory reproductive success. In other words, there are processes that occur after mating that can influence the quantity and quality of offspring that a male sires. In particular, after mating or spawning, there is variation between males in (i) their ability of their sperm to get to the right place at the right time for fertilization (and in the number of sperm that do so); (ii) the quality and quantity of offspring their mates produce; and (iii) whether and when their mates remate with another male, which generally has negative impacts on the first male's future fertilization success. Several factors have been associated with variation in male post-copulatory reproductive success (MPCRS). For example, male morphological and/or behavioral traits have been associated with sperm use patterns in several species of arthropods (e.g., Arnqvist and Danielsson 1999; Edvardsson and Arnqvist 2000; Bloch Qazi 2003; House and Simmons 2003; Sirot et al. 2007; Wojcieszek and Simmons 2011; Hotzy et al. 2012; Rowe and Arnqvist 2012). The relative contributions of female- and male-mediated processes to this variation in MPCRS are difficult to disentangle. Yet, there is evidence to suggest that cryptic female choice (CFC) contributes to some of this variation (e.g., Eberhard 1996; Edvardsson and Arnqvist 2000; Bloch Qazi 2003; Fedina and Lewis 2004).

In addition to male morphological and behavioral traits, male-produced seminal fluid proteins (Sfps) also have been associated with all three aforementioned contributors to variation in MPCRS in arthropods (reviewed in Eberhard 1996; Gillott 2003; Avila et al. 2011). In the mid-1990s, Cordero and Eberhard proposed that many features of Sfps (e.g., the ability of some to get into circulation, the dose dependency in responses to them, and the existence of redundant functions among them) are the expected consequences of evolution by CFC (Cordero 1995; Eberhard and Cordero 1995; Eberhard 1996). They argued for the importance of integrating evolutionary and physiological approaches in the study of Sfps (Eberhard and Cordero 1995). Since that time, many researchers have pursued such an integration through a variety of methods including investigating the mechanisms by which Sfps act in females (e.g., Heifetz and Wolfner 2004; Yang et al. 2008; Yapici et al. 2008; Häsemeyer et al. 2009; Rogers et al. 2009; Fricke et al. 2013; Haussmann et al. 2013; Rubinstein and Wolfner 2013; Heifetz et al. 2014; Rezával et al. 2014), testing for positive Darwinian selection on Sfps (e.g., Aguadé 1998, 1999; Begun et al. 2000; Swanson et al. 2001; Haerty et al. 2007; Almeida and DeSalle 2008; Walters and Harrison 2010; Mancini et al. 2011; Wong et al. 2012; Wong and Rundle 2013; Boes et al. 2014), comparing the rates of Sfp evolution across mating systems that differ in the strength of sexual selection (e.g., Walters and Harrison 2010), conducting experimental evolution studies (e.g., Wong and Rundle 2013), and testing how indicators of sperm competition affect Sfp production and allocation (e.g., Wigby et al. 2009; Sirot et al. 2011). Together, these studies provide a rich new resource for evaluating the potential for CFC to act on Sfps.

In this chapter, we use studies in *Drosophila melanogaster* and other Diptera (as available) to review new evidence for interactions between CFC and Sfps. Following the ideas laid out by Eberhard (1996), we first consider evidence for Sfps as a mechanism to overcome CFC. Then, we explore the hypothesis proposed by Cordero (1995) and Eberhard and Cordero (1995) that females could be using Sfps as criteria for CFC.

13.2 Female Mechanisms to Influence MPCRS

Females have mechanisms to efficiently carry out steps in reproduction (i.e., sperm movement into, maintenance in, and release from storage; egg production, release, and movement; mating behaviors) and to obtain the nourishment needed for these. Some of these mechanisms can be used to increase the quality or quantity of the offspring sired by a particular male.

13.2.1 Sperm Storage, Maintenance, and Release

Female dipterans store sperm in specialized organs of one or two types: seminal (or ventral) receptacles (also called the fertilization chamber; Twig and Yuval 2005) and spermathecae. The numbers and size of these organs (even the presence of two types) vary among taxa (e.g., Pitnick et al. 1999; Arthur et al. 2008) and sometimes within taxa (e.g., Ward et al. 2008; Schäfer et al. 2013). Their functions also may differ (e.g., long-term sperm maintenance versus short-term storage for fertilization; Otronen 1997; Fritz and Turner 2002; Twig and Yuval 2005; Manier et al. 2010; Lüpold et al. 2013; Pérez-Staples et al. 2014). Sperm may get into these storage organs through direct placement by the males, through self-propulsion, and/or through assistance with movement, or by chemo-attraction from female sources (e.g., Hosken et al. 1999; Ilango and Lane 2000). There is some evidence from *D. melanogaster* that the female plays an active role in sperm storage, including in the sperm-entry and sperm-exit phases. First, a female whose nervous system has been genetically transformed to maleness is impaired in sperm storage (Arthur et al. 1998). Second, the accumulation of sperm near the openings of the spermathecal glands suggests that those glands may be producing a substance that attracts sperm (Heifetz and Rivlin 2010). Third, uterine contractions occur after mating, and these contractions appear to be important to expose the openings of the sperm storage organs so that sperm can enter (Adams and Wolfner 2007); it is also possible that these contractions help to move the long sperm (characteristic of many *Drosophila* species) toward those openings. Fourth, in the first two days after mating, normal females switch from storing sperm to releasing those gametes from storage so that fertilization can occur. In females that do not produce eggs (but are otherwise normal), the timing of this switching is delayed, suggesting that female egg production affects the timing of transition between sperm storage stages (Bloch Qazi and Wolfner 2006). Fifth, mated *D. melanogaster* females also temporarily release sperm from storage during subsequent matings (Manier et al. 2010, 2013b). Sixth, there is evidence to suggest that female *D. melanogaster* bias sperm storage against males carrying bacterial infections (Radhakrishnan and Fedorka 2012).

Females of other dipteran species across several families (e.g., Dryomyzidae; Piophilidae; Scathophagidae; Tephritidae; Ulidiidae) also appear to be able to control the movement of sperm into storage. Studies in several species demonstrated that sperm storage patterns are altered when the female nervous system is incapacitated (either through severing neurons or through anesthetizing females; e.g., Hellriegel and Bernasconi 2000; Pérez-Staples et al. 2010). Control of sperm storage by females also has been suggested by studies of innervation and morphology of the female reproductive tract (e.g., Hosken et al. 1999; Fritz 2002; Fritz and Turner 2002). Furthermore, there is evidence that females can actively prevent sperm from getting into storage by expelling them before they are stored (Otronen and Siva-Jothy 1991; Bonduriansky et al. 2005; Brunel and Rull 2010; Manier et al. 2010; Rodriguez-Enriquez et al. 2013).

Females also have mechanisms to preserve or promote viability of sperm in storage. In the sperm storage organs, sperm can receive secretions that nurture them and preserve their viability. In *D. melanogaster*, sperm are thought to be released for fertilization from the seminal receptacle first, with the spermathecae being the sites of long-term sperm storage (Pitnick et al. 1999, 2009; see also Manier et al. 2010). The situation differs in other *Drosophila* species (Manier et al. 2013a, b). Perhaps related to this long-term storage function, *D. melanogaster* spermathecae are surrounded by secretory cells that provide molecules that nourish and support the sperm. In the absence of these female-derived secretions, sperm storage is compromised, as is the motility of stored sperm (even in the seminal receptacle; Schnakenberg et al. 2011; Sun and Spradling 2013; see also Anderson 1945; Allen and Spradling 2008). Findings of secretory glands surrounding the spermatheca of other Diptera suggest that females also play a role in maintenance of sperm in storage in these species (e.g., the dung fly, *Scathophaga stercoraria*, Hosken et al. 1999; Caribbean fruit fly, *Anastrepha suspensa*, Fritz and Turner 2002). Consistent with this hypothesis, the secretions from these glands are more abundant in inseminated females than in virgin females in *A. suspensa* (Fritz and Turner 2002).

Sperm in storage must be released in order to fertilize eggs. It is not known what prompts the release of sperm from storage. Compared to the situation in mice and people, where typically only one or a few sperm fertilize the egg even though 10^7–10^8 sperm enter the female, sperm utilization is quite efficient in *D. melanogaster*. More than 40 % of stored sperm succeed in fertilizing eggs (reviewed in Bloch Qazi et al. 2003), and females modulate sperm release in response to availability of egg laying substrates (Bloch Qazi and Hogdal 2010). Since the openings of the *D. melanogaster* sperm storage organs are near where the egg's micropyle will be when the egg comes to rest in the uterus (or bursa), sperm are released close to where they need to move to fertilize the egg. In contrast to *D. melanogaster*, in *S. stercoraria*, the openings of the sperm storage organs are located away from where the egg's micropyle will be within the bursa, but females appear to have a temporary holding area for sperm that is closer to the micropyle (Arthur et al. 2008). In *S. stercoraria*, sperm release is much less efficient than in *D. melanogaster* (ranging from 1.3 to 4.2 % for the first and second clutches laid after a single mating), and females adjust the rate of release in response to sperm stores (Sbilordo et al. 2009).

Although mechanisms that regulate the release of sperm from storage are not known in any dipteran species, data on two species further support the idea that females play an active role in regulating sperm release from storage. In *D. melanogaster*, two neuromodulators, octopamine and tyramine, are required in females for efficient release of sperm from storage (Avila et al. 2012). Experimentally induced loss of octopamine slows sperm release from the seminal receptacle (only), whereas loss of both octopamine and tyramine impairs sperm release from seminal receptacle and the spermathecae. Female *D. melanogaster* also have a constriction at the proximal end of their seminal receptacle that might provide a site for controlling sperm entry into and release from storage (Heifetz and Rivlin 2010). In *A. suspensa*, females have a valve-like structure at the base of their spermatheca (where it connects with the spermathecal duct) with a pore through which sperm move and which can be opened and closed (Fritz and Turner 2002).

In addition to participating in the storage of sperm from a single mate, the female plays an active role in sperm competition. Initial studies in *D. melanogaster* showed that the female's genotype influences the relative success of two different genotype males in sperm competition (Clark and Begun 1998; Clark et al. 1999). To identify the female components in this interaction, Chow et al. (2013) exploited natural variation. They tested sperm competition parameters in reciprocal matings with flies from a standard strain mated to flies from *Drosophila* lines that each carried a distinct, naturally derived genome (Ayroles et al. 2009). They found great variation in sperm competition parameters across the lines—particularly due to the female's genotype. They then performed association studies for single nucleotide polymorphisms (SNPs) in females that associated with the variation in sperm competition parameters. They found many genes whose alleles, in females, correlated with the differences in sperm competition parameters. Interestingly, of the top 33 genome wide association (GWAS) hits, 15 were genes known to function in the nervous system; some of these encoded ion channels and others were necessary for nervous system development or function. This finding, along with the involvement of neuromodulatorsin sperm release from storage as described above (Avila et al. 2012), argues against the idea that the female has a purely passive role in sperm competition, serving as an “arena” for sperm to “fight it out” (see also Manier et al. 2010) or providing a spermicidal environment to select the best sperm (Holman and Snook 2008). Instead, the genetic data suggest that the female participates actively in regulating the outcome of sperm competition (for example potentially by moving the sperm to useful sites or by controlling their release rate). Future studies of these genes’ functions, and their sites of action, will elucidate how the female actively participates in the control of sperm storage and release dynamics.

13.2.2 Nutrition, Egg Production, and Release

The female also can influence MPCRS through her ability to control egg production, in all its steps: oogenesis, ovulation (release of eggs from the ovary), movement of eggs through the reproductive tract, and oviposition. In some dipterans, the ovaries contain all stages of oogenesis, and eggs are ovulated and laid at a low level even without mating (e.g., *D. melanogaster*). In others, females require a specific stimulus (e.g., a blood meal in anautogenous mosquitoes) to complete oogenesis (Clements 2000). But, in both cases, mechanisms within the female control her oogenesis rate. These mechanisms in turn are affected by the female’s hormone levels (Kelly et al. 1987), condition (e.g., Drummond-Barbosa and Spradling 2001), and environment (reviewed in Baldini et al. 2013; Hansen et al. 2014). For example, *D. melanogaster* female nutrition contributes greatly to oogenic rate and success (Drummond-Barbosa and Spradling 2001; Barnes et al. 2008), and the female controls when and where the eggs are laid, assessing her environment and seeking an appropriate deposition site (e.g., Yang et al. 2008; Dweck et al. 2013; Kacsoh et al. 2013). Environmental assessment by the female also plays a role in egg deposition in *S. stercoraria*, where dung pat age and temperature affect the number of eggs that females lay (Demont et al. 2012). Similarly, in the tephritid fruit fly, *Anastrepha ludens*, host firmness and prior oviposition experience influence clutch size (Díaz-Fleischer and Aluja 2003a, b). In another example, chemical cues as well as predator presence affect the oviposition patterns of mosquitoes (e.g., Bentley and Day 1989), again indicating that the female assesses her environment or situation as part of determining her oviposition pattern. Presumably in all of these cases, the female’s hormonal or neuromodulatory condition is modulated in response to the environmental condition that she detects. A female’s level of oogenesis/ovulation, and her decisions and success in oviposition obviously affect her mate’s reproductive success and could provide a mechanism by which females bias the competitive reproductive success of their mates.

Egg production in female insects is regulated by hormones, particularly juvenile hormone (Kelly et al. 1987; Hansen et al. 2014) and ecdysone (e.g., Baldini et al. 2013). In addition, neuromodulators in the female regulate aspects of the egg production process; for example, octopamine regulates ovulation in *D. melanogaster*. Mutants that fail to synthesize octopamine (or one of its receptors OAMB) do not ovulate (Monastirioti et al. 1996; Han et al. 1998; Lee et al. 2003, 2009; Monastirioti 2003). In vitro and in vivo studies indicate roles for octopamine in regulating muscle contraction/relaxation at the ovaries and the oviduct to cause ovulation (Middleton et al. 2006; Rodríguez-Valentín et al. 2006; Rubinstein and Wolfner 2013). Other molecules, such as the neuromodulator tyramine (Cole et al. 2005) and p24-family intracellular trafficking proteins (Carney and Taylor 2003; Bartoszewski et al. 2004), are required for egg movement through the reproductive tract. The involvement of these molecules in the egg production process provides means by which the female can control her egg production, and thus potential steps that can be modulated for CFC. For example, multiple neuromodulators' release/reuptake changes along the female reproductive tract post-mating (Heifetz et al. 2014). These changes in release/reuptake generate unique combinations of neuromodulators at each reproductive tract region and time, which in turn is suggested to coordinate functions such as egg release, sperm release, and egg movement along the reproductive tract. This mechanism could allow the female some control over the efficiency of reproductive events. Secretions from the spermathecal secretory cells also play a role in ovulation and egg movement. Without those secretory cells, ovulation is impaired (Sun and Spradling 2013) and fertilized eggs can be retained so long within the female that they hatch, resulting in live birth of larvae (Schnakenberg et al. 2011).

13.2.3 Remating Patterns

Whether and when a mated female remates will have a large effect on the reproductive success of her previous mate(s): if a female is more likely to remate, or remates sooner, the first male's sperm are more likely to be displaced or encounter competition. In many dipteran species, males are not able to physically force matings, because mating requires the female to position her ovipositor in an appropriate position (e.g., Sepsid flies, Eberhard 2002). In these instances, we can infer that female behavior is involved in determining whether and when remating occurs. In some species, females actively seek out males by, for example, visiting leks of calling males (e.g., Shelly 1990; Shelly and Kaneshiro 1991; Pie 1998).

In some dipteran species, mated females appear reluctant to remate; they will move away from courting males, extrude or position their ovipositor in such a way that prevents copulation, kick away males, and/or move their body back and forth rapidly in response to a mounting male. However, despite this apparent resistance (see Baena and Eberhard 2007), some females of these species will mate with more than one male over their lifetime (e.g., Clark and Begun 1998; Harshman and Clark 1998; Clark et al. 1999; Tripet et al. 2003; Helinski et al. 2012b; Markow et al. 2012). There is evidence from *D. melanogaster* that remating is under control, in part, of the female's nervous system. Certain neurons that innervate the female's reproductive tract control her remating propensity (Häsemeyer et al. 2009; Yang et al. 2009; Rezával et al. 2014). Moreover, female octopamine levels also impact post-mating behaviors and signals from mating act through octopamine signaling to induce the post-mating rejection responses (Rezával et al. 2014).

13.3 Seminal Fluid Proteins: Male-Derived Modulators of MPCRS

As described above, female Diptera have mechanisms through which they can regulate their egg production, sperm storage/release, and/or remating frequency. Presumably there has been selection for females to express these processes at levels that are the most advantageous to their lifetime reproductive success. But females' optimal levels of reproductive processes may not be as optimal from the point of view of their mates. For example, egg production at a rate that allows optimal female survival and somatic processes might be lower than the highest possible rate; however, higher rates of egg production and oviposition may increase the male's reproductive success. Whereas it might be advantageous to females to be receptive to numerous matings (Arnqvist and Nilsson 2000; Jennions and Petrie 2000; Zeh and Zeh 2003), from the male's perspective, any additional mating by his mate is problematic, as it will result in competition with his sperm (or his progeny). Further, females may bias fertilizations in favor of the sperm of a subset of their mates, thus creating a conflict with the males whose sperm are not favored (Eberhard 1996). This reproductive conflict of interest may explain the evolution of male mechanisms to co-opt female control over reproductive processes. These mechanisms can be physical (such as a male's aedeagus scraping sperm out of the storage organs in Odonata; Waage 1979) or based on sperm numbers (e.g., Parker 1990; Manier et al. 2010), but many are chemical: during mating, a male transfers proteins and other molecules in his seminal fluid, and receipt of these molecules modulates female physiology and behavior. For example, in *D. melanogaster*, 208 seminal proteins (Sfps) are known. Functions in regulating females' egg production, sperm storage and sperm competition, longevity, feeding and excretion, and remating receptivity have been identified for several insects, through genetic studies supplemented by evolutionary or molecular approaches (see Avila et al. 2011 for review). In species of *Anopheles* mosquitoes, specific seminal molecules have, similarly, been associated with affecting females' sperm storage and egg development (Rogers et al. 2009; Baldini et al. 2013). In species of *Aedes* mosquitoes, seminal molecules in aggregate are associated with changes in feeding, egg-laying behavior, flight behavior, and mating behavior (reviewed in Gillott 2003); but the association with specific male molecules has not been determined for most of the post-mating responses, with the exception of hormones transferred by males to females which are associated with increased egg development (Baldini et al. 2013; Clifton et al. 2014).

A few examples of the roles of specific *D. melanogaster* seminal proteins in MPCRS are given as follows.

13.3.1 Sperm Storage, Maintenance, and Release

Some of the seminal proteins that enter the female along with sperm are associated with the movement of sperm into storage (Bertram et al. 1996; Neubaum and Wolfner 1999; Bloch Qazi and Wolfner 2003; Avila and Wolfner 2009), and others appear to be necessary for efficient release of sperm from storage (Ravi Ram and Wolfner 2009; Avila et al. 2010; LaFlamme et al. 2012; Findlay et al. 2014). For example, the seminal glycoprotein Acp36DE is essential for sperm entry into storage; sperm storage is impaired by 50–90 % if it is not provided by the male (Bertram et al. 1996; Neubaum and Wolfner 1999; Bloch Qazi and Wolfner 2003). There are some suggested mechanisms that underlie this association between sperm storage and Acp36DE. *Drosophila* sperm are very long (e.g., Pitnick et al. 1995) and cannot self-propel into storage. They form a sperm mass within the female's uterus. After mating, the uterus opens up in a series of shape changes that may push the sperm up toward the openings of the sperm storage organs (Adams and Wolfner 2007); these morphological changes open the entries to the sperm storage organs as well. Acp36DE, which is found in the anterior mating plug (Bertram et al. 1996), is necessary for completion of these uterine shape changes (Avila and Wolfner 2009); without it the uterus opens only partially, thus preventing the majority of sperm from being stored. How these Acp36DE-associated uterine shape changes come about is not known, but it is likely to involve some of the female mechanisms (e.g., neural, endocrine, or morphological) that were described in Sect. 13.2.

No seminal protein has yet been associated with the maintenance of sperm in storage. However, several seminal proteins have been identified through genetic studies to be necessary for efficient release of sperm from storage (Ravi Ram and Wolfner 2009; Avila et al. 2010; LaFlamme et al. 2012; Findlay et al. 2014). These proteins function in a network. Eight seminal proteins, including four predicted proteases or protease homologs (which could act to inhibit proteolysis), two predicted lectins and two cysteine-rich secretory proteins (CRISPs) from the seminal fluid, act in a pathway that binds a ninth seminal protein, the 36-amino acid peptide "sex peptide" (SP) to sperm. By being bound to sperm, SP can be retained within the mated female (Peng et al. 2005). Its C-terminal active region is gradually released from sperm by proteolytic cleavage (Peng et al. 2005). If SP is not bound to sperm, then sperm are released more slowly than normal (Avila et al. 2010). Thus, the male provides proteins that are associated with the rate of release of his sperm from the female's sperm storage organs. By binding SP to sperm, so that its active C-terminal region is only slowly released, the male network proteins cause the SP's effects to last for a long time (~10 days of post-mating, or as long as the female still contains sperm in storage; see below for more associations between SP and female post-mating responses; Peng et al. 2005). It is not known whether it is the male or the female who provides the protease that releases the SP C-terminal piece from sperm.

13.3.2 Nutrition, Egg Production, and Release

Drosophila melanogaster seminal proteins from the male also alter feeding and excretory behavior in mated females, thus affecting nutritional status and steps in egg production. For example, the SP just mentioned above increases feeding in females (Carvalho et al. 2006) and changes their excretory behavior, slowing excretion, and changing the density of the excreta (Cognigni et al. 2011; Apger-McGlaughon and Wolfner 2013). Details of how SP action causes this effect are not clear, but its sperm binding and release through the pathway noted above are important for this action.

In addition to the effects of Sfps on feeding, at least two seminal proteins (plus the network noted in Sect. 13.3.1) are required for the changes in egg production to occur (Chen et al. 1988; Aigaki et al. 1991; Herndon and Wolfner 1995; Heifetz et al. 2000; Chapman et al. 2001, 2003; Liu and Kubli 2003). SP stimulates egg production, apparently by stimulating oogenesis to proceed beyond an early vitellogenic stage (Soller et al. 1997). The mechanisms underlying this effect are not yet fully understood but there are two, not mutually exclusive, pathways. The first pathway involves SP binding to receptors, one of which is a G-protein-coupled receptor (SPR, which is also a receptor for myoinhibitory peptides (MIPs); Yapici et al. 2008; Kim et al. 2010; Poels et al. 2010; Isaac et al. 2014; see also Haussmann et al. 2013) and then acting through the nervous system to stimulate egg production (Ottiger et al. 2000; Häsemeyer et al. 2009; Yang et al. 2009). Although we do not know how SP binds to its receptors, we do know that the portion of SP that stimulates egg production is the C-terminal region that is released from sperm-bound SP and that also modulates sperm release and feeding/excretion behavior (Peng et al. 2005). The second pathway by which SP could affect egg production derives from the fact that the SP can increase production of a particular form of Juvenile Hormone (JHBIII) (Bontonou et al. (2015); Moshitzky et al. 1996). JH stimulates oogenesis (Kelly et al. 1987), so the increase in JHBIII occasioned by SP can potentially be involved in the increase in oogenic rate after mating. The portion of SP that induces the increase in JHBIII is its N-terminal region (Moshitzky et al. 1996). Since that region is not released from sperm (Peng et al. 2005) and thus is only available during the few hours post-mating that SP that is not bound to sperm remains in the female (Pilpel et al. 2008), any direct effects of SP on JHBIII levels are likely to be short term.

Another seminal protein, ovulin, is also essential in the egg production process (Herndon and Wolfner 1995; Heifetz et al. 2000, 2005; Rubinstein and Wolfner 2013). This 264 amino acid prohormone-like protein is associated with a stimulation of ovulation in mated females (Monsma et al. 1988; Heifetz et al. 2000; Rubinstein and Wolfner 2013). Ovulin, which does not bind to sperm, is only detectable in females for a few hours after mating (Monsma et al. 1990). Thus, its action is thought to be to cause ovulation of mature eggs that were already present in the female when she mated, thereby also allowing the increase in oogenesis that is mediated by SP (Chapman et al. 2001). The association between ovulin and ovulation is mediated through octopaminergic signaling in the female (Rubinstein and Wolfner 2013). The presence of ovulin increases the synaptic strength of octopaminergic neurons that innervate the reproductive tract. Increased octopaminergic signaling results in relaxation of oviduct musculature (and contraction of muscles around the ovaries) for ovulation.

13.3.3 Remating Patterns

Changes in a female's remating propensity are triggered by a number of components that she receives from her mate in *Drosophila*. First, a contact pheromone (7-tricosene) that rubs off from the male appears to make the female less attractive to other males (Ferveur and Sureau 1996). Second, a mating plug protein (PEB-2) is also associated with lower remating propensity (Bretman et al. 2010), although the mechanism for its effect is unknown. The association between these male donations and female attractiveness and remating is short lived (hours). Yet decreased female probability of remating persists for many days (as long as she still is storing sperm, thus ~10 days). This decreased remating is also mediated by the SP's C-terminal region (Peng et al. 2005), again acting through reproductive tract neurons, the SP receptor, and potentially other receptors (Yapici et al. 2008; Häsemeyer et al. 2009; Yang et al. 2009; Haussmann et al. 2013; Rezával et al. 2014).

13.4 Female Use of Sfps for Exerting CFC

The interface of male reproductive molecules such as Sfps with female molecules and pathways can provide opportunities for the female to exert choice by controlling the action of Sfps. At present, we are not aware of direct demonstration of such female use of Sfps, but we present several examples in which such a scenario could be operating.

13.4.1 Regulating Duration and Rate of Sfp Effects

13.4.1.1 Ejaculate Ejection

One mechanism by which females could control the duration and strength of Sfp effects is by ejecting Sfps and sperm before they reach their targets. As described in Sect. 13.2, ejaculate expulsion has been reported in several species of Diptera. In the species for which this process has been described so far, the expelled ejaculate is a sperm-containing droplet or mass (Bonduriansky et al. 2005; Brunel and Rull 2010; Manier et al. 2010, 2013b). The non-sperm components that are expelled along with the sperm have not yet been reported. However, it is likely that Sfps also are within the ejected mass; this can be tested by using anti-Sfp antibodies (e.g., Ravi Ram et al. 2005). If this is the case, females could modulate the effects of Sfps by adjusting the timing of ejaculate expulsion.

In addition to controlling the strength and duration of Sfps effects, ejaculate expulsion provides females with the opportunity to consume the ejaculate. It is possible that females could redirect the action of Sfps to their own advantage by expelling them from their reproductive tract and re-targeting their effects through ingestion. Ejaculate consumption has been reported in two families of Diptera (Piophilidae and Ulidiidae; Bonduriansky et al. 2005; Brunel and Rull 2010; Rodriguez-Enriquez et al. 2013). In a carrion fly species, *Prochyliza xanthostoma*, one effect of ejaculate consumption is oviposition stimulation, but only in well-fed females, suggesting that the effect is not mediated through provision of nutrients (Bonduriansky et al. 2005). If this is an Sfp-mediated effect, females of this species could presumably modulate the level and timing of oviposition stimulation by adjusting ejaculate consumption.

13.4.1.2 Processing of Ovulin and Acp36DE

Females could also modulate the duration and rate of Sfp effects by affecting Sfp processing. The seminal proteins ovulin and Acp36DE (which regulate ovulation and sperm entry into storage, respectively) enter the female as full-length proteins but rapidly undergo proteolytic processing within the female's reproductive tract. The purpose of this processing is as yet unclear. There are some hints from ectopic expression studies that the processing might be releasing more-active sub-regions of the Sfps from less-active precursors (Heifetz et al. 2005; Avila and Wolfner, in preparation). However, it is also possible that the processing initiates degradation of these Sfps.

Ovulin and Acp36DE are processed by a shared proteolytic pathway (Park and Wolfner 1995; Ravi Ram et al. 2006; LaFlamme et al. 2012, 2014). The pathway initiates in the male during mating, with the activation of a trypsin-family protease produced in the male accessory glands (the same tissue that makes ovulin and Acp36DE). The activated trypsin then activates an accessory-gland produced astacin-family metalloprotease, while the proteins are moving through the male. However, ovulin and Acp36DE do not get processed until they are within the female. In males, ejaculates were isolated in the absence of females, the first steps of ovulin processing occur, but the processing cannot complete. Acp36DE's single proteolytic cleavage does not occur in these isolated ejaculates. Thus, complete processing of at least these two Sfps requires contributions from the female as well as the male; the nature of the female's contributions is currently unknown. These results are interesting to consider from a CFC perspective. Whether or not the cleavage of ovulin and Acp36DE activates their action or accelerates their degradation, the essential role of female components in it suggests that the female can control its rate—thus controlling the rate with which her ovulation or sperm storage is stimulated (or limiting the extent of that stimulation). It will be fascinating to identify the female components and to determine their roles and evolutionary characteristics.

13.4.1.3 Degradation and Proteolytic Cleavage of SP

Analogous to the case of ovulin and Acp36DE processing, females could influence the rate and duration of post-mating responses by affecting the proteolytic cleavages that control SP's stability or that release it from sperm. Some of the SP that enters the female during mating enters her circulatory system (Pilpel et al. 2008), from where it is capable of influencing her egg production and receptivity behaviors (Aigaki et al. 1991). This fraction of the SP that is in circulation is degraded within hours, by proteases in the female's hemolymph (Pilpel et al. 2008). Analogous to the arguments made above for ovulin and Acp36DE processing, one can imagine opportunities for CFC in terms of the female's rate of processing SP. For example, the post-mating increase in JHBIII is induced by the N-terminal portion of SP (Moshitzky et al. 1996), which is only available while there is SP in circulation (Peng et al. 2005). Therefore, if a female increases the degradation rate of circulating SP, she can limit how long her JHBIII titers will be induced by the male post-mating.

As described in Sect. 13.3, some of the SP's effects on females are prolonged because a fraction of the transferred SP binds to stored sperm and is thereby stably retained for days within the female. A protease cleaves this bound SP at a trypsin site near SP's N-terminus, releasing SP's C-terminal portion, which is then available to modulate the female's feeding, egg production, receptivity, and release of sperm from storage (Peng et al. 2005; Avila et al. 2010; Apger-McGlaughon and Wolfner 2013). The source of the protease that cleaves SP is not known. If the protease is made by the female, this would provide an opportunity for her to exert CFC: rapid cleavage could release large amounts of SP quickly, causing high short-term stimulation of egg production and sperm release and limiting long-term responses such as decreased remating and sustained increased egg production. Alternatively, slower proteolysis could result in slower rates of egg production and sperm release, but longer persistence of effects on remating and sustained egg production. Therefore, by regulating the rate of SP cleavage from sperm, females could potentially control the rate and duration of several SP-mediated effects. Moreover, even if the protease is provided by the male, protease inhibitors in the female's sperm storage organs (Allen and Spradling 2008; Prokupek et al. 2008, 2009) could potentially regulate its activity and thus the rate of release of SP.

13.4.1.4 Evolution of a Predicted Proteolysis-Regulating Protein Family

The possibility of a female controlling the proteolysis rate of male-derived proteins has the potential to result in a cascade of evolutionary change. For example, the evolutionary dynamics of a serine-type protease family in *Drosophila* could be consistent with such a model (Sirot et al. 2014). The *D. melanogaster* genome encodes three closely related serine-type proteases—two expressed in the female reproductive tract and the third in the male. Evolutionary studies indicate that the ancestral state was to have a single female-expressed gene. This gene duplicated twice, and one of the duplicates evolved expression limited to the male accessory glands in a subset of *Drosophila* species including *D. melanogaster*. The protein product of this gene is a Sfp. Interestingly, the two female proteins promote female remating. In contrast, transfer of the male protein tends to reduce female remating. One could hypothesize that the original gene was able to control remating rates (possibly through control of processing of reproductive proteins), and that the male co-option of the later duplicate allowed males to interfere with this mechanism of female control, suggesting an intersexual arms race.

13.4.2 Modulating Sensitivity to Sfps

Females may be able to modulate their response to Sfps by adjusting the expression or sensitivity of Sfp receptors. Such changes to the *D. melanogaster* SP receptor, for example, could result in modulation of many SP-related processes, including egg production, remating, and release of sperm from storage (e.g., Yapici et al. 2008; Avila et al. 2015). The fact that Sfps (and mating) induce changes in levels and release of female neuromodulators (Heifetz et al. 2014) and that both SP and ovulin act through pre-existing neuromodulator pathways in the female also potentially provides opportunity for CFC (Rubinstein and Wolfner 2013, 2014; Rezával et al. 2014). If a female could modulate her response to Sfps for example, by controlling the rate of octopamine vesicle release, or of development of increased synaptic strength post-mating (Kapelnikov et al. 2008; Rubinstein and Wolfner 2013), that might allow her to adjust her responses to the Sfps from her mate, depending on the male with whom she mated.

13.5 Molecular Signatures of a Battle for Control

In the preceding sections, we reviewed recent evidence from Diptera for female and male mechanisms of controlling female reproductive processes that affect MPCRS. The perspective taken in these sections was that CFC is part of an intersexual battle for control over these processes. Eberhard (1996) argued that several characteristics of Sfps are expected consequences of this battle and thus of CFC. These characteristics include: (i) the ability of Sfps to leave the female's reproductive tract and move through the circulatory system to other parts of the body; (ii) redundancy of Sfp function either with other Sfps or with other male-mediated processes (e.g., physical stimulation from copulation); (iii) graded female response based on Sfp quantity;

(iv) interspecific divergence in Sfp sequences; (v) stimulation of female responses that increase the reproductive success of the Sfp-supplying male; and (vi) mimicry of female messenger molecules by Sfps. Below, we offer new perspectives on a subset of these characteristics using evidence published since the release of Eberhard's book. In the years since Eberhard proposed this hypothesis, sequence-based, molecular, and genetic studies have provided evidence for each of these characteristics.

13.5.1 Ability of Sfps to Enter the Female Circulatory System

In Diptera, some Sfps are found in female hemolymph or in tissues outside of the reproductive tract (e.g., Monsma et al. 1990; Pitnick et al. 1997; Lung and Wolfner 1999; Ravi Ram et al. 2005; Pilpel et al. 2008) or are suggested to enter the hemolymph based on radiotracer studies (e.g., Markow and Ankney 1984; Radhakrishnan et al. 2008) or on the occurrence of traumatic insemination (e.g., Kamimura 2007). In *D. melanogaster*, introduction of SP into the hemolymph has been shown to be sufficient to induce changes in egg production and receptivity (Chen et al. 1988; Aigaki et al. 1991) and two other Sfps with identified effects on female physiology or behavior (ovulin: Herndon and Wolfner 1995; Heifetz et al. 2005; Acp62F: Mueller et al. 2008) have been found in the hemolymph of mated female flies. The presence of these latter two Sfps in the hemolymph has been traced to their movement through the intima of the posterior wall of the vagina (Monsma et al. 1990; Lung and Wolfner 1999). Interestingly, this area of the vagina is unusual in that it is devoid of circular muscles, which may improve its permeability. Further, its permeability to Sfps appears to be transient and selective—certain Sfps move into the hemolymph, whereas others, even ones in the same size range, are detectable only in the reproductive tract (Lung and Wolfner 1999; Ravi Ram et al. 2005). But, is it in the male's or the female's interest (or both) that these proteins enter her circulation? At this point, the answer to this question is not known. This process could be an example of the invasiveness of Sfps proposed by Eberhard (1996). However, as he points out, invasiveness is difficult to distinguish from exploitation of Sfps by females, which is suggested in this case by the selective and temporary passage of particular Sfps. Further complicating matters, the evolution of invasiveness may have been followed by the evolution of exploitation (Eberhard 1996). In any case, it is clear that some Sfps in *D. melanogaster* can access the hemolymph, as predicted by the Eberhard model.

13.5.2 Redundancy of Sfp Function

Redundancy of Sfp function either with other Sfps or with other male-mediated processes (e.g., physical stimulation from copulation or contact pheromones) is an expected signature of an evolutionary battle between males and females for control over female reproduction (Eberhard 1996). There is good evidence for the possibility of redundancy among Sfps. For example, >15 % of *D. melanogaster* Sfps appear to be or have gene-duplicates (e.g., Findlay et al. 2008) that might be expected to have redundant functions; an example is the predicted *D. melanogaster* seminal lectins CG1656 and CG1652 (e.g., Ravi Ram and Wolfner 2009). Beyond just this species, seminal proteins in *Drosophila* and other animals often include multiple members of the same protein family. For example, this is the case with trypsins and other proteases (reviewed in LaFlamme and Wolfner 2013). In addition to expected functional redundancy based on sequence comparisons, detailed analyses of Sfp function in *D. melanogaster* have revealed that seminal proteins may contribute to stimulating the same MCPRS without being truly (chemically) redundant. For example, receipt of ovulin and SP increases egg production, but these two Sfps act at different stages in the egg production process (see Sect. 13.3; Herndon and Wolfner 1995; Heifetz et al. 2000; Chapman et al. 2003; Liu and Kubli 2003). That multiple Sfps could act in the same pathway, but at different stages (thus, not being technically redundant), is also a predicted pattern resulting from ongoing conflict between males and females over control. We see several such cases in *D. melanogaster*. The localization or apparent activation of particular Sfps can require the involvement of multiple other Sfps, but at different steps in the process. For example, as described in Sect. 13.3, binding of SP to sperm requires eight other Sfps and the proteolytic processing of ovulin requires two other Sfps (Ravi Ram et al. 2006; Ravi Ram and Wolfner 2007b; LaFlamme et al. 2012, 2014; Findlay et al. 2014). It is possible that the involvement of each new Sfp in promoting the action of another Sfp (in these cases, SP or ovulin) is an evolutionary response to a corresponding step in which females regained control over reproduction. Consistent with this hypothesis, evolutionary analyses of the genes in the SP-binding pathway show that the timing of origin of the most recent additions corresponds with a change from more frequent to less frequent remating along the *Drosophila* phylogeny (Findlay et al. 2014). Notably, involvement of multiple Sfps in the pathway of a single Sfp is not an expected outcome of female exploitation of Sfps for triggering reproductive processes. Finally, Sfps can be redundant with other male contributions, such as the case described above in Sect. 13.3, in which a cuticular pheromone and two protein components of seminal fluid (PEB-2 and SP) all decrease female attractiveness to males, although on different timescales (Ferveur and Sureau 1996; Chapman et al. 2003; Liu and Kubli 2003; Bretman et al. 2010). Thus, sequence and functional analyses in *D. melanogaster* have identified multiple instances of redundancy (or similarity) of Sfp function, consistent with Eberhard's predictions.

13.5.3 Interspecific Divergence in Sfp Sequences

Post-copulatory sexual selection is likely to result in rapid interspecific divergence of Sfps since different mutations will arise and be favored under different conditions (Eberhard and Cordero 1995; Eberhard 1996). There is ample evidence that, across a wide range of taxa, Sfps do tend to diverge more quickly on averge than non-Sfp genes (e.g., Aguadé 1998, 1999; Begun et al. 2000; Swanson et al. 2001; Haerty et al. 2007; Almeida and DeSalle 2008; Mancini et al. 2011; Wong et al. 2012; Wong and Rundle 2013; Boes et al. 2014). However, this pattern is consistent not only with CFC and sexually antagonistic coevolution, but also with male–male competition and other forms of selection such as host–pathogen interactions (Clark et al. 2006). We recently reviewed evidence for this and other predicted evolutionary patterns that would be predicted if Sfps are evolving through conflict over control of reproduction (Sirot et al. 2014). One pattern of particular interest from the perspective of CFC is the predicted outcome of correlated rates of evolution of male and female reproductive molecules. If males and females are competing for control over female reproductive processes, we would expect not only rapid and divergent evolution between species, but also correlated rates of evolution between male and female proteins that interact or affect the same process (Clark et al. 2009). An example that supports this prediction comes from *D. melanogaster* in which the male and female proteins that affect the long-term post-mating response pathway have correlated rates of evolution (Findlay et al. 2014). In sum, although there is evidence that Sfps are rapidly evolving, it is difficult to be certain whether that is driven by CFC.

13.5.4 Mimicry of Female Messenger Molecules

Eberhard (1996) argued that similarities between male seminal products and messenger molecules produced by the female are the predicted consequences of selection for males' ability to overcome female control of reproduction-related processes. Such patterns are also predicted as a result of selection for females to exploit the male ejaculate for costly products that stimulate the female's reproductive processes. There is increasing evidence that such similarities do exist and come about in at least two different ways. In some instances, both males and females produce the same proteins—transferred as Sfps in the case of the male. For example, *Aedes albopictus* males transfer the neuropeptide adipokinetic hormone-1 (AKH-1) to females during mating (Boes et al. 2014). AKH-1 is also produced by female mosquitoes and mobilizes the release of carbohydrates from the fat body into the hemolymph in some species (Kaufmann and Brown 2008; Kaufmann et al. 2009). In a similar example, *D. melanogaster* male and female reproductive organs synthesize and secrete glucose dehydrogenase (GLD). Genetic experiments suggest that this enzyme is important for sperm storage and utilization, and that GLD contributions from both male and female are important in fertility (Iida and Cavener 2004). In addition to proteins, males also transfer non-proteinaceous messenger molecules, such as hormones, that are also produced by the female. For example, in the mosquito *Aedes aegypti*, males transfer juvenile hormone (Clifton et al. 2014) and, in *Anopheles gambiae*, males transfer 20-hydroxyecdysone (Baldini et al. 2013), both of which are known to play critically important roles in females including promoting egg development. Interestingly, in the case of *An. gambiae*, the pathways by which male- and female-derived 20-hydroxyecdysones affect egg development appear to be at least partially independent, with the male-derived hormone acting through interactions with a mating-induced protein produced in the female reproductive tract (Baldini et al. 2013).

In addition to males transferring the same proteins that are synthesized by females, mimicry of female messenger models could involve male transfer of proteins that are similar but not identical to those produced by females. This appears to be the case for the family of closely related serine-like protease proteins discussed in Sect. 13.4.1.4 in which two members of the family are expressed exclusively or primarily in the female reproductive tract, whereas the third member is expressed exclusively in the male accessory glands and is a Sfp. Evolutionary and gene expression analyses are consistent with the hypothesis that this Sfp evolved through "co-option" of a duplicate of an ancestral female protein (Sirot et al. 2014). Therefore, since Eberhard (1996), there is even more evidence for ejaculate products that mimic female messenger molecules; however, there is no direct evidence thus far of whether those products are maintained through selection on males to circumvent female control mechanisms.

13.6 CFC on Sfps

In the previous sections, we considered Sfps as a male-derived mechanism for overcoming CFC. Another manner in which CFC and Sfps could interact is if females are using Sfps as criteria for biasing sperm use toward particular mates. In this scenario, CFC would be acting directly on Sfps. If CFC is acting on Sfps, there would need to be variation between males in the quantity or quality (e.g., primary sequence) of Sfps transferred to females. Further, there would be an association between this variation and differences in MPCRS. Finally, for CFC to result in the evolution of Sfps, some of the variation in Sfp quantity or quality must be heritable. Below, we evaluate the evidence relevant to these predictions for Sfp quantity and quality separately and then consider how CFC might be differentiated from other forms of selection.

13.6.1 Sexual Selection on Sfp Quantity

The prediction that there are consistent differences between males in the quantity of Sfps transferred to females has not been tested directly, but there is evidence to suggest that this might be the case. Both Sfp gene expression and protein levels (measured directly or by using accessory gland size as a proxy) in the male accessory glands can vary among males raised under the same conditions in *Drosophila* and in *Aedes* mosquitoes (Smith et al. 2009; Wigby et al. 2009; Ayroles et al. 2011; Fedorka et al. 2011; Alfonso-Parra et al. 2014; LaFlamme and Wolfner, unpublished data). Further, males from different *D. melanogaster* genetic isolines (derived from natural populations) vary consistently in the amount of Sfps transcribed or of proteins transferred to females (Fiumera et al. 2005; Smith et al. 2009; Smith et al., unpublished data). Variation in Sfp quantity also appears to have a heritable component, at least in *D. melanogaster*. Evidence for this assertion comes from a study in which male accessory glands responded to selection for large size and showed a concomitant increase in the production of both SP and ovulin (Wigby et al. 2009). As the Sfps of more dipteran species are identified and techniques for quantifying proteins (e.g., Bantscheff et al. 2007, 2012) become more affordable and accessible, it will be important to directly test the prediction of consistent differences between males in the amount of Sfps transferred.

It is interesting to note that, in addition to the inter-male variation in Sfp quantity, there was also a similar scale of variation between multiple ejaculates of the same male when mated in rapid succession to different females (Smith et al., unpublished data). Further, the quantity of Sfp transferred to females can vary depending on a number of extrinsic factors such as social environment, female size, and female mating status (Wigby et al. 2009; Sirot et al. 2011; Wigby et al., in preparation). Therefore, in addition to males varying in their basal amount of Sfps transferred, the quantity of transferred Sfps can vary across time with mating experience, social experience, and mate qualities.

As described in Sect. 13.3, the importance of Sfps to MPCRS has been established in *D. melanogaster* through studies in which individual Sfps have been either eliminated (or knocked down) in males or added to virgin females through ectopic expression or injection. Genetic variation is likely to result in more subtle differences in Sfp amount. Effects of Sfps on MPCRS also are detectable at this level in Diptera such as mosquitoes and tephritid flies, in experiments in which virgin females were injected with different amounts of male accessory gland homogenates (reviewed in Eberhard 1996; see also Jang 1995; Radhakrishnan and Taylor 2007; Helinski et al. 2012a). However, the relationship between Sfp dose and female post-mating responses associated with MPCRS does not increase linearly and then reach an asymptote in all cases, as had been previously predicted (Eberhard 1996). For example, in some dipteran species, females injected with intermediate amounts of AG homogenate (within the predicted range of normal transfer) respond most similarly to mated females, whereas females injected with low or high amounts respond more like virgin females (Jang 1995; Radhakrishnan and Taylor 2007). Together, the evidence from all-or-none and dose-dependent injection studies suggests that variation in Sfp quantity is positively associated with variation in MCPRS, but not at high levels that are likely to be outside of the natural range of Sfp transfer.

Studies of associations between MPCRS and Sfp quantity in transferred ejaculates (as opposed to injected ones) can also be used to evaluate the evidence for dose dependency. In *D. melanogaster*, artificial selection on male accessory gland size resulted in larger accessory glands that produced higher quantities of the two Sfps measured (SP and ovulin) than the accessory glands of control line males (Wigby et al. 2009). When placed with competitor males, large-AG males sired relatively more offspring than control line males. Although Sfp transfer and MPCRS could not be measured in the same females (because females must be killed to measure Sfp transfer), the data in Wigby et al. (2009) suggest that variation in the quantity of Sfp produced is positively associated with variation in MPCRS. However, in a study of SP variation across different *D. melanogaster* isolines derived from a wild population, no association was found between isoline averages for quantity of SP transferred and subsequent number of eggs laid (Smith et al. 2012; see also Fiumera et al. 2005). The differences between the dose dependency related results of these two studies could be due to different methodologies or to the different female responses measured.

13.6.2 Sexual Selection on Sfp Quality

Sexual selection could also act on variation in Sfp quality. Such variation could arise due to polymorphism in the amino acid sequence of the proteins (or to production of different isoforms) or through post-translational modifications (e.g., glycosylation and phosphorylation) since such differences would result in the integration of different resources into the protein. Consistent with the hypothesis that sexual selection could be acting on variation in Sfp quality, Sfps have high levels of amino acid polymorphisms (i.e., higher than non-Sfps) in several *Drosophila* species (e.g., Clark et al. 1995; Begun et al. 2000; Tsaur et al. 2001; Kern et al. 2004; Schully and Hellberg 2006). Some of these polymorphisms have been associated with the variation in measures of MPCRS (Clark et al. 1995; Fiumera et al. 2005, 2006, 2007; Chow et al. 2010; Zhang et al. 2013). For example, a single substitution of isoleucine for serine in *D. melanogaster* ovulin is associated with reduced levels of P2 (measured as proportion of progeny sired by the second of two males to mate with a female; Fiumera et al. 2005) and a single substitution of alanine for threonine in another Sfp (Mst57Dc) in this species is associated with differences in female fecundity (Fiumera et al. 2007). Curiously, associations have also been found between MPCRS and SNPs in Sfp genes that do not change the amino acid sequence of the protein (Fiumera et al. 2005, 2006, 2007; Chow et al. 2010). These associations could be due to effects of the polymorphisms on transcription rates or mRNA stability or an association with a polymorphism in another part of the same gene, a nearby gene, or a gene in linkage disequilibrium with the investigated gene. Further studies are needed to determine the causal relationships, if any, between variation in Sfp sequences and variation in MPCRS.

Variations in amino acid sequence could affect the cost of Sfp synthesis through differential needs for raw materials and or production mechanics (Bragg and Wagner 2009; Smith and Chapman 2010) and males could vary in their ability to withstand these costs (e.g., Smith and Greig 2010). Essential amino acids that are only derived through food sources and are limited in the environment could be costly to acquire and thus be a rate-limiting step for translation of particular Sfps. Their use for high levels of Sfp production could also limit their use in production of other proteins. Costs based on amino acid composition are not limited to those incurred in obtaining essential amino acids: There could be differences in the amount of energy it takes for an animal to produce different amino acids. For example, in bacteria, the number of ATPs required for synthesis of tryptophan is over six times higher than that required for the synthesis of glycine (Akashi and Gojobori 2002). Further, amino acid composition can affect translation rate, protein stability, and degradation rate (Yutani et al. 1977, 1987). Variation in amino acid sequence can also result in different patterns of post-translational modifications. Together, these variations could result in proteins of differing quality from the same or very similar DNA sequences. Males might then opt to produce different qualities of Sfps based on their condition or their ability to acquire limited resources for protein synthesis.

13.6.3 Predicted Evolutionary Patterns of CFC on Sfps

If females are exerting CFC on Sfp quantity, we would expect there to be selection on males to allocate more (or better) Sfps to females in situations in which there is actual or perceived competition from other males for fertilization of the female's eggs. Consistent with this prediction, *D. melanogaster* males transfer higher quantities of both ovulin and SP when in the presence of other males (Wigby et al. 2009). Yet, when *D. melanogaster* males mate with a recently mated female, they actually reduce the amount of ovulin transferred (Sirot et al. 2011). This pattern is not expected if CFC selects for greater Sfp transfer, but is expected if the male is exploiting his predecessor's ovulin (Hodgson and Hosken 2006). It will be important to use both modeling and empirical approaches in the future to make and test predictions, respectively, about expected Sfp allocation patterns if females are discriminating among males based on Sfp quantity.

In the case of Sfp quality, we might expect different patterns of Sfp evolution depending on the consistency of female choice. The relative quality of different Sfp variants may be consistent for all females if, for example, there is a particular essential amino acid that is always scarce in the environment and Sfp variants differ in the inclusion of this amino acid. In this case, females may favor sperm of males with Sfp variants that are richer in this amino acid, as an indicator of a heritable trait such as foraging ability (Debelle et al. 1989; Kent et al. 2009). We would expect CFC to result in strong directional selection for a particular Sfp variant. Such selection should result in selective sweeps leading to the fixation of particular variants within a population, unless the gene has pleiotropic effects or is in linkage disequilibrium with other genes of important effect. Evidence from interspecific comparisons across closely related *Drosophila* species suggests that such selective sweeps are common in Sfp genes (e.g., Aguadé 1998, 1999; Begun et al. 2000; Wagstaff and Begun 2005, 2007; Almeida and DeSalle 2008; Findlay et al. 2008, 2009; Wong et al. 2012; Wong and Rundle 2013).

In contrast, if assessment of relative quality of Sfp variants depended on variable female condition and/or genotype, we would expect high frequencies of within-population polymorphisms in Sfps. For example, particular combinations of Sfp and interacting female-derived protein variants (e.g., receptors or members of a protein complex) could be more compatible than others. More compatible combinations may result in more pronounced female post-mating responses which, in turn, could bias paternity patterns toward males possessing particular variants. Indeed, male competitive fertilization success (P1 and P2) varies among different females in *D. melanogaster* (Clark et al. 1999; Bjork et al. 2007). The hypothesis that Sfp variant quality depends on female genotype is further supported by the finding that of an association between a male's P1 (the proportion of offspring sired by the first male to mate when two males mated with the same female) and his SP genotype depends on the SPR genotype of his mate (Clark et al. 1999; Chow et al. 2010). Further, the finding of high levels of Sfp polymorphisms is consistent with the hypothesis that CFC acts on Sfp quality but that females vary in their assessment of quality.

13.6.4 Differentiating CFC on Sfps from Other Forms of Selection

In some cases, selection acting through female choice can be difficult to distinguish from selection acting through male–male competition or sexually antagonistic coevolution (e.g., Eberhard 1996, 2015, Chap. 1). This can be true for pre-mating processes, but can be especially problematic for processes that occur during or after mating because those processes are more difficult to observe (Birkhead 1998). The criteria for demonstrating CFC have been the subject of debate (Birkhead 2000; Eberhard 2000; Kempenaers et al. 2000; Pitnick and Brown 2000). There is consensus that mechanisms must exist by which females could bias the paternity of their mates (e.g., sperm management, control over timing of egg production, and remating). As described in Sect. 13.2 and elsewhere (Eberhard 1996), there is strong evidence that female Diptera have such mechanisms. A second set of criteria that has been proposed for inferring CFC is demonstrable variation in a measure of MPCRS that is attributable to male genotype, female genotype, and an interaction between male and female genotypes (Birkhead 1998, 2000; Pitnick and Brown 2000; but see Eberhard 2000; Kempenaers et al. 2000). Although this set of criteria is consistent with CFC, it does not definitively distinguish CFC from male–male competition. Nevertheless, in *D. melanogaster*, these criteria are met by several measures of MPCRS including the number of offspring produced, female remating rate, P1 and P2 (Clark et al. 1995, 1999; Clark and Begun 1998; Bjork et al. 2007; Chow et al. 2010).

Further, there is evidence to suggest that the associations between these measures and the genotypes of the interacting flies could be mediated through variation in SP and SPR. An association study found that there are several DNA sequence polymorphisms in both genes that are associated with P1, female remating rate, and offspring production (Chow et al. 2010). Further, there are significant SP × SPR (i.e., male × female) interaction effects on P1 (Chow et al. 2010). Together, the evidence from studies of *Drosophila* suggests that the requirements are met for CFC to be acting on male Sfps, although much of this evidence is also consistent with other potential mechanisms of selection.

13.7 Conclusions and Future Directions

In conclusion, studies from a range of species have demonstrated that female Diptera have mechanisms for controlling processes that affect MPCRS (e.g., sperm storage and release) and that could be used to exert CFC. Further, it is clear from data for *D. melanogaster*, and suggested by evidence from other species, that male Diptera also have mechanisms for influencing female processes that affect the male's reproductive success; the Sfps on which we focused in this article are one such mechanism. However, what is still unknown is whether Sfps are purely a mechanism of male manipulation by which males co-opt female pathways to affect female physiology and behavior, or whether Sfps also provide a resource that females can exploit (beyond just the amino acids that are provided by the male) to adjust their own reproductive processes to exert CFC or as criteria on which to base their CFC. In other words, "who's zoomin' who" and how can we know?

One approach for dissecting the nature of potential interactions between Sfps and CFC is to test the effects of manipulating female perception of male phenotype on the fate of Sfps and of Sfp-mediated changes in the females (e.g., see Edvardsson and Arnqvist (2000) for this type of study on CFC and copulatory courtship). To conduct such investigations, one would first need to find a manipulable male phenotypic trait that is associated with a Sfp-mediated process. One would then compare the fate of Sfps and the Sfp-mediated process in females that mated with control males versus with males that had a manipulated version of the trait. If the fate of Sfps or Sfp-mediated processes in females mated to these two types of male differed, that would support the hypothesis that females are using Sfps as tools for affecting their own reproductive processes.

As a hypothetical example, let us use sex combs (bristle-like structures found on the forelegs of males) in *Drosophila bipectinata*. In this species, P2 increases linearly with the size of the male's sex combs (Polak and Simmons 2009). For the sake of this example, let us assume that there is a "sex peptide" in *D. bipectinata* that acts similarly to the one described in *D. melanogaster* (Chen et al. 1988) and that impacts P2 (Chow et al. 2010). Then, a test of the CFC/Sfp interaction axis could involve comparing P2 in matings of females with control males versus genetically identical males whose sex comb bristles had been shortened through microscale laser ablation (Polak and Rashed 2010). Alternatively, one could compare P2 between control females and females in which structures that sense the sex combs are blocked (see Eberhard 2015, Chap. 1). Differences between either of these two groups of females in P2 would suggest that the relationship between P2 and sex comb size is mediated, at least in part, by female control. If such differences are seen, further comparisons between the two types of mated female could assess whether the P2 differences reflected differences in fate of SP, expression of SPR, and rate of neuromodulator release. Findings of differences between the two groups of females in these measures would suggest that the female control is through female use of Sfps as tools for affecting their own reproductive processes. Before such studies can be conducted, however, we need to know (i) what aspects of MPCRS are mediated by Sfps and (ii) what manipulable male phenotypic traits are associated with Sfp-mediated aspects of MPCRS.

In this chapter, we also suggest that females could be basing CFC decisions on Sfp quantity or quality. This hypothesis is difficult to test as altering Sfp quantity or quality (e.g., amino acid sequence) could affect not only female preference for the male but also the strength of action of the Sfps. Further, since males produce many different Sfps (e.g., over 200 in *D. melanogaster*; Ravi Ram and Wolfner 2007a; Findlay et al. 2008, 2009; Avila et al. 2011), females could be basing choice on the quality or quantity of a subset of these proteins. Nevertheless, comparing MPCRS between males with experimentally altered Sfp amino acid sequences would allow us to test whether the associations between Sfp polymorphisms and MPCRS (described in Sect. 13.6) are due to the Sfp polymorphisms themselves or to associated changes in other proteins. The recent development of the CRISPR/Cas9 technique for genome editing (Jinek et al. 2012; see Carroll 2014 for review) and its applicability across a wide range of organisms including Diptera (e.g., Bassett et al. 2013; Gratz et al. 2013a, b) will make such experiments feasible to carry out. The CRISPR/Cas9 method allows deletion of any endogenous gene or its replacement with an allele of the experimenter's design. Therefore, using this method, it should be possible to generate dipteran males that produce different levels of a given Sfp (from zero, to levels above normal), or that express mutant forms of that Sfp, altered in specific amino acids of interest. Those males can then be used in tests of MPCRS.

The interactions seen today between males and females are also likely the consequence of an evolutionary to-and-fro in which a contribution or effect from one sex was responded to over evolutionary time by the other. This can mean that effects of CFC and of Sfps are inextricably combined by this point. Thus, it may be very difficult to tease apart which one precisely is acting in a given situation, or even whether or how each is involved in driving the rapid evolution of the primary sequences of some reproductive proteins. Our own view is that both male and female "manipulation" is probably going on, and that components originally selected to "zoom" one sex may, in fact, be co-opted by that sex to "zoom" right back.

Acknowledgments We are grateful to F. Avila, M. Bloch Qazi, A. Wong, and two anonymous reviewers for valuable comments on this chapter and to Anita Aisenberg and Alfredo Peretti for the opportunity to write it. We appreciate support from NIH grants R01-HD038921 (to MFW), R01-HD059060 (to MFW and A.G. Clark), R01-AI095491 (to MFW and L.C. Harrington), and R15-ES020051 (to A. Fiumera).

References

Adams EM, Wolfner MF (2007) Seminal proteins but not sperm induce morphological changes in the *Drosophila melanogaster* female reproductive tract during sperm storage. J Insect Physiol 53:319–331

Aguadé M (1998) Different forces drive the evolution of the Acp26Aa and Acp26Ab accessory gland genes in the *Drosophila melanogaster* species complex. Genetics 150:1079–1089

Aguadé M (1999) Positive selection drives the evolution of the Acp29AB accessory gland protein in *Drosophila*. Genetics 152:543–551

Aigaki T, Fleischmann I, Chen PS, Kubli E (1991) Ectopic expression of sex peptide alters reproductive behavior of female *Drosophila melanogaster*. Neuron 7:557–563

Akashi H, Gojobori T (2002) Metabolic efficiency and amino acid composition in the proteomes of *Escherichia coli* and *Bacillus subtilis*. Proc Natl Acad Sci USA 99:3695–3700

Alfonso-Parra C, Avila FW, Deewatthanawong P, Sirot LK, Wolfner MF, Harrington LC (2014) Synthesis, depletion and cell-type expression of a protein from the male accessory glands of the dengue vector mosquito *Aedes aegypti*. J Insect Physiol 70:117–124

Allen AK, Spradling AC (2008) The Sf1-related nuclear hormone receptor Hr39 regulates *Drosophila* female reproductive tract development and function. Development 135:311–321

Almeida FC, DeSalle R (2008) Evidence of adaptive evolution of accessory gland proteins in closely related species of the *Drosophila repleta* group. Mol Biol Evol 25:2043–2053

Anderson RC (1945) A study of the factors affecting fertility of lozenge females of *Drosophila melanogaster*. Genetics 30:280

Apger-McGlaughon JL, Wolfner MF (2013) Post-mating change in excretion by mated *Drosophila melanogaster* females is a long-term response that depends on sex peptide and sperm. J Insect Physiol 59:1024–1030

Arnqvist G, Danielsson I (1999) Copulatory behavior, genital morphology, and male fertilization success in water striders. Evolution 53:147–156

Arnqvist G, Nilsson T (2000) The evolution of polyandry: multiple mating and female fitness in insects. Anim Behav 60:145–164

Arthur BI, Hauschteck-Jungen E, Nothiger R, Ward PI (1998) A female nervous system is necessary for normal sperm storage in *Drosophila melanogaster*: a masculinized nervous system is as good as none. Proc R Soc London, Ser B 265:1749–1753

Arthur BI, Sbilordo SH, Pemberton AJ, Ward PI (2008) The anatomy of fertilization in the yellow dung fly *Scathophaga stercoraria*. J Morphol 269:630–637

Avila FW, Wolfner MF (2009) Acp36DE is required for uterine conformational changes in mated *Drosophila* females. Proc Natl Acad Sci USA 106:15796–15800

Avila FW, Mattei AL, Wolfner MF (2015) Sex peptide receptor is required for the release of stored sperm by mated *Drosophila melanogaster* females. J Insect Physiol 14(76):1–6. doi:10.1016/j.jinsphys.2015.03.006. [Epub ahead of print]

Avila FW, Ravi Ram K, Bloch Qazi MC, Wolfner MF (2015) Sex peptide is required for the efficient release of stored sperm in mated *Drosophila* females. Genetics 186:595–600

Avila FW, Sirot LK, LaFlamme BA, Rubinstein CD, Wolfner MF (2011) Insect seminal fluid proteins: identification and function. Annu Rev Entomol 56:21–40

Avila FW, Qazi MCB, Rubinstein CD, Wolfner MF (2012) A requirement for the neuromodulators octopamine and tyramine in *Drosophila melanogaster* female sperm storage. Proc Natl Acad Sci USA 109:4562–4567

Ayroles JF, Carbone MA, Stone EA, Jordan KW, Lyman RF, Magwire MM et al (2009) Systems genetics of complex traits in *Drosophila melanogaster*. Nat Genet 41:299–307

Ayroles JF, Laflamme BA, Stone EA, Wolfner MF, Mackay TFC (2011) Functional genome annotation of *Drosophila* seminal fluid proteins using transcriptional genetic networks. Genet Res 93:387–395

Baena ML, Eberhard WG (2007) Appearances deceive: female "resistance" behaviour in a sepsid fly is not a test of male ability to hold on. Ethol Ecol Evol 19:27–50

Baldini F, Gabrieli P, South A, Valim C, Mancini F, Catteruccia F (2013) The interaction between a sexually transferred steroid hormone and a female protein regulates oogenesis in the malaria mosquito *Anopheles gambiae*. PLoS Biol 11(10):e1001695

Bantscheff M, Schirle M, Sweetman G, Rick J, Kuster B (2007) Quantitative mass spectrometry in proteomics: a critical review. Anal Bioanal Chem 389:1017–1031

Bantscheff M, Lemeer S, Savitski MM, Kuster B (2012) Quantitative mass spectrometry in proteomics: critical review update from 2007 to the present. Anal Bioanal Chem 404:939–965

Barnes AI, Wigby S, Boone JM, Partridge L, Chapman T (2008) Feeding, fecundity and lifespan in female *Drosophila melanogaster*. Proc R Soc London, Ser B 275:1675–1683

Bartoszewski S, Luschnig S, Desjeux I, Grosshans J, Nüsslein-Volhard C (2004) *Drosophila* p24 homologues eclair and baiser are necessary for the activity of the maternally expressed Tkv receptor during early embryogenesis. Mech Dev 121:1259–1273

Bassett AR, Tibbit C, Ponting CP, Liu JL (2013) Highly efficient targeted mutagenesis of *Drosophila* with the CRISPR/Cas9 system. Cell Rep 4:220–228

Begun DJ, Whitley P, Todd BL, Waldrip-Dail HM, Clark AG (2000) Molecular population genetics of male accessory gland proteins in *Drosophila*. Genetics 156:1879–1888

Bentley MD, Day JF (1989) Chemical ecology and behavioral aspects of mosquito oviposition. Annu Rev Entomol 34:401–421

Bertram MJ, Neubaum DM, Wolfner MF (1996) Localization of the *Drosophila* male accessory gland protein Acp36DE in the mated female suggests a role in sperm storage. Insect Biochem Mol Biol 26:971–980

Birkhead TR (1998) Cryptic female choice: criteria for establishing female sperm choice. Evolution 52:1212–1218

Birkhead TR (2000) Defining and demonstrating postcopulatory female choice—again. Evolution 54:1057–1060

Bjork A, Starmer WT, Higginson DM, Rhodes CJ, Pitnick S (2007) Complex interactions with females and rival males limit the evolution of sperm offence and defence. Proc R Soc London, Ser B 274:1779–1788

Bloch Qazi MC (2003) A potential mechanism for cryptic female choice in a flour beetle. J Evol Biol 16:170–176

Bloch Qazi MC, Hogdal L (2010) Hold on: females modulate sperm depletion from storage sites in the fly *Drosophila melanogaster*. J Insect Physiol 56:1332–1340

Bloch Qazi MC, Wolfner MF (2003) An early role for the *Drosophila melanogaster* male seminal protein Acp36DE in female sperm storage. J Exp Biol 206:3521–3528

Bloch Qazi MC, Wolfner MF (2006) Emergence of sperm from female storage sites has egg-influenced and egg-independent phases in *Drosophila melanogaster*. Biol Lett 2:128–130

Bloch Qazi MC, Heifetz Y, Wolfner MF (2003) The developments between gametogenesis and fertilization: ovulation and female sperm storage in *Drosophila melanogaster*. Dev Biol 256:195–211

Boes KE, Ribeiro JMC, Wong A, Harrington LC, Wolfner MF, Sirot LK (2014) Identification and characterization of seminal fluid proteins in the Asian Tiger mosquito *Aedes albopictus*. PLoS Negl Trop Dis 8:e2946

Bonduriansky R, Wheeler J, Rowe L (2005) Ejaculate feeding and female fitness in the sexually dimorphic fly *Prochyliza xanthostoma* (Diptera: Piophilidae). Anim Behav 69:489–497

Bontonou G, Shaik HA, Denis B, Wicker-Thomas C (2015) Acp70A regulates *Drosophila* pheromones through juvenile hormone induction. Insect Biochem Mol Biol 56:36–49

Bragg JG, Wagner A (2009) Protein material costs: single atoms can make an evolutionary difference. Trends Genet 25:5–8

Bretman A, Lawniczak MKN, Boone J, Chapman T (2010) A mating plug protein reduces early female remating in *Drosophila melanogaster*. J Insect Physiol 56:107–113

Brunel O, Rull J (2010) The natural history and unusual mating behavior of *Euxesta bilimeki* (Diptera: Ulidiidae). Ann Entomol Soc Am 103:111–119

Carney GE, Taylor BJ (2003) Logjam encodes a predicted EMP24/GP25 protein that is required for *Drosophila* oviposition behavior. Genetics 164:173–186

Carroll D (2014) Genome engineering with targetable nucleases. Annu Rev Biochem 83:409–439

Carvalho GB, Kapahi P, Anderson DJ, Benzer S (2006) Allocrine modulation of feeding behavior by the sex peptide of *Drosophila*. Curr Biol 16:692–696

Chapman T, Herndon LA, Heifetz Y, Partridge L, Wolfner MF (2001) The Acp26Aa seminal fluid protein is a modulator of early egg hatchability in *Drosophila melanogaster*. Proc R Soc London, Ser B 268:1647–1654

Chapman T, Bangham J, Vinti G, Seifried B, Lung O, Wolfner MF et al (2003) The sex peptide of *Drosophila melanogaster*: female post-mating responses analyzed by using RNA interference. Proc Natl Acad Sci USA 100:9923–9928

Chen PS, Stumm-Zollinger E, Aigaki T, Balmer J, Bienz M, Böhlen P (1988) A male accessory gland peptide that regulates reproductive behavior of female *D. melanogaster*. Cell 54:291–298

Chow CY, Wolfner MF, Clark AG (2010) The genetic basis for male × female interactions underlying variation in reproductive phenotypes of *Drosophila*. Genetics 186:1355-U1451

Chow CY, Wolfner MF, Clark AG (2013) Large neurological component to genetic differences underlying biased sperm use in *Drosophila*. Genetics 193:177–185

Clark AG, Begun DJ (1998) Female genotypes affect sperm displacement in *Drosophila*. Genetics 149:1487–1493

Clark AG, Aguadé M, Prout T, Harshman LG, Langley CH (1995) Variation in sperm displacement and its association with accessory gland protein loci in *Drosophila melanogaster*. Genetics 139:189–201

Clark AG, Begun DJ, Prout T (1999) Female × male interactions in *Drosophila* sperm competition. Science 283:217–220

Clark NL, Aagaard JE, Swanson WJ (2006) Evolution of reproductive proteins from animals and plants. Reproduction 131:11–22

Clark NL, Gasper J, Sekino M, Springer SA, Aquadro CF, Swanson WJ (2009) Coevolution of interacting fertilization proteins. PLoS Genet 5:e1000570

Clements AN (2000) The biology of mosquitoes. Chapman and Hall, London

Clifton ME, Correa S, Rivera-Pérez C, Nouzova M, Noriega FG (2014) Male *Aedes aegypti* mosquitoes use JH III transferred during copulation to influence previtellogenic ovary physiology and affect the reproductive output of female mosquitoes. J Insect Physiol 64:40–47

Cognigni P, Bailey AP, Miguel-Aliaga I (2011) Enteric neurons and systemic signals couple nutritional and reproductive status with intestinal homeostasis. Cell Metab 13:92–104

Cole SH, Carney GE, McClung CA, Willard SS, Taylor BJ, Hirsh J (2005) Two functional but noncomplementing *Drosophila* tyrosine decarboxylase genes: distinct roles for neural tyramine and octopamine in female fertility. J Biol Chem 280:14948–14955

Cordero C (1995) Ejaculate substances that affect female insect reproductive physiology and behavior—honest or arbitrary traits. J Theor Biol 174:453–461

Debelle JS, Hilliker AJ, Sokolowski MB (1989) Genetic localization of *foraging* (*for*)—a major gene for larval behavior in *Drosophila melanogaster*. Genetics 123:157–163

Demont M, Martin OY, Bussiere LF (2012) Wild yellow dung fly females may not select sperm based on dung pat microclimate but could nevertheless benefit from polyandry. Evol Ecol 26:715–731

Díaz-Fleischer F, Aluja M (2003a) Clutch size in frugivorous insects as a function of host firmness: the case of the tephritid fly *Anastrepha ludens*. Ecol Entomol 28:268–277

Díaz-Fleischer F, Aluja M (2003b) Influence of conspecific presence, experience, and host quality on oviposition behavior and clutch size determination in *Anastrepha ludens* (Diptera: Tephritidae). J Insect Behav 16:537–554
Drummond-Barbosa D, Spradling AC (2001) Stem cells and their progeny respond to nutritional changes during *Drosophila* oogenesis. Dev Biol 231:265–278
Dweck HKM, Ebrahim SAM, Kromann S, Bown D, Hillbur Y, Sachse S et al (2013) Olfactory preference for egg laying on citrus substrates in *Drosophila*. Curr Biol 23:2472–2480
Eberhard WG (1996) Female control: sexual selection by cryptic female choice. Princeton University Press, Princeton
Eberhard WG (2000) Criteria for demonstrating postcopulatory female choice. Evolution 54:1047–1050
Eberhard WG (2002) The function of female resistance behavior: intromission by male coercion vs. female cooperation in sepsid flies (Diptera: Sepsidae). Rev Biol Trop 50:485–505
Eberhard WG (2015) Cryptic female choice and other types of post-copulatory sexual selection. In: Peretti AV, Aisenberg A (eds) Cryptic female choice in arthropods. Springer, Cham
Eberhard WG, Cordero C (1995) Sexual selection by cryptic female choice on male seminal products—a new bridge between sexual selection and reproductive physiology. Trends Ecol Evol 10:493–496
Edvardsson M, Arnqvist G (2000) Copulatory courtship and cryptic female choice in red flour beetles *Tribolium castaneum*. Proc R Soc London, Ser B 267:559–563
Fedina TY, Lewis SM (2004) Female influence over offspring paternity in the red flour beetle *Tribolium castaneum*. Proc R Soc London, Ser B 271:1393–1399
Fedorka KM, Winterhalter WE, Ware B (2011) Perceived sperm competition intensity influences seminal fluid protein production prior to courtship and mating. Evolution 65:584–590
Ferveur JF, Sureau G (1996) Simultaneous influence on male courtship of stimulatory and inhibitory pheromones produced by live sex-mosaic *Drosophila melanogaster*. Proc R Soc London, Ser B 263:967–973
Findlay GD, Yi XH, MacCoss MJ, Swanson WJ (2008) Proteomics reveals novel *Drosophila* seminal fluid proteins transferred at mating. PLoS Biol 6:1417–1426
Findlay GD, MacCoss MJ, Swanson WJ (2009) Proteomic discovery of previously unannotated, rapidly evolving seminal fluid genes in *Drosophila*. Genome Res 19:886–896
Findlay GD, Sitnik JL, Wang W, Aquadro CF, Clark NL, Wolfner MF (2014) Evolutionary rate covariation identifies new members of a protein network required for *Drosophila melanogaster* female post-mating responses. PLoS Genet 10:e1004108
Fiumera AC, Dumont BL, Clark AG (2005) Sperm competitive ability in *Drosophila melanogaster* associated with variation in male reproductive proteins. Genetics 169:243–257
Fiumera AC, Dumont BL, Clark AG (2006) Natural variation in male-induced 'cost-of-mating' and allele-specific association with male reproductive genes in *Drosophila melanogaster*. Philos Trans R Soc London, Ser B 361:355–361
Fiumera AC, Dumont BL, Clark AG (2007) Associations between sperm competition and natural variation in male reproductive genes on the third chromosome of *Drosophila melanogaster*. Genetics 176:1245–1260
Fricke C, Green D, Mills WE, Chapman T (2013) Age-dependent female responses to a male ejaculate signal alter demographic opportunities for selection. Proc R Soc London, Ser B 280:20130428
Fritz AH (2002) A single, abdominal ganglion in *Anastrepha suspensa* (Diptera: Tephritidae) and its innervation of the female sperm storage organs. Ann Entomol Soc Am 95:103–108
Fritz AH, Turner FR (2002) A light and electron microscopical study of the spermathecae and ventral receptacle of *Anastrepha suspensa* (Diptera: Tephritidae) and implications in female influence of sperm storage. Arthropod Struct Dev 30:293–313
Gillott C (2003) Male accessory gland secretions: modulators of female reproductive physiology and behavior. Annu Rev Entomol 48:163–184

Gratz SJ, Cummings AM, Nguyen JN, Hamm DC, Donohue LK, Harrison MM, Wildonger J, O'Connor-Giles KM (2013a) Genome engineering of *Drosophila* with the CRISPR RNA-guided Cas9 nuclease. Genetics 194:1029–1035

Gratz SJ, Wildonger J, Harrison MM, O'Connor-Giles KM (2013b) CRISPR/Cas9-mediated genome engineering and the promise of designer flies on demand. Fly 7:249–255

Haerty W, Jagadeeshan S, Kulathinal RJ, Wong A, Ram KR, Sirot LK et al (2007) Evolution in the fast lane: rapidly evolving sex-related genes in *Drosophila*. Genetics 177:1321–1335

Han KA, Millar NS, Davis RL (1998) A novel octopamine receptor with preferential expression in *Drosophila* mushroom bodies. J Neurosci 18:3650–3658

Hansen IA, Attardo GM, Rodriguez SD, Drake LL (2014) Four-way regulation of mosquito yolk protein precursor genes by juvenile hormone-, ecdysone-, nutrient-, and insulin-like peptide signaling pathways. Front Psychol 5:103

Harshman LG, Clark AG (1998) Inference of sperm competition from broods of field-caught *Drosophila*. Evolution 52:1334–1341

Häsemeyer M, Yapici N, Heberlein U, Dickson BJ (2009) Sensory neurons in the *Drosophila* genital tract regulate female reproductive behavior. Neuron 61:511–518

Haussmann IU, Hemani Y, Wijesekera T, Dauwalder B, Soller M (2013) Multiple pathways mediate the sex-peptide-regulated switch in female *Drosophila* reproductive behaviours. Proc R Soc London, Ser B 280

Heifetz Y, Rivlin PK (2010) Beyond the mouse model: using *Drosophila* as a model for sperm interaction with the female reproductive tract. Theriogenology 73:723–739

Heifetz Y, Wolfner MF (2004) Mating, seminal fluid components, and sperm cause changes in vesicle release in the *Drosophila* female reproductive tract. Proc Natl Acad Sci USA 101:6261–6266

Heifetz Y, Lung O, Frongillo EA, Wolfner MF (2000) The *Drosophila* seminal fluid protein Acp26Aa stimulates release of oocytes by the ovary. Curr Biol 10:99–102

Heifetz Y, Vandenberg LN, Cohn HI, Wolfner MF (2005) Two cleavage products of the *Drosophila* accessory gland protein ovulin can independently induce ovulation. Proc Natl Acad Sci USA 102:743–748

Heifetz Y, Lindner M, Garini Y, Wolfner MF (2014) Mating regulates neuromodulator ensembles at nerve termini innervating the *Drosophila* reproductive tract. Curr Biol 24:731–737

Helinski MEH, Deewatthanawong P, Sirot LK, Wolfner MF, Harrington LC (2012a) Duration and dose-dependency of female sexual receptivity responses to seminal fluid proteins in *Aedes albopictus* and *Ae. aegypti* mosquitoes. J Insect Physiol 58:1307–1313

Helinski MEH, Valerio L, Facchinelli L, Scott TW, Ramsey J, Harrington LC (2012b) Evidence of polyandry for *Aedes aegypti* in semifield enclosures. Am J Trop Med Hyg 86:635–641

Hellriegel B, Bernasconi G (2000) Female-mediated differential sperm storage in a fly with complex spermathecae, *Scatophaga stercoraria*. Anim Behav 59:311–317

Herndon LA, Wolfner MF (1995) A *Drosophila* seminal fluid protein, Acp26Aa, stimulates egg laying in females for 1 day after mating. Proc Natl Acad Sci USA 92:10114–10118

Hodgson DJ, Hosken DJ (2006) Sperm competition promotes the exploitation of rival ejaculates. J Theor Biol 243:230–234

Holman L, Snook RR (2008) A sterile sperm caste protects brother fertile sperm from female-mediated death in *Drosophila pseudoobscura*. Curr Biol 18:292–296

Hosken DJ, Meyer EP, Ward PI (1999) Internal female reproductive anatomy and genital interactions during copula in the yellow dung fly, *Scathophaga stercoraria* (Diptera: Scathophagidae). Can J Zool 77:1975–1983

Hotzy C, Polak M, Ronn JL, Arnqvist G (2012) Phenotypic engineering unveils the function of genital morphology. Curr Biol 22:2258–2261

House CM, Simmons LW (2003) Genital morphology and fertilization success in the dung beetle *Onthophagus taurus*: an example of sexually selected male genitalia. Proc R Soc London, Ser B 270:447–455

Iida K, Cavener DR (2004) Glucose dehydrogenase is required for normal sperm storage and utilization in female *Drosophila melanogaster*. J Exp Biol 207:675–681

Ilango K, Lane RP (2000) Coadaptation of male aedeagal filaments and female spermathecal ducts of the Old World Phlebotomine sand flies (Diptera: Psychodidae). J Med Entomol 37:653–659

Isaac RE, Kim YJ, Audsley N (2014) The degradome and the evolution of *Drosophila* sex peptide as a ligand for the MIP receptor. Peptides 53:258–264

Jang EB (1995) Effects of mating and accessory gland injections on olfactory-mediated behavior in the female mediterranean fruit fly, *Ceratitis capitata*. J Insect Physiol 41:705–710

Jennions MD, Petrie M (2000) Why do females mate multiply? A review of the genetic benefits. Biol Rev 75:21–64

Jinek M, Chylinski K, Fonfara I, Hauer M, Doudna JA, Charpentier E (2012) A programmable dual-RNA-guided DNA endonuclease in adaptive bacterial immunity. Science 337:816–821

Kacsoh BZ, Lynch ZR, Mortimer NT, Schlenke TA (2013) Fruit flies medicate offspring after seeing parasites. Science 339:947–950

Kamimura Y (2007) Twin intromittent organs of *Drosophila* for traumatic insemination. Biol Lett 3:401–404

Kapelnikov A, Rivlin PK, Hoy RR, Heifetz Y (2008) Tissue remodeling: a mating-induced differentiation program for the *Drosophila* oviduct. BMC Dev Biol 8:114

Kaufmann C, Brown MR (2008) Regulation of carbohydrate metabolism and flight performance by a hypertrehalosaemic hormone in the mosquito *Anopheles gambiae*. J Insect Physiol 54:367–377

Kaufmann C, Merzendorfer H, Gade G (2009) The adipokinetic hormone system in Culicinae (Diptera: Culicidae): molecular identification and characterization of two adipokinetic hormone (AKH) precursors from *Aedes aegypti* and *Culex pipiens* and two putative AKH receptor variants from *A. aegypti*. Insect Biochem Mol Biol 39:770–781

Kelly TJ, Adams TS, Schwartz MB, Birnbaum MJ, Rubenstein EC, Imberski RB (1987) Juvenile hormone and ovarian maturation in the Diptera—a review of recent results. Insect Biochem 17:1089–1093

Kempenaers B, Foerster K, Questiau S, Robertson BC, Vermeirssen ELM (2000) Distinguishing between female sperm choice versus male sperm competition: a comment on Birkhead. Evolution 54:1050–1052

Kent CF, Daskalchuk T, Cook L, Sokolowski MB, Greenspan RJ (2009) The *Drosophila foraging* gene mediates adult plasticity and gene-environment interactions in behaviour, metabolites, and gene expression in response to food deprivation. PLoS Genet 5:e1000609

Kern AD, Jones CD, Begun DJ (2004) Molecular population genetics of male accessory gland proteins in the *Drosophila simulans* complex. Genetics 167:725–735

Kim YJ, Bartalska K, Audsley N, Yamanaka N, Yapici N, Lee JY et al (2010) MIPs are ancestral ligands for the sex peptide receptor. Proc Natl Acad Sci USA 107:6520–6525

LaFlamme BA, Wolfner MF (2013) Identification and function of proteolysis regulators in seminal fluid. Mol Reprod Dev 80:80–101

LaFlamme BA, Ram KR, Wolfner MF (2012) The *Drosophila melanogaster* seminal fluid protease "seminase" regulates proteolytic and post-mating reproductive processes. PLoS Genet 8:e1002435

LaFlamme B, Avila F, Michalski K, Wolfner M (2014) A *Drosophila* protease cascade member, seminal metalloprotease-1, is activated stepwise by male factors and requires female factors for full activity. Genetics 196(4):1117–1129

Lee HG, Seong CS, Kim YC, Davis RL, Han KA (2003) Octopamine receptor OAMB is required for ovulation in *Drosophila melanogaster*. Dev Biol 264:179–190

Lee HG, Rohila S, Han KA (2009) The octopamine receptor OAMB mediates ovulation via Ca2+/calmodulin-dependent protein kinase II in the *Drosophila* oviduct epithelium. PLoS ONE 4:e4716

Liu HF, Kubli E (2003) Sex-peptide is the molecular basis of the sperm effect in *Drosophila melanogaster*. Proc Natl Acad Sci USA 100:9929–9933

Lung O, Wolfner MF (1999) *Drosophila* seminal fluid proteins enter the circulatory system of the mated female fly by crossing the posterior vaginal wall. Insect Biochem Mol Biol 29:1043–1052

Lüpold S, Pitnick S, Berben KS, Blengini CS, Belote JM, Manier MK (2013) Female mediation of competitive fertilization success in *Drosophila melanogaster*. Proc Natl Acad Sci USA 110:10693–10698

Mancini E, Tammaro F, Baldini F, Via A, Raimondo D, George P et al (2011) Molecular evolution of a gene cluster of serine proteases expressed in the *Anopheles gambiae* female reproductive tract. BMC Evol Biol 11:72

Manier MK, Belote JM, Berben KS, Novikov D, Stuart WT, Pitnick S (2010) Resolving mechanisms of competitive fertilization success in *Drosophila melanogaster*. Science 328:354–357

Manier MK, Belote JM, Berben KS, Lüpold S, Ala-Honkola O, Collins WF et al (2013a) Rapid diversification of sperm precedence traits and processes among three sibling *Drosophila* species. Evolution 67:2348–2362

Manier MK, Lüpold S, Belote JM, Starmer WT, Berben KS, Ala-Honkola O et al (2013b) Postcopulatory sexual selection generates speciation phenotypes in *Drosophila*. Curr Biol 23:1853–1862

Markow TA, Ankney PF (1984) *Drosophila* males contribute to oogenesis in a multiple mating species. Science 224:302–303

Markow TA, Beall S, Castrezana S (2012) The wild side of life *Drosophila* reproduction in nature. Fly 6:98–101

Middleton CA, Nongthomba U, Parry K, Sweeney ST, Sparrow JC, Elliott CJ (2006) Neuromuscular organization and aminergic modulation of contractions in the *Drosophila* ovary. BMC Biol 4:17

Monastirioti M (2003) Distinct octopamine cell population residing in the CNS abdominal ganglion controls ovulation in *Drosophila melanogaster*. Dev Biol 264:38–49

Monastirioti M, Linn CE, White K (1996) Characterization of *Drosophila* tyramine beta-hydroxylase gene and isolation of mutant flies lacking octopamine. J Neurosci 16:3900–3911

Monsma SA, Wolfner MF (1988) Structure and expression of a *Drosophila* male accessory gland gene whose product resembles a peptide pheromone precursor. Genes Dev 2:1063–1073

Monsma SA, Harada HA, Wolfner MF (1990) Synthesis of two *Drosophila* male accessory gland proteins and their fate after transfer to the female during mating. Dev Biol 142:465–475

Moshitzky P, Fleischmann I, Chaimov N, Saudan P, Klauser S, Kubli E et al (1996) Sex-peptide activates juvenile hormone biosynthesis in the *Drosophila melanogaster* corpus allatum. Arch Insect Biochem Physiol 32:363–374

Mueller JL, Linklater JR, Ram KR, Chapman T, Wolfner MR (2008) Targeted gene deletion and phenotypic analysis of the *Drosophila melanogaster* seminal fluid protease inhibitor Acp62F. Genetics 178:1605–1614

Neubaum DM, Wolfner MF (1999) Mated *Drosophila melanogaster* females require a seminal fluid protein, Acp36DE, to store sperm efficiently. Genetics 153:845–857

Otronen M (1997) Sperm numbers, their storage and usage in the fly *Dryomyza anilis*. Proc R Soc London, Ser B 264:777–782

Otronen M, Siva-Jothy MT (1991) The effect of postcopulatory male behavior on ejaculate distribution within the female sperm storage organs of the fly *Dryomyza anilis* (Diptera: Dryomyzidae). Behav Ecol Sociobiol 29:33–37

Ottiger M, Soller M, Stocker RF, Kubli E (2000) Binding sites of *Drosophila melanogaste*r sex peptide pheromones. J Neurobiol 44:57–71

Park M, Wolfner MF (1995) Male and female cooperate in the prohormone-like processing of a *Drosophila melanogaster* seminal fluid protein. Dev Biol 171:694–702

Parker GA (1990) Sperm competition games—raffles and roles. Proc R Soc London, Ser B 242:120–126
Peng J, Chen S, Büsser S, Liu H, Honegger T, Kubli E (2005) Gradual release of sperm bound sex-peptide controls female postmating behavior in *Drosophila*. Curr Biol 15:207–213
Pérez-Staples D, Weldon CW, Radhakrishnan P, Prenter J, Taylor PW (2010) Control of copula duration and sperm storage by female Queensland fruit flies. J Insect Physiol 56:1755–1762
Pérez-Staples D, Cordova-Garcia G, Aluja M (2014) Sperm dynamics and cryptic male choice in tephritid flies. Anim Behav 89:131–139
Pie MR (1998) Lek behavior as the mating strategy of *Setellia* sp. (Diptera: Richardiidae). J Insect Behav 11:823–832
Pilpel N, Nezer I, Applebaum SW, Helfetz Y (2008) Mating-increases trypsin in female *Drosophila* hemolymph. Insect Biochem Mol Biol 38:320–330
Pitnick S, Brown WD (2000) Criteria for demonstrating female sperm choice. Evolution 54:1052–1056
Pitnick S, Spicer GS, Markow TA (1995) How long is a giant sperm? Nature 375:109
Pitnick S, Spicer GS, Markow T (1997) Phylogenetic examination of female incorporation of ejaculate in *Drosophila*. Evolution 51:833–845
Pitnick S, Markow T, Spicer GS (1999) Evolution of multiple kinds of female sperm-storage organs in *Drosophila*. Evolution 53:1804–1822
Pitnick S, Wolfner MF, Suarez S (2009) Sperm-female interactions. In: Birkhead TR DJHaSP (ed) Sperm biology: an evolutionary perspective. Academic Press, London, pp 247–304
Poels J, Van Loy T, Vandersmissen HP, Van Hiel B, Van Soest S, Nachman RJ et al (2010) Myoinhibiting peptides are the ancestral ligands of the promiscuous *Drosophila* sex peptide receptor. Cell Mol Life Sci 67:3511–3522
Polak M, Rashed A (2010) Microscale laser surgery reveals adaptive function of male intromittent genitalia. Proc R Soc London, Ser B 277:1371–1376
Polak M, Simmons LW (2009) Secondary sexual trait size reveals competitive fertilization success in *Drosophila bipectinata* Duda. Behav Ecol 20:753–760
Prokupek A, Hoffmann F, Eyun SI, Moriyama E, Zhou M, Harshman L (2008) An evolutionary expressed sequence tag analysis of *Drosophila* spermatheca genes. Evolution 62:2936–2947
Prokupek AM, Kachman SD, Ladunga I, Harshman LG (2009) Transcriptional profiling of the sperm storage organs of *Drosophila melanogaster*. Insect Mol Biol 18:465–475
Radhakrishnan P, Fedorka KM (2012) Immune activation decreases sperm viability in both sexes and influences female sperm storage. Proc R Soc London, Ser B 279:3577–3583
Radhakrishnan P, Taylor PW (2007) Seminal fluids mediate sexual inhibition and short copula duration in mated female Queensland fruit flies. J Insect Physiol 53:741–745
Radhakrishnan P, Nair S, Raftos D, Taylor PW (2008) Transfer and fate of male ejaculate in female Queensland fruit flies. Physiol Entomol 33:302–309
Ravi Ram K, Wolfner MF (2007a) Seminal influences: *Drosophila* Acps and the molecular interplay between males and females during reproduction. Integr Comp Biol 47:427–445
Ravi Ram K, Wolfner MF (2007b) Sustained post-mating response in *Drosophila melanogaster* requires multiple seminal fluid proteins. PLoS Genet 3:2428–2438
Ravi Ram K, Wolfner MF (2009) A network of interactions among seminal proteins underlies the long-term postmating response in *Drosophila*. Proc Natl Acad Sci USA 106:15384–15389
Ravi Ram K, Ji S, Wolfner MF (2005) Fates and targets of male accessory gland proteins in mated female *Drosophila melanogaster*. Insect Biochem Mol Biol 35:1059–1071
Ravi Ram K, Sirot LK, Wolfner MF (2006) Predicted seminal astacin-like protease is required for processing of reproductive proteins in *Drosophila melanogaster*. Proc Natl Acad Sci USA 103:18674–18679
Rezával C, Nojima T, Neville MC, Lin AC, Goodwin SF (2014) Sexually dimorphic octopaminergic neurons modulate female postmating behaviors in *Drosophila*. Curr Biol 24:725–730

Rodriguez-Enriquez CL, Tadeo E, Rull J (2013) Elucidating the function of ejaculate expulsion and consumption after copulation by female *Euxesta bilimeki*. Behav Ecol Sociobiol 67:937–946

Rodríguez-Valentín R, López-González I, Jorquera R, Labarca P, Zurita M, Reynaud E (2006) Oviduct contraction in *Drosophila* is modulated by a neural network that is both octopaminergic and glutamatergic. J Cell Physiol 209:183–198

Rogers DW, Baldini F, Battaglia F, Panico M, Dell A, Morris HR et al (2009) Transglutaminase-mediated semen coagulation controls sperm storage in the malaria mosquito. PLoS Biol 7:e1000272

Rowe L, Arnqvist G (2012) Sexual selection and the evolution of genital shape and complexity in water striders. Evolution 66:40–54

Rubinstein CD, Wolfner MF (2013) *Drosophila* seminal protein ovulin mediates ovulation through female octopamine neuronal signaling. Proc Natl Acad Sci USA 110:17420–17425

Rubinstein CD, Wolfner MF (2014) Reproductive hacking: a male seminal protein acts through intact reproductive pathways in female *Drosophila*. FLY 8:80–85

Sbilordo SH, Schäfer MA, Ward PI (2009) Sperm release and use at fertilization by yellow dung fly females (*Scathophaga stercoraria*). Biol J Linn Soc 98:511–518

Schäfer MA, Berger D, Jochmann R, Blanckenhorn WU, Bussiere LF (2013) The developmental plasticity and functional significance of an additional sperm storage compartment in female yellow dung flies. Funct Ecol 27:1392–1402

Schnakenberg SL, Matias WR, Siegal ML (2011) Sperm-storage defects and live birth in *Drosophila* females lacking spermathecal secretory cells. PLoS Biol 9:e1001192

Schully SD, Hellberg ME (2006) Positive selection on nucleotide substitutions and indels in accessory gland proteins of the *Drosophila pseudoobscura* subgroup. J Mol Evol 62:793–802

Shelly TE (1990) Observations on the lek behavior of *Drosophila imparisetae* (Diptera: Drosophilidae) in Hawaii. J Kans Entomol Soc 63:652–655

Shelly TE, Kaneshiro KY (1991) Lek behavior of the oriental fruit fly, *Dacus dorsalis*, in Hawaii (Diptera: Tephriditae). J Insect Behav 4:235–241

Sirot LK, Brockmann HJ, Lapointe SL (2007) Male postcopulatory reproductive success in the beetle, *Diaprepes abbreviatus*. Anim Behav 74:143–152

Sirot LK, Wolfner MF, Wigby S (2011) Protein-specific manipulation of ejaculate composition in response to female mating status in *Drosophila melanogaster*. Proc Natl Acad Sci USA 108:9922–9926

Sirot LK, Findlay GD, Sitnik JL, Frasheri D, Avila FW, Wolfner MF (2014) Molecular characterization and evolution of a gene family encoding both female- and male-specific reproductive proteins in *Drosophila*. Mol Biol Evol 31(6):1554–1567

Smith DR, Chapman MR (2010) Economical evolution: microbes reduce the synthetic cost of extracellular proteins. MBio 1(3):pii: e00131–10

Smith C, Greig D (2010) The cost of sexual signaling in yeast. Evolution 64:3114–3122

Smith DT, Hosken DJ, Ffrench-Constant RH, Wedell N (2009) Variation in sex peptide expression in *D. melanogaster*. Genet Res 91:237–242

Smith DT, Sirot LK, Wolfner MF, Hosken DJ, Wedell N (2012) The consequences of genetic variation in sex peptide expression levels for egg laying and retention in females. Heredity 109:222–225

Soller M, Bownes M, Kubli E (1997) Mating and sex peptide stimulate the accumulation of yolk in oocytes of *Drosophila melanogaster*. Eur J Biochem 243:732–738

Sun J, Spradling AC (2013) Ovulation in *Drosophila* is controlled by secretory cells of the female reproductive tract. eLife 2: e00415

Swanson WJ, Clark AG, Waldrip-Dail HM, Wolfner MF, Aquadro CF (2001) Evolutionary EST analysis identifies rapidly evolving male reproductive proteins in *Drosophila*. Proc Natl Acad Sci USA 98:7375–7379

Tripet F, Toure YT, Dolo G, Lanzaro GC (2003) Frequency of multiple inseminations in field-collected *Anopheles gambiae* females revealed by DNA analysis of transferred sperm. Am J Trop Med Hyg 68:1–5

Tsaur SC, Ting CT, Wu CI (2001) Sex in *Drosophila mauritiana*: a very high level of amino acid polymorphism in a male reproductive protein gene, Acp26Aa. Mol Biol Evol 18:22–26

Twig E, Yuval B (2005) Function of multiple sperm storage organs in female Mediterranean fruit flies (*Ceratitis capitata*, Diptera: Tephritidae). J Insect Physiol 51:67–74

Waage JK (1979) Dual function of the damselfly penis—sperm removal and transfer. Science 203:916–918

Wagstaff BJ, Begun DJ (2005) Comparative genomics of accessory gland protein genes in *Drosophila melanogaster* and *D. pseudoobscura*. Mol Biol Evol 22:818–832

Wagstaff BJ, Begun DJ (2007) Adaptive evolution of recently duplicated accessory gland protein genes in desert *Drosophila*. Genetics 177:1023–1030

Walters JR, Harrison RG (2010) Combined EST and proteomic analysis identifies rapidly evolving seminal fluid proteins in *Heliconius* butterflies. Mol Biol Evol 27:2000–2013

Ward PI, Wilson AJ, Reim C (2008) A cost of cryptic female choice in the yellow dung fly. Genetica 134:63–67

Wigby S, Sirot LK, Linklater JR, Buehner N, Calboli FCF, Bretman A et al (2009) Seminal fluid protein allocation and male reproductive success. Curr Biol 19:751–757

Wojcieszek JM, Simmons LW (2011) Male genital morphology influences paternity success in the millipede *Antichiropus variabilis*. Behav Ecol Sociobiol 65:1843–1856

Wong A, Rundle H (2013) Selection on the *Drosophila* seminal fluid protein Acp62F. Ecol Evol 3:1942–1950

Wong A, Turchin M, Wolfner MF, Aquadro CF (2012) Temporally variable selection on proteolysis-related reproductive tract proteins in *Drosophila*. Mol Biol Evol 29:229–238

Yang CH, Belawat P, Hafen E, Jan LY, Jan YN (2008) *Drosophila* egg-laying site selection as a system to study simple decision-making processes. Science 319:1679–1683

Yang CH, Rumpf S, Xiang Y, Gordon MD, Song W, Jan LY et al (2009) Control of the post-mating behavioral switch in *Drosophila* females by internal sensory neurons. Neuron 61:519–526

Yapici N, Kim YJ, Ribeiro C, Dickson BJ (2008) A receptor that mediates the post-mating switch in *Drosophila* reproductive behaviour. Nature 451:33-U31

Yutani K, Ogasahara K, Sugino Y, Matsushiro A (1977) Effect of a single amino acid substitution on stability conformation of a protein. Nature 267:274–275

Yutani K, Ogasahara K, Tsujita T, Sugino Y (1987) The dependence of conformational stability on the hydrophobicity of the amino acid residue in a series of variant proteins substituted by each of the 20 amino acids at a unique position of the tryptophan synthase alpha subunit. Protein Eng 1:249–249

Zeh JA, Zeh DW (2003) Toward a new sexual selection paradigm: polyandry, conflict and incompatibility. Ethology 109:929–950

Zhang R, Clark AG, Fiumera AC (2013) Natural genetic variation in male reproductive genes contributes to nontransitivity of sperm competitive ability in *Drosophila melanogaster*. Mol Ecol 22:1400–1415

Chapter 14
An Integrative View of Postcopulatory Sexual Selection in a Soldier Fly: Interplay Between Cryptic Mate Choice and Sperm Competition

Flavia Barbosa

Abstract Cryptic female choice (CFC) can occur in the same species in conjunction with other postcopulatory processes, such as sperm competition (SC) and cryptic male choice (CMC). However, each of these processes has been mostly studied in isolation. Little is known about how they interact with each other and how this interplay affects the role they play in sexual selection. This chapter addresses the interplay between CFC, CMC, and SC in the soldier fly, *Merosargus cingulatus* (Diptera: Stratiomyidae). Soldier flies mate at oviposition sites; both sexes mate multiply and males perform copulatory courtship. Here, we describe the different postcopulatory processes that occur in this species: Females use control of oviposition timing as a CFC mechanism. Males show cryptic preferences for larger females and respond to SC cues by increasing copulation duration. Male reproductive success is, therefore, a result of a number of factors: oviposition behavior, copulatory courtship, copulation duration, female size, and male density at the oviposition site. The interaction between CFC and other postcopulatory processes sheds light on its potential impact on sexual selection and evolution.

14.1 Introduction

Postcopulatory sexual selection (hereby defined as sexual selection occurring during or after copulation) is a major evolutionary engine that can promote morphological and behavioral changes in polygamous species. It has been described in a variety of taxa (Eberhard 1996; Simmons and Siva-Jothy 1998). Postcopulatory sexual selection has both an intrasexual and an intersexual component: sperm

---

F. Barbosa (✉)
Institut de Recherche Sur La Biologie de L'Insecte (IRBI),
Université François Rabelais, Tours, France
e-mail: flavia.barbosa@univ-tours.fr

A.V. Peretti and A. Aisenberg (eds.), *Cryptic Female Choice in Arthropods*,
DOI 10.1007/978-3-319-17894-3_14

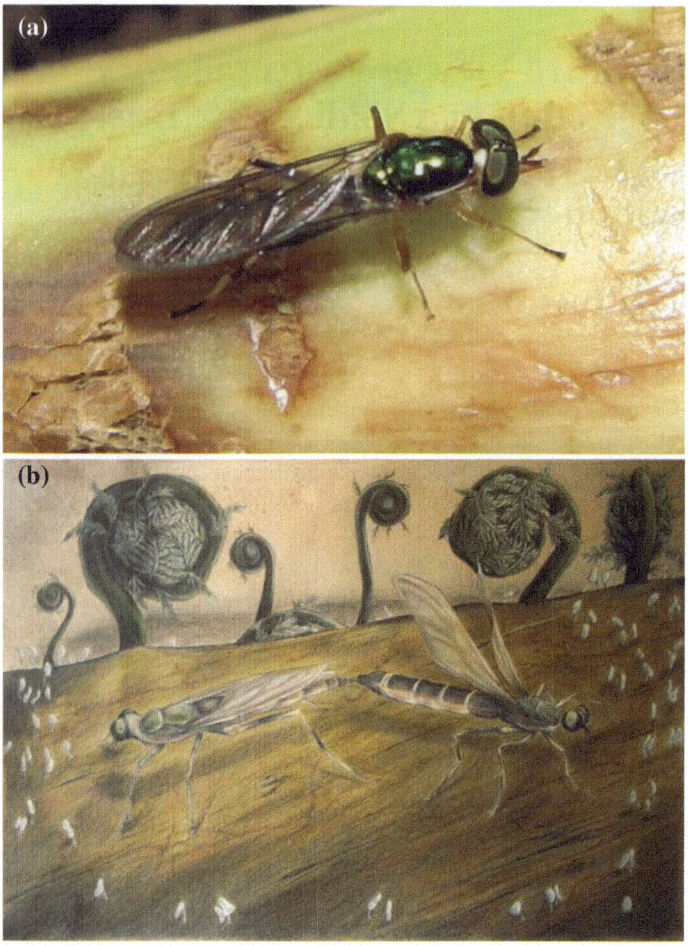

Fig. 14.1 The soldier fly *Merosargus cingulatus*. **a** A male at an oviposition site (photo by F. Barbosa). **b** Drawing of a mating pair (drawing by A. Hallett)

competition (SC) and cryptic mate choice, respectively. Sperm competition theory was first proposed over four decades ago (Parker 1970) and was quickly accepted as an important and nearly ubiquitous process shaping male physiology, morphology, and behavior in animals (Parker 1970; Birkhead and Møller 1998; Simmons 2001). Cryptic female choice (CFC) was proposed later (Thornhill 1983), but received considerably less attention, and its role in sexual selection was considered controversial for several years (Birkhead 1998, 2000; Kempenaers et al. 2000). As a result, research on CFC lagged behind in comparison with that on SC. Nonetheless, the importance of CFC in sexual selection is now widely acknowledged, and a large body of empirical work demonstrates its importance across taxa. More recently, it was suggested that males can also perform postcopulatory mate choice by selectively allocating resources to preferred females (Bonduriansky 2001), a process known as cryptic male choice (CMC).

Because postcopulatory sexual selection acts on both males and females, multiple postcopulatory processes can be at play in a given species. Still, while these processes have been extensively studied and empirically demonstrated, they have traditionally been examined in isolation from one another. We thus lack information on how SC and cryptic mate choice may interact and affect the reproductive success of each sex. In order to have a comprehensive view of the role, postcopulatory processes play in sexual selection, an integrative approach where they are investigated in the same species is necessary. In this chapter, I apply this comprehensive view by describing the interplay between SC, CFC, and CMC and their effect on reproductive success, using the soldier fly *Merosargus cingulatus* as a case study (Fig. 14.1). Specifically, I (1) describe some aspects of the life history and mating system of the soldier fly while exploring their effect on sexual selection; (2) summarize the evidence for CFC, SC, and CMC in this species; (3) explore how these processes interact with one another and affect reproductive success; and (4) suggest future studies integrating the multiple components of postcopulatory sexual selection.

14.2 The Soldier Fly: Natural History and Mating Behavior

Merosargus is a Neotropical genus which encompasses over one hundred species (James and McFadden 1971). These species are morphologically and ecologically diverse, being commonly found in a variety of habitats, such as primary and secondary tropical rainforest, suburban residential areas, and city parks (Woodley 2001). Even though this is a diverse and abundant genus, very little is known about its biology and natural history. The larvae are detritivores and feed on a variety of decomposing plant matter. Several species develop in flower parts of *Heliconia* plants (Seifert and Seifert 1976, 1979), while others are commonly found in a variety of decomposing fruits, recently cut grass and fallen stems of succulents; a few species develop in animal fecal matter (Woodley 2001). Adults aggregate around these larval substrates, where females oviposit and males often defend territories. Very little is known about the foraging habits of the adults.

In *M. cingulatus*, copulations occur at oviposition sites, the larval development substrate. This species develops in decomposing vegetable matter such as rotting fruit. Territories consist of small patches over which the territory holder flies continuously during the times of the day females search for oviposition sites (approximately between 9 AM and 4 PM). Males will defend their territory from any fly that approaches, including heterospecifics (*F. Barbosa*, personal observation). They do so by chasing the intruder; if the intruder is a conspecific, the interaction

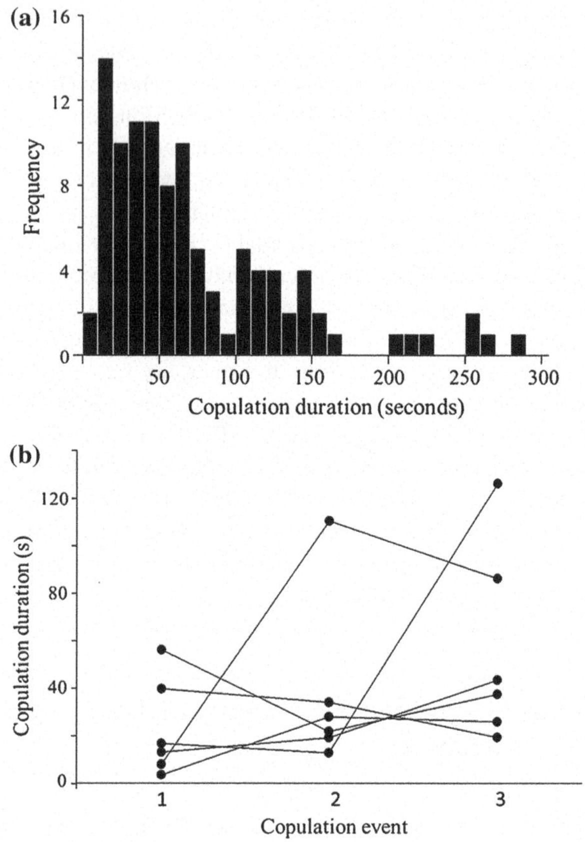

Fig. 14.2 Copulation duration varies both among and within males in the soldier fly. **a** Frequency distribution of copulation durations of 104 mating pairs measured in the field. **b** Variation in copulation duration within males: duration of 3 different copulations of the same male, measured from field recordings of marked-and-released individuals

often escalates. First, the males collide against each other; then, one of the males may use his legs to capture the other one and carry him away from the territory. He then releases that male and returns (*F. Barbosa*, personal observation).

Oviposition sites attract large numbers of males and females. Male territories are relatively small, with neighboring males typically about 10–15 cm apart. The territory holder attempts to grab and copulate with any female that flies near his territory. Females do not appear to have any opportunity to choose a mate, although it is possible that they can observe males from a distance before approaching. However, when they arrive at the oviposition site, they do not appear to approach an individual male, but the substrate. Females also do not seem to resist mating once they are grasped by a male. It is unclear whether females come to the territories primarily for mating or for oviposition. There appears to be no precopulatory signaling or courtship, but males perform courtship throughout the copulation. Copulatory courtship involves an alternation of two distinct behaviors: Males tap the female's abdomen with their hind legs and wave their hind legs in the air. It lasts for the entire duration of copulation, and although there is considerable variation in copulation duration (Fig. 14.2), there is very little variation in the rate of tapping and waving behaviors (Barbosa 2009).

After mating, females usually lay eggs deep in the rotting vegetable matter, in or close to the male's territory. If a female leaves a male's territory without ovipositing, she will likely mate again before having the chance to oviposit elsewhere. This is due to the fact that male density at oviposition sites is high. If a female approaches a different oviposition site, she will likely be detected by a male, who will then grab her and mate (Barbosa 2009). Observations of marked animals show that both males and females mate multiply in the field (F. Barbosa, personal observation).

Several features of the reproductive behavior of soldier flies suggest that postcopulatory sexual selection plays a major role in sexual selection in this species, making it well suited for a comprehensive study of the interplay between SC and cryptic mate choice. First, males do not court females before mating, and precopulatory mate choice does not appear to occur (although females may have preferences for certain oviposition sites). However, males do court females during copulation. Copulatory courtship is generally associated with CFC, since it occurs after copulation has begun and therefore cannot play a role in attracting mates (Eberhard 1994). Second, there is considerable variation in copulation duration (and subsequently, in courtship duration) both between and within males. In other words, males may be selectively allocating resources (time and energy spent mating and courting) during copulation, a behavior associated with SC and CMC.

Soldier flies also present the advantage of being a well-suited study species for field experiments. It is relatively easy to manipulate and test field-collected individuals under natural or seminatural conditions, which allows for studies that address their behavior in nature. Postcopulatory sexual selection is virtually a laboratory phenomenon, since the vast majority of what we know about both SC and cryptic mate choice comes from studies conducted entirely in laboratories. Sexual selection can play out very differently in the field than it does in the laboratory (Jennions and Petrie 1997); still, very little is known about the effects of postcopulatory sexual selection on natural populations.

14.3 Cryptic Female Choice by Female Control of Oviposition Timing in a Soldier Fly

There are numerous ways through which females can bias fertilization toward preferred males; in fact, over 20 potential mechanisms of CFC have been proposed (Eberhard 1996). Although not all of these mechanisms have been empirically demonstrated, there is the potential for them to be widespread among species where females mate multiply. Female soldier flies mate several times during their lifetime (often multiple times in the same day) and oviposit after most, but not all, copulations. This means there is the potential for CFC by female control of oviposition timing: A female may choose not to oviposit after mating with a less preferred male, a behavior which may decrease that male's fertilization success.

In this section, I review the evidence supporting the hypothesis that CFC occurs through control of oviposition timing in *M. cingulatus*. Two critical predictions of this hypothesis were tested: (1) That females are less likely to oviposit immediately after mating with a less preferred male and (2) that failure to immediately oviposit after mating results in lower reproductive success for the male she just mated with. Since copulatory courtship is associated with selection by CFC, this work focused on female preference for this behavior (Barbosa 2009).

The first prediction was tested through a field experiment where I compared the oviposition behavior of females that mated with males of two groups: individuals who did not perform copulatory courtship and individuals who did. Piles of rotting fruit were set up in the field to attract soldier flies. Males were captured and uniquely marked with acrylic paint on their thorax and were then assigned to either a control or a treatment group. The treatment males had black acrylic paint applied to their hind legs. Once the paint dried, it made their legs stiff and prevented normal movement during copulatory courtship. The control males were marked and handled like the treatment males, but did not have paint applied to their legs, allowing for normal movement during copulatory courtship.

All males were released and most of them returned to their territories. Those who did were observed until they obtained a mate; their copulations were video-recorded and the female behavior after copulation was observed. None of the treatment males performed copulatory courtship; and, supporting the first prediction, none of their mates oviposited after mating, flying away as soon as copulation ended. On the other hand, all of the control males performed copulatory courtship and all of their mates oviposited (difference in oviposition behavior of females with control versus manipulated males: Fisher's exact test, $p < 0.005$). Female oviposition behavior was the only difference detected in copulations with males of the two groups: There were no visible differences in male mating behavior other than the presence of copulatory courtship, and mean copulation duration did not differ between groups either. These results confirm the prediction that females would be less likely to oviposit after mating with a less preferred male. However, to demonstrate that this female behavior is indeed a mechanism of CFC, a second prediction must be validated: This female behavior must affect male reproductive success.

To determine whether failure to immediately oviposit by the female results in lower reproductive success for the male she just mated with, the pattern of sperm precedence in soldier flies was investigated. Female oviposition timing is especially important for male reproductive success in species where there is last male sperm precedence (i.e., where the last male to mate with the female fertilizes most of her eggs). When that is the case, if a female fails to oviposit after mating with a male and remates with a different male, the first male will fertilize very few of her eggs.

Field-collected individuals of unknown mating history were allowed to mate, and females were allowed to oviposit. AFLP markers were then used for paternity analysis of the offspring. Although female mating history was not established, the last male she mated with was known. It was therefore possible to calculate the

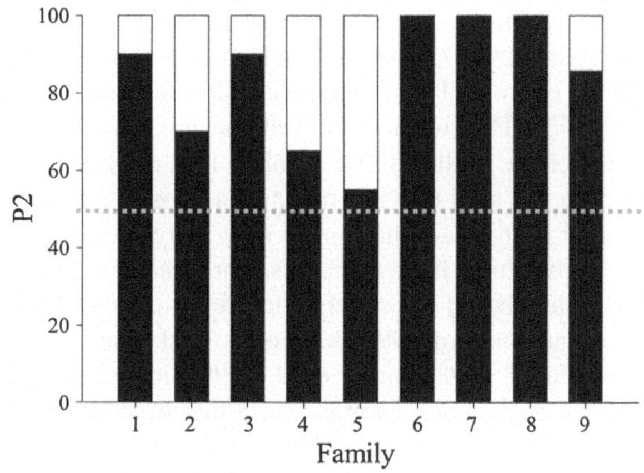

Fig. 14.3 P_2, or percentage of offspring assigned to the last male to mate with each female. The *dotted line* shows the P_2 value of 0.5

percentage of offspring sired by the last male. The average P_2, or percentage of offspring sired by the last male, was 0.839 (range 0.55–1; Fig. 14.3). This demonstrates that there is last male sperm precedence in *M. cingulatus*.

In summary, female soldier flies are less likely to oviposit after mating with males that do not perform copulatory courtship. This female behavior results in lower reproductive success for the male she just mated with. Since oviposition is under female control, the behavior of female *M. cingulatus* suggests that CFC is at play. However, as discussed in the introduction of this book, it is necessary to demonstrate that females benefit from this behavior to rule out SAC and demonstrate CFC. This would be impossible to do in this case, and therefore, we cannot discard SAC as a hypothesis to explain this female behavior, even though CFC appears to play an important role.

The experimental design of this study does not rule out the additional possibility that females fail to oviposit in the absence of copulatory courtship because sperm transfer may not happen in that case. Copulatory courtship has been shown to be crucial for males to achieve complete penetration and sperm transfer in other species (Tallamy et al. 2002), but it is unknown whether this is the case for soldier flies. However, whether or not females are responding to the lack of courtship per se, or to other aspects of the copulation, the observed female behavior is still under female control, suggesting CFC by female control of oviposition timing.

14.4 Sperm Competition (SC) and Cryptic Male Choice (CMC) in the Soldier Fly

Much of the framework of sexual selection theory comes from the idea of differential parental investment between males and females, which is derived from the fact that gamete expenditure is much higher for females than for males (Trivers 1972). However, it is now generally acknowledged that reproduction can also be quite costly for males, in terms of sperm production (Dewsbury 1982; Tang-Martinez and Ryder 2005), ejaculate and spermatophore making (Wagner 2005; Sakaluk 1985; Fox et al. 1995; Oberhauser 1988), time and energy spent in courtship and copulation (Prestwich 1994; Vehrencamp et al. 1989), and direct benefits supplied to the female (Tallamy 2000; Vahed 1998; Gross and Sargent 1985).

With the idea that reproduction is costly for males, more recent research has focused on the fact that males may increase their reproductive success by selectively allocating their resources. For instance, sperm competition theory predicts that males should selectively allocate resources (Gage 1991; Cook and Gage 1995; Gage and Barnard 1996; Marconato and Shapiro 1996; Kvarnemo and Simmons 1999; Fuller 1998; Parker et al. 1999), increasing their investment when the risk of SC is high. For example, males that transfer more sperm when the probability of SC is high have increased fertilization success (Parker et al. 1996, 1997; Wedell et al. 2002; Parker and Pizzari 2010). Males may respond in multiple ways to increased risks of SC: They may prolong copulations, which may increase sperm transfer, increase the transfer of seminal fluids that affect female future receptivity, or prevent the female from remating by mate guarding (Simmons 2001). Males may also change courtship intensity (Lorch et al. 1993; Andrés and Cordero Rivera 2000; Tompkins and Hall 1981; Friberg 2006), or the quality of nuptial gifts (Sakaluk 1985; Fox et al. 1995; Oberhauser 1988; Vahed 1998; Wagner 2005).

When resources limit male reproductive success, males are expected to allocate more resources to higher quality females, a process known as CMC (Bonduriansky 2001). Males have been shown to allocate more resources to females with traits that correlate with fecundity in a number of species (Gage and Barnard 1996; Gage 1998; Parker et al. 1999; Engqvist and Sauer 2001; Reinhold et al. 2002; Xu and Wang 2009; Cornwallis and O'Connor 2009; Yasui 1996; Wedell 1998).

Copulation duration in the soldier fly appears to be under male control and is remarkably variable, in among-male and within-male copulation duration (Fig. 14.2). One prediction from sperm competition theory is that males will increase copulation duration when male density at the oviposition site is high (since males may use density as a cue to assess SC risk). One prediction from CMC is that males will increase copulation duration when mating with larger females, since larger females are more fecund (Fig. 14.4). Here, I review evidence showing that male soldier flies adjust copulation duration in response to both SC risk and to variation in female quality (Barbosa 2011).

To test the hypothesis that males allocate resources according to the risk of SC, a field experiment was conducted where males were introduced into an enclosure that contained oviposition substrate. To adjust perceived SC, males were either by themselves or with four other males. These constituted a low and a high male density treatment, respectively. Males had an acclimation period of 1 h, during which they established territories in the oviposition substrate. After acclimation, the four extra males in the high-density treatment were removed. A female was then introduced into the enclosure for both treatments (the females were not exposed to different male densities). The pair was allowed to mate and copulations were video-recorded. Copulation duration and intensity of copulatory courtship were measured from the videos.

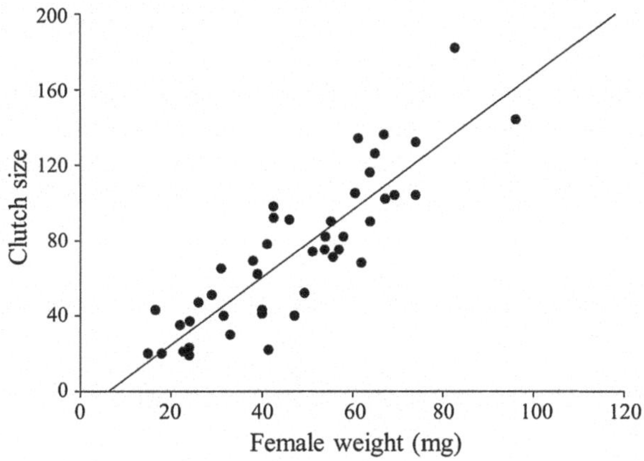

Fig. 14.4 Number of eggs present in female reproductive tract as a function of her weight ($R^2 = 0.76$, $P < 0.001$)

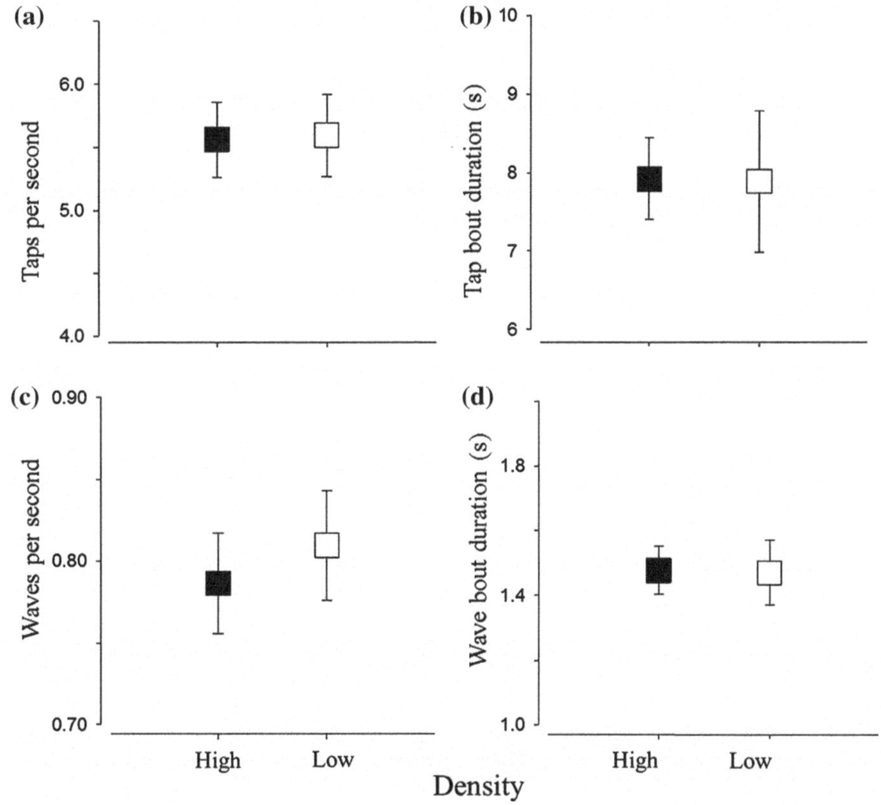

Fig. 14.5 Copulatory courtship behavior of males in high (*black*)- and low (*white*)-density treatments: rate of taps (**a**), mean duration of tap bouts (**b**), rate of waves (**c**), and mean duration of wave bouts (**d**). No statistical difference between the behaviors of the two treatment groups was found (Mann–Whitney U-test, all *p* values >0.5). Error bars represent one standard error

Although it is possible that males respond to SC risk by changing some aspect of their copulatory courtship behavior, our results showed that this is not the case for soldier flies: Different components of male courtship behavior were measured for both groups, but no significant differences were found (Fig. 14.5). However, copulations were significantly longer when male density was higher at the oviposition sites (Fig. 14.6).

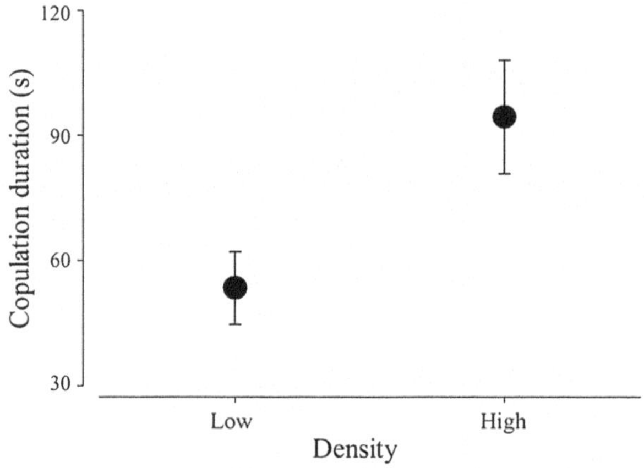

Fig. 14.6 Mean copulation duration in high- and low-density oviposition sites (Mann–Whitney U-test, $p = 0.012$). Error bars represent one standard error

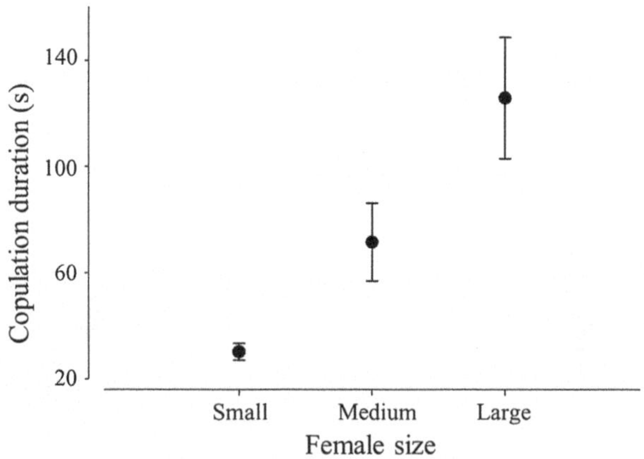

Fig. 14.7 Mean copulation duration of males mating with females of three different size groups. Copulation duration was significantly different between the three female groups (Kruskal–Wallis, $P < 0.0005$). Pairwise comparisons revealed that copulation duration was significantly different after correction for multiple comparisons with both sequential Bonferroni and Sidak adjustment methods between small versus medium and small versus large females (Mann–Whitney U-test, $P = 0.013$, $P < 0.0005$, respectively) and marginally significant between medium versus large females (Mann–Whitney test, $P = 0.057$). Error bars represent one standard error

To test the hypothesis of CMC, a similar experiment was conducted. Males were introduced alone to the enclosure with oviposition substrate and allowed to establish territories. Then, a female belonging to one of three size categories was introduced (small, medium, and large, which corresponded to the bottom, middle, and top third of the size distribution of previously measured field-caught females). Copulation duration was found to be strongly influenced by female size: Copulations with the largest females were on average 4 times longer than with the smallest females (Fig. 14.7).

These results highlight that both SC and CMC are important agents affecting sexual selection in the soldier fly. Males change copulation duration in response to cues of SC risk and to variation in female quality. These results illustrate the interaction between two sources of postcopulatory selection in the same trait, copulation duration.

14.5 Copulation Duration, Sperm Competition Cues, and Fertilization Success

It has been well established that males selectively allocate resources in response to SC in order to achieve higher reproductive success. However, although male behavioral responses to SC have been extensively documented, very few studies have attempted to directly demonstrate the effects of these responses on male fitness (but see Bretman et al. 2009). In this section, I review evidence that shows that in the soldier fly, male behavioral response to SC results in higher reproductive success (Barbosa 2012). Two different hypotheses of how males can increase their reproductive success were tested. First, males may stimulate females to lay a larger number of eggs by increasing copulation duration, either by transferring ejaculate components that stimulate oviposition or by influencing female behavior through prolonged courtship. Second, males may fertilize a larger number of a female's eggs by prolonging copulations, either by transferring more sperm or accessory gland secretions, or through CFC for traits associated with longer copulations.

The effects of male selective resource allocation were tested in a field experiment where males were submitted to treatments that simulated different levels of SC risk: high and low male density at the oviposition site. They were then mated to either a small or a large female, resulting in a 2 × 2 design. Copulations were video-recorded and females were allowed to oviposit. To check whether males stimulate females to lay more eggs, clutch size was measured, as well as the percentage of mature eggs in a female's reproductive tract that were laid. The second part was done by dissecting females and counting the mature eggs that were left in their ovaries after they oviposited. To test whether longer copulations resulted in higher fertilization success for males, the egg clutches were allowed to develop into larvae, and AFLP profiles were used to determine the percentage of the offspring that were fertilized by the experimental male.

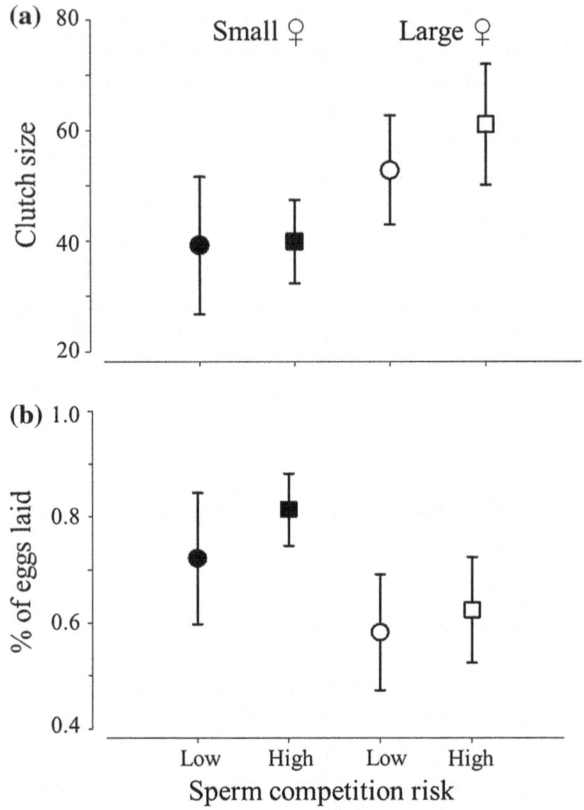

Fig. 14.8 Female oviposition rate was not affected when females (*small*, *black* or *large*, *white*) were mated with males under high (*squares*) and low (*circles*) simulated sperm competition. **a** Clutch size of females who mated with males from both treatments. A 2 × 2 ANOVA on square-root-transformed clutch size showed no effect of density ($F_{1,36} = 0.121$, $p = 0.730$) or female size ($F_{1,36} = 2.991$, $p = 0.092$), and there was no interaction between density and female size ($F_{1,36} = 0.001$, $p = 0.972$). **b** The percentage of mature eggs laid by females mated with males from both treatments. A 2 × 2 ANOVA on arcsine-transformed percentage of eggs showed no effect of density ($F_{1,36} = 0.548$, $p = 0.464$). There was an effect of female size on the percentage of eggs laid with smaller females tending to lay a larger percentage of their available eggs ($F_{1,36} = 5.287$, $p = 0.027$), but no interaction between density and female size ($F_{1,36} = 0.474$, $p = 0.469$). *Bars* represent one standard error

Simulated SC risk did not affect clutch size, nor the percentage of eggs laid (Fig. 14.8). Not surprisingly, female size affected oviposition behavior: Larger females laid more eggs and smaller females laid a larger percentage of their available eggs. However, there was no effect of male density at the oviposition site on female oviposition behavior. On the other hand, fertilization success was influenced by simulated SC risk: When SC was higher, males increased copulation duration and fertilized a higher percentage of a female's eggs than when SC was

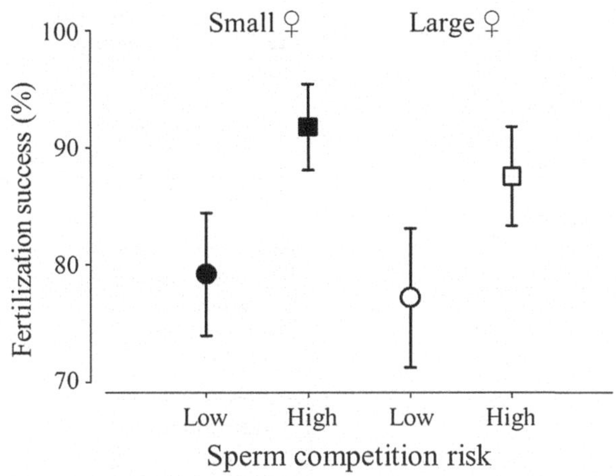

Fig. 14.9 Male fertilization success under high (*squares*) and low (*circles*) simulated sperm competition when mating with small (*black*) and large (*white*) females. A 2 × 2 ANOVA on arcsine-transformed fertilization success revealed a main effect of sperm competition risk ($F_{1,31} = 5.758$, $p = 0.023$). There was no effect of female size ($F_{1,31} = 0.195$, $p = 0.662$) and no interaction between sperm competition and female size ($F_{1,31} = 0.067$, $p = 0.797$)

low (Fig. 14.9). These results demonstrate that male behavioral response to the risk of SC results in differential fertilization success. Males increased copulation duration under high simulated SC and, as a result, fertilized on average 14.4 % more of a female's clutch than males under low simulated SC. This shows a direct link between male selective resource allocation and fitness.

14.6 Conclusions and Future Directions

CFC has been mainly studied in isolation from other postcopulatory processes, but the evidence presented here suggests that CFC is not the only mechanism affecting the outcome of sexual selection. Rather, CFC likely occurs in conjunction with SC and CMC in numerous species, and one may expect these processes to act in concert to determine the reproductive success of males and females. In this case study, I have compiled evidence that demonstrates that male reproductive success in the soldier fly is a result of the interaction between CFC, SC, and CMC. Since females mate multiply, male fitness depends on his fertilization success. Fertilization success is affected by female oviposition timing, which is potentially a mechanism of CFC in this species and which is in turn affected by male copulatory courtship behavior. Fertilization success also depends on copulation duration, which is under male control. Males adjust copulation duration based on female size (CMC) and SC risk at the oviposition site (Fig. 14.10). This interplay between postcopulatory mechanisms is most likely not a unique attribute of soldier flies. Multiple

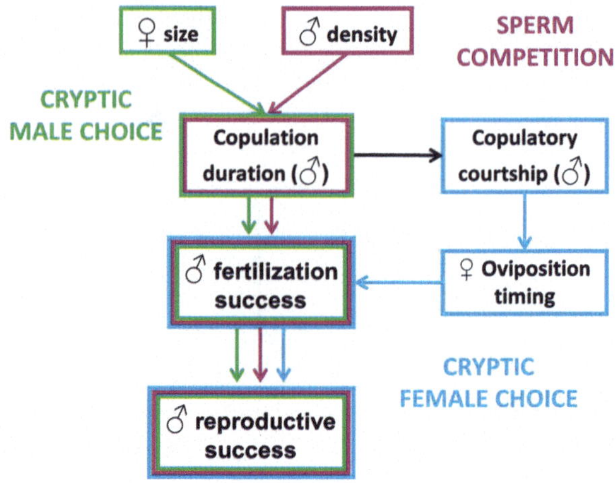

Fig. 14.10 Summary of the effects of cryptic female choice, sperm competition, and cryptic male choice on male reproductive success in the soldier fly. Different colors denote the effects of different postcopulatory processes (*blue* cryptic female choice; *pink* sperm competition; *green* cryptic male choice)

postcopulatory mechanisms are likely to occur together in many other species as well, and investigating the co-occurrence of and interplay between these processes is an important future direction to the field.

Copulatory courtship and copulation duration appear to be key traits in determining male reproductive success in soldier flies. These two traits happen to be highly correlated, since males court during the entire duration of copulation and there is very little variation in courtship rate. Therefore, it may be challenging to tease apart the effect of each component on male fitness. While females respond to the presence of courtship by adjusting oviposition timing, the specific mechanism that leads to higher fertilization success after longer copulations in this species remains unknown, as well as whether this effect is a result of longer copulations or longer courtships.

Longer copulations may result in higher fertilization success for males as a consequence of increased sperm and seminal fluid transfer. However, it is unknown whether there is a positive relationship between copulation duration and ejaculate transfer in soldier flies. Males have been shown to adjust ejaculate size in response to SC in several species (Gage 1991; Wedell and Cook 1999; Engqvist 2007; Ramm et al. 2009), and there are numerous examples that show that larger ejaculates result in higher fertilization success (Simmons 1987; Simmons et al. 1996; Arnqvist and Danielsson 1999).

Alternatively, higher fertilization success may be the result of CFC. Females may bias fertilization toward males that prolong copulations (or courtship) by controlling sperm transfer and storage. Empirical work on other species shows that females bias fertilization toward males that perform more intense copulatory courtship (Tallamy et al. 2002, 2003; Fedina and Lewis 2004; Edvardsson and Arnqvist 2000), and this may be the case of the soldier fly as well.

The field of CFC has vastly advanced in the past decade, and a large body of work supports CFC as a widespread mechanism across taxa. Still, the strength of

CFC in driving evolutionary processes remains relatively unknown. But measuring the strength of all postcopulatory sexual selection processes is needed to fully understand their evolutionary importance (Simmons and Beveridge 2010). Attention has also been drawn to the interaction between precopulatory and postcopulatory selection (Fedina and Lewis 2008; Rebar et al. 2011). With this case study, I highlight that a full understanding of CFC requires us to not only study it in isolation, but to develop experiments that tease apart and quantify the various components of postcopulatory sexual selection at play between males and females.

Acknowledgments I thank Anita Aisenberg and Alfredo Peretti for inviting me to write this chapter, Darren Rebar and an anonymous reviewer for useful comments on a previous draft, and Allysa Hallett for allowing me to use her drawing. Financial support was provided by NSF Postdoctoral Fellowship in Biology FY 2012 number 1202761.

References

Andrés JA, Cordero Rivera A (2000) Copulation duration and fertilization success in a damselfly: an example of cryptic female choice? Anim Behav 59(4):695–703

Arnqvist G, Danielsson I (1999) Postmating sexual selection: the effects of male body size and recovery period on paternity and egg production rate in a water strider. Behav Ecol 10(4):358–365

Barbosa F (2009) Cryptic female choice by female control of oviposition timing in a soldier fly. Behav Ecol 20(5):957–960.

Barbosa F (2011) Copulation duration in the soldier fly: the roles of cryptic male choice and sperm competition risk. Behav Ecol 22(6):1332–1336

Barbosa F (2012) Males responding to sperm competition cues have higher fertilization success in a soldier fly. Behav Ecol 23(4):815–819.

Birkhead T (1998) Cryptic female choice: criteria for establishing female sperm choice. Evolution 52(4):1212–1218

Birkhead T (2000) Defining and demonstrating postcopulatory female choice–again. Evol Int J Org Evol 54(3):1057–1060

Birkhead TR, Møller AP (1998) Sperm competition and sexual selection. Academic Press, Massachusetts

Bonduriansky R (2001) The evolution of male mate choice in insects: a synthesis of ideas and evidence. Biol Rev 76(3):305–339

Bretman A, Fricke C, Chapman T (2009) Plastic responses of male *Drosophila melanogaster* to the level of sperm competition increase male reproductive fitness. Proc Biol Sci/R Soc 276(1662):1705–1711.

Cook PA, Gage MJ (1995) Effects of risks of sperm competition on the numbers of eupyrene and apyrene sperm ejaculated by the moth *Plodia interpunctella* (Lepidoptera: Pyralidae). Behav Ecol Sociobiol 36(4):261–268

Cornwallis CK, O'Connor EA (2009) Sperm: seminal fluid interactions and the adjustment of sperm quality in relation to female attractiveness. Proc R Soc B Biol Sci 276(1672):3467–3475

Dewsbury DA (1982) Ejaculate cost and male choice. Am Nat 119:601–610

Eberhard WG (1994) Evidence for widespread courtship during copulation in 131 species of insects and spiders, and implications for cryptic female choice. Evolution 48(3):711–733

Eberhard WG (1996) Female control: sexual selection by cryptic female choice. Princeton University Press, New Jersey

Edvardsson M, Arnqvist G (2000) Copulatory courtship and cryptic female choice in red flour beetles *Tribolium castaneum*. Proc Biol Sci/R Soc 267(1443):559–563.

Engqvist L (2007) Male scorpionflies assess the amount of rival sperm transferred by females' previous mates. Evol Int J Org Evol 61(6):1489–1494.
Engqvist L, Sauer KP (2001) Strategic male mating effort and cryptic male choice in a scorpion-fly. Proc Biol Sci/R Soc 268(1468):729–735.
Fedina TY, Lewis SM (2004) Female influence over offspring paternity in the red flour beetle *Tribolium castaneum*. Proc Biol Sci/R Soc 271(1546):1393–1399.
Fedina TY, Lewis SM (2008) An integrative view of sexual selection in *Tribolium* flour beetles. Biol Rev Camb Philos Soc 83(2):151–171.
Fox CW, Hickman DL, Raleigh EL, Mousseau TA (1995) Paternal investment in a seed beetle (Coleoptera: Bruchidae): influence of male size, age, and mating history. Ann Entomol Soc Am 88(1):100–103
Friberg U (2006) Male perception of female mating status: its effect on copulation duration, sperm defence and female fitness. Anim Behav 72(6):1259–1268.
Fuller RC (1998) Sperm competition affects male behaviour and sperm output in the rainbow darter. Proc R Soc Lond B Biol Sci 265(1413):2365–2371
Gage A, Barnard C (1996) Male crickets increase sperm number in relation to competition and female size. Behav Ecol Sociobiol 38(5):349–353
Gage MJ (1991) Risk of sperm competition directly affects ejaculate size in the Mediterranean fruit fly. Anim Behav 42(6):1036–1037
Gage MJG (1998) Influences of sex, size, and symmetry on ejaculate expenditure in a moth. Behav Ecol 9(6):592–597
Gross MR, Sargent RC (1985) The evolution of male and female parental care in fishes. Am Zool 25(3):807–822
James MT, McFadden MW (1971) The genus *Merosargus* in Middle America and the Andean subregion (Diptera: Stratiomyidae). Melanderia 7:1–76.
Jennions MD, Petrie M (1997) Variation in mate choice and mating preferences: a review of causes and consequences. Biol Rev 72(2):283–327
Kempenaers B, Foerster K, Questiau S, Robertson B, Vermeirssen ELM (2000) Distinguishing between female sperm choice versus male sperm competition: a comment on Birkhead. Evol Int J Org Evol 54(3):1050–1052
Kvarnemo C, Simmons LW (1999) Variance in female quality, operational sex ratio and male mate choice in a bushcricket. Behav Ecol Sociobiol 45(3):245–252
Lorch PD, Wilkinson GS, Reillo PR (1993) Copulation duration and sperm precedence in the stalk-eyed fly *Cyrtodiopsis whitei* (Diptera: Diopsidae). Behav Ecol Sociobiol 32(5):303–311
Marconato A, Shapiro D (1996) Sperm allocation, sperm production and fertilization rates in the bucktooth parrotfish. Anim Behav 52(5):971–980
Oberhauser KS (1988) Male monarch butterfly spermatophore mass and mating strategies. Anim Behav 36(5):1384–1388
Parker G, Ball M, Stockley P, Gage M (1996) Sperm competition games: individual assessment of sperm competition intensity by group spawners. Proc R Soc Lond B Biol Sci 263(1375):1291–1297
Parker G, Ball M, Stockley P, Gage M (1997) Sperm competition games: a prospective analysis of risk assessment. Proc R Soc Lond B Biol Sci 264(1389):1793–1802
Parker GA (1970) Sperm competition and its evolutionary consequences in the insects. Biol Rev 45(4):525–567
Parker GA, Pizzari T (2010) Sperm competition and ejaculate economics. Biol Rev 85(4):897–934
Parker GA, Simmons LW, Stockley P, McChristie DM, Charnov EL (1999) Optimal copula duration in yellow dung flies: effects of female size and egg content. Anim Behav 57(4):795–805
Prestwich KN (1994) The energetics of acoustic signaling in anurans and insects. Am Zool 34(6):625–643
Ramm SA, McDonald L, Hurst JL, Beynon RJ, Stockley P (2009) Comparative proteomics reveals evidence for evolutionary diversification of rodent seminal fluid and its functional significance in sperm competition. Mol Biol Evol 26(1):189–198.

Rebar D, Zuk M, Bailey NW (2011) Mating experience in field crickets modifies pre- and postcopulatory female choice in parallel. Behav Ecol 22(2):303–309.
Reinhold K, Kurtz J, Engqvist L (2002) Cryptic male choice: sperm allocation strategies when female quality varies. J Evol Biol 15(2):201–209
Sakaluk SK (1985) Spermatophore size and its role in the reproductive behaviour of the cricket, *Gryllodes supplicans* (Orthoptera: Gryllidae). Can J Zool 63(7):1652–1656
Seifert RP, Seifert FH (1976) A community matrix analysis of *Heliconia* insect communities. Am Nat 110(973):461–483
Seifert RP, Seifert FH (1979) A Heliconia insect community in a Venezuelan cloud forest. Ecology 60(3):462–467
Simmons L (1987) Sperm competition as a mechanism of female choice in the field cricket, *Gryllus bimaculatus*. Behav Ecol Sociobiol 21(3):197–202
Simmons L, Siva-Jothy M (1998) Sperm competition in insects: mechanisms and the potential for selection. In: Sperm competition and sexual selection . Academic Press. pp 341–434
Simmons L, Stockley P, Jackson R, Parker G (1996) Sperm competition or sperm selection: no evidence for female influence over paternity in yellow dung flies *Scatophaga stercoraria*. Behav Ecol Sociobiol 38(3):199–206
Simmons LW (2001) Sperm competition and its evolutionary consequences in the insects. Princeton University Press, New Jersey
Simmons LW, Beveridge M (2010) The strength of postcopulatory sexual selection within natural populations of field crickets. Behav Ecol 21(6):1179–1185.
Tallamy DW (2000) Sexual selection and the evolution of exclusive paternal care in arthropods. Anim Behav 60(5):559–567
Tallamy DW, Darlington MB, Pesek JD, Powell BE (2003) Copulatory courtship signals male genetic quality in cucumber beetles. Proc Biol Sci/R Soc 270(1510):77–82.
Tallamy DW, Powell BE, McClafferty JA (2002) Male traits under cryptic female choice in the spotted cucumber beetle (Coleoptera: Chrysomelidae). Behav Ecol 13(4):511–518
Tang-Martinez Z, Ryder TB (2005) The problem with paradigms: Bateman's worldview as a case study. Integr Comp Biol 45(5):821–830
Thornhill R (1983) Cryptic female choice and its implications in the scorpionfly *Harpobittacus nigriceps*. Am Nat 122(6): 765–788
Tompkins L, Hall JC (1981) The different effects on courtship of volatile compounds from mated and virgin *Drosophila* females. J Insect Physiol 27(1):17–21
Trivers R (1972) Parental investment and sexual selection. In: Campbell B (ed) Sexual selection and the descent of man 1871–1971. Aldine, Chicago, pp 136–179
Vahed K (1998) The function of nuptial feeding in insects: a review of empirical studies. Biol Rev 73(1):43–78
Vehrencamp SL, Bradbury JW, Gibson RM (1989) The energetic cost of display in male sage grouse. Anim Behav 38(5):885–896
Wagner WE (2005) Male field crickets that provide reproductive benefits to females incur higher costs. Ecol Entomol 30(3):350–357
Wedell N (1998) Sperm protection and mate assessment in the bushcricket *Coptaspis* sp. Anim Behav 56(2):357–363
Wedell N, Cook P (1999) Strategic sperm allocation in the small white butterfly *Pieris rapae* (Lepidoptera: Pieridae). Funct Ecol 13(1):85–93
Wedell N, Gage MJG, Parker GA (2002) Sperm competition, male prudence and sperm-limited females. Trends Ecol Evol 17(7):313–320
Woodley NE (2001) A world catalog of the Stratiomyidae: (Insecta: Diptera), vol 11. North American Dipterists' Society, USA
Xu J, Wang Q (2009) Male moths undertake both pre- and in-copulation mate choice based on female age and weight. Behav Ecol Sociobiol 63(6):801–808.
Yasui Y (1996) Males of a mite, *Macrocheles muscaedomesticae*, estimate a female's value on the basis of her age and reproductive status. J Insect Behav 9(3):517–524

Chapter 15
Species-Specific Behavioral Differences in Tsetse Fly Genital Morphology and Probable Cryptic Female Choice

R.D. Briceño and W.G. Eberhard

Abstract A long-standing mystery in morphological evolution is why male genitalia tend to diverge more rapidly than other structures. One possible explanation of this trend is that male genitalia function as "internal courtship devices," and are under sexual selection by cryptic female choice (CFC) to induce female responses that improve the male's chances of fathering her offspring. Males of closely related species, which have species-specific genital structures, are thought to provide divergent stimulation. Testing this hypothesis has been difficult; the presumed genital courtship behavior is hidden from view inside the female; appropriate experimental manipulations of male and female genitalia are often technically difficult and seldom performed; and most studies of how the male's genitalia interact with those of the female are limited to a single species in a given group, thus limiting opportunities for comparisons of species-specific structures. In this chapter, we summarize data from morphological, behavioral, and experimental studies of six species in the tsetse fly genus *Glossina*, including new X-ray recordings that allowed visualization of events inside the female during real time. Species-specific male genital structures perform dramatic, stereotyped, rhythmic movements, some on the external surface of the female's abdomen and others within her reproductive tract. Counting conservatively, a female *Glossina* may sense stimuli from the male's genitalia at up to 8 sites on her body during some stages of copulation. As predicted by CFC theory, these movements differ among closely related species;

R.D. Briceño (✉) · W.G. Eberhard
Escuela de Biología, Universidad de Costa Rica, Ciudad Universitaria, Costa Rica
e-mail: rbriceno@biologia.ucr.ac.cr

W.G. Eberhard
e-mail: william.eberhard@gmail.com

W.G. Eberhard
Smithsonian Tropical Research Institute, Panama, Republic of Panama

W.G. Eberhard
Biological Sciences and Museum of Natural History, Louisiana State University, Baton Rouge, USA

A.V. Peretti and A. Aisenberg (eds.), *Cryptic Female Choice in Arthropods*,
DOI 10.1007/978-3-319-17894-3_15

some of the species-specific male genital structures that are apparently designed to stimulate the female move with different rhythms against different portions of the female's body in different species. In no case does female morphology show signs of counter adaptations to avoid or reduce male stimulation or to fit mechanically with male structures, as predicted by some alternative hypotheses to explain rapid divergent evolution of genitalia; for most male structures, the corresponding portion of the female is featureless and uniform in different species. Experimental modifications of one species-specific genital structure (the cercus), and of the possible female sense organs in the portion of the female that this male structure contacts during copulation, elicited female reproductive responses in two species (reductions in sperm transport, ovulation, and resistance to further copulations) that could result in cryptic female choice favoring this male structure.

15.1 Introduction

The male genitalia of animals with internal fertilization tend to diverge rapidly, probably because they are under one or more types of sexual selection (Eberhard 1985, 1996, 2009; Hosken and Stockley 2003; Leonard and Córdoba-Aguilar 2010). This pattern of rapid divergence is amply demonstrated in the taxonomic literature on many groups of animals; the morphological differences between the male genitalia of closely related species are often greater than those between other body parts of the same species. The ways in which sexual selection has acted on male genitalia function are much less studied. One major hypothesis is that they are used as internal courtship devices by males to stimulate females. The male's genitalia are thought to be under selection to induce the female to use the male's sperm to fertilize her eggs, and thus to exercise cryptic female choice in his favor (Eberhard 1985). Testing this hypothesis and the several alternatives (see Eberhard, this volume; also Discussion below) has been difficult, however, because the presumed genital courtship behavior during copulation is generally hidden from view inside the female.

Nearly, all studies of the possible functions of genital structures have relied on static views of the form of the male's genitalia, and their positions within the female (for Diptera, see references to studies of 43 species in Table 1 of Eberhard 2004a, also Briceño et al. 2007).

This chapter summarizes data from morphological studies of the erect male genitalia during copulation, video recordings of genital movements both outside and inside females, and experimental modifications of the species-specific details of male genital structures and of female sense organs in the areas that are contacted by these structures in different species of the tsetse fly genus *Glossina*. Because of the abundant background data that are available due to the medical and economic importance of tsetse flies, these studies provide one of the most extensive comparative views of the functional genital morphology ever published. Tsetse studies are also unique in providing the first direct behavioral observations in an arthropod of how male genital structures move within the female.

15.2 Background: The Natural History of *Glossina*

Glossina (tsetse) flies are important vectors of diseases of humans and domestic animals in sub-Saharan Africa (Gooding and Krafsur 2005; Feldman et al. 2005), and their taxonomy, biogeography, habitat preferences, genome composition, and reproductive physiology have been studied intensively (e.g., Newstead et al. 1924; Buxton 1955; Mulligan 1970; Tobe and Langley 1978; Gillot and Langley 1981; Gooding and Krafsur 2005; International *Glossina* Genome Initiative 2014). The genus is composed of approximately 20 species. Their reproduction is unusual. Females ovulate only a single egg at a time; it is fertilized and then retained in her modified oviduct (the "uterus"); the larva hatches here and feeds on material secreted by the female's milk gland until it is mature. A single larva is raised at a time, and only finally emerges from the female when it is ready to pupate. A female can raise several larvae (up to an estimated 11 cycles in field-collected females of *G. pallidipes* and *morsitans*) (Hargrove 2012) in a lifetime in the field.

Both males and females feed by sucking blood from vertebrates (generally, but not always, large mammals) (Hoppenheit et al. 2013), and males apparently lurk near hosts to grab females in the air when they come to feed. *Glossina* females are probably effectively isolated from heterospecific crossings by different diurnal activity cycles, habitats, and geographic ranges, and by species-specific surface hydrocarbons that allow males to distinguish the sex and species identity of females prior to copulation (Huyton et al. 1980; Wall and Langley 1993). Although females may occasionally be subject to brief chases or strikes by heterospecific males, they are apparently not normally subject to intromission attempts by cross-specific males in nature. Dissections of field-captured females of *G. pallidipes* and *morsitans morsitans* indicate that females do not begin mating until they are several days old and that some females mate more than once (Hargrove 2012). Some female *G. palpalis* also mate more than once in the field (Squire 1951). Multiple matings by females are reproductively significant, because both first and second males sometimes sired offspring in twice-mated females of *G. morsitans* (Kawooya 1977).

Females need to feed several times to rear a single larva, so there are opportunities for a female to copulate repeatedly. Males seize and attempt to copulate with objects coated with species-specific female surface hydrocarbons (Wall and Langley 1993). Copulation is long (normally about 30–120 min, but rarely up to 24 h Saunders 1970), and in at least some species, it is so long that the male's genitalia are designed to allow the female to defecate during copulation, thus avoiding the danger of her intestinal tract becoming plugged (Pollock 1974). Sperm transfer occurs just before separation, at the end of copulation (Jaensen 1979a, b). The male first constructs a spermatophore which is surrounded by a diffuse male accessory gland product and whose tip is inserted into the lower portion of the common spermathecal duct; sperm is then transferred into the spermatophore and (in smaller quantities) up the spermathecal duct (Pollock 1970, 1974).

Females probably participate actively in moving the sperm up the spermathecal ducts and into the spermathecae, as is the general rule in insects (summaries in Eberhard 1996; Simmons 2001): Spermatophores that are discarded by females commonly contained "considerable" amounts of sperm (Pollock 1974 on *G. austeni*); and there is a negative effect on sperm transfer to the spermathecae of blinding female sense organs to male stimuli (Briceño and Eberhard 2009a, b on *G. pallidipes* and *G. morsitans*). Away from hosts, tsetse flies are rare, and the densities of their populations are quite low (Feldmann et al. 2005). The flies are relatively large (>1 cm long; their bite is painful) and are robust and can survive rough handling, facilitating experimental manipulations.

The phylogenies of both the bacterial gut symbionts (which may have been necessary to permit *Glossina* as well as related hippoboscoid flies to evolve to feed on vertebrate blood—McAlpine 1989), and also of the ribosomal ITS-2 sequence, as well as other, morphological traits and habitat preferences of the flies indicate that there are three subgenera: The *morsitans* and *palpalis* subgenera are more closely related to each other than they are to the *fusca* group (Newstead et al. 1924; Potts 1970; Aksoy et al. 1997; Chen et al. 1999).

15.3 Copulatory Courtship Behavior

Male *G. pallidipes* performed six highly repetitive male behavior patterns during copulation that are likely to stimulate the female (Jaensen 1979a, b; Briceño et al. 2007): "Peep" (a high pitched whine produced by rapid vibration of the male's wings while they were folded over his abdomen; "wing buzz" (both wings were repeatedly brought forward to extend laterally and buzzed for about 0.5 s; usually the male made two wing-forward movements in each burst of buzzing); "raised legs II"(both middle legs were raised dorsally and anteriorly at the initiation of a bout of wing buzzing); "rub with legs I" (the front legs rubbed or tapped repeatedly on the pronotum and/or dorsal surface of the head of the female for about 0.5 s/burst); "rub with legs II" (the middle legs rubbed or tapped repeatedly on the sides of the female's thorax, her head or her abdomen); and "rub with legs III" (the hind legs rubbed or tapped repeatedly on the ventral surface of the female's abdomen). None of these male behaviors had any obvious mechanical consequences for copulation (e.g., they did not help the male hold onto the female); they appear to be designed instead to stimulate the female during copulation. They resembled the copulatory courtship behavior of many other insect species (Eberhard 1994, 1996).

15.4 Clasping Male Genitalia

15.4.1 Morphology of Male Clasping Genitalia

Several male genital structures have long been known to have species-specific forms (Fig. 15.1) and are well illustrated in the taxonomic literature on *Glossina*

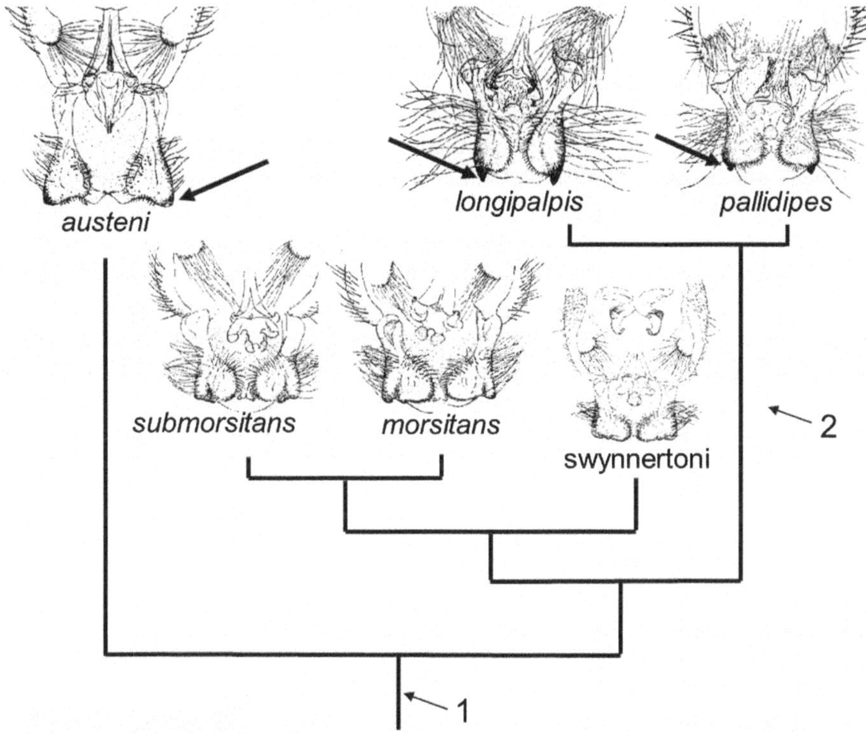

Fig. 15.1 Drawings of the male genitalia (cerci) of *Glossina* species in the *morsitans* subgenus (from Newstead et al. 1924) superimposed on the phylogenetic relations between these species (after Chen et al. 1999). At *1*, there were lateral cercal teeth but they were not elongate or strengthened (*heavy arrow* to *austeni*). At *2* these teeth, whose function was studied experimentally in *G. pallidipes* by Briceño and Eberhard (2009a), were elongate and strengthened (*heavy arrows* in *longipalpis* and *pallidipes*) (changes deduced from outgroup comparisons with the other two subgenera of *Glossina*) (from Briceño and Eberhard 2009a)

(summary in Potts 1970). The male genitalia are highly derived with respect to those of other flies, and the homologies of some structures are not clear (D.M. Wood, pers. comm.). Most taxonomic illustrations of male genitalia include only their positions at rest, rather than to the everted positions that they assume during copulation. Thus, several of the male sclerites and processes long known to differ between species were not understood with respect to their mechanical relationships to each other and to the female during copulation until recently.

We have observed copulating pairs of six species *G. pallidipes*, G. *morsitans*, and *G. swynnertoni* of the *morsitans* subgenus; *palpalis* and *fuscipes* of the *palpalis* subgenus; and *G. brevipalpis* of the *fusca* subgenus under a dissecting microscope; in addition, we have flash-frozen pairs and then dissected them (Briceño et al. 2007, 2015; Briceño and Eberhard 2009a, b), Pollock (1974) sectioned more than 70 pairs of *G. austeni* (in the *morsitans* subgenus) that were flash frozen during copulation. The following description of clasping in the best-studied species, *G. pallidipes*, applies in at least in its major details to all seven species (and probably for species throughout the genus).

The male genitalic structures of *Glossina* can be classified in two categories: those which function to grip or press against the external surface of the female's abdomen—the cercus, the fifth abdominal sternite, and the inferior clasper of the male (we include male's modified sternite 5 as part of his genitalia, though strictly speaking this is not correct); and those inserted into her genital tract—the phallobase or phallosome (e.g., Patton 1936; Pollock 1974) (with its associated structures) (Fig. 15.2). There are species-specific traits on both sets of male structures, though the morphology of the gripping structures is better studied (they are larger and are more easily visible when the male genitalia are not everted). The mechanical interactions of the male's grasping structures with the female are also better understood (Figs. 15.2 and 15.3) than those of his intromittent genitalia.

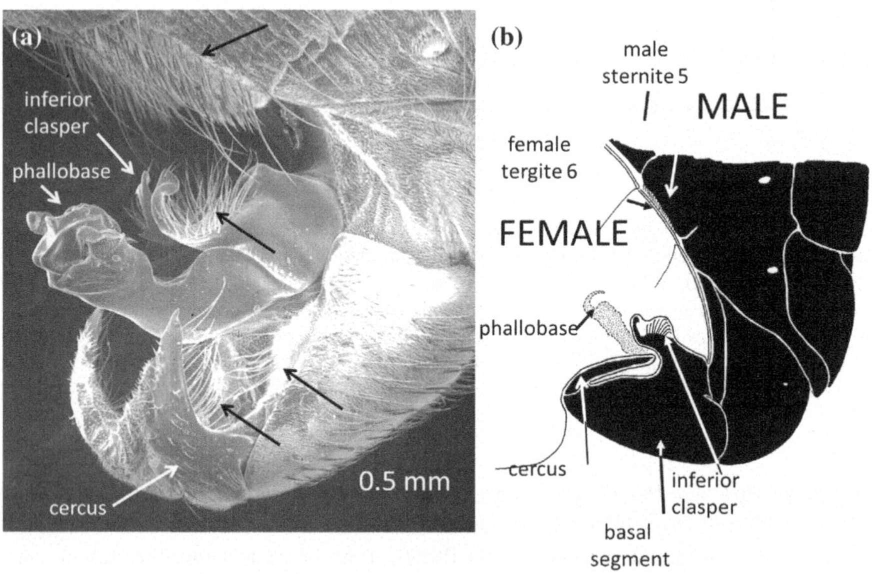

Fig. 15.2 **a** SEM portrait of the everted male genitalia of *G. fuscipes*. The *heavy arrows* indicate groups of setae that rub against the female when the male squeezes her with his cerci. **b** Schematic representation of the male (*black*) and female genitalia of *G. fuscipes* during copulation. The male's phallobase is deep inside the female's vagina, his tightly folded cerci pinch the ventral wall of the female's abdomen tightly, the dense, stout setae on his sternite 5 press on her tergite 6, and the curved processes of the his inferior claspers and their brush of setae press on the external surface of her abdomen just ventral to the posterior tip of her tergite 7 (the tips of the inferior clasper setae were not observed directly; probably they are curved as pictured, given their lengths, the tight quarters, and the substantial pressures that males apply when squeezing females, as evidenced by deformation of the ventral surface of the female's abdomen) (from Briceño et al. 2015)

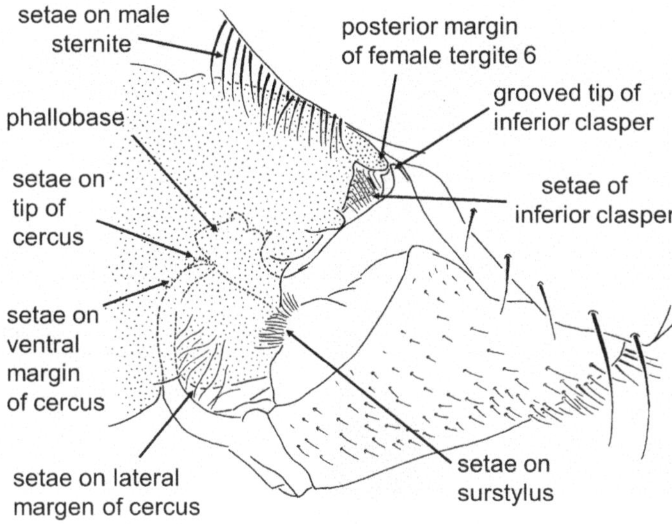

Fig. 15.3 Diagrammatic view of the genitalia of a copulating pair of *G. pallidipes* showing multiple points where apparently stimulatory male structures contact and move against the female. Six groups of male setae that press or rub against the female during copulation, as well as the inferior clasper contact with the posterior margin of her tergite and his intromittent phallobase, are all indicated (from Briceño et al. 2015)

The male's cerci and his sternum 5 clasp the external surface of the female in a pliers-like grip. The plate-like cerci articulate basally with the thick "basal segment," which houses powerful muscles (Fig. 15.4). The tips of the cerci press against the membranous ventral surface near the tip of the female's abdomen, causing it to invaginate sharply (Fig. 15.2). The other arm of the "pliers" is the highly setose ventral surface of the male's sternite 5 (which is sexually dimorphic); it presses on dorsal surface of tergite 6 or 7 of the female abdomen. Studies of *G. pallidipes* (Briceño et al. 2007) showed that the genitalic clasp and the subsequent movements of the male genitalia cause five different groups of setae on the male genitalia and a further group on his sternite 5 to rub or brush against the external surface of the female's abdomen (Fig. 15.3).

There are minor differences between the subgenera in the positions of the male's genitalia on the female during copulation. Males in the *morsitans* subgenus were positioned slightly more posteriorly on the female. In *G. brevipalpis*, G. *palpalis*, and *G. fuscipes*, the cerci folded tightly against the ventral surface of the basal segment, making an angle of substantially less than 90° with the basal segment (Fig. 15.2), while in the *morsitans* subgenus this angle was closer to 90° (Fig. 15.3).

Details of the contact between the inferior claspers of the male genitalia and the surface of the female's abdomen during copulation also varied, and the members of the *morsitans* subgenus again differed from the other species. The inferior claspers of the *morsitans* subgenus pressed on the tip of tergite 6 itself, and the tergite tip fit into a groove in the inferior clasper (Fig. 15.3), while the inferior

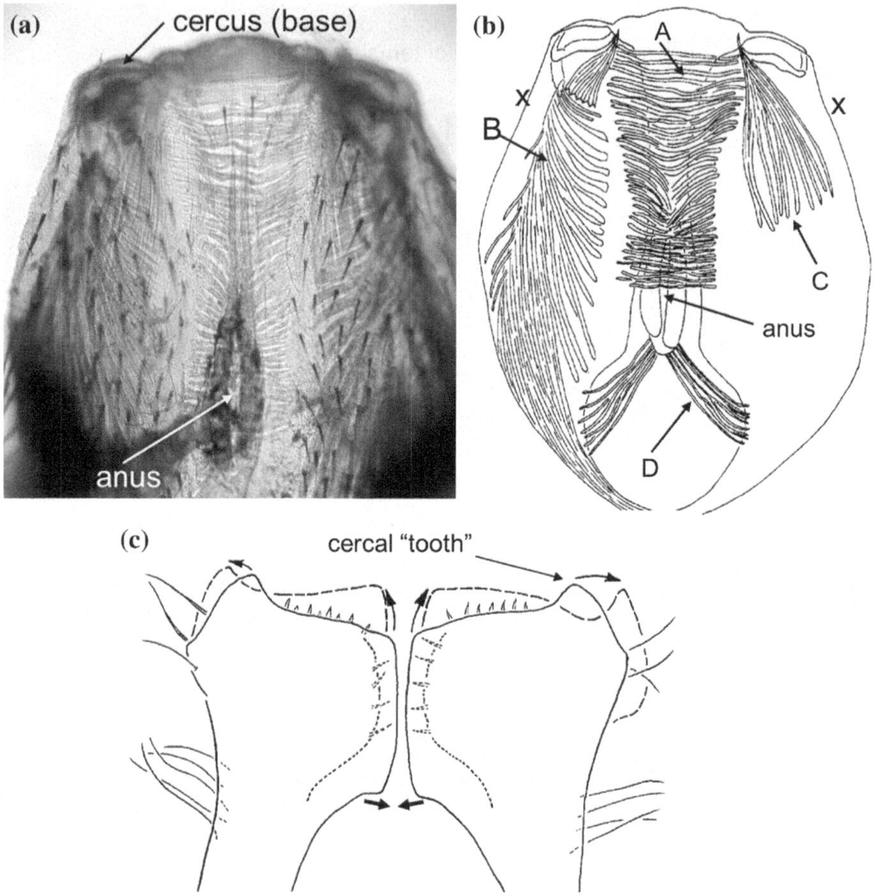

Fig. 15.4 The powerful muscles of the male genitalia of *G. pallidipes* provide indirect evidence for further movements that could not be directly observed during copulation. When the dorsal wall of the male's basal segment is made transparent (**a**), several groups of muscle fibers are visible (*A*–*D* in **b**). The most massive group of fibers (*B*) apparently serves to move the two cerci, as shown diagrammatically in (**c**) (artificially squeezing the cerci at points "x" produced the rocking motion shown in **c**). Rocking movements would cause each cercal tooth to scrape against or stretch the abdominal cuticle of the female during copulation (the tips of the cerci are at the bottom in **a** and **b**, at the top in **c**) (after Briceño et al. 2007)

claspers and their dense setae in the males of the *palpalis* and *fusca* subgenera pressed on the membranous body wall of the female, directly ventral to the tip of her abdomen (tergite 7) (Fig. 15.2) (Briceño et al. 2015). In *brevipalpis*, in contrast, each inferior clasper bears a flat plate-like extension and lacks strong setae; the medial surfaces of these planar processes pressed on the female's body wall just ventral to the tip of her tergite 7 (Briceño et al. 2015).

There are no external differences in the females that corresponded to any of these differences in the morphology and positions of the male genitalia. The ventral surface of her abdomen that was grasped by the cerci, the tip of her tergites 6 and 7, and the membranous area just ventral to the tergite that were contacted by the male inferior clasper are all featureless and at least superficially similar in the different species, except for a transverse membranous cleft dorsal to the vulva of *G. austeni* into which the inferior claspers fit (Pollock 1974).

15.4.2 Behavior of Male Grasping Genitalia. Observations of the Behavior of Cerci and Associated Structures

Direct observations under the dissecting microscope showed that the cerci squeeze the distal tip of the female's abdomen rhythmically in long bursts of activity (lasting many minutes). These squeezes exert substantial force on the female. They cause the membranous ventral surface of her abdomen to invaginate sharply (Fig. 15.2) and bend the lower portion of her internal reproductive tract sharply (Vanderplank 1947; Pollock 1974; Briceño et al. 2007); in *G. palpalis*, the male cerci produce small areas of abrasion on the membranous female abdomen (Squire 1951). Another indication of strong forces is that the distal margins of the cerci in field-collected males sometimes have a "distinctly chipped or worn appearance" in *G. morsitans* (Newstead et al. 1924, p. 38).

The squeezing behavior of *G. pallidipes* is highly structured, with long rhythmic series that vary consistently at different stages of copulation (Briceño et al. 2007). Species apparently differ in the relative strengths and durations of squeezes (Table 15.1). Although different strengths were only classified in loose, probably somewhat imprecise categories, and although the squeezes at different stages of copulation were combined, some differences were so clear that they surely reflect real behavioral differences between species. For instance, both the strong and very strong squeezes by *G. pallidipes* were much longer than those of *G. fuscipes*; small squeezes were common in *G. pallidipes*, rare in *G. brevipalpis* and absent in *G. fuscipes* and *palpalis*.

Artificial manipulation of the cerci of *G. pallidipes* by squeezing their bases together gently with a forceps showed that the two cerci could rock against each other (Fig. 15.4c); this movement would cause the cercal teeth to scrape against or stretch the female's abdominal cuticle during copulation. These movements were not verified by direct observations (as can be seen in Fig. 15.3, the cercal tips are out of view in the deep fold in the female's abdomen), but the heavy musculature in the basal segment (muscles A, B, C in Fig. 15.4b) and the sustained, rhythmic narrowing movements of the basal segment during copulation in *G. pallidipes* imply that such movements do occur (Briceño et al. 2007).

X-ray videos (Briceño et al. 2010, Fig. 15.8) of all species of the *morsitans* subgenus consistently revealed a pair of retractable spines arising near the bases of the cerci that pinched the surface of the female's abdomen against the cerci, just posterior to the distal tips of the male's cerci, causing a small, rounded portion of the female's abdomen to be pinched off between the tip of the spine and the tip of the male's cercus (Briceño et al. 2015). This pinch was constant, and the spine did not move, except when the male extended his cerci and relaxed his squeeze on the

Table 15.1 Durations (in seconds) of different types of squeezing movements by the male's cerci in five species of *Glossina*

	G. pallidipes	*G. morsitans morsitans*	*G. morsitans centralis*	*G. brevipalpis*	*G. fuscipes*	*G. palpalis*
Relative strength of the squeeze						
Very strong	$62.6 \pm 44.1a_1a_2a_3a_4$ (256)	$26.6 \pm 27.4a_1a_5a_6a_7b_1$ (376)	$13.5 \pm 16.7a_2a_5a_8a_9a_{10}$ (195)	$6.3 \pm 5.97a_3a_6a_8a_{11}a_{12}$ (280)	$3.92 \pm 21.4a_4a_7a_9a_{11}a_{13}$ (51)	$131.1 \pm 13.0b_1b_2a_{10}a_{12}a_{13}$ (13)
Strong	$50.4 \pm 53.5a_1c_1a_2a_3$ (119)	$53.5 \pm 33.0a_1a_4a_5a_6$ (50)	$16.5 \pm 15.1c_1a_4a_7a_8$ (18)	–	$0.64 \pm 0.47a_2a_5a_7a_9$ (409)	$4.47 \pm 6.67a_3a_6\ a_8a_9$ (279)
Medium	$22.0 \pm 9.0a_1a_2$ (112)	$6.7 \pm 6.2a_1$ (20)	–	$5.6 \pm 3.7a_2$ (218)	–	–
Mean number of sequences	$14.8 \pm 8.9b_1a_1$ (55)	$9.4 \pm 9b_1$ (20)	–	$7.8 \pm 5.1a_1$ (211)	–	–
Small	0.17 ± 0.03a1a2 (150)	0.12 ± 0.16a1a3a4 (150)	0.14 ± 0.13a3 (50)	$0.23 \pm 0.04a_2a_4$ (45)	–	–
Mean number of sequences	33.6 ± 20.3a1a_2 (160)	$27.6 \pm 26.1a_1a_3$ (288)	$30.6 \pm 14.9a_4$ (160)	$3.1 \pm 1.1a_2a_3a_4$ (30)	–	–
N (pairs)	22	19	9	5	4	5

The relative strengths of squeezes were determined by qualitative judgements of how deeply the male cerci folded the ventral surface of the female's abdomen inward. Numbers in parentheses refer to the number of squeezes measured. The values in the same row with the same letter and number (e.g., a_1, a_2) were significantly different with Mann–Whitney U tests (a = $p < 0.001$; b = $p < 0.01$; c = $p < 0.05$)

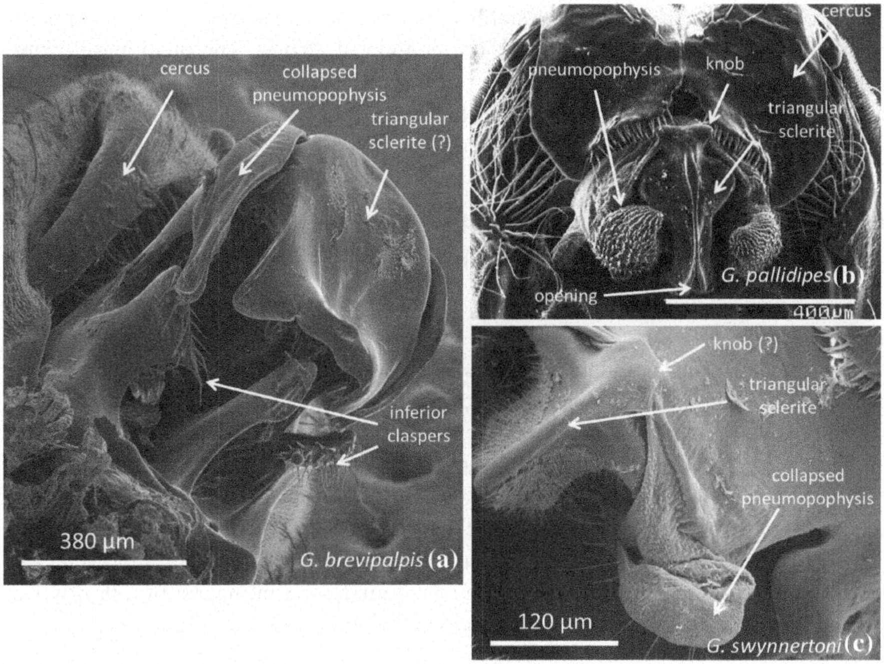

Fig. 15.5 SEM portraits of selected aspects of the distal portion of the phallobase in *G. brevipalpis* (**a**), *G. pallidipes* (**b**), and *G. swynnertoni* (**c**) that illustrate types of inter-specific differences. *G. pallidipes* and *G. swynnertoni* are in the same subgenus, and some homologous though differently shaped structures are recognizable (e.g., the triangular sclerite and the knob); both species have pneumopophyses, but those of *G. pallidipes* are heavily spined while those of *G. synnertoni* are nearly smooth, at least in the distal portion. In *G. brevipalpis*, which is a member of a different subgenus, some traits, such as the triangular plate and knob are absent or so different in form that they are not easily recognizable; other homologous traits, such as the pneumopophyses, are recognizable but have different forms and lack spines. In **b**, the cerci of *G. pallidipes* press forcefully against the curved surface of the basal portion of the phallobase (the female tissue which was between these male structures in the copulating pair—abdominal cuticle plus one wall of the oviduct—has been removed) (after Briceño et al. 2015)

female's abdomen, when the spines were withdrawn toward the bases of the cerci. Despite the consistency with which these pinches were seen in the X-ray videos, they never occurred in flash-frozen pairs, and indeed, no such pinches were seen in the extensive study of *G. pallidipes* (Briceño et al. 2007). Presumably the spines were withdrawn when the copulating pair was frozen.

In X-ray videos of two species, *G. pallidipes* and *G. brevipalpis*, the tips of the cerci sometimes repeatedly flexed posteriorly (with respect to the female) with small amplitude movements. Because the tips of the cerci were near the basal, cylindrical portion of the phallobase during copulation (Fig. 15.5b), the effect of these flexions was to rub or knead the female tissue between the male's cerci and his phallobase.

One type of movement seen in X-ray videos of one pair of *G. swynnertoni* was not observed in any other species. A pointed sclerite between the male's cercus and his phallobase rhythmically stabbed the female, probably on the membrane or a sclerite just ventral to her vulva (Briceño et al. 2015). The phallobase of this male was not inserted in the female's vagina; it was not clear whether it had been withdrawn after a previous insertion or had not yet been inserted into the female.

15.5 Deducing the Functions of Clasping Genitalia from Morphology and Behavior

The morphological designs of genital structures and their positions with respect to the female during copulation suggested that several non-intromittent structures function to stimulate the female (Briceño et al. 2007). Six different groups of modified setae on and near the male genitalia of *G. pallidipes* rub directly against particular sites on the female's body during the squeezing behavior that accompanies clasping (Fig. 15.3) (Briceño et al. 2007). Differences in the designs of these setae correlate with the probable force with which they press on the female and the probable sensitivity of the female surfaces that they contact (e.g., the setae are robust where the male presses strongly on a female abdominal tergite, thin where they brush against female membranes) (Briceño et al. 2007).

15.6 Experimental Tests of the Functions of Male Genital Structures

Experimental modifications utilizing two techniques demonstrated that in *G. pallidipes* and *G. morsitans* two species-specific structures on the male cerci stimulate virgin females to ovulate, to transport sperm, and to refrain from remating following copulation (Briceño and Eberhard 2009a, b). In some pairs, the male cerci were modified experimentally by cutting off the large teeth on the cerci (see arrows in Fig. 15.1); control males were restrained as if to produce the experimental modifications but were left untreated. In other pairs, the male was left intact, but any sense organs in the area of the female's abdomen that is contacted by the male during copulation were blocked (by coating them with nail polish) or inactivated (by pressing a red hot needle briefly (<1 s) against her abdomen) (in both experiments adjacent segments were modified in similar ways in control females).

Sperm transfer to the spermathecae and ovulation were checked by dissecting females 9–10 days following copulation. Sperm transfer was estimated by the degree of filling of the spermathecae (Fig. 15.6). Sperm are thought to be deposited by the male in a spermatophore at the mouth of the common spermathecal duct (as in other *Glossina*); spermathecal filling (or lack of filling) may be influenced by female processes (sperm transport, sperm dumping), or by sperm motility). Ovulation was verified by checking whether there was a developing larva in the uterus 9–10 days later. Female receptivity to further mating was tested by placing the female in a glass vial

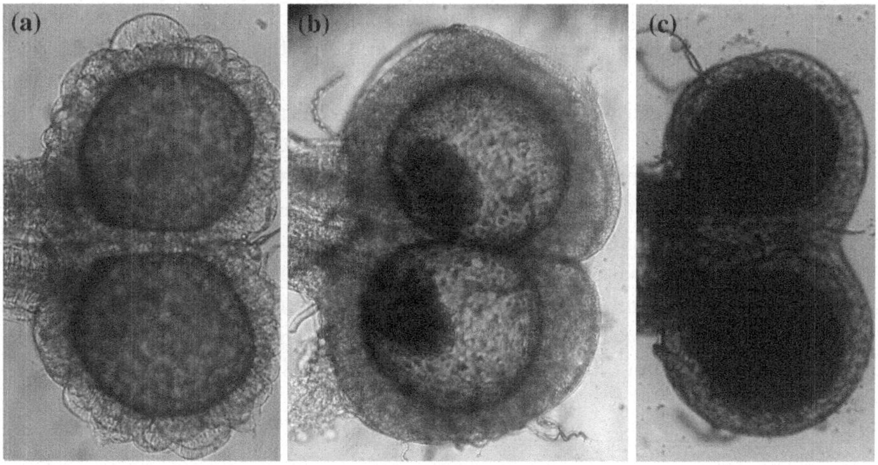

Fig. 15.6 Different degrees of filling of the spermathecae of *G. pallidipes*: **a** empty; **b** 15–20 % full; **c** full (after Briceño and Eberhard 2009a)

for 3 min with a 7-day-old virgin male. Remating tests were conducted (until the female accepted a mating) on each of the first 11 days after copulation, and then on days 25, 50, and 75 post-mating. All males attempted to mate in these trials.

As predicted by CFC theory, these manipulations of both the males and the females induced female reproductive responses that would likely produce biases against the male's chances of paternity: Ovulation and sperm storage decreased, while female remating increased. The fact that experimental blocking of female sense organs elicited the same responses as modifying the males showed that responses to altered male genital morphology were due to changes in tactile stimuli received by the female from the male's genitalia, rather than to other possible changes in the males that resulted from alterations of their genitalia. Experimental modifications of females also revealed a previously unappreciated female response. When unable to sense male contact, females prevented clasping, by keeping their wings folded posteriorly (an occasional response noted previously in intact females by Squire 1951). In combination with previous studies of tsetse reproductive physiology, these data constitute the most complete experimental confirmation that sexual selection (probably by CFC—see below) acts on the stimulatory properties of male genitalia (Briceño and Eberhard 2009a, b).

In similar experiments, in which the strong setae on the male sternite 5 of *G. pallidipes* were covered with nail polish, the likelihood that the female would ovulate was not affected, but the likelihood that she would have sperm in her spermathecae decreased (Briceño and Eberhard 2009a). The reciprocal experiment, in which the male was left unaltered, but nail polish was applied to the dorsal surface of the female tergite which is contacted by male sternite 5 during copulation, gave similar results: The fraction of females with sperm in the spermathecae decreased. In addition, ovulation decreased significantly (Briceño and Eberhard 2009a).

In one additional, unpublished experiment, nail polish was applied to the pneumopophyses of *G. pallidipes* (inflatable sacs on the male phallobase—see below), smoothing their surfaces but allowing at least some expansion. This treatment resulted in reduced sperm transfer to the female's spermathecae (7 of 12 females mated to experimental males lacked sperm in their spermatheca 9–10 days after copulation; 2 of 19 females mated to control males lacked sperm; $chi^2 = 4.27$, $p = 0.04$.). The roles of the two sexes in these experiments remain undetermined, however. It is possible, for instance, that expansions of the pneumopophyses help drag the male's phallobase deeper into the female or position it correctly to transfer sperm, but they might also serve to stimulate the female and induce a response favorable to the male's reproduction (or both). The significance of these results regarding possible cryptic female choice is thus uncertain.

15.7 Intromittent Male Genitalia

15.7.1 Morphology of Intromittent Male Genitalia

Additional species-specific male genital structures of *Glossina* are inserted into the female's vagina. The intromittent male genital structure is approximately cylindrical phallosome, which bears several structures that have been less completely characterized (see, however, Patton 1936). All five species examined had a pair of small membranous sacs (pneumopophyses); their sizes and shapes differed; and some but not others were covered with short spines (Fig. 15.5) (Briceño et al. 2015). The positions of the two pneumopophyses varied even on the two sides of the phallobase of the same male in copulating specimens (e.g., Fig. 15.5b), so these sacs are mobile during copulation (see also direct observations below). All species also had a sclerite at the tip of the phallobase; in some, it was more or less triangular, with the distal portion produced into a point bearing an opening through which sperm are probably transferred (Fig. 15.5b). This opening contained a small folded sac that, in some specimens of *G. pallidipes* that were flash frozen during copulation, was everted into the mouth of the female's spermathecal duct (Briceño et al. 2007). In *G. austeni*, the male phallobase is inserted into the lower portion of the common spermathecal duct (Pollock 1974). Taxonomic studies that have focused on the phallobase have documented substantial inter-specific variation (Newstead et al. 1924; Patton 1936).

15.7.2 Behavior of Intromittent Male Genitalia

The male's genitalia are hidden inside the female during copulation, so their behavior is difficult to study and is poorly understood. There are nevertheless two sources of data. One technique involved removing the male's head, positioning him immobile with his cerci lifted to expose his phallobase under a dissecting microscope, and then eliciting movements of his genitalia by gentling deflecting the setae on his interior claspers (stimulation of these setae in intact males failed to evoke genital behavior). Movements of genital sclerites and the pneumopophyses were videotaped through the microscope. A second more direct technique was to make X-ray recordings of movements that occur inside the female during copulation (Briceño et al. 2010, in press). For obtaining X-ray videos, virgin male and female flies were placed together in small chambers; recording began approximately 30 s or less after the male had mounted the female and seized her abdomen with his genitalia. Intromission was not always immediate (as also determined in some pairs of *G. austeni* and *pallidipes* by interrupting copulations—Pollock 1974; Briceño et al. 2007). The process of intromission, which, judging by some other fly species (Eberhard 2002), may involve female movements that expose her genital opening, has not been studied.

Each technique has limitations (behavior outside the female is not necessarily the same as that inside; and the resolution of the X-ray recordings was limited by the size of the genitalia and differences in X-ray opacity). It was clear, however, that the male genitalia move actively inside the female during copulation. Some movements are clearly rhythmic and are not mechanically necessary to bring the phallobase to the probable site of sperm deposition; instead, they seem likely to stimulate the female (Briceño et al. 2010, in press).

15.7.3 Pneumopophyses

When headless males of all five species were stimulated, some everted their intromittent genitalia and repeatedly inflated and collapsed their pneumopophyses rapidly. These inflations were directed laterally and basally. The shapes and positions of the fully inflated pneumopophyses appeared to be consistent intra-specifically and to vary between species (Fig. 15.7; Briceño et al. 2015). The pneumopophyses could not be seen in the X-ray videos of copulation; they were visible when the male genitalia were outside the female, but not when they were inside her. The mechanical result of pneumopophysis inflation inside the female is presumably either to brace the male's phallobase within the female's vagina, or to push it deeper inward; both types of movement seem likely to stimulate the female via stretch receptors (if they are present) in the walls of the oviduct.

15.7.4 Distal Triangular Sclerite of Phallobase

In headless males, the distal sclerite periodically swung about 180° so that its distal end projected distally. In all species, extension of this sclerite occurred only while the pneumopophyses were extended. Similar movements, though only

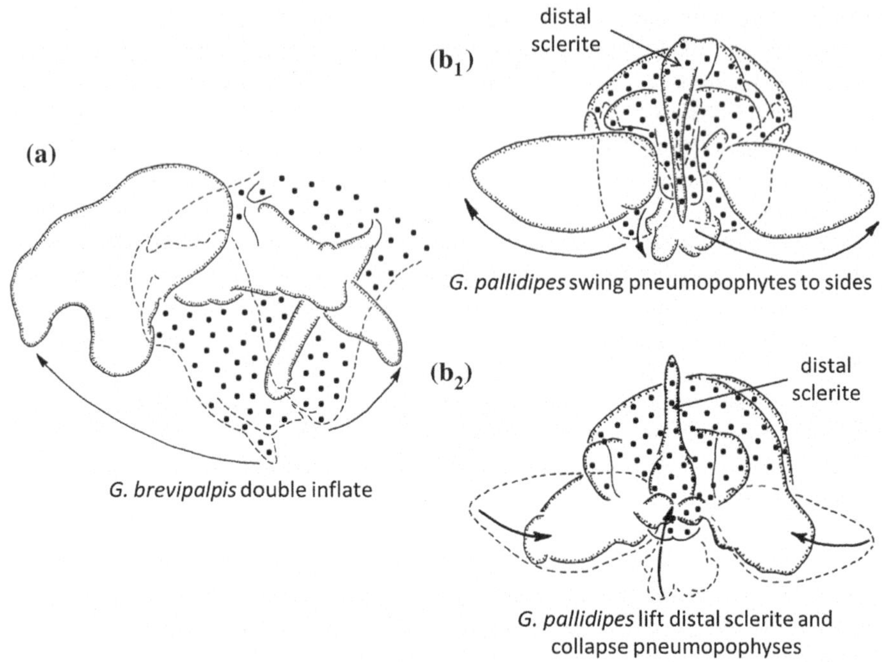

Fig. 15.7 Schematic drawings of different movements and shapes of the inflated pneumopophyses of *G. brevipalpis* (**a**) and *G. pallidipes* (**b**). The stippled portions show the basiphallus; the *curved arrows* indicate the movements of the pneumopophyses immediately preceding the positions shown in the drawing (**a** in lateral view, $\mathbf{b_1}$ and $\mathbf{b_2}$ in caudal view). The distal sclerite of *G. pallidipes* also moved just prior to $\mathbf{b_2}$ (*G. brevipalpis* from Briceño et al. 2015; *G. pallidipes* from Briceño et al. 2010)

incomplete in scope and less clearly resolved, may have occurred sometimes when the phallobase was at the apex of a thrust in the X-ray videos (Briceño et al. 2015). It is possible that these movements exert strong forces. In *G. austeni*, the "hook-like aedeagus" (perhaps homologous with the distal sclerite) is provided with powerful muscles and appears to push strongly against the female's genital papilla on the wall of the uterus (Pollock 1974).

15.7.5 Phallobase as a Unit

The phallobase did not move as a unit in headless males, but it made long series of thrusting movements as a unit inside the female in the X-ray video recordings of five species (Fig. 15.8) (Briceño et al. 2015). Some patterns of phallobase movements in X-ray recordings, including the thrusting movements of the entire phallobase, as well as an unidentified sclerite, pulsing movements of the phallobase,

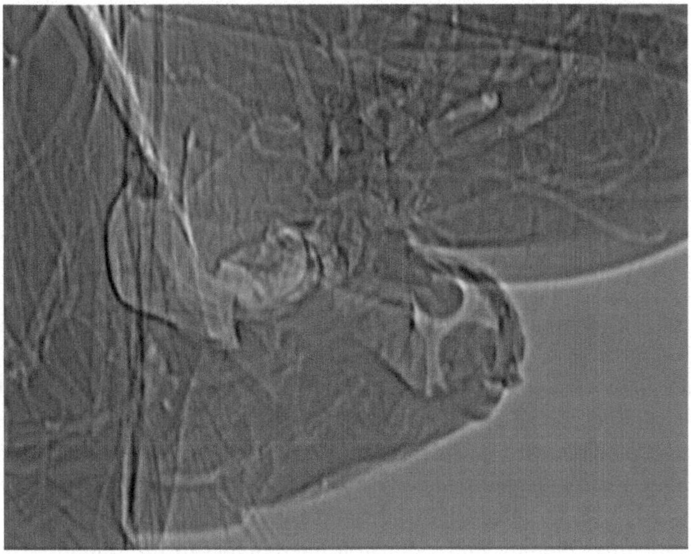

Fig. 15.8 An X ray image of a copulating pair of *Glossina pallidipes* (male shaded pink) showing the male's intromittent phallobase inside the female. The outlines of the phallobase were determined by analysis of movement patterns during thrusting behavior (see Briceño et al. 2015)

and withdrawal of the phallobase during strong cercal squeezes, and rubbing movements of the tips of the cerci against the central cylinder of the phallobase, were seen in some species but not others (Briceño et al. 2015). There were also some probable species-specific differences in the durations of individual thrusts of the phallobase; for example, the mean thrust duration was 0.28 ± 0.16 s in *G. pallidipes*, but only 0.04 ± 0.01 s in *G. brevipalpis*) (Briceño et al. 2015). The small sizes of some samples and their incomplete coverage of the entire copulation period preclude confident conclusions regarding the absence of particular types of movements in any given species; nevertheless, it appears that there are differences between species in relative frequencies.

Thrusting movements caused the dorsal surface of the basal portion of the phallobase to press and move structures on the female's external surface near her vulva, such as the sternal plate just ventral to her anus. In *G. pallidipes,* these movements presumably caused the arc of long setae on the male's inferior claspers to press deeper into or against the membranous groove around the anal plates and the sternal plate (this contact occurred in flash-frozen pairs) (Briceño et al. 2007). Thrusting movements of the basal portion of the phallobase also applied pressure to the curved projections of the inferior claspers of *G. palpalis*, which in turn pressed against the female. The rhythmic thrusting movements (about 1 every 3 s) of a long thin, hinged structure in *G. palpalis* caused the tips of the inferior claspers to rub across a surface (probably the intersegmental membrane of the female) (Briceno et al. in press). The phallobase was also temporarily withdrawn entirely from inside the female for periods of up to 49 s during especially strong cercal squeezes in *G. pallidipes* and *morsitans* (Briceño et al. 2015).

15.8 Discussion

15.8.1 Stimulation of the Female: Implications from Morphological Designs and Behavior

Several kinds of data confirmed two cryptic female choice predictions for *Glossina*: Males stimulate females during copulation; and the stimuli received by the female diverge rapidly over evolutionary time. Even though in external view the abdomen, legs, wings, and head of a male *Glossina* tsetse fly move only sporadically during copulation, the female receives a continuous barrage of stimuli from his genitalia, which move actively and in a sustained manner, both inside and on the external surface of her body. Externally, the male's cerci pinch and squeeze the ventral surface of the tip of her abdomen rhythmically. The squeezing movements are not mechanically necessary to restrain the female. They cause one patch of setae on the male's sternum and five other patches on his genitalia to rub against the female (Fig. 15.3). Internally, the male's phallobase also executes sustained, rhythmic thrusting, pulsing and extension movements in the female's vagina and her oviduct; there are also probable movements of his pneumopophyses, the distal triangular sclerite of his phallobase and a long, thin sclerite, and rhythmic movements of the basal portion his phallobase also push against other parts of the female. In addition to these genital movements, the male executes bursts of other rhythmic movements of his legs and wings, brushing, tapping, and vibrating other parts of her body during copulation (Jaensen 1979a, b; Briceño et al. 2007; Briceño and Eberhard 2009a, b). A female *Glossina* may sense stimuli from the male at up to ten sites on her body at once during some moments of copulation!

Most of these stimulatory male structures and their movements are not utilitarian, in the sense that they are not needed to hold or to penetrate the female. Even the thrusting movements with the phallobase did not result in progressively deeper penetration in the species observed with the X-ray technique (Briceño et al. 2015). In *G. austeni*, the phallobase has also already arrived at the mouth of the spermathecal duct only half way through copulation (Pollock 1974). It seems inevitable, however, that the friction of the male's intromittent genitalia with the walls of the vagina and the oviduct must stimulate the female. In at least three of the four species in the *morsitans* subgenus (including *G. pallidipes*), the male's intromittent genitalia are periodically withdrawn completely or nearly completely from the female's vagina when the male's cerci begin to squeeze strongly on the ventral surface of her abdomen and are then reinserted after the cercal squeeze ends.

The possibility that the movements of the pneumopophyses function to stimulate the female is supported by the reduction in sperm transfer when the pneumopophyses of *G. pallidipes* were covered with nail polish (above). In addition, the likelihood that copulation would induce ovulation was nearly extinguished when the pneumopophyses of *G. morsitans* were cauterized (Dodd 1973). The possibility that sperm transfer by the male (rather than some female response to the male) was altered by these pneumopophysis treatments was not checked, however.

Only a few of the male structures that contact the external surface of the female's abdomen are also mechanically necessary to restrain the female and hold her abdomen in a position that permits him to insert his phallobase. It is highly likely that the male's powerful grasp of the tip of the female's abdomen between his cerci and his fifth sternite functions to restrain the female. And it could be argued that the dense setae on his fifth sternite function as anti-slip devices that make his grip more effective. But the complex squeezing and flexing behavior of his cerci, the species-specific teeth on the tips of his cerci, and the multiple patches of setae on his genitalia at points where they rub on the female during the squeezing movements, all seem useless in terms of mechanical restraint. The groove in the inferior clasper that receives the edge of female tergite 6 in species of the *morsitans* group also seem superfluous with respect to a hold-fast function, as the cerci hold the female in a powerful, apparently unbreakable grip. The inferior claspers of *G. palpalis* (Squire 1951) and *G. austeni* (Pollock 1974) have been hypothesized to lift the female sternum and expose her vulva. But even if this occurs, the differences in their widely divergent forms seem mechanically irrelevant, given the relative uniformity of female morphology.

In sum, the functions of nearly all of the different types of genital movements and many of the setose genital structures described here are probably not directly related to restraining the female or to transferring sperm; in contrast, their designs seem appropriate to stimulate the female.

It is important to emphasize the fragmentary nature of our observations of phallobase behavior, even in the one species for which there is an appreciable sample size (*G. pallidipes*). Copulation in *Glossina* lasts up to two hours, while the observations with X-rays were limited to a few minutes because the X-rays damaged the animals (Socha et al. 2007). Judging by the clear differences in the copulatory courtship behavior of the legs and wings of male *G. pallidipes* that occur during the course of a normal copulation (Jaensen 1979a, b; Briceño et al. 2007), as well as in the patterns of squeezing behavior of the male's cerci (Briceño et al. 2007), the behavior of the phallobase may also vary during the course of a copulation. Such changes seem particularly likely if, as appears to be the case, the male sometimes succeeds in grasping the female's abdomen with his cerci, but is not immediately able to introduce his phallobase into the female's vagina. In *G. austeni* pairs that were flash frozen about half way through copulation and then sectioned, the male seemed to exert outward rather than inward force on some portions of the female's reproductive tract, and the distal portion of his basiphallus seemed to "grasp" a projection of the uterus (the genital papilla) (Pollock 1974). We suspect that the otherwise unique rhythmic stabbing movements observed in *G. swynnertoni* (Briceño et al. 2015) were attempts by the male to achieve intromission and that we may have failed to see similar movements in the other species because we did not happen to record pairs that were in that stage of copulation. It has long been known that some "copulations" in *Glossina* do not result in sperm transfer (Buxton 1955; Pinhão and Grácio 1973; Pollock 1974). Another stage that may show different behavior patterns is ejaculation (there was very distinctive copulatory courtship behavior in *G. pallidipes* during ejaculation—Jaensen 1979a, b;

Briceño et al. 2007); this stage was probably not included in any of the X-ray recordings. In summary, our list of positions and movements of the phallobase is likely to be incomplete.

15.8.2 Female Responses to Male Stimuli

If cryptic female choice is the reason why male genital morphology and behavior are elaborate and species-specific in *Glossina*, then it should be true that females respond to the male stimulation by altering their reproductive processes in ways that improve the male's chances of paternity. This prediction was tested experimentally in two species for two of the several structures mentioned above, but the results clearly confirmed the prediction. Both removing the male cercal tooth and smoothing his setae on sternite 5 and incapacitating female sense organs in the areas in which they are likely to be stimulated by these structures elicited female responses that were likely to decrease the male's chances of paternity: reduced sperm transfer to her spermathecae; reduced ovulation; and greater receptivity to subsequent male mating attempts (Briceño and Eberhard 2009a, b).

Some female responses were greater when the possible female sense organs in the area contacted by the male were blocked than when the male structure was altered. This is not surprising, because the female alterations were likely to eradicate rather than reduce stimulation from the male structure and may also have reduced or eliminated stimuli from male squeezing behavior.

It is revealing to place these observations of the triggering of ovulation in the context of previous detailed research on female reproductive physiology in *Glossina*. Saunders and Dodd (1972) concluded, on the basis of extensive experiments, that ovulation was not triggered by transfer of sperm, deposition of the spermatophore in the female, secretions of the male's testes, his accessory glands or his ejaculatory duct, or by humeral factors from the spermathecae of inseminated females. They speculated, by elimination, that stimuli received during copulation must induce ovulation. These physiological studies thus reinforce the likely importance of the stimuli from the male's genitalia in inducing ovulation.

15.8.3 Species Specificity and Evolutionary Transitions

One prediction of cryptic female choice is that different aspects of the behavior of genitalia that are presumed to be used as internal courtship devices, such as the squeezing movements of the cerci, the thrusting movements, the inflations of pneumopophyses, and the flexion of the triangular sclerite, are likely to differ among even these closely related species (as, of course, is typical of courtship behavior in general). As far as the available data go, it appears that this prediction is fulfilled and that many of the stimuli delivered by male *Glossina* differ in different species. As shown above and by Briceño et al. (in press), there are many differences in both the morphology and the behavior of the male genitalia of different species of *Glossina*. The morphological differences were partially documented in earlier taxonomic studies, while the differences in their behavior and physical interactions with the female's genitalia have been documented only recently, and much less completely. Differences are especially clear in the stimuli from the cerci and the inferior claspers.

It is possible to trace some of the probable evolutionary transitions of morphological and behavioral genital traits we have described, using the phylogenetic relations of species in *Glossina*. Several genital traits that are shared between species in the *palpalis* and *fusca* subgenera, but not with those in the *morsitans* subgenus are probably synapomorphies in the *morsitans* group. These putatively derived traits of the *morsitans* subgenus include the central joining of the cerci (as argued by Potts 1970), pressing the female tergite 6 rather than tergite 7 with the male's modified sternum 5 during copulation, folding the cerci less sharply to make a relatively large angle (approximately 90°) with the basal segment during copulation, stabbing the ventral surface of the female's abdomen so as to fold off a small round mass of female tissue near the tip of the cercus during copulation, and pressing the inferior clasper against the posterior tip of the female's tergite 6 rather than against the membranous surface that is directly ventral to this tip.

Male stimulation of the female in the *G. morsitans* spp. lineage during copulation thus appears to have become more concentrated on female tergite 6. Experimental covering of the female tergite of *G. pallidipes* confirmed that stimulation of this tergite during copulation induces the female to ovulate, to store sperm, and to resist further copulation (Briceño and Eberhard 2009a) and that it induces sperm storage in *G. morsitans* females (Briceño and Eberhard 2009b). The positions and forms of large cercal teeth have also changed in this lineage (Fig. 15.1), and experiments in two species have shown that stimulation from these teeth increase ovulation, sperm storage and sexual receptivity responses in ways that could improve the male's chances of paternity.

15.8.4 Alternative Explanations for the Evolution of Genital Morphology and Behavior in Glossina

15.8.4.1 Lock and Key

Three other commonly cited hypotheses to explain rapid divergent evolution of animal genitalia can be examined in light of the data presented here (see Eberhard, this volume). The species isolation hypothesis proposes that divergence in male genital morphology is driven by selection on females to avoid receiving sperm from heterospecific males. The avoidance mechanism could be via a mechanical lock-and-key mechanisms, in which heterospecific male genitalia are mechanically excluded from those of the female ("mechanical lock and key"), or via female discrimination against heterospecific males on the basis of the stimuli that she receives from their genitalia ("sensory lock and key") (see Eberhard 1985, Simmons 2014). The mechanical version is clearly contradicted as an explanation for the species-specific forms of the male cercus and inferior clasper in *Glossina*. Both the male cerci and the inferior claspers contact featureless, smooth membranous portions of the female abdomen that are mechanically incapable of excluding the genitalia of heterospecific males; the female structures show no perceptible differences between species. Thus, female morphology clearly does not mechanically exclude the species-specific cerci and inferior claspters of heterospecific males, and the mechanical lock-and-key hypothesis cannot explain their rapid divergence.

Observations of the morphology of the female's common oviduct, where the male's phallobase is inserted, are less complete. The lining of the female's reproductive tract is relatively soft rather than rigid, so mechanical exclusion seems unlikely a priori. There is, however, is a sclerite in the *fusca* group (the "signum") on the inner surface of a thick symmetrical, transparent, gelatinous structure, the "genital fossae," and the signum exhibits considerable specific diversity of form (Newstead et al. 1924; Patton 1936); the functions of the signum and the genital fossae are apparently unknown.

The mechanical lock-and-key hypothesis also has problems explaining the inter-specific differences in phallobase morphology, because the phallobase slides back and forth within the female; at least during early stages of copulation, and thus does not have a fine physical mesh with the female, at least along a large portion of the lower portion of her reproductive tract. Physical exclusion of heterospecific male phallobases seems unlikely, at least in early stages of copulation. Rejection of mechanical lock and key on the basis of the apparent lack of a female "lock" is less certain for the phallobase, however, because it is possible that the available observations may have missed positions assumed by the male at critical moments (e.g., intromission, ejaculation, or spermatophore deposition). There is wide inter-specific variation in the female signum and the male phallobase (e.g., Patton 1936), and their coupling (if it occurs) remains to be described. The X-ray recordings did not permit detailed analyses of how male genitalia were positioned with respect to details the internal morphology of the female vagina and uterus. A further reason to doubt mechanical lock-and-key arguments for the male's intromittent genitalia is that they fail to explain the elaborate, highly repeated, and divergent stimulatory movements of the clasping and of the intromittent portions of the male's genitalia.

A final reason to doubt both mechanical and sensory lock-and-key hypothesis is that, as explained above, *Glossina* species are probably effectively isolated by differences in diurnal activity cycles, habitat, geographic range, and species-specific surface hydrocarbons that allow males to distinguish the sex and species identity of females prior to copulation (Huyton et al. 1980; Wall and Langley 1993) (occasional crosses do occur between subspecies—Curtis et al. 1980). This likely reproductive isolation also implies that the experimentally demonstrated female responses at later stages of copulation to the male's lateral cercal teeth and to his sternite 5 are unlikely to represent adaptations to avoid cross-specific pairing.

In sum, the mechanical lock-and-key hypothesis can be confidently discarded for some species-specific aspects of male genitalia, and could be incorrect for all of them. There is also evidence, though less extensive, against the sensory lock-and-key hypothesis.

15.8.4.2 Sperm Competition

The males of some animals directly influence the fate of sperm from other males, for instance by removing them, diluting them, killing them, flushing them from the female, etc. (e.g., Simmons 2001; Leonard and Córdoba-Aguilar 2010). Such sperm competition may possibly occur in *Glossina*, but none of the copulatory courtship behavior, male genital morphology, or male genital behavior described here has any obvious relation with the sperm of previous or subsequent males. Thus, sperm competition, as usually understood (e.g., Simmons 2001), seems unlikely to explain their evolution.

15.8.4.3 Sexually Antagonistic Coevolution

A third alternative hypothesis to explain the rapid divergent evolution of the behavioral and morphological traits of *Glossina* described here is sexually antagonistic coevolution (SAC). Male–female conflict of interests could result in coevolutionary races between male traits that increase the male's chances of paternity but at the same time reduce the female's ability to reproduce, and female defenses against these traits that increase her ability to resist or overcome the male-imposed damage (Chapman et al. 2003; Arnqvist and Rowe 2005). One version of SAC is that the species-specific male traits function to physically or mechanically overcome species-specific female defensive morphological traits (e.g., Alexander et al. 1997). This version can be confidently rejected for the male cerci and inferior claspers of *Glossina* for the same reasons discussed above for lock and key. In essence, there are simply no defensive female structures in the areas contacted by these species-specific male structures of *Glossina*. The only complementary female morphological trait is the transverse cleft dorsal to the vulva, into which the male's inferior claspers fit in *G. austeni* (Pollock 1974). And this cleft would appear to aid rather than impede the possible lifting action of the inferior claspers on the female's sternal plate that may uncover her vulva. The genus *Glossina* can thus be added to the list of groups in which female morphology does not show the species-specific defensive traits that are predicted by SAC to be common in combination with rapid divergent evolution of male genital structures (Eberhard 2004a, b).

These arguments regarding the mechanical effects of male genitalia can also be applied, though with less confidence, to the inflatable sacs or pneumopophyses of the male phallobase. These sacs were present in all species but showed differences between species in both morphology and behavior. One possible SAC-related function of the movements of the pneumopophyses, suggested by observations of other insects, is that they (at least those with teeth) rub holes in the lining of the female's vagina (Merritt 1989; Crudgington and Siva-Jothy 2000; Flowers and Eberhard 2006). This could allow male seminal products to escape into her body cavity (Gillot and Langley 1981) where they could induce female responses such as sperm transport and resistance to further copulations (Riemann and Thorson 1969; Chen 1984). Previous studies of the reproductive physiology of *G. morsitans* argue strongly against this alternative, however; introduction of male seminal products into the female's body cavity did not affect ovulation, insemination or remating (Saunders and Dodd 1972; Gillot and Langley 1981).

A second version of the SAC hypothesis emphasizes conflict over stimulation rather than mechanical coupling. Females might defend against damaging male effects by means of species-specific differences in the nervous system, rather than differences in external morphology. Changes in a female's nervous system could make her less likely to respond to male manipulation and might be less costly than changes in her morphology. This version of the SAC hypothesis could explain the rapid divergence in both the morphology and the behavior of male *Glossina* genitalia and is compatible with the lack of species-specific morphological differences in the portions of the female that are contacted by the male. It supposes that male stimulation constitutes a sort of a sensory trap (Holland and Rice 1998; Córdoba-Aguilar 2005; Arnqvist 2006), in which females "cannot help themselves" from responding to the male's stimulation.

This argument supposes that females have not been able to solve the problem of being overly responsive to the male and thus damaging their own reproductive prospects, due to natural selection that favors the female responses in other contexts. But in contrast to other sensory traps, in which female responsiveness to male stimuli is maintained by natural selection (Christy 1995; Córdoba-Aguilar 2005), this SAC argument seems unlikely for *Glossina*. There is no obvious reason why the thresholds for female response to male stimuli in *Glossina* would be constrained by natural selection. That is, a female *Glossina* would seem to be free to adjust her degree of sensitivity to male post- or syn-copulatory stimulation so that her responses are in accord with her own reproductive interests.

This stimulatory version of SAC also depends on the supposition that the effects of the species-specific aspects of the male morphology, which at least in *G. pallidipes* and *G. morsitans* include increased sperm transport, increased probability of ovulation, and increased resistance to further copulations, damage female reproductive output. Specifically, the female must lose future offspring due to the increase in sperm transport that is induced by the male, the increased probability of ovulation that is induced by the male, and her decreased chance of copulations with other males that result from her increased resistance to further copulations. And all three of these losses must be larger than the possible gains she could obtain through the increased abilities of her sons to induce these responses in females of the following generation under the SAC hypothesis. There is no evidence, however, to support any of these SAC suppositions in *Glossina* (though it must be noted that we know of no studies that could have tested any of these possibilities).

All of the female responses to males in *Glossina* that have been documented are likely to be favored by natural selection on females—to store and utilize the gametes received from the current male and to avoid the likely costs of additional copulations with subsequent males. This does not mean that it is not possible that the disadvantages posited by SAC exist, but that there is no empirical support for invoking them.

In sum, the SAC hypothesis cannot be definitively ruled out. But it depends on several less than certain suppositions, for which there is little or no evidence. As is typical (Popper 1970), advances in science often involve gradually discarding hypotheses that require more and more post hoc adjustments to comply with the accumulating data. Use of Occam's razor suggests that SAC is less appealing than CFC as a general explanation for genital evolution in *Glossina*.

15.8.5 Limits of the Techniques Used

This chapter shows that a combination of different techniques was needed to gain understanding of the genital behavior in tsetse flies. These included direct detailed observation of external events (Briceño et al. 2007), dissection of flash-frozen pairs, direct observation of the genitalia of isolated, headless males, and indirect observation of mating with X-ray images. Each technique has advantages and disadvantages, and none gives a complete view. External events can be observed without altering the flies' behavior during the entire 30–120 min copulations, but they fail to reveal internal events. Flash-frozen pairs give detailed snap shots of the positions of different sclerites at particular stages during copulation, but cannot document their movements and may sometimes be incomplete; for instance, the basal spines that were seen with X-rays (Briceño et al. 2015) were missed in an earlier study of *G. pallidipes* (Briceño et al. 2007). The genitalia may be induced to move (as in the headless males of thus study), and these movements can be observed in detail, but the males performed only a subset of the behavior patterns that they are capable of executing (e.g., the rhythmic thrusting movements were entirely absent, eversion of the basal spines near the cerci did not occur). The X-ray video images revealed movements that were otherwise hidden from view, but resolution of details was not possible, some structures such as the pneumonophyses were not visible, and only short segments of copulation behavior could be recorded before the X-rays damaged the flies (Socha et al. 2007). And of course, cyclotron facilities where such X-ray recordings are possible are not widely available. We believe that it is very likely that further details of genital behavior remain to be discovered in *Glossina*.

The ability to visualize internal movements using X-ray images opens a new field of study, the behavior of genitalia during copulation. Results may help illuminate the functional morphology of puzzlingly elaborate genital structures and may provide both additional taxonomic characters for distinguishing closely related species, and tests of sexual selection theory.

Acknowledgments We acknowledge collaboration and assistance from A.S. Robinson, D. Wegrzynek, E. Chinea-Cano, T. Dos Santos Rolo, A. Markowicz, C. Streli, P. Wobrauschek, and the staff of the Forschungszentrum Karlsruhe, in particular Alexander Rack, Timm Weitkamp, Patrik Vagovic, and Tilo Baumbach. We thank the International Atomic Energy Agency for the use of flies and facilities, Rudolf Boigner and Carmen Marin for help in rearing flies, Anita Aisenberg, Yoshitaka Kamimura and an unusually thorough anonymous reviewer for helpful comments on the manuscript, D. Monty Wood for advice on morphological homologies, Maribelle Vargas for help obtaining SEM images, and Jorge Hendrichs, Andrew Parker and Marc Vreysen for logistic support. The IAEA, STRI, and the Universidad de Costa Rica provided financial support.

References

Aksoy S, Chen X, Hypsa V (1997) Phylogeny and possible transmission routes of midgut-associated endosymbionts of tsetse (Diptera: Glossinidae). Ins Mol Biol 6:183–190

Alexander RD, Marshall DC, Cooley JR (1997) Evolutionary perspectives on insect mating. In: Choe J, Crespie B (eds) The evolution of mating systems in insects and arachnids. Cambridge University Press, Cambridge, pp 4–31

Arnqvist G (2006) Sensory exploitation and sexual conflict. Philos Trans Roy Soc B 361: 375–386

Arnqvist G, Rowe L (2005) Sexual conflict. Princeton University Press, New Jersey

Briceño RD, Eberhard WG (2009a) Experimental modifications of male genitalia confirm cryptic female choice theory of genital evolution. J Evol Biol 22:1516–1525

Briceño RD, Eberhard WG (2009b) Experimental demonstration of possible cryptic female choice on male tsetse fly genitalia. J Ins Physiol 55:989–996

Briceño RD, Eberhard WG, Robinson A (2007) Copulation behavior of *Glossina pallidipes* (Diptera: Muscidae) outside and inside the female, and genitalic evolution. Bull Ent Res 97:1–18

Briceño RD, Wegrzynek D, Chinea-Cano E, Eberhard WG, dos Santos-Rolo T (2010) Movements and morphology under sexual selection: tsetse fly genitalia. Ethol Ecol Evol 22:385–391

Briceño RD, Eberhard WG, Chinea-Cano E, Wegrzynek D, dos Santos Rolo T (2015) Species specific behavioral differences in male tsetse fly genitalia behavior during copulation, with a discussion of genitalic evolution in *Glossina*. Ethol Ecol Evol

Buxton PA (1955) The natural history of tsetse flies. HK Lewis, London

Chapman T, Arnqvist G, Bangham J, Rowe L (2003) Sexual conflict. Trends Ecol Evol 18:41–47

Chen PS (1984) The functional morphology and biochemistry of insect male accessory glands and their secretions. Ann Rev Ent 29:233–255

Chen X, Li S, Aksoy S (1999) Concordant evolution of a symbiont with its host insect species: molecular phylogeny of genus *Glossina* and its bacteriome-associated endosymbiont, *Wigglesworthia glossinidia*. J Mol Evol 48:49–58

Christy JH (1995) Mimicry, mate choice, and the sensory trap hypothesis. Am Nat 146:171–181

Córdoba-Aguilar A (2005) Possible coevolution of male and female genital form and function in a calopterygid damselfly. J Evol Biol 18:132–137

Crudgington HS, Siva-Jothy MJ (2000) Genital damage, kicking and early death. Nature 407:855–856

Curtis CF, Langley PA, Trewern MA (1980) X-Chromosome involvement in male hybrid sterility from *Glossina morsitans* sub-species crosses. Heredity 45:405–410

Dodd CHW (1973) Control of the reproductive cycle in tsetse flies (*Glossina* spp.). PhD thesis, University Edinburgh

Eberhard WG (1985) Sexual selection and animal genitalia. Harvard University Press, Cambridge

Eberhard WG (1994) Evidence for widespread courtship during copulation in 131 species of insects and spiders, and implications for cryptic female choice. Evolution 48:711–733

Eberhard WG (1996) Female control: sexual selection by cryptic female choice. Princeton University Press, New Jersey
Eberhard WG (2002) The function of female resistance behavior: intromission by male coercion vs. female cooperation in sepsid flies (Dipera: Sepsidae). Rev Biol Trop 50:485–505
Eberhard WG (2004a) Male-female conflict and genitalia: failure to confirm predictions in insects and spiders. Biol Rev 79:121–186
Eberhard WG (2004b) Rapid divergent evolution of sexual morphology: comparative tests of antagonistic coevolution and traditional female choice. Evolution 58:1947–1970
Eberhard WG (2009) Static allometry and animal genitalia. Evolution 63(1):48–66
Feldmann U, Dyck VA, Mattioli RC, Jannin J (2005) Potential impact of tsetse fly control involving the sterile insect technique. In: Robinson AS, Dyck VA, Hendrichs J (eds) Sterile insect technique principles and practice in area-wide integrated pest management. Springer, The Netherlands, pp 701–723
Flowers RW, Eberhard WG (2006) Fitting together: copulatory linking in some tropical Chrysomelidae. Rev Biol Trop 54:829–842
Gillot C, Langley PA (1981) The control of receptivity and ovulation in the tsetse fly, *Glossina morsitans*. Physiol Ent 6:269–281
Gooding RH, Krafsur ES (2005) Tsetse genetics: contributions to biology, systematic, and control of tsetse flies. Ann Rev Entomol 50:101–123
Hargrove JW (2012) Age-specific changes in sperm levels among female tsetse flies (*Glossina* spp.) with a model for the time course of insemination. Physiol Entomol 37:278–290
Holland B, Rice WR (1998) Chase-away sexual selection: antagonistic seduction versus resistance. Evolution 52:1–7
Hoppenheit A, Bauer B, Steuber S, Terhalle W, Diall O, Zessin KH, Clausen PH (2013) Multiple host feeding in *Glossina palpalis gambiensis* and *Glossina tachinoides* in Southeast Mali. Med Vet Entomol 27:222–225
Hosken D, Stockley P (2003) Sexual selection and genital evolution. Trends Ecol Evol 19:87–93
Huyton PM, Langley PA, Carlson AD, Schwartz M (1980) Specificity of contact sex pheromones in tsetse flies, *Glossina* spp. Physiol Entomol 5:253–264
International Glossina Genome Initiative (169 authors) (2014) Genome sequence of the tsetse fly (*Glossina morsitans*): vector of African trypanosomiasis. Science 344:380–386
Jaensen T (1979a) Mating of males of *Glossina pallidipes* Austen (Diptera: Glossinidae). Bull Ent Res 69:573–588
Jaensen T (1979b) Mating behaviour of *Glossina pallidipes* Austen (Diptera: Glossinidae): duration of copulation, insemination and fecundity. Entomol Exp Appl 26:1–12
Kawooya J (1977) Demonstration of multiple insemination in females of *Glossina morsitans* Westwood, using a phenotypic marker Ocra. J Appl Entomol 84:321–327
Leonard J, Córdoba-Aguilar A (2010) The evolution of primary sexual characters in animals. Oxford University Press, New York
McAlpine JF (1989) Phylogeny and classification of the Muscomorpha. In: McAlpine JF, Wood DM (eds) Manual of nearctic Diptera. Monograph 32, vol 3. Research Branch Agriculture Canada, pp 1397–1518
Merritt D (1989) The morphology of the phallosome and accessory gland material transfer during copulation in the blowfly, *Lucilia cuprina* (Insecta, Diptera). Zoomorphology 108:359–366
Mulligan HW (1970) The African trypanosomiases. Allen and Unwin, London
Newstead R, Evans AM, Potts WH (1924) Guide to the study of tsetse-flies. Hodder and Stoughton Limited, London
Patton WS (1936) Studies on the higher Diptera of medical and veterinary importance. A revision of the species of the genus *Glossina* Wiedemann based on a comparative study of male and female terminalia. Ann Trop Med Paras 30:71–89
Pinhão RC, Grácio AJS (1973) The degree of spermathecal impregnation and the number of matings in *Glossina austeni*. Ann Inst Higiene Med Trop 1:103–106

Pollock JN (1970) Sperm transfer by spermatophore in *Glossina austeni* Newstead. Nature 225:1063–1064
Pollock JN (1974) Anatomical relations during sperm transfer in *Glossina austeni* Newstead (Glossinidae, Diptera). Trans Roy Ent Soc Lond 125:489–501
Popper K (1970) Objective knowledge: an evolutionary approach. Oxford University Press, New York
Potts WH (1970) Systematics and identification of *Glossina*. In: Mulligan HW (ed) The African trypanosomiases. Allen & Unwin, London, pp 243–273
Riemann JG, Thorson BJ (1969) Effect of male material on oviposition and mating by female house flies. Ann Ent Soc Am 62:828–834
Saunders DS (1970) Reproduction of *Glossina*. In: Mulligan W (ed) The African trypanosomiases. Allen & Unwin, London, pp 327–344
Saunders DS, Dodd CHW (1972) Mating, insemination, and ovulation in the tsetse fly, *Glossina morsitans*. J Ins Physiol 18:187–198
Simmons LW (2001) Sperm competition and its evolutionary consequences in the insects. Princeton University Press, New Jersey
Simmons LW (2014) Sexual selection and genital evolution. Austral Entom 53:1–17
Socha JJ, Westneat MW, Harrison JF, Waters JS, Lee WK (2007) Real-time phase-contrast x-ray imaging: a new technique for the study of animal form and function. BMC Biol 5:6
Squire FA (1951) Observations on mating scars in *Glossina palpalis* (RD) Bull Ent Res 42:601–604
Tobe SS, Langley PA (1978) Reproductive physiology of *Glossina*. Ann Rev Entomol 23:283–307
Vanderplank FL (1947) Experiments in the hybridization of tsetse-flies (*Glossina*, Diptera) and the possibility of a new method of control. Trans Roy Ent Soc Lond 98:1–18
Wall R, Langley PA (1993) The mating behavior of tsetse flies (Glossina): a review. Physiol Entomol 18:211–218

Chapter 16
Evaluating Cryptic Female Choice in Highly Promiscuous *Tribolium* Beetles

Tatyana Y. Fedina and Sara M. Lewis

Abstract *Tribolium* flour beetles have been an important model organism for the study of postcopulatory sexual selection. These beetles are representative of many insects that live in high-density aggregations and therefore experience frequent mating opportunities. However, unlike many such insects, *Tribolium* females are extremely polyandrous and seemingly promiscuous as they engage in frequent matings without any apparent pre-mating courtship ritual. Instead, intrasexual competition as well as intersexual mate/paternity choices occurs primarily during and after copulation. Here, we review the evidence for at least four cryptic female choice (CFC) mechanisms in *Tribolium* that may affect how offspring paternity is apportioned among a female's mates. By comparing live to freshly euthanized females (both readily mated by males), studies reveal that during mating, females reject spermatophore transfer or limit sperm numbers transferred by inferior males. Re-mating with additional males is another CFC mechanism, and mated *Tribolium* females will more readily accept a spermatophore from males that are more attractive than their previous mate. Finally, *Tribolium* females may also use the timing of spermatophore ejection after mating to bias offspring paternity. *Tribolium* life history and mating system traits (including long adult life, continuous egg-laying, and frequent re-mating) suggest that females are unlikely to be harmed by multiple matings. Indeed, there exists little experimental evidence for the cost of polyandry to females, and the *Tribolium* mating system appears to lack many elements that would indicate sexually antagonistic coevolution. The multiple CFC mechanisms demonstrated in *Tribolium* may stem from females' inability to effectively assess males before mating. Such mechanisms may provide an effective

T.Y. Fedina (✉)
Department of Molecular and Integrative Physiology, University of Michigan, Ann Arbor, USA
e-mail: tfedina@umich.edu

S.M. Lewis
Department of Biology, Tufts University, Medford, USA
e-mail: sara.lewis@tufts.edu

A.V. Peretti and A. Aisenberg (eds.), *Cryptic Female Choice in Arthropods*,
DOI 10.1007/978-3-319-17894-3_16

strategy for females to adjust their paternity decisions at different stages during and after mating, as they receive more information about their current and future mates and about their environment.

16.1 Conditions Favoring Extreme Female Promiscuity

Differential investment in gametes and/or parental care between the sexes in polygamous species inevitably creates sexual conflict over reproduction. However, sexual antagonism does not necessarily lead to evolution of traits in one sex that harm the other sex (Pizzari and Snook 2003; Eberhard 2004; Chapman 2006). Recent theoretical and empirical studies suggest that under certain life histories, sexual conflict can be alleviated by increased female polyandry (see Sect. 16.1.5 below). Because environment, life history, and mating system of an organism are intimately interconnected in defining the evolution of sexually selected traits, in this section we will describe natural history of *Tribolium* flour beetles that might have generated their highly polyandrous (mating with multiple males) and seemingly promiscuous (indiscriminate) mating system with no elaborate pre-mating interactions. The rest of the chapter will discuss the mechanisms of cryptic female choice (CFC) that have evolved under this mating system.

16.1.1 Habitat, Food, Life History, and Population Cycles

Tribolium flour beetles are cosmopolitan pests that inhabit dried foodstuffs, including cereal products and stored grain. They have been associated with human food stores for at least 4000 years, and their ancestors were probably living off decomposing wood and other decaying organic matter, supplemented with fungi and bacteria (Hinton 1942; Pray and Goodnight 1995; Levinson and Levinson 1995). Unlike many other holometabolous insects, the juvenile stages of *Tribolium* share habitat and food with adults.

Stored foodstuffs are nutritionally rich sources of food able to maintain several generations of flour beetles on the same food patch. *Tribolium* is an income breeder (i.e., reproduction is based on energy acquired during the reproductive period; Stearns 1992). Within 4 days after adult eclosion, females start laying eggs each weighing only 2 % of female body (Bernasconi et al. 2006). Depending on environmental conditions and population density, females with continuous access to males lay on average 10–15 eggs per day for 3–6 months, after which oviposition rates gradually decline. This amounts to female lifetime fecundity of about 500–1500 eggs, which develop into adults within 32 days (Sokoloff 1974; Mertz 1975; Sonleitner 1978). Egg-laying is continuous and evenly distributed over this fertile period, with eggs laid singly rather than in batches. In addition to reproduction, natural *Tribolium* populations can increase through migration.

In food storage facilities, flour beetles tend to aggregate rather than distribute themselves randomly (Sokoloff 1974; Subramanyam and Harein 1990; Trematerra and Sciarretta 2004). Living in aggregations is beneficial for them because adults produce and release into the media noxious defensive compounds, benzoquinones (Sokoloff 1974). Such conditioning makes the media unpalatable for other consumers and repellent to potential vertebrate and invertebrate predators and microorganisms (Yezerski et al. 2007). Aggregation behavior is promoted by male-produced aggregation pheromone (see Sect. 16.1.3). As populations increase, however, the food becomes less nutritious and accumulates toxic concentrations of waste products and benzoquinones (Duehl 2011). Such conditions induce adult dispersal via flight (Zyromska-Rudzka 1966a, b; Sokoloff 1974; Ziegler 1978).

In spite of the benefits of communal defense, sharing habitat and food by adults and juveniles creates competition and leads to cannibalism of up to 99 % of immobile life stages (eggs and pupae) by mobile stages (larvae and adults) at high population densities (Sokoloff 1974). *Tribolium* juvenile stages are also more sensitive to catastrophic habitat destruction, shortages of food, and conditioning of the media (Sokoloff 1974). Higher mortality risk of juveniles compared to adults is predicted to promote long adult life span (Stearns 1992), and in fact, a distinguishing feature of *T. castaneum* is their long life span (over 1 year) with continuous egg laying over their several months of fertile period (Sokoloff 1974; Nilsson et al. 2002).

Because average adult life span equals up to 10 full generation cycles, individual beetles are likely to encounter a range of density-dependent environmental conditions over their lifetime. As a consequence, many of *Tribolium* life-history traits exhibit considerable phenotypic plasticity (Sokoloff 1974; Nilsson et al. 2002; Lewis et al. 2012). Thus, many male life history and some reproductive performance traits (e.g., mating rate, insemination success) are dependent on food quality (Lewis et al. 2012), and even minimal conditioning of the medium decreases female oviposition rate by up to 50 % (Park 1936; Prus 1961; Sokoloff 1974; Sonleitner and Guthrie 1991). CFC is also likely to respond to changing environmental/population conditions. One can predict, for instance, that at high population densities when females are laying significantly fewer eggs, they might be more discerning in their choice of sires. However, these kinds of questions have not been addressed experimentally so far.

16.1.2 Benefits and Costs of Female Polyandry

In addition to long adult life span and continuous egg-laying, another distinguishing feature of *Tribolium* beetles is continuous female receptivity to mating. These females can be fertilized as early as 3 h after eclosion, even though they do not start laying eggs until 4 days posteclosion (Dawson 1964). They can also store and use sperm from a single mating for up to four months, laying over 700 fertile eggs during this period (Bloch Qazi et al. 1996). Despite such abilities, these females appear to be continuously receptive, mating with up to 12 different males per hour in various experimental settings (Park 1933; Sokoloff 1974; Pai and Yan 2003b; Pai et al. 2007). Even though these are probably overestimates due to most studies using previously isolated males having unlimited access to virgin females, *T. castaneum* natural mating rates are still extremely high (Wool 1967) and must be explained by some other net benefits besides sperm replenishment.

It is commonly accepted that polyandry is maintained because its benefits to females outweigh the costs. Among the benefits identified for flour beetles are increased female fecundity (Park 1933; Lewis and Austad 1994) and the probability of successful insemination (Pai et al. 2005). Notably, some of these may come from higher ejaculate quantities transferred, because females who mate repeatedly with the same male also produce more offspring (Nilsson et al. 2002). Females might also directly benefit from polyandry if they receive nutrition from male ejaculate. Such benefits have been studied in the mealworm beetle *Tenebrio molitor*, a tenebrionid relative that shares many similarities in ecology and reproductive biology (see Section 16.3.6). Worden and Parker (2001) manipulated both the number of mates and feeding regimens of *T. molitor* females and found that females did not gain nutritional benefits from multiple matings. The observation that *Tribolium* females quickly expel spermatophores after mating (Bloch Qazi et al. 1996; Fedina 2007; Sect. 16.3.3) further supports the conclusion that these females are also unlikely to draw any nutrition from male ejaculates.

In addition to direct benefits, polyandrous *Tribolium* females may enjoy several indirect genetic benefits. These include higher larval viability of offspring (Pai et al. 2005; but see Pai and Yan 2003b), higher insemination success of sons (Pai and Yan 2002b), and higher egg viability in grand-offspring (Pai and Yan 2002b). A recent study has also found significant father–son heritability for sperm offense, reproductive success, and longevity (Lewis et al. 2012), suggesting that females could benefit from mate choice. And in fact, sons from polyandrous mothers showed higher sperm offense ability (Bernasconi and Keller 2001). Such transgenerational effects may also be due to epigenetic inheritance, maternal investment, and cryptic male choice (Mills and Moore 2004; Zeh and Zeh 2008; Kelly and Jennions 2011); however, these are impossible to disentangle from indirect genetic benefits based on the experimental designs used in these studies. Furthermore, since flour beetles exhibit inbreeding depression (Gaur and Rao 1997; Pray and Goodnight 1995; Wade et al. 1996; Wool and Mendlinger 1981), CFC may also help ameliorate negative consequences of inbreeding. Indeed, a recent study has shown that inbred *T. castaneum* females that were selected via brother–sister mating for 8 generations evolved to be more promiscuous, accepting more males per unit time (Michalczyk et al. 2011a). Notably, even in the absence of CFC, females could benefit from polyandry through reduced inbreeding of their grand-offspring (Cornell and Tregenza 2007).

Studies have identified costs of polyandry in comparison with enforced monandry in *T. castaneum*; these include lower fitness of daughters (Pai and Yan 2002b; Pai et al. 2007) and reduced sperm defense capacity of sons (Bernasconi and Keller 2001). Using more realistic comparison of different degrees of polyandry, a recent study observed a negative correlation between the number of mates (non-selected) and short-term fitness of females from female-biased selection lines, and no correlation between these parameters for females from male-biased selection lines (Michalczyk et al. 2011b). While the study does suggest that polyandrous females may enjoy protection from the potential negative effects of multiple matings (thus revealing genetic "load" of polyandry), it does not show that this protection is costly and that *coevolved* males would harm females having such protection.

Another potential negative consequence of multiple mating is transmission of pathogenic microorganisms during copulation; however, such transmission has not been reported in *T. castaneum* (Sokoloff 1974). Finally, if the costs to females of resistance to male mating attempts are higher than the costs of submission, extreme mating frequencies can also be explained by convenience polyandry (reviewed in Kvarnemo and Simmons 2013; Parker and Birkhead 2013). The cost of mating itself in *Tribolium* does not appear to be high. Thus, no effect on life span of increased polyandry was observed for females selected for low mating frequencies (Michalczyk et al. 2011b). Long life with frequent mating by itself suggests the absence or buffering of detrimental mating effects. Furthermore, as discussed below (Sect. 16.4), *T. castaneum* seem to lack identifiable analogs of harmful *Drosophila* male sex peptide and its female response elements.

16.1.3 Evidence for Pre-mating Female Assessment of Long-Range Male Pheromones

In several *Tribolium* species, adult males produce a highly volatile compound, 4,8-dimethyldecanal (DMD), which is attractive to both sexes (Arnaud and Haubruge 2002; Suzuki 1980; Suzuki et al. 1987; Suzuki and Sugawara 1979). Although DMD is commonly known as an aggregation pheromone, the potential exists for *Tribolium* females to use it in mate assessment (Levinson and Levinson 1995). Thus, significant variation has been observed in DMD production among males and for the same individuals over time (Bloch Qazi et al. 1998b). Furthermore, males fed low-nutrition diet showed a threefold reduction in their daily DMD production (Ming and Lewis 2010), suggesting that females could use DMD as a condition-dependent indicator of male quality.

However, females were not preferentially attracted to filter paper disks carrying odors from high-nutrition males (Ming and Lewis 2010), and existing evidence for sexual dimorphism in response to synthetic DMD is conflicting (Levinson and Mori 1983; Obeng-Ofori and Coaker 1990; Verheggen et al. 2007). In the former and numerous other studies, males' total olfactory cues have been collected onto filter paper disks over many days, thus reflecting a composite and time-integrated chemical profile. Such disks are typically used in olfactory bioassays to measure female attraction in pitfall arenas (Boake and Wade 1984; Boake 1985, 1986;

Bloch Qazi et al. 1998b; Wang 1992; Lewis and Austad 1994; Pai and Yan 2003a). One such study found that the long-range olfactory attractiveness was not correlated with the reproductive performance of male progeny, and therefore, females are unlikely to use these chemical signals as a basis for their choice (Boake 1986). In another study, males that were more attractive did gain higher paternity success when mating with previously mated females (Lewis and Austad 1994); however, higher attractiveness in this case may just be a correlate of male overall health and vigor, and not necessarily per se the determinant of male reproductive success. Therefore, current data do not support the idea that male long-range pheromones (DMD, in particular) affect female choice. In addition, the results of these and similar bioassays should be interpreted with caution for two reasons. First, because flour beetles release noxious benzoquinones when disturbed, female net response will represent the attraction to pheromones integrated with aversion to benzoquinones; how much of the latter is released depends on how males are handled. Second, most bioassays measure the responses of virgin females that are more motivated to mate and therefore less discriminating than mated beetles (Jennions and Petrie 2000).

Finally, under high population density, male–female encounters may essentially be random, with potential female pre-mating choice restricted to a brief close-range mate assessment based on low-volatility and contact pheromones, such as cuticular hydrocarbons (Howard and Blomquist 2005). Few hydrocarbons have been found in several *Tribolium* species (Markarian et al. 1978; Lockey 1978), and some (e.g., 1-pentadecene, 1-tetradecene, and 1-hexadecene) have been assigned pheromonal function for *T. confusum* (Verheggen et al. 2007). These and potentially other still unidentified compounds might be responsible for male attractiveness together with DMD, or they may act sequentially at different stages of intersexual interactions. However, information is currently lacking on the role of such cues in *Tribolium* mate choice.

16.1.4 Lack of Pre-mating Interactions

T. castaneum beetles lack any elaborate pre-mating interactions between the sexes. Upon encounter, a male generally mounts the female immediately and tries to achieve intromission with the extruded aedeagus while rapidly rubbing his legs along female's sides (Fig. 16.1). During such attempts, females may allow or refuse intromission by extruding or retracting their ovipositor (Wojcik 1969) or by dislodging a mounted male (Pai and Yan 2003b). When observed in a petri dish, only about half of all male copulation attempts lead to successful intromission, independent of female mating status (Lewis and Iannini 1995). Since in a majority of cases it is the female who dislodges the male (rather than the male dismounting voluntarily), females may thereby be exercising mate choice at this stage. However, it is currently unknown what male traits females might be assessing that could affect the likelihood of successful intromission. Male contact pheromones

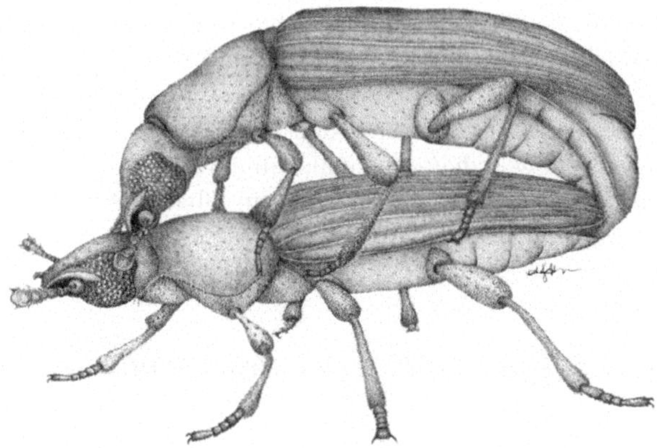

Fig. 16.1 Copulatory position of *Tribolium castaneum* flour beetles, with male mounted dorsally on female (drawing by April Hobart). Scale bar, 3 mm. From: *Biol Rev* 83:151–171

(cuticular hydrocarbons) and copulatory leg-rubbing are the likely candidates, as some evidence exists for their involvement in female choice at later stages (Sects. 16.2.2 and 16.2.3).

We offer several non-mutually exclusive explanations for the absence of apparent pre-mating interactions in flour beetles. First, at high population densities, the chance of forfeiting a mating to another male increases dramatically, so males will be selected to avoid pre-mating interactions and to monopolize females immediately upon encounter (Thornhill and Alcock 1983). Another reason for the absence of pre-mating interactions might be the lack of room and light for behavioral displays as well as excessive interference from conspecifics for any chemical, acoustic, or tactile displays, especially at high population densities. Such interference also makes pre-mating assessment of males by females error-prone, and several other insects mating under similar conditions also lack pre-mating interactions (Wojcik 1969; Ashworth 1993). In fact, under such conditions, an even better strategy for a male would be to mount any beetle and leave sex discrimination for later stages of copulation. Indeed, *T. castaneum* males are known for their sexually indiscriminant mating attempts and high frequencies of homosexual copulations (LeVan et al. 2009). Such seemingly maladaptive behavior has also been reported for other animals that mate in high-density aggregations (Thornhill and Alcock 1983; Scharf and Martin 2013).

Thus, we conclude that for *Tribolium,* the arena for male courtship and mate assessment has largely shifted from the more traditional pre-mating stage to copulatory and postcopulatory stages. Consequently, high female mating frequencies may be explained by the need to copulate in order to assess potential mates. If this is true, we should expect higher rates of insemination failure and higher variability in male paternity success in *Tribolium* compared to insects known for their elaborate pre-mating interactions (e.g., *Drosophila*). This will happen because in the absence of pre-mating interactions, females are expected to reject inferior males (and their ejaculates) via CFC mechanisms during and after mating. Existing data agree with the above hypothesis; thus, depending on conditions, only about

50–90 % of copulation attempts result in offspring production in *T. castaneum* (Bloch Qazi et al. 1996; Fedina and Lewis 2006; Lewis et al. 2012; Tyler and Tregenza 2013) compared to almost 100 % fertilization success for *Drosophila* males (e.g., Pischeda and Rice 2012; *pers. obs.*). In the following sections, we will discuss how *Tribolium* females might assess males during copulation, and we will describe specific CFC mechanisms that allow these females to prevent inferior males from siring many offspring.

16.1.5 Effect of Extreme Promiscuity on Sexual Selection

In polyandrous species, males are predicted to increase allocation to copulatory and postcopulatory competitive traits in the face of increased competition for mates and their gametes. In some cases, male competitive traits may evolve that are detrimental to females; this is presumably followed by the evolution of female counter-adaptations to reduce the harm. Therefore, polyandry may lead to sexually antagonistic coevolution resulting in net cost to females (Rice 1998; Arnqvist and Rowe 2005). Despite theoretically sound considerations and a number of empirical examples, the level of sexual antagonism in nature can be overestimated, and its effect on females assumed detrimental in the absence of any measures of female total fitness (Eberhard 2004; Chapman 2006; Peretti and Cordoba-Aguilar 2007; Priest et al. 2008; Eberhard this volume). Furthermore, higher levels of polyandry are often assumed to produce higher levels of sexual antagonism. However, recent theoretical studies suggest that under increasing polyandry, the return to males of further ejaculate allocation (and of any allocation to costly courtship and competitive traits) starts to decline, particularly if sperm precedence is determined by female anatomy and physiology rather than male competitive traits (Kvarnemo and Simmons 2013; Parker and Birkhead 2013; Shuster et al. 2013).

Sexual selection/conflict intensity is decreased under such conditions because males cannot monopolize females and their gametes as effectively, and this decreases variation in fitness among males. Monopolization of females and their eggs is more achievable in species with shorter life spans, where females lay eggs in batches or lay many eggs over short periods of time (Kvarnemo and Simmons 2013). In such species, males evolve traits that displace or otherwise disable the stored sperm from female's previous matings and/or traits that temporarily prevent females from mating with other males, while simultaneously increasing their egg-laying rate. In such species, males have a lot to gain from allocation to ejaculate and any form of mate guarding, and females have a lot to lose if the monopolizing male is not the ideal sire for their offspring, especially if mating has harmful side effects (e.g., in *Drosophila*). In such species, pre-mating assessment becomes more important for females.

Thus, sexual conflict is alleviated in species like *Tribolium*, where females lay eggs continuously over relatively long reproductive life span at the same time accepting multiple matings. Frequent promiscuous matings by females have been noted for a number of arthropods and other animals (e.g., Pomfret and Knell 2008—dung beetles; Panova et al. 2010—*Littorina* snails; Blyth et al. 2006—coelopid flies; Brouwer et al. 2011—fairy wrens), and different explanations have been given. Explanations aside, however, extreme polyandry must decrease sexual selection/conflict intensity in these species. For example, the analysis of 14 different species of dung beetles showed that males from more crowded populations with female-biased operational sex ratios have lost their horns, indicating lower sexual selection intensity (Pomfret and Knell 2008).

16.2 Cryptic Female Choice During Copulation

Throughout this chapter, we adhere to Thornhill and Alcock's (1983) definition of CFC as female-controlled processes occurring during and/or after copulation that bias offspring paternity toward some of her mates at the expense of other mates (see also the introductory chapter of this book). Interactions between the sexes during copulation have the potential to influence subsequent offspring paternity of polyandrous females, even in species with an extensive repertoire of pre-mating interactions (Eberhard 1991, 1996). Because flour beetles lack obvious pre-mating interactions, they may consequently rely to a greater extent on mate assessment during copulation. Copulatory interactions may involve sperm competition as well as CFC that can mediate or alter the outcomes of male actions (Eberhard 1996; Simmons 2001). Because such processes are likely to be influenced by both sexes, experimental manipulation of female control is necessary to partition male *versus* female influence; alternatively, some studies have relied on statistical partitioning of male *versus* female effects after crossing different male and female strains (Arnqvist 2014). Here, we will review existing evidence for CFC during copulation in flour beetles, including female influence over insemination success, spermatophore placement, and the amount of sperm transferred and stored.

16.2.1 Female Reproductive Anatomy and Male Spermatophore

T. castaneum males transfer sperm into the lower portion of female reproductive tract, called the *bursa copulatrix* (Fig. 16.2a, b). Short-term sperm storage, lasting about 1 week, occurs in the anterior part of the bursa, which can be closed off from the lower bursa by a muscular sphincter. Sperm in the bursa is used for fertilizations first (Droge-Young, *pers. comm.*). Longer term sperm storage, lasting several months, takes place within the female spermatheca, a structure consisting of several long, narrow tubules connected to the anterior bursa through a common duct (Surtees 1961; Fedina and Lewis 2004). Close to the spermatheca lies a

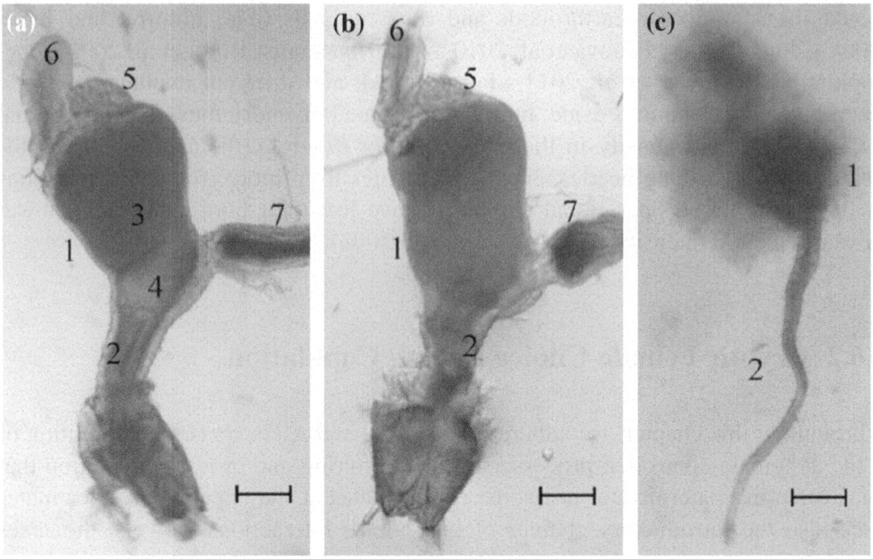

Fig. 16.2 Spermatophore transfer in *T. castaneum*: female dissections immediately following naturally terminated copulations: **a** and **b** progressive stages of spermatophore inside the female *bursa copulatrix* and **c** spermatophore dissected out of the female bursa. The scale bars are equal to 100 μm. *1* Spermatophore sac in the anterior bursa, *2* spermatophore tail in the posterior bursa, *3* and *4* anterior and posterior spermatophore sac, *5* spermatheca, *6* spermathecal gland, and *7* common oviduct. From: *J Insect Physiol* 53:93–98

prominent spermathecal gland, which releases its secretions into the anterior bursa through a duct passing through a sclerotized O-shaped ring. The function of this gland is currently unknown (Sect. 16.4)

Tribolium males transfer their sperm and seminal fluid within a so-called spermatophore (Fig. 16.2), a proteinaceous structure manufactured by two pairs of male accessory glands. Spermatophore production, histochemistry, and behavior have been particularly well described in *Tenebrio* mealworm beetles, whose initially invaginated spermatophore everts to release sperm once it is inside the female bursa (e.g., Gadzama and Happ 1974; Grimnes et al. 1986; Paesen et al. 1996). The spermatophore of *T. castaneum* is structurally and behaviorally similar to that of *T. molitor,* but it ruptures and releases sperm more quickly after transfer (Bloch Qazi et al. 1996; Fedina 2007); this may guard against sperm displacement in the face of frequent female re-mating. The *T. castaneum* spermatophore is also transferred to females as an invaginated tube that becomes filled with sperm during transfer, and this allows for female manipulation of the numbers of sperm transferred (Sect. 16.2.4).

16.2.2 Male Copulatory Courtship and Female Quiescence Behavior

In *T. castaneum*, copulations generally last from 0.5 to 2 min, but some may continue for over 30 min (Sokoloff 1974; Bloch Qazi et al. 1996; Haubruge et al. 1999; Edvardsson and Arnqvist 2000; Lewis 2004). Such extreme variability is in agreement with the idea that females assess mates during copulation (although it may also reflect differences in male motivation or spermatophore transfer efficiency). This contrasts with *D. melanogaster*, where following extensive pre-mating interactions, copulations almost invariably last around 20 min, time affected mostly by male (Economos et al. 1979; Taylor et al. 2013; Crickmore and Vosshall 2013; and references therein). In *T. castaneum,* only female copulatory behaviors were found to predict the likelihood of sperm transfer and the number of sperm transferred (Bloch Qazi 2003). In addition, copulations with dead females (which provide no resistance) are about 20 % shorter than copulations with live females (Fedina and Lewis 2008). Both these facts as well as the fact that females are more likely than males to terminate copulations suggest that it is primarily females that control copulation duration in this species, taking variable time for mate assessment and CFC. How might *Tribolium* females assess mate quality during copulation?

In many tenebrionid beetles, males perform stereotypic leg movements during mating (Wojcik 1969; Carazo et al. 2004). *T. castaneum* males rapidly rub their legs back and forth against the sides of the female's body in short bouts interspersed with periods of inactivity (Fig. 16.1, Bloch Qazi 2003). This eventually leads to females exhibiting quiescence behavior: They stop moving and assume a characteristic posture (Bloch Qazi 2003). This female behavior is associated with spermatophore transfer: If copulations are interrupted during quiescence, 60 % of them fail to result in spermatophore transfer, compared to only 20 % failure when copulations were allowed to proceed naturally or interrupted after full quiescence (Bloch Qazi 2003). Female quiescence may therefore indicate female acceptance of male spermatophore.

It has been suggested that some characteristics of male leg-rubbing may communicate male quality to females and thereby affect female CFC decisions. Several studies have tested this hypothesis in *T. castaneum*, but have reached different conclusions. Thus, Edvardsson and Arnqvist (2000, 2005) observed a positive correlation between male leg-rubbing rate and paternity share when males mated with previously mated females. However, this correlation by itself does not imply that female choice is based on male leg-rubbing. In fact, when in the same study males' legs were shortened to alter female perception of leg-rubbing, no changes in either second-male paternity share or number of offspring were detected for manipulated compared to control males. Furthermore, in these studies, male leg-rubbing behavior was measured in real time by direct human observation, which seems likely to be error-prone due to the speed and complexity of this behavior. Using a microscope, the total number of rubbing bouts (each comprised of several leg swings) by each leg was counted, then rubbing rate was determined after dividing by copulation duration (no measurements were made of bout duration). A more comprehensive study was performed using video analysis (Bloch Qazi 2003), which found that neither the rate nor total duration of leg-rubbing predicted second-male paternity share. In this study, though, leg-rubbing was only assessed during quiescence; at this point, the female has already made a decision to accept the male's spermatophore, so this may not be the stage of copulation most relevant to CFC. Finally, Fedina and Lewis (2006) analyzed video-recorded behaviors *before* the onset of female quiescence while also manipulating male condition by starvation. They found that only rubbing frequency (number of swings/sec summed over all leg) was decreased in starved compared to fed males, suggesting that this measure may communicate male condition to females. However, rubbing frequency did not affect male paternity share, while the two other measures (% time spent rubbing and rubbing asymmetry, which were both unaffected by starvation) were positively associated with paternity share. This suggests that females either do not use rubbing frequency to assess male condition, or that male current condition is of no interest to them. The same study noted that males intensified their leg-rubbing in response to female movement and that percentage of time spent leg-rubbing was positively associated with the probability of insemination failure. These latter observations suggest that male leg-rubbing may merely encourage females to remain stationary (perhaps, via some mechanism of sensory exploitation), thus facilitating sperm transfer. This study also included an additional analysis that used male leg-rubbing rate as defined by Edvardsson and Arnqvist (2000), yet found that it was unrelated to the male's paternity share (Fedina and Lewis 2006). Furthermore, male leg-rubbing does not change in response to variation in rearing density (Edvardsson and Arnqvist 2006), or induced parasitism (Pai and Yan 2003a), and showed no heritability for sons or genetic benefits in terms of progeny survival (Edvardsson and Arnqvist 2006). Therefore, taken together the current evidence does not support the idea that *T. castaneum* females use male leg-rubbing to assess mate quality and manipulate paternity share.

Another way for females to assess male quality during copulation could be through close-range chemical cues, such as cuticular hydrocarbons (Howard and Blomquist 2005). However, the role of such chemical cues in *Tribolium* mate choice has been largely unstudied. *T. castaneum* males have paired setiferous glands located on the tarsi of their front legs that produce copious waxy secretions that accumulate as males get older and are attractive to both sexes (Faustini et al. 1982). The function of these glands is still unclear, but they are not necessary for long-distance attraction (Bloch Qazi et al. 1998b). Their proximity to the female's chemosensory apparatus during copulation makes them good candidates for use in close-range sexual interactions. Additional studies are needed to identify and manipulate specific chemical cues that might be used by females for mate assessment during copulation.

16.2.3 Female Control Over Spermatophore Transfer

In *T. castaneum,* spermatophore transfer fails in up to 15 % of copulations with virgin females (Bloch Qazi et al. 1996) and up to 50 % of copulations with previously mated females (Tyler and Tregenza 2013). Females were shown to play an active role in allowing spermatophore transfer in experiments that manipulated male phenotypic quality by starvation (Fedina and Lewis 2006). These experiments leveraged the fact that males will readily mate and transfer sperm to females that have been freshly killed (using ethyl acetate vapors). Males behave similarly toward live or dead females at least in the beginning of copulation when male behaviors are not yet modified by the presence or absence of female response (Fedina and Lewis 2006). After starvation for 7 days, males lose 10–15 % of their body mass. Interestingly, when mating with dead females, these starved males were just as likely to transfer a single spermatophore as non-starved males are, suggesting that male starvation per se does not affect spermatophore transfer ability (Fedina and Lewis 2006). When mating with live females, however, these phenotypically inferior males were 20 % less likely to transfer spermatophore than fed males (Fig. 16.3). This result demonstrates that live females exert control over spermatophore transfer and more often reject spermatophores from inferior males. Notably, this was not an absolute response, as 70 % of inferior males still managed to transfer spermatophores to live females. However, these were no-choice experiments that used virgin females who are likely more motivated to mate and may accept low-quality males more often than already mated females. Furthermore, *T. castaneum* females can afford partially effective responses, because they can "update" their decisions via additional CFC mechanisms (see next sections).

Another potential CFC mechanism that has been investigated is female control over where the male spermatophore is placed within the female reproductive tract. Dissections of females immediately after mating revealed that the sperm-containing sac is positioned in female *bursa copulatrix* in one of two distinct locations: either above or below the sphincter that separates the anterior and posterior bursa (Fig. 16.4). Because common oviduct enters right below this sphincter, an egg passing down the oviduct

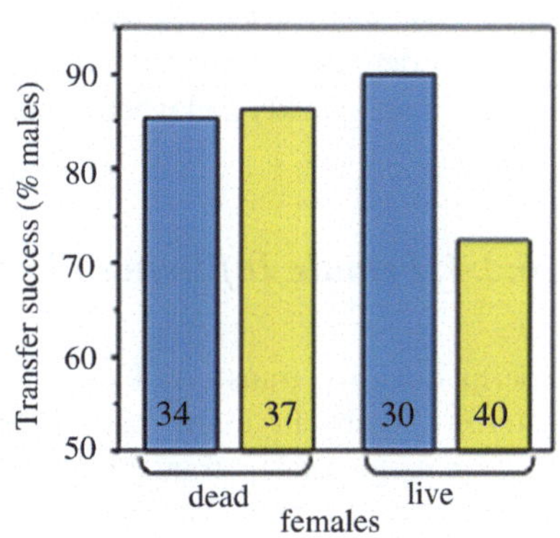

Fig. 16.3 Percentage of fed (*blue bars*) versus starved (*yellow bars*) *T. castaneum* males that successfully transferred spermatophores to dead or to live females. Sample sizes are given inside bar. From: *Behav Ecol Sociobiol* 60:844–853

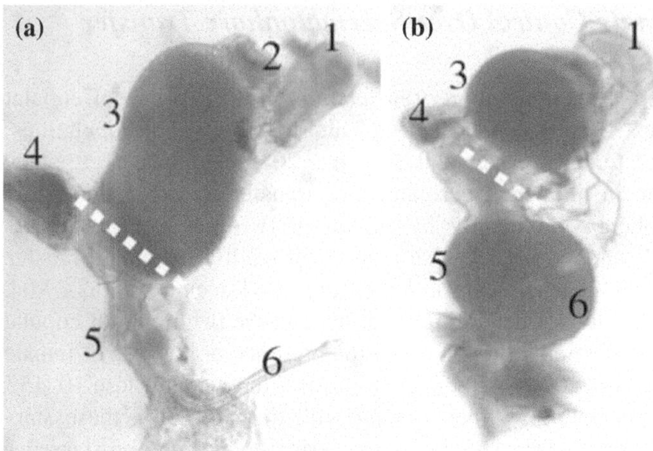

Fig. 16.4 Differences in male spermatophore positioning within the *T. castaneum* female reproductive tract (*white dotted line* separates anterior from posterior bursa). **a** "Successful" positioning: The mating male's spermatophore sac is visible as a dark mass filling the female's anterior bursa, with spermatophore tail located in posterior bursa. **b** "Unsuccessful" positioning: Mating male's spermatophore sac and tail are both located in female posterior bursa, separated by muscular sphincter from the anterior bursa, where sperm from a previous mate is visible as a dark mass. *1* Spermathecal gland, *2* spermatheca, *3* anterior bursa, *4* common oviduct, *5* posterior bursa, and *6* spermatophore tail. *Behav Ecol Sociobiol* (2006) 60:844–853

would likely displace any sperm located in posterior bursa. Additionally, spermatophores positioned in posterior bursa would be more vulnerable to displacement by the female's next mate. Since females lay eggs and mate every few hours, spermatophore positioning could therefore be an important determinant of how much sperm from each mating gets stored and subsequently used for fertilizations. Fedina and Lewis (2006) detected no effects of male treatment (starved vs. fed), female treatment (live vs. dead), or male copulatory courtship behaviors on spermatophore positioning. However, in these experiments, the females were pre-mated to multiple males 24 h before the experimental mating, by which point any previous male spermatophores would already have been expelled. Relative spermatophore placement is likely to be more important when two males' spermatophores overlap inside the female bursa. Thus, it would be interesting to determine whether females can control spermatophore position when they re-mate more quickly and whether such placement predicts male paternity share.

16.2.4 Female Influence Over Sperm Quantity Transferred

Sperm quantity transferred during copulation is the main determinant of male paternity share in many taxa (Parker 1970; Eberhard 1996; Simmons 2001). *T. castaneum* males transfer 0.5–3.0 $\times$ 10^5 sperm during a single copulation (Arnaud et al. 2001; Bloch Qazi et al. 1998a, 1996; Fedina and Lewis 2006).

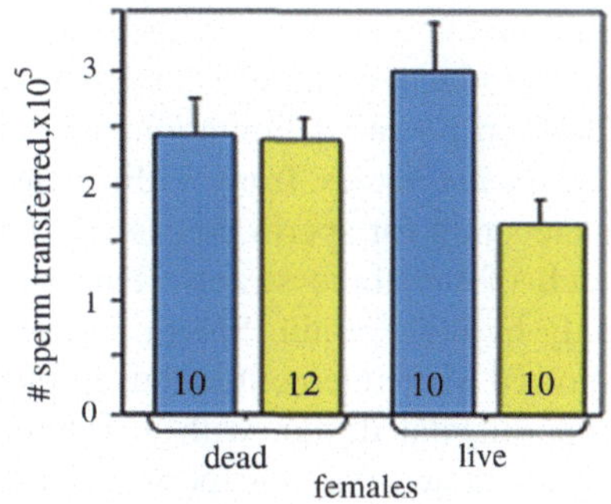

Fig. 16.5 Sperm quantity (mean + SE) transferred per spermatophore by either fed (*blue bars*) or starved (*yellow bars*) *T. castaneum* males that were randomly assigned to mate with dead or live females. Sample sizes are shown inside the bars. From: *J Insect Physiol* 53:93–9

Because this number does not depend on male body mass or copulation duration (Bloch Qazi et al. 1996), but instead increases with longer female quiescence (Bloch Qazi 2003), this female behavior appears critical for successful sperm transfer. When males mate in rapid succession, the number of sperm they transfer decreases up to threefold (Bloch Qazi et al. 1996), and this substantially diminishes males' paternity share in matings with previously inseminated females (Lewis 2004). This suggests that females could potentially influence paternity by restricting the numbers of sperm transferred during mating. Such female control is made possible because in *T. castaneum,* sperm is not prepackaged into the spermatophore before mating, but rather is injected into it during copulation. Fedina (2007) demonstrated that females use control over sperm quantity transferred as a CFC mechanism, using a 2-factorial mating design in which males of high and low phenotypic quality (fed vs. starved) were mated to either live (full control) or dead (no control) females. Starved males transferred about half as much sperm as fed males when mating with live females (Fig. 16.5). When mating with dead females, however, starved males transferred the same number of sperm as fed males. Thus, female control rather than male condition is responsible for restricting the number of sperm injected into the spermatophore by inferior males.

We have seen that during copulation, females exercise choice over spermatophore transfer itself, as well as over the quantity of sperm transferred per spermatophore, and that longer quiescence during copulation is reflective of female favoring the male through both of these mechanisms. It remains unknown, however, whether females use spermatophore positioning inside their bursa as another CFC mechanism and how they perceive male condition.

16.3 Postcopulatory Cryptic Female Choice

16.3.1 Inferring CFC from MxF Statistical Interactions

One experimental approach that has been widely used to demonstrate the existence of CFC is crossing a panel of genetically similar groups of males and females (e.g., strains) in all possible combinations. If a statistical interaction between male and female genotypes in their effects on paternity, fecundity, and other measures of reproductive output is observed, it suggests the existence of CFC (see Arnqvist 2014 on potential limitations of this approach). Pai and Yan (2002) competed *T. castaneum* males from 3 different wild-type strains against a standard marker male strain for sperm defense (P_1) and offense (P_2) when mating to females from each of the three strains. This study found that sperm offense was influenced only by male strain. Sperm defense, however, was affected by female strain and showed MxF interaction, thus implicating females in biasing paternity. In a similar experimental design with 3 different *T. castaneum* strains, Nilsson et al. (2003) observed no effect on sperm defense, but sperm offense was significantly affected by both male and female genotype, as well as their interaction. In addition, this study found that copulation duration was significantly affected by genotype of each sex and their interaction, suggesting CFC during copulation. Discrepancies between results of these two studies may be due to different strains being used, or to differences in experimental design; Pai and Yan (2002) employed 24-h cohabitation with each male as mating regime for a female, while Nilsson et al. (2003) observed a single mating with each male. The former mating regimen results in multiple ejaculates being transferred to females, and this alters sperm precedence compared to when each male is restricted to a single mating (reviewed in Fedina and Lewis 2008).

Although the above experiments provide evidence for female involvement in determining sperm precedence, they shed no light on the mechanisms by which females exercise their choices. Determining precisely how CFC is achieved requires a targeted examination of each hypothesized mechanism, accompanied by manipulation of female control.

16.3.2 Sperm Movement into Storage

In *T. castaneum*, sperm are quickly released from the spermatophore and begin moving into the female spermatheca even before copulation ends. Under normal conditions, this process is completed within one hour, when about 4 % (=4000–5000) of all transferred sperm is moved into spermatheca (Bloch Qazi et al. 1996). When females were anaesthetized with carbon dioxide for 30 min following mating (this abrogates female muscular activity, but leaves sperm motility unaffected), there was an 11-fold reduction in the number of sperm that moved into the female spermatheca. In addition, live females stored ~6.5 times more sperm than dead females, with greater variation (Bloch Qazi et al. 1998a). Thus, muscular activity within the female reproductive tract is required for sperm storage. Such active sperm transport could represent yet another CFC mechanism if females were shown to move more sperm into spermatheca when mating with higher quality males. Fedina and Lewis (2006) found suggestive evidence for differential sperm storage, as fed males not only transferred more sperm compared to starved males, but also a higher percentage of their sperm ended up in storage. However, while these results suggest that females might differentially store sperm from superior males, it is also possible that fed males are better able to transfer and move their sperm into storage. Manipulations of female control over storage (e.g., using CO_2) along with male quality are needed to demonstrate that females can use this mechanism to bias male paternity.

In addition to females actively transporting certain males' sperm into storage, the volume or shape of sperm storage compartments may represent another CFC mechanism, if it makes some males more successful at siring offspring than others. *T. castaneum* females show high variation in spermathecal morphology (tubule number) and volume (Fig. 16.6, Fedina and Lewis 2004, Bernasconi et al. 2006). Since females do not seem to have sphincters or muscular control over individual tubules, they are unlikely to partition sperm from different males as has been suggested for some flies (Eberhard 1996). However, a negative

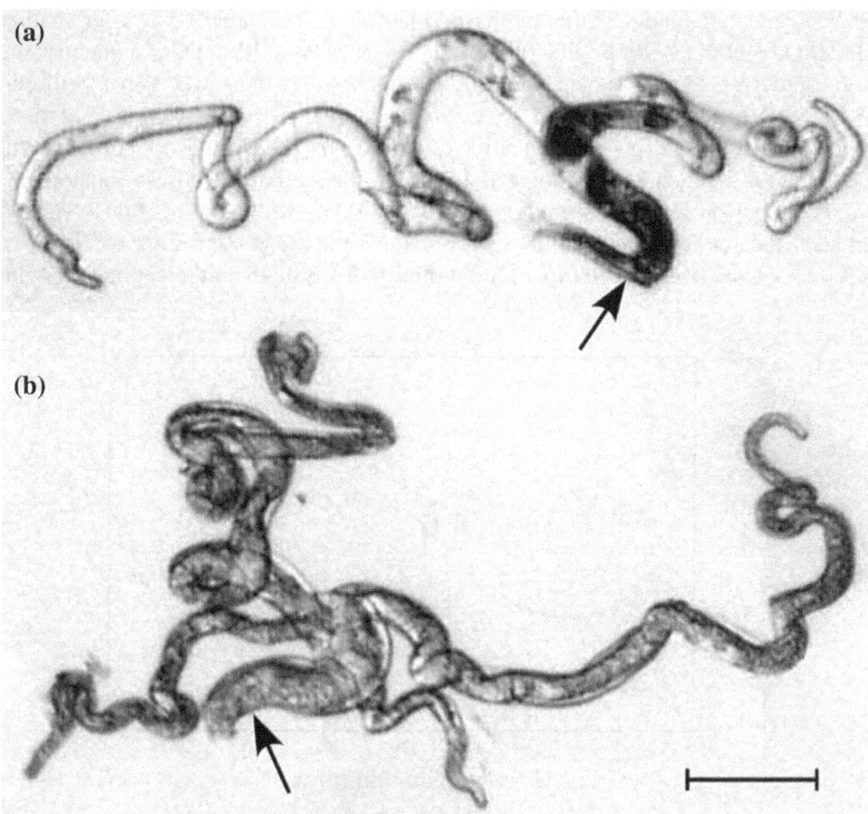

Fig. 16.6 Natural variation in spermathecal morphology of *T. castaneum* females: representative examples of **a** simple spermathecal structure with two primary tubules and **b** complex spermathecal structure with four primary tubules. Primary tubules are connected to the spermathecal duct (indicated by *arrow*), which opens into the anterior *bursa copulatrix* (not shown). Scale bar, 50 μm. From: *Proc R Soc Lond* B 271:1393–1399

correlation was found between second-male paternity success and spermathecal volume (Fedina and Lewis 2004; but see Bernasconi et al. 2006); this may reflect increased storage and/or reduced displacement of first-male sperm in females with larger spermathecae.

16.3.3 Spermatophore Ejection After Mating

Following successful insemination, females can further affect sperm fate via spermatophore/sperm ejection as shown for some other beetles, flies, and dragonflies (Eberhard 1996, Cordoba-Aguilar 2006, Lüpold et al. 2011). In *T. castaneum*, ~4 % of total transferred sperm is stored in the female spermatheca, while another 10 % is transiently stored in the anterior bursa; the remaining sperm (~85 %) are expelled together with spermatophore remnants soon after mating (Bloch Qazi et al. 1996, Fedina 2007, Fedina and Lewis 2007). Consequently, the more quickly a female expels a male's spermatophore, the fewer sperm will end up in storage. Previously mated *T. castaneum* females frequently expelled spermatophores within a few minutes after re-mating, and these still contained sperm (Fedina 2007). When kept in flour after mating, the median time to spermatophore expulsion for previously virgin females was >15 h, while for previously mated females, the median spermatophore expulsion time was ~2 h (Fig. 16.7). Thus, if females selectively expel spermatophores more quickly after copulating with

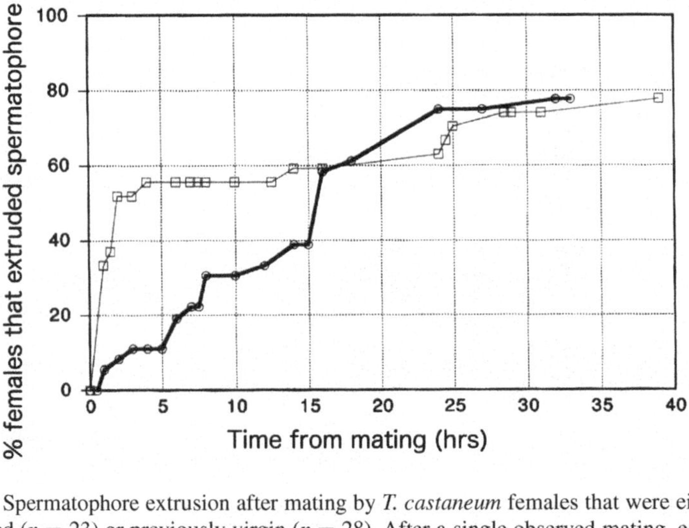

Fig. 16.7 Spermatophore extrusion after mating by *T. castaneum* females that were either previously mated ($n = 23$) or previously virgin ($n = 28$). After a single observed mating, each female was placed in fine flour that was sifted at various intervals (ranging from 0.5 to 9 h) to check for any eggs and spermatophores that had been deposited. Virgin females retained male spermatophores longer than mated ones: For virgin females, the median retention time was >15 h, compared to 2 h for previously mated females

inferior males, it could provide another CFC mechanism. This potential mechanism is currently under investigation (Droge-Young, *pers. comm.*), and preliminary observations have shown that spermatophore retention time predicts the relative representation of a focal male sperm in female bursa (which is in turn proportional to short-term paternity success of the male) and that spermatophores are expelled after longer delay following copulations with bigger males.

16.3.4 Potential for CFC During Sperm Storage and Use

In other taxa, females have been shown to interact with male sperm and ejaculate components and to favor some males over others via molecular/cellular interaction mechanisms (Andersson and Simmons 2006; Evans and Sherman 2013). These mechanisms have not been explicitly studied in flour beetles; however, some progress has been made in that direction (see Sect. 16.4). Furthermore, conspecific sperm precedence has been observed in crosses between *T. castaneum* and *T. freemani* (Wade et al. 1994). Following interspecific matings, females produce normal numbers of hybrid (sterile) offspring; however, after females mate with both a conspecific and a heterospecific male, they produce over 99 % of conspecific progeny, independent of male order (Robinson et al. 1994; Fricke and Arnqvist 2004). Which mechanisms of CFC might be responsible for this conditional reproductive isolation remains to be explored.

16.3.5 Cryptic Female Choice via Re-mating

Following sperm transfer and storage, females can further affect male paternity by re-mating with a different male (Eberhard 1996; Simmons 2001). Because of high *T. castaneum* mating rates and continuous egg-laying, male paternity success will depend on the female's previous mating history as well as her future mating trajectory (Lewis and Jutkiewicz 1998; Lewis et al. 2005). *T. castaneum* is characterized by the last male precedence, in which the last male to mate with a female fertilizes the majority of her eggs over the next few days (Lewis and Austad 1990; reviewed in Fedina and Lewis 2008). Because many potential mates are typically available, females could potentially manipulate offspring paternity by re-mating with a higher quality male. Nilsson et al. (2003) examined re-mating delay by *T. castaneum* females in reciprocal pairings between three wild-type strains and detected significant male-by-female genotype interaction. This implies that females may control their re-mating interval in order to bias paternity. Further support for this CFC mechanism is that when *T. castaneum* females mated sequentially to two males differing in their attractiveness, they were twice as likely to accept a spermatophore from a more attractive male following an initial mating with a less attractive male than in the reciprocal mating order (Fedina and Lewis 2007).

Relative male attractiveness in this study was assayed by giving each female a choice between three tubes of flour: In two tubes, the flour had been conditioned by two different males, while a third tube contained fresh flour. Only females who entered one of the two male tubes were used; half of them were assigned to mate first with the chosen male, then after 24 h with the non-chosen one, and the remaining females mated in the opposite order. The females were dissected after their second copulation, and the presence of the second-male spermatophore in the female's bursa was recorded as successful transfer (the success of the first mating was judged by the presence of larvae in a vial). Similarly, in *T. molitor,* females were more likely to re-mate with males larger than their first mate, and they also re-mated more quickly after mating with tapeworm-infected male (Worden and Parker 2005).

Re-mating is the final CFC mechanism in the sequence of possible male–female interactions during and after single mating event, and at the same time, it starts another cycle of interactions with a new male. In Sects. 16.2 and 16.3, we have reviewed evidence for at least four distinct CFC mechanisms operating in *T. castaneum* (summarized in Table 16.1). These include: (1) female control over spermatophore acceptance, (2) female control over the number of sperm transferred during mating, (3) female expulsion of spermatophores after mating, and (4) female re-mating with a different male. We discussed the potential for two additional CFC mechanisms that are based on characteristics of the *T. castaneum* female reproductive system: female control over position of the male spermatophore and female control over sperm storage. The existence of multiple CFC mechanisms in *Tribolium* may stem from constraints on females' ability to assess males effectively before mating. These various CFC mechanisms may benefit

Table 16.1 List of *T. castaneum* cryptic female choice mechanisms

Mating stage	CFC mechanism	Evidence	References
Pre-mating	Allowing intromission	Not studied	Lewis and Iannini (1995)
During mating	Allowing spermatophore transfer	Good	Bloch Qazi et al. (Bloch Qazi et al. 1996), Fedina and Lewis (2006), Tyler and Tregenza (2013)
	Control sperm quantity transferred per spermatophore	Good	Bloch Qazi et al. (1996), Bloch Qazi (2003), Bloch Qazi et al. (1998a)
	Spermatophore positioning in female bursa	Requires additional study	Fedina and Lewis (2006)
Postmating	Movement of sperm into storage by a female	Requires additional study	Bloch Qazi et al. (1996), Fedina and Lewis (2006), Bloch Qazi et al. (1998a), Fedina and Lewis (2004)
	Time of spermatophore ejection	Preliminary	Droge-Young E. M. unpublished data
	Re-mating with another male	Good	Fedina and Lewis (2004)

females by allowing them to use sequential cues to evaluate male quality and to adjust their paternity decisions at different stages during and after mating as they receive more information.

16.3.6 Comparative Study of CFC Evolution and Mechanisms

Comparisons among closely related species are a classical approach that has proven valuable in discerning traits' evolution and adaptive significance. *T. castaneum* has several relatively well-studied tenebrionid relatives with distinct differences in anatomy, physiology, life history, and mating systems that may allow inferences about the operation of sexual selection. Thus, a more distant relative of *Tribolium* spp., *T. molitor* is distinguished by larger body size, shorter adult life span, lower tolerance to high densities, much lower female fecundity (Cotton and George 1929; Dick 1937), and more elaborate pre-mating interactions (Happ 1969; August 1971; Obata and Hidaka 1982). These life history and mating system features make mealworm beetles an informative comparison to flour beetles. For example, differences in spermatophore behavior inside female reproductive tract in the two species are suggestive of different selection forces acting at the time of transfer. Thus, in *T. molitor,* spermatophore takes several minutes to open and release sperm, while in *T. castaneum* spermatophore ruptures to release sperm almost immediately after and often even before the end of mating. Such difference has allowed *T. molitor* male to evolve spermatophore incapacitation as a mechanism of sperm offense. In this species, if another male mates with a female within 5 min after previous male, the previous male's spermatophore does not release its sperm whereby completely negating that males' paternity (Drnevich 2003). The ratio between spermathecal volume and the male ejaculate volume also has the potential to influence mating, sperm storage dynamics, and probability of displacement. Thus, in *T. castaneum,* single spermatophore fills the entire female bursa (Fedina and Lewis 2006), and a single ejaculate fills spermatheca to two-thirds of its maximum capacity (Lewis and Jutkiewicz 1998). In contrast, *T. molitor* female bursa can accommodate as much as 12 spermatophores, and spermatheca is not full even after three complete sperm transfers (Drnevich 2003). This may produce the difference in sperm precedence between the species: high last sperm precedence in *T. castaneum* (Fedina and Lewis 2006) and a more evenly distributed paternity by multiple males in *T. molitor* (Drnevich 2003). It would be informative to determine the effects of such profound differences in spermatophore transfer and sperm storage dynamics on CFC and find their connection with the divergent life histories of these beetles.

Another example of an interesting divergence in sexual characters between *Tribolium* species is spermathecal shape, known to influence sperm displacement dynamics. Thus, the *T. castaneum* spermatheca has long and narrow tubules, which promote initial sperm stratification and last-male sperm precedence, followed by gradual rise in paternity by previous males (Fedina and Lewis 2008). In its close relative *T. confusum,* the female spermatheca is a wide U-shaped sac connected to female bursa via a narrow duct (Surtees 1961). The latter shape promotes more uniform initial sperm mixing and different temporal changes in paternity compared to *T. castaneum* (Vardell and Brower 1978). Determining the order of evolutionary transitions in spermathecal shape and other related differences could reveal what might have driven the evolution of such divergent sperm storage structures.

16.4 Latest Developments and Future Directions

In many insects, males' accessory gland substances that are transferred to females during copulation exert a profound effect on female physiology and reproductive behavior, aiding sperm movement into storage, increased ovulation rate, increased refractoriness to re-mating, and decreased life span (Eberhard 1996; Simmons 2001; Chap. 14 of this book). In flour beetles, indirect evidence suggests these male substances might alter female physiology. Thus, the observed male X female interaction effects on female mating rate, reproductive output, and longevity in reciprocal crosses of different wild-type *T. castaneum* strains (Attia and Tregenza 2004; Nilsson et al. 2002) implicates postmating interactions between male ejaculates and female reproductive physiology, and the increase in female reproductive rate in response to elevated mating frequency (Nilsson et al. 2002) indicates the presence of oviposition-boosting substances in male ejaculate.

Tribolium spp. males have two pairs of accessory glands with at least six differentially staining cell types (Sevener et al. 1992; Roberts and Grimnes 1994; Novaczewski and Grimnes 1996). Histological staining in *T. brevicornis*, *T. freemani*, and *T. anaphe* suggests that these glands produce mostly proteins and some carbohydrate-containing molecules; however, detailed molecular characterization of these substances has been lacking until recently. Using a combination of proteomic and genomic methods, South et al. (2011a) identified 14 seminal fluid proteins (13 with male-biased expression) in *T. castaneum*; seven of these proteins had identifiable homologues in other insects, including protease inhibitors and odorant binding proteins. Another study using proteomics but somewhat different criteria for defining male accessory gland proteins identified 13 proteins, 5 of which overlapped with the previous study (Xu et al. 2013). For the majority of these proteins, RNAi knockdown in males had negative effect on female offspring production. However, the strongest negative effect was observed for angiotensin-converting enzyme that maintains sperm structural integrity, and it is still unclear whether any of these proteins target female physiology (Xu et al. 2013). Notably, many of the *Tribolium* male accessory gland proteins did not have identifiable homologs in previously studied model systems (South et al. 2011a; Xu et al. 2013). No homologs of SP have been found in *Tribolium*, but this does not prove its absence since reproductive proteins are known for rapid divergence (Clark et al. 2006). The *Tribolium* homolog of a recently discovered *Drosophila* SP receptor showed no response to *Drosophila* SP when tested ex vivo (Yapici et al. 2008), although this receptor is conserved among insects (Poels et al. 2010). Therefore, the lack of response of *Tribolium* SP receptor homolog to *Drosophila* SP might indicate either the absence of SP in flour beetles, or a requirement for some additional cofactors absent in ex vivo system. In any case, future studies of female molecules responding to male accessory gland proteins should be facilitated by availability of the *T. castaneum* genome (Brown et al. 2003; Wang et al. 2007), as well as the advances made in other model systems (see Chap. 14 in this book).

Another major missing link in understanding of postmating male–female interactions,—not only for *Tribolium* but also for most arthropods,—is a general lack of knowledge on the function of female reproductive glands (Al-Wathiqui et al. 2014). In *Tribolium,* the close proximity of the spermathecal gland to the sites of sperm storage and fertilization suggest a possible role in female sperm choice. The cellular structure of the gland has been studied in both *T. castaneum* (Al-Khalifa 1981) and *T. molitor* (Happ and Happ 1970, 1977), and it has been suggested that its secretions serve a nutritive function for stored sperm. In *D. melanogaster,* spermathecal gland secretions are required for effective sperm storage (Sun and Spradling 2013), and in bees, these secretions maintain stored sperm viability (Den Boer et al. 2009): Both of these physiological functions suggest the potential for female spermathecal glands to play a role in CFC. However, beyond basic anatomical and histological investigations, remarkably little is known about female reproductive gland functions in *Tribolium* and other insects.

16.5 Conclusions

The environment, demographics, and population dynamics of *Tribolium* favor long adult life span and continuous laying of single eggs by females over their long fertility period.

These features prevent males from effectively monopolizing females and their eggs (and, possibly, from evolving female-harming sexually antagonistic adaptations).

At high population density, frequent promiscuous matings are favored by unsuitable conditions for pre-mating choice as well as by the absence of significant costs of polyandry for females.

Pre-mating choice is replaced by mate assessment during and possibly after copulation. *Tribolium* females can bias paternity via several identified CFC mechanisms, including: (1) female control over spermatophore transfer, (2) female control over the number of sperm transferred during mating, (3) female expulsion of spermatophores after mating, and (4) female re-mating with a different male.

Most of these CFC mechanisms are only partially effective individually (i.e., they do not act as all or nothing), but together these mechanisms may benefit females by allowing them to use sequential cues to assess male quality and to adjust their paternity decisions at different stages during and after mating.

Future efforts should focus on identifying male traits subject to female choice and on characterizing postmating interactions between male ejaculate components and female reproductive physiology.

Acknowledgments We thank the editors Anita Aisenberg and Alfredo Peretti for inviting us to write this chapter, as well as for their helpful comments and guidance in writing the chapter. We also thank Margaret Bloch-Qazi and an anonymous reviewer for their valuable comments on an earlier version of this chapter. Finally, we are grateful to Elizabeth Droge-Young for sharing the results of her unpublished work on *Tribolium*.

References

Al-Khalifa MS (1981) The spermathecal structure of the flour beetle *Tribolium castaneum* (Coleoptera: Tenebrionidae). Indian J Zool 9:1–6

Al-Wathiqui N, Lewis SM, Dopman E (2014) Using RNA sequencing to characterize female reproductive genes between Z and E strains of European corn borer moth (*Ostrinia nubilalis*). BMC Genomics 15:189

Alexander P, Barton DHR (1943) The excretion of ethylquinone by the flour beetle. Biochem J 37:463–465

Andersson M, Simmons LW (2006) Sexual selection and mate choice. TREE 21(6):296–302

Arnaud L, Haubruge E (2002) Insecticide resistance enhances male reproductive success in a beetle. Evolution 56:2435–244

Arnaud L, Haubruge E, Gage MJG (2001) Sperm size and number variation in the red flour beetle. Zool J Linnean Soc 133:369–375

Arnqvist G (2014) Cryptic female choice. In: The evolution of insect mating systems. Oxford University Press.

Arnqvist G, Rowe L (2005) Sexual conflict. Princeton University Press, Princeton

Ashworth JR (1993) The biology of *Lasioderma serricorne*. J Stored Prod Res 29(4):291–303

Attia FA, Tregenza T (2004) Divergence revealed by population crosses in the red flour beetle *Tribolium castaneum*. Evol Ecol Res 6:927–935

August CJ (1971) The role of male and female pheromones in the mating behavior of *Tenebrio molitor*. J Insect Physiol 17:739–751

Bernasconi G, Brostaux Y, Meyer EP, Arnaud L (2006) Do spermathecal morphology and inter-mating interval influence paternity in the polyandrous beetle *Tribolium castaneum*? Behaviour 143:643–658

Bernasconi G, Keller L (2001) Female polyandry affects their sons' reproductive success in the red flour beetle *Tribolium castaneum*. J Evol Biol 14:186–193

Bloch Qazi MC (2003) A potential mechanism for cryptic female choice in a flour beetle. J Evol Biol 16:170–176

Bloch Qazi MC, Aprille JR, Lewis SM (1998a) Female role in sperm storage in the red flour beetle, *Tribolium castaneum*. Comp Biochem Physiol A 120:641–647

Bloch Qazi MC, Boake CRB, Lewis SM (1998b) The femoral setiferous glands of *Tribolium castaneum* males and production of the pheromone 4,8-dimethyldecanal. Entomol Exp Appl 89:313–317

Bloch Qazi MC, Herbeck JT, Lewis SM (1996) Mechanisms of sperm transfer and storage in the red flour beetle (Coleoptera: Tenebrionidae). Ann Entomol Soc Am 89:892–897

Blyth JE, Gilburn AS (2006) Extreme promiscuity in a mating system dominated by sexual conflict. J Insect Behav 19(4):447–455

Boake CRB (1985) Genetic consequences of mate choice a quantitative genetic method for testing sexual selection theory. Science 227:1061–1063

Boake CRB (1986) A method for testing adaptive hypotheses of mate choice. Am Nat 127:654–666

Boake CRB, Wade MJ (1984) Populations of the red flour beetle *Tribolium castaneum* (Coleoptera: Tenebrionidae) differ in their sensitivity to aggregation pheromones. Environ Entomol 13:1182–1185

Brouwer L, van de Pol M, Atema E, Cockburn A (2011) Strategic promiscuity helps avoid inbreeding at multiple levels in a cooperative breeder where both sexes are philopatric. Mol Ecol 20(22):4796–4807

Brown SJ, Denell RE, Beeman RW (2003) Beetling around the genome. Genet Res 82:155–161

Carazo P, Sanchez E, Font E, Desfilis E (2004) Chemosensory cues allow male *Tenebrio molitor* beetles to assess the reproductive status of potential mates. Anim Beh 68:123–129

Chapman T (2006) Evolutionary conflicts of interest between males and females. Curr Biol 16(17):R744–R754

Clark NL, Aagaard JE, Swanson WJ (2006) Evolution of reproductive proteins from animals and plants. Reproduction 131(1):11–22

Cordoba-Aguilar A (2006) Sperm ejection as a possible cryptic female choice mechanism in Odonata (Insecta). Physiol Entomol 31:146–153

Cornell SJ, Tregenza T (2007) A new theory for the evolution of polyandry as a means of inbreeding avoidance. Proc R Soc Biol Sci B 274:2873–2879

Cotton RT, George AS (1929) The mealworms. USDA Tech Bull 95:1–37

Crickmore MA, Vosshall LB (2013) Opposing dopaminergic and GABAergic neurons control the duration and persistence of copulation in *Drosophila*. Cell 155(4):881–893

Dawson PS (1964) Age at sexual maturity in female flour beetles, *Tribolium castaneum* and *T. confusum*. Ann Entomol Soc Am 57:1–3

Den Boer SP, Boomsma JJ, Baer B (2009) Honey bee males and queens use glandular secretions to enhance sperm viability before and after storage. J Insect Physiol 55(6):538–543

Dick J (1937) Oviposition in certain Coleoptera. Ann Appl Biol 24:762–796

Drnevich JM (2003) Number of mating males and mating interval affect last-male sperm precedence in *Tenebrio molitor* L. Anim Behav 66:349–357

Duehl AJ, Arbogast RT, Teal PEA (2011) Density-related volatile emissions and responses in the red flour beetle, *Tribolium castaneum*. J Cheml Ecol 37(5):525–532

Eberhard WG (1991) Copulatory courtship and cryptic female choice in insects. Biol Rev Camb Philos Soc 66:1–32

Eberhard WG (1996) Female control: sexual selection by cryptic female choice. Princeton University Press, Princeton

Eberhard WG (2004) Rapid divergent evolution of sexual morphology: comparative tests of antagonistic coevolution and traditional female choice. Evolution 58(9):1947–1970

Economos AC, Miquel J, Binnard R, Kessler S (1979) Quantitative analysis of mating behavior in aging male *Drosophila melanogaster*. Mech Ageing Dev 10(3–4):233–240

Edvardsson M, Arnqvist G (2000) Copulatory courtship and cryptic female choice in red flour beetles *Tribolium castaneum*. Proc Royal Soc Biol Sci B 267:559–563

Edvardsson M, Arnqvist G (2005) The effects of copulatory courtship on differential allocation in the red flour beetle *Tribolium castaneum*. J Insect Behav 18:313–322

Edvardsson M, Arnqvist G (2006) No apparent indirect genetic benefits to female red flour beetles preferring males with intense copulatory courtship. Behav Genet 36:775–782

Evans JP, Sherman CDH (2013) Sexual selection and the evolution of egg-sperm interactions in broadcast-spawning invertebrates. Biol Bull 224(3):166–183

Faustini DL, Post DC, Burkholder WE (1982) Histology of aggregation pheromone gland in the red flour beetle *Tribolium castaneum*. Ann Entomol Soc Am 75:187–190

Fedina TY (2007) Cryptic female choice during spermatophore transfer in *Tribolium castaneum* (Coleoptera: Tenebrionidae). J Insect Physiol 53:93–98

Fedina TY, Lewis SM (2004) Female influence over offspring paternity in the red flour beetle *Tribolium castaneum*. Proc Royal Soc Biol Sci B 271:1393–1399

Fedina TY, Lewis SM (2006) Proximal traits and mechanisms for biasing paternity in the red flour beetle *Tribolium castaneum* (Coleoptera: Tenebrionidae). Behav Ecol Sociobiol 60:844–853

Fedina TY, Lewis SM (2007) Female mate choice across mating stages and between sequential mates in flour beetles. J Evol Biol 20:2138–2143

Fedina TY, Lewis SM (2008) An integrative view of sexual selection in Tribolium flour beetles. Biol Rev 83:151–171

Fricke C, Arnqvist G (2004) Conspecific sperm precedence in flour beetles. Anim Behav 67:729–732

Gadzama NM, Happ GM (1974) The structure and evacuation of the spermatophore of *Tenebrio molitor* L. (Coleoptera: Tenebrionidae). Tissue Cell 6:95–108

Gaur GK, Rao MK (1997) Effects of inbreeding and incrossing on fecundity traits (egg number and hatchability) in *Tribolium castaneum*. India Vet J 74:43–45

Grimnes KA, Bricker CS, Happ GM (1986) Ordered flow of secretion from accessory glands to specific layers of the spermatophore of mealworm beetles: demonstration with a monoclonal antibody. J Exp Zool 240:275–286

Happ GM (1969) Multiple sex pheromones of the mealworm beetle, *Tenebrio molitor* L. Nature 222:180–181

Happ GM, Happ CM (1970) Fine structure and histochemistry of the spermathecal gland in the mealworm beetle, *Tenebrio molitor*. Tissue Cell 2:441–466

Happ GM, Happ CM (1977) Cytodifferentiation in the accessory glands of *Tenebrio molitor*, part 3. Fine structure of the spermathecal accessory gland in the pupa. Tissue Cell 9:711–732

Haubruge E, Arnaud L, Mignon J, Gage MJG (1999) Fertilization by proxy: Rival sperm removal and translocation in a beetle. Proc Royal Soc Biol Sci B 266:1183–1187

Hinton HE (1942) Secondary sexual characters of *Tribolium*. Nature 149:500

Howard RW, Blomquist GJ (2005) Ecological, behavioral, and biochemical aspects of insect hydrocarbons. Annu Rev Entomol 50:371–393

Jennions MD, Petrie M (2000) Why do females mate multiply? A review of the genetic benefits. Biol Rev Camb Philos Soc 75:21–64

Kelly C, Jennions M (2011) Sexual selection and sperm quantity: meta-analyses of strategic ejaculation. Biol Rev 86:863–884

Kvarnemo C, Simmons LW (2013) Polyandry as a mediator of sexual selection before and after mating. Philos Trans R Soc London B Biol Sci 368(1613):20120042

LeVan KE, Fedina TY, Lewis SM (2009) Testing multiple hypotheses for the maintenance of male homosexual copulatory behaviors in flour beetles. J Evol Biol 22(1):60–70

Levinson A, Levinson H (1995) Reflections on structure and function of pheromone glands in storage insect species. Anzeiger fuer Schaedlingskunde Pflanzenschutz Umweltschutz 68:99–118

Levinson HZ, Mori K (1983) Chirality determines pheromone activity for flour beetles. Naturwissenschaften 70(4):190–192

Lewis SM, Austad SN (1990) Sources of intraspecific variation in sperm precedence in red flour beetles. Am Nat 135:351–359

Lewis SM (2004) Multiple mating and repeated copulations: effects on male reproductive success in red flour beetles. Anim Behav 67:799–804

Lewis SM, Austad SN (1994) Sexual selection in flour beetles: the relationship between sperm precedence and male olfactory attractiveness. Behav Ecol 5:219–224

Lewis SM, Iannini J (1995) Fitness consequences of differences in male mating behaviour in relation to female reproductive status in flour beetles. Anim Behav 50:1157–1160

Lewis SM, Jutkiewicz E (1998) Sperm precedence and sperm storage in multiply mated red flour beetles. Behav Ecol Sociobiol 43:365–369

Lewis SM, Kobel A, Fedina T, Beeman RW (2005) Sperm stratification and paternity success in red flour beetles. Physiol Entomol 30:303–307

Lewis SM, Tigreros N, Fedina TY, Ming QL (2012) Genetic and nutritional effects on male traits and reproductive performance in *Tribolium* flour beetles. J Evol Biol 25(3):438–451

Lockey KH (1978) Hydrocarbons of adult *Tribolium castaneum* and *Tribolium confusum* (Coleoptera: Tenebrionidae). Comp Biochem Physiol B 61(3):401–408

Lüpold S, Manier MK, Ala-Honkola O, Belote JM, Pitnick S (2011) Male *Drosophila melanogaster* adjust ejaculate size based on female mating status, fecundity, and age. Behav Ecol 22:184–191

Markarian H, Florentine GJ, Pratt JJ (1978) Quinone production of some species of *Tribolium*. J Insect Physiol 24:785–790

Mertz DB (1975) Senescent decline in flour beetle strains selected for early adult fitness. Physiol Zool 48(1):1–23

Michalczyk L, Millard AL, Martin OY, Lumley AJ, Emerson BC, Chapman T, Gage MJ (2011a) Inbreeding promotes female promiscuity. Science 333(6050):1739–1742

Michalczyk L, Millard AL, Martin OY, Lumley AJ, Emerson BC, Gage MJ (2011b) Experimental evolution exposes female and male responses to sexual selection and conflict in *Tribolium castaneum*. Evolution 65(3):713–724

Mills W, Moore T (2004) Polyandry, life-history trade-offs and the evolution of imprinting at Mendelian loci. Genetics 168:2317–2327

Ming QL, Lewis SM (2010) Pheromone production by male *Tribolium castaneum* (Coleoptera: Tenebrionidae) is influenced by diet quality. J Econ Entomol 103(5):1915–1919

Nilsson T, Fricke C, Arnqvist G (2002) Patterns of divergence in the effects of mating on female reproductive performance in flour beetles. Evolution 56:111–120

Nilsson T, Fricke C, Arnqvist G (2003) The effects of male and female genotype on variance in male fertilization success in the red flour beetle (*Tribolium castaneum*). Behav Ecol Sociobiol 53:227–233

Novaczewski M, Grimnes KA (1996) Histological characterization of the reproductive accessory gland complex of *Tribolium anaphe* (Coleoptera: Tenebrionidae). Tribolium Inf Bull 36:74–78

Obata S, Hidaka T (1982) Experimental analysis of mating behavior in *Tenebrio molitor* (Coleoptera: Tenebrionidae). Appl Entomol Zool 17(1):60–66

Obeng-Ofori D, Coaker TH (1990) Some factors affecting responses of four stored product beetles (Coleoptera: Tenebrionidae and Bostrichidae) to pheromones. Bull Entomol Res 80:433–442

Paesen GC, Feng X, Happ GM (1996) Structure of a D-protein gene and amino-acid sequences of the highly repetitive D-proteins secreted by the accessory glands of the mealworm beetle. Biochim Biophys Acta 1293:171–176

Pai A, Bennett L, Yan G (2005) Female multiple mating for fertility assurance in red flour beetles (*Tribolium castaneum*). Can J Zool 83:913–919

Pai A, Feil S, Yan G (2007) Variation in polyandry and its fitness consequences among populations of the red flour beetle, *Tribolium castaneum*. Evol Ecol 21(5):687–702

Pai A, Yan G (2002a) Female mate choice in relation to heterozygosity in *Tribolium castaneum*. J Evol Biol 15:1076–1082

Pai A, Yan G (2002b) Polyandry produces sexy sons at the cost of daughters in red flour beetles. Proc R Soc Biol Sci B 269:361–368

Pai A, Yan G (2003a) Effects of tapeworm infection on male reproductive success and mating vigor in the red flour beetle, *Tribolium castaneum*. J Parasitol 89:516–521

Pai A, Yan G (2003b) Rapid female multiple mating in red flour beetles (*Tribolium castaneum*). Can J Zool 81:888–896

Panova M, Boström J, Hofving T, Areskoug T, Eriksson A, Mehlig B, Mäkinen T, André C, Johannesson K (2010) Extreme female promiscuity in a non-social invertebrate species. PLoS ONE 5(3):e9640

Park T (1933) Studies in population physiology. II. Factors regulating initial growth of *Tribolium confusum* populations. J Exp Zool 65:17–42

Park T (1936) Studies in population physiology. VI. The effect of differentially conditioned flour upon the fecundity and fertility of *Tribolium confusum* Duv and its populations. J Exp Zool 73:393–404

Parker GA (1970) Sperm competition and its evolutionary consequences in the insects. Biol Rev 45:525–567

Parker GA, Birkhead TR (2013) Polyandry: the history of a revolution. Philos Trans R Soc London B 368(1613):20120335

Peretti AV, Cordoba-Aguilar A (2007) On the value of fine-scaled behavioural observations for studies of sexual coercion. Ethol Ecol Evol 19(1):77–86

Pischedda A, Rice WR (2012) Partitioning sexual selection into its mating success and fertilization success components. PNAS 109(6):2049–2053

Pizzari T, Snook RR (2003) Perspective: sexual conflict and sexual selection: chasing away paradigm shifts. Evolution 57(6):1223–1236

Poels J, Van Loy T, Vandersmissen HP, Van Hiel B, Van Soest S, Nachman RJ, Van den Broeck J (2010) Myoinhibiting peptides are the ancestral ligands of the promiscuous *Drosophila* sex peptide receptor. Cell Mol Life Sci 67(20):3511–3522

Pomfret JC, Knell RJ (2008) Crowding, sex ratio and horn evolution in a South African beetle community. Proc R Soc Biol Sci B 275(1632):315–321

Pray LA, Goodnight CJ (1995) Genetic variation in inbreeding depression in the red flour beetle *Tribolium castaneum*. Evolution 49:176–188

Priest NK, Galloway LF, Roach DA (2008) Mating frequency and inclusive fitness in *Drosophila melanogaster*. Am Nat 171(1):10–21

Prus T (1961) The effect of homotypic and heterotypic conditioning of medium upon the net fecundity of *Tribolium castaneum* Herbst and *T. confusum* Duval. Ekol Polska Ser A 9:245–257

Rice WR (1998) Intergenomic conflict, interlocus antagonistic coevolution, and the evolution of reproductive isolation. In: Endless forms: species and speciation, pp 161–270

Roberts MM, Grimnes KA (1994) Histological evidence for five cell types in the male accessory glands of *Tribolium freemani* (Coleoptera: Tenebrionidae). Tribolium Inf Bull 34:72–74

Robinson T, Johnson NA, Wade MJ (1994) Postcopulatory, prezygotic isolation: Intraspecific and interspecific sperm precedence in *Tribolium* spp., flour beetles. Heredity 73:155–159

Scharf I, Martin OY (2013) Same-sex sexual behavior in insects and arachnids: prevalence, causes, and consequences. Behav Ecol Sociobiol 67:1719–1730

Sevener JD, Dennard NN, Grimnes KA (1992) Histological and histochemical evidence for an additional cell type in the male accessory reproductive glands of *Tribolium brevicornis* (Coleoptera: Tenebrionidae). Tribolium Inform Bull 32:93–95

Shuster SM, Briggs WR, Dennis PA (2013) How multiple mating by females affects sexual selection. Philosl Trans R Soc London B 368(1613):20120046

Simmons LW (2001) Sperm competition and its evolutionary consequences in the insects. Princeton University Press, Princeton

Sokoloff A (1974) The biology of Tribolium with special emphasis on genetic aspects. Oxford University Press, Oxford

Sonleitner FJ (1978) Longevity and oviposition rate in *Tribolium castaneum*; Rubner's hypothesis and natural selection. Tribolium Inf Bull 21:150–154

Sonleitner FJ, Guthrie PJ (1991) Factors affecting oviposition rate in the flour beetle *Tribolium castaneum* and the origin of the population regulating mechanism. Res Popul Ecol (Tokyo) 33:1–12

South A, Sirot LK, Lewis SM (2011) Identification of predicted seminal fluid proteins in *Tribolium castaneum*. Insect Mol Biol 20(4):447–456

Stearns SC (1992) The evolution of life histories. Oxford University Press, Oxford

Subramanyam B, Harein PK (1990) Accuracies and sample sizes associated with estimating densities of adult beetles (Coleoptera) caught in probe traps in stored barley. J Econ Entomol 83:1102–1109

Sun J, Spradling AC (2013) Ovulation in *Drosophila* is controlled by secretory cells of the female reproductive tract. eLife 2013, 2:e00415.

Surtees G (1961) Spermathecal structure in some Coleoptera associated with stored products. Proc R Entomol Soc Lond A 36:144–152

Suzuki T (1980) 4-8-dimethyldecanal the aggregation pheromone of the flour beetles *Tribolium castaneum* and *Tribolium confusum* (Coleoptera: Tenebrionidae). Agricultl Biol Chem 44:2519–2520

Suzuki T, Nakakita H, Kuwahara Y (1987) Aggregation pheromone of *Tribolium freemani* Hinton (Coleoptera: Tenebrionidae) I. Identification of the aggregation pheromone. Appl Entomol Zool 22:340–347
Suzuki T, Sugawara R (1979) Isolation of an aggregation pheromone from the flour beetles *Tribolium castaneum* and *Tribolium confusum*. Appl Entomol Zool 14:228–230
Taylor ML, Evans JP, Garcia-Gonzalez F (2013) No evidence for heritability of male mating latency or copulation duration across social environments in *Drosophila melanogaster*. PLoS ONE 8(10):e77347
Thornhill R, Alcock J (1983) The evolution of insect mating systems. Harvard University Press, Cambridge
Trematerra P, Sciarretta A (2004) Spatial distribution of some beetles infesting a feed mill with spatio-temporal dynamics of *Oryzaephilus surinamensis*, *Tribolium castaneum* and *Tribolium confusum*. J Stored Prod Res 40:363–377
Tyler F, Tregenza T (2013) Why do so many flour beetle copulations fail? Entomol Exp Appl 146(1):199–206
Vardell HH, Brower JH (1978) Sperm precedence in *Tribolium confusum* (Coleoptera: Tenebrionidae). J Kansas Entomol Soc 51:187–190
Verheggen F, Ryne C, Olsson POC, Arnaud L, Lognay G, Högberg HE, Persson D, Haubruge E, Löfstedt C (2007) Electrophysiological and behavioral activity of secondary metabolites in the confused flour beetle, *Tribolium confusum*. J Chem Ecol 33:525–539
Wade MJ, Patterson H, Chang NW, Johnson NA (1994) Postcopulatory, prezygotic isolation in flour beetles. Heredity 72:163–167
Wade MJ, Shuster SM, Stevens L (1996) Inbreeding: its effect on response to selection for pupal weight and the heritable variance in fitness in the flour beetle, *Tribolium castaneum*. Evolution 50:723–733
Wang OT (1992) Experimental assessment of male attractiveness female choice or Hobson's choice? Am Nat 139:433–441
Wang L, Wang S, Li Y, Paradesi MSR, Brown SJ (2007) BeetleBase: the model organism database for *Tribolium castaneum* Nucl Acids Res 35 (suppl 1): D476–D479
Wojcik DP (1969) Mating behavior of 8 stored-product beetles (Coleoptera: Dermestidae, Tenebrionidae, Cucujidae, and Curculionidae). Florida Entomol 52:171–197
Wool D (1967) Some observations on mating frequencies in *Tribolium castaneum* strains. Tribolium Inf Bull 10:182–186
Wool D, Mendlinger S (1981) Genetic and ecological consequences of subdivision in *Tribolium castaneum* populations. J Anim Ecol 50:421–434
Worden BD, Parker PG (2001) Polyandry in grain beetles, *Tenebrio molitor*, leads to greater reproductive success: material or genetic benefits? Behav Ecol 12:761–767
Worden BD, Parker PG (2005) Females prefer noninfected males as mates in the grain beetle *Tenebrio molitor*: evidence in pre- and postcopulatory behaviours. Anim Behav 70:1047–1053
Xu J, Baulding J, Palli SR (2013) Proteomics of *Tribolium castaneum* seminal fluid proteins: identification of an angiotensin-converting enzyme as a key player in regulation of reproduction. J Proteomics 78:83–93
Yapici N, Kim YJ, Ribeiro C, Dickson BJ (2008) A receptor that mediates the post-mating switch in *Drosophila* reproductive behaviour. Nature 451(7174):33–37
Yezerski A, Ciccone C, Rozitski J, Volingavage B (2007) The effects of a naturally produced benzoquinone on microbes common to flour. J Chem Ecol 33(6):1217–1225
Zeh JA, Zeh DW (2008) Maternal inheritance, epigenetics and the evolution of polyandry. Genetica 134:45–54
Ziegler JR (1978) Dispersal and reproduction in *Tribolium*: the influence of initial density. Environ Entomol 7:149–156
Zyromska-Rudzka H (1966a) Abundance and emigrations of *Tribolium* in a laboratory model. Ekol Pol, Ser A 14:491–518
Zyromska-Rudzka H (1966b) An appreciation of the laboratory set for investigating the emigratory activity of *Tribolium*. Ekol Pol, Ser A 14:125–132

Chapter 17
Female Choice in Social Insects

Boris Baer

Abstract The mating biology of eusocial insects, being the ants, bees, wasps, and termites, is truly amazing as a number of reproductive traits have evolved in these species that are not or rarely found in other species, such as the absence of remating later in life, prolonged sperm storage, and extreme levels of queen fertility. Kin selection is recognized as a driving force shaping these insect societies and their reproductive biology, selecting for high relatedness among helpers, and limiting the number of fathers contributing to offspring. The study of the mating biology of social insects received remarkably little scientific attention, despite the fact that mating behavior can provide a mechanism through which high relatedness can be achieved. As a consequence, our current knowledge about the presence or absence of sexual selection including female choice remains poorly investigated. In this chapter, I provide a theoretical introduction to female choice in social insects, arguing that in the absence of female remating later in life and exceptional high demands for large numbers of viable sperm, queens should express male choice throughout all steps of the mating process. I then discuss some examples from the recent literature that provide empirical evidence for female choice (precopulatory and cryptic choice) and develop a number of questions and hypotheses that can be addressed in the future.

17.1 Introduction

Standing in a dense and humid rainforest of French Guyana, I spotted a hill that was not overgrown with vegetation and therefore offered a better view through the impenetrable ground vegetation. This was welcome given I tried to spot

B. Baer (✉)
Centre for Integrative Bee Research (CIBER), The University of Western Australia, 6009 Crawley, Australia
e-mail: boris.baer@uwa.edu.au

A.V. Peretti and A. Aisenberg (eds.), *Cryptic Female Choice in Arthropods*,
DOI 10.1007/978-3-319-17894-3_17

red-handed tamarins (*Saguinus midas*) in the canopy some 20–30 m above me. However, I should have been warned that there was a reason for the absence of vegetation on that hill, as within minutes, I found myself covered in thousands of angry ants, some of them heavily armed with razor-sharp mandibles, all crawling up my boots and trousers and every single one dedicated to their last drop of hemolymph to fight me off that hill. By setting my foot onto this mature colony of *Atta* leaf-cutting ants, I had triggered an alarm that spread rapidly. Such colonies grow to the size of a family home, containing up to 8 million workers and surviving in the wild for 20 years or more (Weber 1972). This experience of interacting with a fully grown social insect organism was further elaborated a couple of weeks later when our field station was raided by a colony of *Eciton* army ants. These ants maintain no permanent nest structures but build temporary bivouacs made by workers for brood rearing, reaching colony sizes of up to 20 million individuals in some African species (Hölldobler and Wilson 1990; Raignier and van Boven 1955). Any form of resistance would have been futile as each worker is armed with pointy mandibles as well as a stinger, and I was therefore defeated once more.

Amazingly, in both of these cases, I was confronted with the offspring of a single reproducing female, known as the queen. These animals are well hidden and protected by their societies, because they are the sole reproductive individuals in the colony and therefore highly valuable, as they cannot be replaced in most species. This cryptic lifestyle of queens is typical for all eusocial ants, bees, wasps, and termites. Yet these queens represent the pinnacle of social evolution, as their reproductive potential is key for their social lifestyle. The life histories of social insect queens are quite variable between species, and as already indicated above, often truly spectacular as their reproductive traits are either unique or have evolved to spectacular extremes. These traits determine colony success, which depends on maintaining a large number of helpers, all originating from fertilized eggs. As queens dominate reproduction in these societies, fathers have adopted a very cryptic lifestyle. In the case of the hymenopteran social insects (ants, bees and wasps)—fathers are only present as stored sperm within the queen (Boomsma and Ratnieks 1996).

In this chapter, I focus on social insect queens and explore what we know about these extraordinary animals' potential to determine or bias paternity after copulations and in the absence of males, especially whether and if so cryptic female choice is present. The study of the reproductive biology of social insects received scientific attention only recently, but males and their reproductive agendas were studied in more detail, for reviews see for example (Baer 2003, 2005, 2011; Boomsma et al. 2005; Hölldobler and Bartz 1985). As I point out, social insects offer unique opportunities to investigate cryptic female choice (CFC), because the available theoretical framework of inclusive fitness theory allows to develop specific predictions and hypotheses, and newly available knowledge and technologies offer ample opportunities for experimental work to address them.

I start this chapter with a brief introduction to social insects and their reproductive biology. Because the mating biologies can differ substantially between social insect species, this introduction presents a very generalized overview describing reproductive traits and behaviors as found in a majority of species. Such a general overview is important for the following section presenting a theoretical framework of why CFC is expected to be important in social insects and discussing the available empirical support found in the literature. I will also point out research areas and questions that should receive further attention in the future to stimulate further research using social insects to study CFC.

17.2 Social Insects

Eusociality is defined by the presence of a division of labor, cooperative brood care, and overlapping generations within a colony (Wilson 1971). It is widespread in bees and wasps, and present in all known species of ants and termites. Darwin was puzzled by the presence of a worker caste in these species (Rubenstein 2012) that does not reproduce but altruistically helps raising non own offspring. However, it was not until Hamilton (1964) and Trivers and Hare (1976) formulated the necessary theoretical framework of inclusive fitness and kin selection that a powerful evolutionary explanation became available to explain altruistic helping. In essence, the incentive for helping increases with increasing relatedness between a helper and the individual receiving help, thereby increasing inclusive fitness of the helper. The development of kin selection theory triggered a substantial body of theoretical work to identify conflicts arising through cooperation in these societies (e.g., see West et al. 2002; Bourke and Franks 1995; Queller 2003), including reproductive conflicts among colony members (Ratnieks et al. 2006) and their potential consequences for social evolution. These contributions stimulated a wealth of empirical work to test ideas derived from theory (see, e.g., (Bourke and Franks 1995; Foster and Ratnieks 2000). As a consequence, sociobiology became and remains a vibrant field of research (Wilson 2000). Its output of research is comparable to another field receiving broad scientific attention, being sexual selection, which studies biases in paternity contributions to explain fitness consequences of male–male competition and female choice (Baer 2014). Despite the common interest of both fields in the genetic makeup of offspring, kin selection research developed independently from research conducted on sexual selection (Boomsma 2007). The reasons being that studies on kin selection were more concerned with the consequences of paternity distributions (Baer 2014), whereas sexual selection research such as CFC focused on explaining how paternity distributions are generated and their effects on the evolution of individual life history traits.

17.3 The Reproductive Biology of Social Insect Queens

Social insects are characterized by the presence of extreme levels of reproductive skew, where one or very few females (normally referred to as queens, sometimes as gamergates) monopolize reproduction. In the hymenopteran social insects, queen development is typically initiated during the egg or larval period and is dependent on environmental factors that trigger elevated juvenile hormone levels (Penick et al. 2012), such as the amount of food provided to a larva (Alford 1975). In honeybees, the provisioning of royal jelly to 3 day old larvae initiates modifications in DNA methylation that triggers the relevant developmental pathways for queen development (Maleszka 2008). When virgin queens hatch, they are supported by their sister workers until they reach sexual maturity. They then leave the colony to take part in nuptial flights to choose mates and copulate before founding a new colony. With some known exceptions such as honeybees or swarm-founding epiponine wasps [see (Ratnieks et al. 2006) and references therein], queens do not return to their maternal colony. Instead, they go through a phase of solitary living, which can be of substantial length and can include periods of hibernation, dispersion, or foraging (Alford 1975). In many species, only a single queen initiates a new colony, but multiple foundresses have been reported in some wasps, ants, and termites (Schmid-Hempel and Crozier 1999; Atkinson and Adams 1997). During that time, queens also perform worker tasks such as foraging (Pollock et al. 2012; Hölldobler and Wilson 1990), symbiont cultivation (Fernandez-Marin et al. 2004), brood care, or nest defense (Fig. 17.1). Colony foundation is the most critical time

Fig. 17.1 A leaf-cutter ant queen (*Atta colombica*) photographed after her nuptial flight, while initiating a new colony. The queen has dug a first initial brood chamber 20–30 cm below the surface and has started to metabolize her flight muscles to lay eggs and maintain a small initial fungus garden. A colony can reach sexual maturity within 5–7 years, containing around 7 million workers and producing next generations of thousands of males and gynes during each mating season. Picture by the author

period in the life of a queen and is accompanied by extreme levels of queen mortality (Diehl-Fleig 1995; Baer et al. 2006; Schmid-Hempel 1998). Queens need to maintain their reproductive potential by continuously producing eggs as well as by keeping sperm alive and viable within their spermatheca. Associated sperm storage costs can be substantial and trade off with other life history traits. In the leaf-cutter ant *A. colombica*, for example, queens that mate more often or store higher numbers of sperm during their nuptial flight have a reduced capacity to up-regulate their immune system during colony foundation, which makes them susceptible to infections (Baer et al. 2006).

As soon as a first generation of helpers emerges, queens become reproductive organs within a larger "superorganism." They are then responsible to deliver large numbers of fertilized eggs to build and maintain a colony's worker force. This requires queens of some species to be spectacularly fertile and able to lay hundreds to thousands of eggs per day. In honeybees, for example, queens can lay up to 2000 eggs per day, the equivalent of their own body weight (Maleszka 2008). Once colonies have reached their mature size, queens produce new generations of sexual offspring. In the hymenopteran social insects, queens control the fertilisation process and therefore, the number of males (unfertilized eggs) and queens or workers (fertilized eggs) produced (Heimpel and de Boer 2008). Reproductive conflicts arise between the queen and her workers over the sex ratio in sexual offspring (Ratnieks et al. 2006; Boomsma 1996; Tsuji 1996), and workers sometimes modify primary sex ratios in their own interest, for example, by eating queen- or male-destined eggs/larvae (Sundström 1994). In some species, workers also kill the queen and replace her with one of their sisters or start to lay their own, unfertilized eggs (Winston 1991; Alford 1975; Foster and Ratnieks 2000).

Social insect queens are therefore characterized by astonishing levels of lifetime fecundity. Contrary to other animals, reproduction and longevity are positively correlated in social insects (Heinze et al. 2013). Honeybee queens are only marginally larger than workers, but can live up to 8 years and produce around 1.7 million fertilized eggs (Baer 2005). In the fungus growing ant *A. colombica*, queens initially store up to 450 million sperm, allowing them to maintain colonies for decades and consisting of several million workers (Baer et al. 2006; Weber 1972). Army ants seem to hold the current world record, with queens of some species storing up to 1 billion sperm and fertilizing 250 million eggs (Kronauer 2009). Such continuous high levels of female fecundity are truly spectacular, but, as already mentioned, are all achieved during a single round of sperm acquisition early in the life of these animals.

In summary, reproductive queens are found in low frequencies in insect societies, but they are key individuals initiating new colonies and producing most if not all offspring. They evolved a number of spectacular adaptations to achieve astounding levels of fertility, and elaborations in reproductive traits as found between species are key determinants of their eusocial lifestyles.

17.4 Why Social Insect Queens Should Be Choosy

As already pointed out, queens of social ants, bees, and wasps preform only a single round of mate choice and sperm acquisition early in life and never remate once they have started to lay eggs (Boomsma et al. 2005; Baer 2011). In the majority of hymenopteran species, males die during or shortly after copulation and only survive as stored sperm inside their mate. As a consequence, sperm rather than egg number limits the size and longevity of their societies. Termites provide an exception because males survive alongside the queen as kings and continuously remate with them to replenish sperm supplies (Hartke and Baer 2011). However, as for the ants, bees, and wasps, termites are also closed genetic systems where no additional genetic contributions are typically accepted after an initial round of mate choice (Boomsma et al. 2005). Such "marriages for live," combined with the observation that only a single or a few males sire offspring in many species (Boomsma 2009; Hughes et al. 2008), determine the genetic architecture of the colony before workers are produced. Single or highly skewed paternity distributions as found in many social insects are expected from kin selection, because they maximize relatedness among helpers and thereby increase the incentive of helping (Jaffe et al. 2012). Consequently, a queen's decision with whom she mates, the number of mating partners she chooses to copulate with, and the amount of sperm she stores from each of her mating partners are of paramount importance defining the success of the later emerging society. There is indeed ample empirical evidence that these mating decisions of queens can have dramatic fitness consequences, see Table 17.1 for some examples. As inferior mating decisions such as inbreeding (Armitage et al. 2010) cannot be corrected later in life, queens are expected to be more than passive ejaculate recipients and extremely choosy in order to identify preferred or high-quality males and/or to discriminate against unwanted males or their ejaculates. Consequently, some of the spectacular reproductive characteristics found in some social insects are expected to represent evolutionary end points that evolved through continuous rounds of (cryptic) female choice.

As in other animals, female choice of social insect queens can occur both precopulatory as mate choice or postcopulatory as CFC. There is some empirical evidence for mate choice in social insects, for example, based on secondary sexual male traits (Izzo and Tibbetts 2012), or females resisting copulations or sometimes even killing males (Baer 2003). Overall, very few studies investigated precopulatory female choice, but typical indicators for the presence of female choice such as elaborate secondary sexual male traits seem mostly absent, but see (Izzo and Tibbetts 2012) for an exception. This lack of empirical work on precopulatory female choice is partially caused by experimental limitations, because mate choice and copulations in social insects are often difficult to observe in the field or in the laboratory (Baer 2003). However, precopulatory female choice could in fact be less important in species where queens participate in short nuptial flights and are exposed to various environmental risks such as predation/parasitism or adverse

Table 17.1 Life history and fitness consequences resulting from mate choice decisions of social insect queens

Trait	Effect	Description	References
Mate choice	Incompatibly	Sperm can induce an immune response in queens and increase queen mortality	Greeff and Schmid-Hempel (2008)
Mate choice	Inbreeding	Queens mating with related males produce diploid males, which have no or reduced fitness	Armitage et al. (2010),; Beye et al. (2003); Gerloff et al. (2003)
Mate choice	Division of labour	Patriline specific variation in task performance among workers increases overall colony fitness	Oldroyd and Fewell(2007)
Mate choice	Worker production	Queens need to mate with two male genotypes to produce workers and queens	Ashe and Oldroyd (2002)
Mate choice	Hibernation success	Sperm of different sire groups (brothers) differentially affect queen hibernation success, longevity and fitness	Baer and Schmid-Hempel (2005)
Mate choice	Parasite susceptibility	Patrilines differ in disease susceptibility	Baer and Schmid-Hempel (2003)
Mate choice	Inbreeding	Worker homozygosity results in reduced production of sexuals	Haag-Liautard et al. (2009)
Mating costs	Disease transmission	Infected males transfer pathogens to the female during copulation	Greeff and Schmid-Hempel (2008); Yue et al. (2007)
Polyandry	Sperm storage costs	Queens mating too often and/or storing too much sperm have lower immunity	Baer et al. (2006); Greeff and Schmid-Hempel (2008)
Polyandry	Genetic diversity	Genetically diversity among workers reduces parasitism	Baer and Schmid-Hempel (1999); Tarpy (2003); Hughes et al. (2003)
Polyandry	Half sib conflicts	Polyandry results in conflicts over reproduction and sex ratios	Baer and Schmid-Hempel (2001)
Polyandry	Sperm number	Queens storing more sperm produce larger colonies or colonies that survive longer	Tschinkel (1987a, b)
Polyandry	Sex ratio	Polyandry reduces queen-worker conflicts over sex ratios	Sundström (1994)
Polyandry	Colony performance	Colonies with increased genetic variation in workers grow faster	Wiernasz et al. (2004)

climatic conditions (Fig. 17.2) (Hölldobler and Wilson 1990). As a consequence, queens mate quickly and rather indiscriminately with males to collect ejaculates and perform mate choice postcopulatory during the sperm storage process in less dangerous environments. A precondition for postcopulatory CFC is polyandry, i.e., females mating with different males. Analyses of queen copulation frequencies show that polyandry is more widespread in social insects than indicated from paternity analyses in worker offspring (Baer 2011; Boomsma and Ratnieks 1996;

Fig. 17.2 The reproductive biology of honeybees has been intensively studied, partially because of commercial interests for breeding programs and the development of an artificial insemination technique. Honeybee queens perform one or very few nuptial flight(s) and mate in quick succession with a large number of males (drones). In *A. mellifera*, only a very small fraction of 3–5 % of sperm initially received by the queen will be stored in the spermatheca, a process lasting up to 40 h. Picture taken from the footage of the theatrical documentary "More than Honey," for video footage see the movie trailer at http://www.ciber.science.uwa.edu.au/blog/?page_id=121

Jaffe et al. 2012). This implies the presence of postcopulatory mechanisms that reduce and/or bias paternities (Jaffe et al. 2012). CFC of social insect queens could therefore provide the proximate mechanisms to explain the ultimate mismatch between observed queen mating frequencies and paternity. A first step to test this idea is to look for empirical evidence in the published literature.

17.5 Evidence for Cryptic Queen Choice

A search using Web of Science in March 2014 using "CFC" and "social insect" as search parameters resulted in a list of only 12 papers, 7 of which I (co)author. This illustrates that CFC has basically not been investigated in social insects, despite its predicted impact on eusocial living. However, there are a number of studies available that investigated the reproductive biology of social insects and provided some evidence for the presence of cryptic queen choice, although these findings were not necessarily discussed in that context.

In his influential book, Eberhard (1996) listed a number of mechanisms of CFC and I selected a subset of those traits, which seemed relevant for social insects, together with supporting evidence found in the literature which is summarized in Table 17.2. Although this list is unlikely to be complete and some of these observations might be more convincing than others, it nevertheless provides very encouraging evidence to justify further research.

17.5.1 Morphologically Based Cryptic Female Choice

Eberhard (1996) pointed out that insect females are generally in control of sperm migration and transport within their bodies and that the relevant morphological structures facilitating these processes are also used for CFC. Based on our current knowledge, this could also be the case in social insects. The sexual organs of queens (and males) are often morphologically complex. They contain structures such as valves, sperm pumps, or narrow ducts for the movement of ejaculates, as well as organs for the temporal storage of sperm prior to transfer to the spermatheca. Although these structures can be expected to have evolved through natural selection to maximize sperm acquisition and storage efficiency, they also offer queens the possibility for CFC. For example, ejaculates are often not directly transferred to the spermatheca, which would be the most efficient mechanism to transfer male gametes to a female storage organ. Instead, they are initially received and temporarily stored in other parts of the queen's reproductive tract, for example, in the bursa copulatrix or the lateral oviducts (Baer 2003, 2005, 2011). Honeybee queens can actively close both the sting chamber and the bursa copulatrix (Baer 2005; Dade 1962), and they need to actively contract their bursa after having received an individual ejaculate in order to transfer sperm into their lateral oviducts (Koeniger and Koeniger 1991). These specialized reproductive organs could therefore enable honeybee queens to reject entire ejaculates or parts thereof and offer opportunities to study CFC in more detail in the future. A similar mechanism seems to be present in leaf-cutter ants, where males have no physical access to the female's sexual tract and queens can close the entrance to their sexual organs with a muscle (Baer and den Boer unpublished data).

In species where queens receive more sperm form their mate(s) than required to fill the spermatheca, excess sperm is dumped (Robertson 1995; Baer 2005; Woyke 1983). The process of sperm storage in the *Apis mellifera* occurs over a period of 40 h. During this process, ejaculates are moved back from the lateral oviducts into the bursa copulatrix through muscular contractions, and some sperm is transferred to the spermatheca. However, more than 95 % of the sperm initially received is expelled, and observed mating frequencies are substantially higher than the number of fathers found in offspring (Baer 2005). The spermathecal duct of honeybee queens is a narrow tube surrounded by muscular tissue (Bresslau 1905; Snodgrass 1984), which provides queens with control over the amount of sperm passing through. Furthermore, a morphological structure present between the spermathecal duct and the spermatheca, known as Bresslau's sperm pump (Bresslau 1905), is believed to control access of sperm into and out of the spermatheca. However, apart from a detailed description more than 100 years ago, we still lack experimental work to understand its relevance for CFC. Our present knowledge about the mating biology of honeybee queens indicates that they might be able to manipulate ejaculates in multiple ways and during every stage of the mating process: (1) while receiving them, (2) while transporting them to the lateral oviducts, (3) during the storage process, or (4) during fertilization. This suggests that observed

Table 17.2 A subset of mechanisms of cryptic female choice taken from Eberhard (Eberhard 1996) and empirical observations supporting their presence in social insects

Mechanism	Empirical observations	Species	Key References
Inter-ejaculate biases	Queens neutralise sperm damaging effects of seminal fluid proteins of rival males	Ants	den Boer et al. (2010)
	Paternity skew is low in highly polyandrous species and high in queens with low mating frequencies	Ants, Bees, Wasps	Jaffe et al. (2012)
Intra-ejaculate biases	Queens store a non-random sample of longer or shorter sperm	Bees	Baer et al. (2003)
Failure to store sperm of some males	Queens dump sperm during the storage process	Bees	Ruttner (1956)
	Spermathecal duct has a valve, pump or constriction to control access to spermatheca during sperm storage process or egg fertilisation	Ants, Bees	Oppelt and Heinze(2007), Bresslau (1905)
Discarding sperm of some males	Observed queen mating frequencies are higher than molecular studies identifying the number of contributing fathers	Ants, Bees	Baer (2011), Keller and Passera (1992), Oberstadt and Heinze (2003), Boomsma and Ratnieks (1996)
Biased use of stored sperm	Some patrilines are over represented in sexual (queen) offspring	Bees	Moritz et al. (2005)
	Queens produce (some) queen offspring asexually	Ants	Doums et al. (2013)
Choice of sperm that reaches the egg	Queens control the fertilisation process and thereby male contributions to sexual offspring	Ants, Bees, Wasps	Boomsma (1996)
Premature interruption of copulation	Males are killed by queens during copula	Ants	Monnin and Peeters (1998), Allard et al. (2007)
	Queens bite males to terminate copulation	Ants	Keller and Passera (1992)
	Queen movements interrupt copulation	Ants	Kronauer and Boomsma (2007)
Denial of deeper genital access	Queens close opening to genital tract, denying ejaculate access to the females sexual tract	Ants, Bees	Baer (unpublished data)
	Queens relocate ejaculates of previous copulations within their sexual tract	Bees	Sauter et al. (2001)
	Spermathecal ducts are long and narrow and are beyond access of male genitalia	Bees	Schoeters and Billen (2000), Dade (1962)
Reduce number of offspring	Female bias sexual offspring towards non fertilised eggs (males)	Bees	Beekman and Van Stratum (1998)
Re-mating with additional males	Queens receive sufficient sperm from a single copulation to fill spermatheca, but continue to re-mate with additional males	Ants, Bees	Tarpy (2003); Hughes and Boomsma (2005)

paternities are the result of a complex interplay between different mechanisms, which each represents a different level of female choice.

In bumblebees, males transfer an ejaculate consisting of sperm, seminal fluid, and a mating plug into the female's bursa copulatrix (Duvoisin et al. 1999; Brown and Baer 2005). Sperm is placed at the entrance of the spermathecal duct (Duvoisin et al. 1999) from where it is transferred to the spermatheca. Bumblebee queens also possess a long and narrow spermathecal duct, so the process of storing the sperm into the spermatheca takes up considerably longer than copulation itself (Duvoisin et al. 1999). If bumblebee queens accept additional matings, the first ejaculate is pushed further up into the lateral oviducts (Sauter et al. 2001). Consequently, a queen's decision to delay remating might influence a first male's contribution to stored sperm. Interestingly, bumblebee queens seem to possess the necessary morphological structures to store displaced ejaculates. Using morphological structures to delay sperm storage seems also present in the ant *Leptothorax gredleri*, where the spermathecal duct of the queen is so narrow that sperm have to pass one by one into the spermatheca (Oppelt and Heinze 2007), which takes several hours after mating. Interestingly, only a single male is found to sire worker offspring in this species despite queens mating with up to four males. The same is found in the Argentine ant *Linepithema humile*, where queens mate with multiple males, but only a single male sires offspring (Keller and Passera 1992). These examples illustrate that paternity distributions in offspring are not reliable predictors of queen mating frequencies (Baer 2011), and CFC could explain mismatches between observed and detected numbers of copulations.

Queens of ants, bees, and wasps can determine the sex of their offspring, because they control whether an egg is fertilized or not. Spermathecal ducts or sperm pumps/valves are expected to control this process, which can already be defined as a form of CFC during egg fertilization. Because males gain direct fitness only by siring queen offspring, reproductive conflicts emerge between the queen and her mate(s) over the sex ratio in sexual offspring (Boomsma 1996). Males prefer a highly queen-based sex ratio, whereas queens prefer an equal investment into reproductive offspring. Sex rations vary greatly between social insect species, indicating that such reproductive conflicts are resolved differently depending on the species and it's mating system. Furthermore, queens can produce virgin queens asexually in some species and thereby manipulate male fitness, for example, in the ants *Cataglyphis cursor* (Doums et al. 2013) or *Platythyrea punctata* (Kellner and Heinze 2011). Similarly, in the little fire ant, queens use sperm to produce workers only, but virgin queens develop without any genetic contributions of males (Fournier et al. 2005). Although these examples might not be seen as classical cases of CFC, the conflict between the queen and her mate(s) results in manipulations of a male's reproductive success. The power to determine the caste of offspring in the absence of males benefits the queens but reduces male fitness. It would therefore be interesting to investigate whether paternity contributions differ in sexual offspring compared to worker offspring, i.e., whether some fathers are more likely to sire queens than others and whether queens can control paternal representation in their sexual offspring. Very little empirical work has been conducted so far to quantify this. In honeybees, queens are reared from rare "royal" subfamilies (Moritz et al. 2005), indicating that some fathers are more likely to sire virgin queens than others. However, because honeybee workers can influence the fate of a fertilized egg, further research is needed to understand the influence of queens versus workers over caste fate. In general, future work is needed to quantify whether paternity contributions differ between worker and sexual offspring and whether workers, which carry paternal genes, manipulate caste determination and paternity in their fathers' interest.

In summary, there is good evidence that social insect queens are able to manipulate the process of sperm storage using multiple morphological structures within their sexual tracts. Additionally, storing and using sperm can take up considerable time (Oppelt and Heinze 2007; Reichardt and Wheeler 1996; Woyke 1983; Duvoisin et al. 1999) providing queens with the necessary time window to perform CFC.

17.5.2 Molecular Based Cryptic Female Choice

Apart from morphological structures, queens could also use molecules present in various glandular secretions to bias paternity. Queens have a number of glands associated with their sexual tract (Snodgrass 1984; Janet 1904), but very little is known about these secretions or their influence on ejaculates or paternity. As sperm becomes increasingly dependent on the queen's support, they can be compared to endosymbionts (Baer et al. 2009). A queen's power over sperm fate could also be used for CFC, if the amount or compositions of these secretions are modified. The spermathecal gland secretions of honeybee queens are provided to store sperm (Klenk et al. 2004) and are biochemically complex (Baer et al. 2009). They contain proteins that are very efficient at keeping sperm alive (den Boer et al. 2009), but seem to have a variety of additional functions, some of which could be linked to CFC, such as for examples proteins with cytotoxic and signaling functions or chaperons (Baer et al. 2009). The protein composition of spermathecal fluid changes substantially once sperm has become stored, indicating that queens interact differentially with newly arriving compared to stored sperm (Baer et al. 2009). Interestingly, sperm respond to these changes in their host environment as well, as indicated by substantial proteomic differences between stored and ejaculated sperm (Poland et al. 2011).

In highly polyandrous species, ejaculates of multiple males co-occur in the queen's sexual tract resulting in sperm competition. This is the case in honeybees such as *A. mellifera* as well as in several leaf-cutter ants such as *A. colombica* and *A. echinatior*. In these three species, sperm competition occurs in the form of sperm incapacitation, where seminal fluid proteins kill sperm of rival males (den Boer et al. 2010). However, this is not necessarily in the interest of the queen, especially if insufficient numbers or damaged sperm become stored and compromise her fecundity. As expected, *A. colombica* queens use secretions from their spermathecal glands to neutralize sperm incapacitation and proteins are known to be the molecules responsible for this effect (unpublished data). As the queen controls the release of secretions from her glands into the spermatheca, she can influence sperm competition and thereby manipulate paternity of males. *A. colombica* therefore provides another intriguing example that paternity in social insects seems determined by multiple traits, which evolved under postcopulatory sexual selection.

As the number of sequenced social insect genomes is substantially growing, research can now take full advantage of state-of-the-art–*omics* technologies that become increasingly united as part of systems biology. These techniques allow the detection of a large numbers of molecules as well as their abundance in samples of interest. Furthermore, bioinformatics can assign detected molecules to biochemical networks, offering detailed insights into their biological functions on the phenotypic level. These techniques are therefore highly promising tools for future research to study reproductive traits such as CFC, which were so far challenging to address because they occur within the sexual tract of an individual on a very small scale.

17.5.3 Ultimate Consequences of Cryptic Queen Choice

Postcopulatory female manipulations of paternity will finally determine the frequency of fathers in offspring (Eberhard 1996). The relative contributions of different fathers to offspring (paternity skew) can vary quite substantially in social insects. In polyandrous ants, bees, and wasps where queens mate only with one or two males, paternities are normally highly biased towards one male (Jaffe et al. 2012). Inclusive fitness of helpers is therefore maximized in these species, as predicted from kin selection theory. However, if paternity skew is high in social insects, species currently described as monandrous based on molecular paternity analyses might still be polyandrous. Queen multiple mating might therefore be even more common than acknowledged so far, and CFC could provide the necessary morphological or molecular mechanisms to reduce the number of mates down to a single father. Obviously, polyandrous queens with single paternity such as the previously mentioned ants *C. cursor* (Doums et al. 2013) and *P. punctata* (Kellner and Heinze 2011) would be primary target species for future research to unravel the mechanisms by which single fathers are determined.

In species where queens mate with a large number of males, paternity skew becomes increasingly equalized (Jaffe et al. 2012). This seems driven by a number of well-documented fitness benefits gained from increased genetic diversity among helpers (see Table 17.1, Baer and Schmid Hempel 1999; Tarpy 2003; Hughes and Boomsma 2005), which are specifically important in large and long-lived insect societies. Obviously, manipulations of ejaculates that bias or equalize paternity skew is in the interest of queens, as her choice of the number of fathers and their individual contributions define both the level of conflicts in her worker offspring, as well as the potential benefits gained through genetic diversity. Social insects might therefore represent a group of insects where selection on female choice might in fact be stronger than male–male competition, an idea that should certainly be investigated in the future. If true, the most successful societies on earth would be characterized by the presence of an astonishing dominance of female power, both over the reproductive process and during the phase of later social living.

17.6 Conclusions

CFC is admittedly poorly investigated in social insects, because research conducted so far was mostly guided by questions derived from kin selection theory. However, as the study of social insect reproduction received increasing scientific attention over recent years, new findings also provided first evidence for the presence of sexual selection in these species. Theoretical considerations predict that social insect queens are choosy and manipulate paternities in their own interest because they need to store large numbers of high-quality sperm that can only be acquired once in a lifetime. Empirical data support this idea, because social insect queens possess morphological structures as well as secreted molecules that both seem involved in CFC. Cryptic queen choice is not only expected from theoretical considerations, but could also provide an explanation for the observed mismatch between queen mating frequencies and the numbers of fathers and their relative abundance in offspring. Consequently, social insects offer exciting opportunities to study the interplay between sexual and kin selection, especially since methodological and technological progress offers rather spectacular opportunities for future experimental work.

Acknowledgments I thank Alfredo Peretti and Anita Aisenberg for their invitation to write this review and their continuous help throughout the publication process. I also thank Juergen Heinze and Barbara Baer-Imhoof whose detailed comments substantially improved the manuscript. This work was supported by a Future Fellowship (DP0770050) and a Discovery grant (DP0878107) offered by the Australian Research Council.

References

Alford DV (1975) Bumblebees. Davis-Poynter, London

Allard D, Gobin B, Billen J (2007) Timing of sperm transfer in *Diacamma pallidum*. Physiol Entomol 32:382–387

Armitage S, Boosma J, Baer B (2010) Diploid male production in a leaf-cutting ant. Ecol Entomol 35:175–182

Ashe A, Oldroyd B (2002) Genetic determination of caste in harvester ants. Trends Ecol Evol 17:448–449

Atkinson L, Adams ES (1997) The origins and relatedness of multiple reproductives in colonies of the termite *Nasutitermes corniger*. Proc R Soc B-Biol Sci 264:1131–1136

Baer B (2003) Bumblebees as model organisms to study male sexual selection in social insects. Behav Ecol Sociobiol 54:521–533

Baer B (2005) Sexual selection in Apis bees. Apidologie 36:187–200
Baer B (2011) The copulation biology of ants (Hymenoptera: Formicidae). Myrmecol. News 14:55–68
Baer B (2014) Sexual selection in social insects. In: Shuker D, Simmons LW (eds) The evolution of insect mating systems. Oxford University Press, Oxford
Baer B, Schmid-Hempel P (1999) Experimental variation in polyandry affects parasite loads and fitness in a bumble-bee. Nature 397:151–154
Baer B, Schmid-Hempel P (2001) Unexpected consequences of polyandry for parasitism and fitness in the bumblebee, *Bombus terrestris*. Evolution 55:1639–1643
Baer B, Schmid-Hempel P (2003) Bumblebee workers from different sire groups vary in susceptibility to parasite infection. Ecol Lett 6:106–110
Baer B, Schmid-Hempel P (2005) Sperm influences female hibernation success, survival and fitness in the bumblebee *Bombus terrestris*. Proc R Soc B 272:319–323
Baer B, Schmid-Hempel P, Hoeg JT, Boosma JJ (2003) Sperm length, sperm storage and mating system characteristics in bumblebees. Insect Soc 50:101–108
Baer B, Armitage SAO, Boosma JJ (2006) Sperm storage induces an immunity cost in ants. Nature 441:872–875
Baer B, Eubel H, Taylor NL, O'Toole N, Millar AH (2009). Insights into female sperm storage from the spermathecal fluid proteome of the honeybee *Apis mellifera*. Genome Biol 10: R67
Beekman M, Van Stratum P (1998) Bumblebee sex ratios: Why do bumblebees produce so many males? Proc R Soc B 265:1535–1543
Beye M, Hasselmann M, Fondrk MK, Page RE, Omholt SW (2003) The gene csd is the primary signal for sexual development in the honeybee and encodes an SR-type protein. Cell 114:419–429
Boomsma JJ (1996) Split sex ratios and queen-male conflict over sperm allocation. Proc R Soc B 263:697–704
Boomsma JJ (2007) Kin selection versus sexual selection: why the ends do not meet. Curr Biol 17:R673–R683
Boomsma JJ (2009) Lifetime monogamy and the evolution of eusociality. Phil Trans R Soc 364:3191–3207
Boomsma JJ, Ratnieks FLW (1996) Paternity in eusocial Hymenoptera. Philos T Roy Soc B 351:947–975
Boomsma JJ, Baer B, Heinze J (2005) The evolution of male traits in social insects. Annu Rev Entomol 50:395–420
Bourke AFG, Franks NR (1995) Social evolution in ants. Princeton University Press, Princeton
Bresslau E (1905) Der Samenblasengang der Bienenkönigin. Zool Anz 29:299–325
Brown MJF, Baer B (2005) The evolutionary significance of long copulation duration in bumblebees. Apidologie 36:157–167
Dade HA (1962) Anatomy and dissection of the honeybee. Bee Research Association, London
Den Boer SPA, Boomsma JJ, Baer B (2009) Honey bee males and queens use glandular secretions to enhance sperm viability before and after storage. J Insect Physiol 55:538–543
Den Boer SPA, Baer B, Boomsma JJ (2010) Seminal fluid mediates ejaculate competition in social insects. Science 327:1506–1509
Diehl-Fleig E (1995) Success of colony establishment by *Acromyrmex striatus* (Roger) (Hymenoptera: Formicidae). Ann Soc Entomol Brasil 24:625–630
Doums C, Cronin AL, Ruel C, Federici P, Haussy C, Tirard C, Monnin T (2013) Facultative use of thelytokous parthenogenesis for queen production in the polyandrous ant *Cataglyphis cursor*. J Evol Biol 26:1431–1444
Duvoisin N, Baer B, Schmid-Hempel P (1999) Sperm transfer and male competition in a bumblebee. Anim Behav 58:743–749
Eberhard WG (1996) Female control: sexual selection by cryptic female choice. Princeton University Press, Princeton
Fernandez-Marin H, Zimmerman JK, Wcislo WT (2004) Ecological traits and evolutionary sequence of nest establishment in fungus-growing ants (Hymenoptera, Formicidae, Attini). Biol J Linn Soc 81:39–48

Foster KR, Ratnieks FL (2000) Facultative worker policing in a wasp. Nature 407:692–693
Fournier D, Estoup A, Orivel J, Foucaud J, Jourdan H, Le Breton J, Keller L (2005) Clonal reproduction by males and females in the little fire ant. Nature 435:1230–1234
Gerloff CU, Ottmer BK, Schmid-Hempel P (2003) Effects of inbreeding on immune response and body size in a social insect, *Bombus terrestris*. Funct Ecol 17:582–589
Greeff M, Schmid-Hempel P (2008) Sperm reduces female longevity and increases melanization of the spermatheca in the bumblebee *Bombus terrestris*. Insect. Soc. 55:313–319
Haag-Liautard C, Vitikainen E, Keller L, Sundström L (2009) Fitness and the level of homozygosity in a social insect. J Evol Biol 22:134–142
Hamilton WD (1964) The genetical evolution of social behavior I & II. J Theor Biol 7(I):1–16; (II): 17–32
Hartke TR, Baer B (2011) The mating biology of termites: a comparative review. Anim Behav 82:927–936
Heimpel GE, De Boer JG (2008) Sex determination in the Hymenoptera. Annu Rev Entomol 53:209–230
Heinze J, Frohschammer S, Bernadou A (2013) Queen life-span and total reproductive success are positively associated in the ant *Cardiocondyla* cf. *kagutsuchi*. Behav Ecol Sociobiol 67:1555–1562
Hölldobler B, Bartz SH (1985) Sociobiology of reproduction in ants. In: Hölldobler B, Lindauer M (eds) Experimental behavioural ecology and sociobiology. Gustav Fischer Verlag, Stuttgart
Hölldobler B, Wilson EO (1990) The ants. Springer, Berlin
Hughes WOH, Boomsma JJ (2005) Genetic diversity and disease resistance in leafcutting ant societies. Evolution 58:1251–1260
Hughes WOH, Sumner S, Van-Borm S, Boomsma JJ (2003) Worker caste polymorphism has a genetic basis in Acromyrmex leaf-cutting ants. Proc Natl Acad Sci USA 100:9394–9397
Hughes WOH, Oldroyd BP, Beekman M, Ratnieks FLW (2008) Ancestral monogamy shows kin selection is key to the evolution of eusociality. Science 320:1213–1216
Izzo AS, Tibbetts EA (2012) Spotting the top male: sexually selected signals in male *Polistes dominulus wasps*. Anim Behav 83:839–845
Jaffe R, Garcia-Gonzalez F, Den Boer SPA, Simmons LW, Baer B (2012) Patterns of paternity skew among polyandrous social insects: what can they tell us about the potential for sexual selection? Evolution 66:3778–3788
Janet C (1904) Anatomie du gaster de la Myrmica rubra. Georges Carré et C. Nuad, Paris
Keller L, Passera L (1992) Mating system, optimal number of mating, and sperm transfer in the Argentine ant *Iridomyrmex humilis*. Behav Ecol Sociobiol 31:359–366
Kellner K, Heinze J (2011) Mechanism of facultative parthenogenesis in the ant *Platythyrea punctata*. Evol Ecol 25:77–89
Klenk M, Koeniger G, Koeniger N, Fasold H (2004) Proteins in spermathecal gland secretion and spermathecal fluid and the properties of a 29 kDa protein in queens of *Apis mellifera*. Apidologie 35:371–381
Koeniger N, Koeniger G (1991) An evolutionary approach to mating behaviour and drone copulatory organs in Apis. Apidologie 22:581–590
Kronauer DJC (2009) Recent advances in army ant biology (Hymenoptera: Formicidae). Myrmecol News 12:51–65
Kronauer DM, Boomsma JJ (2007) Do army ant queens re-mate later in life? Insect Soc 54:20–28
Maleszka R (2008) Epigenetic integration of environmental and genomic signals in honey bees. Epigenetics 3:188–192
Monnin T, Peeters C (1998) Monogyny and regulation of worker mating in the queenless ant *Dinoponera quadriceps*. Anim Behav 55:299–306
Moritz RFA, Lattorff HMG, Neumann P, Kraus FB, Radloff SE, Hepburn HR (2005) Rare royal families in honeybees, *Apis mellifera*. Naturwissenschaften 92:488–491
Oberstadt B, Heinze J (2003) Mating biology and population structure of the ant, *Leptothorax gredleri*. Insect Soc 50:340–345
Oldroyd BP, Fewell JH (2007) Genetic diversity promotes homeostasis in insect colonies. Trends Ecol Evol 22:408–413

Oppelt A, Heinze J (2007) Dynamics of sperm transfer in the ant *Leptothorax gredleri*. Naturwissenschaften 94:781–786
Penick CA, Prager SS, Liebig J (2012) Juvenile hormone induces queen development in late-stage larvae of the ant *Harpegnathos saltator*. J Insect Physiol 58:1643–1649
Poland V, Eubel H, King M, Solheim C, Millar AH, Baer B (2011) Stored sperm differs from ejaculated sperm by proteome alterations associated with energy metabolism in the honeybee *Apis mellifera*. Mol Ecol 20:2643–2654
Pollock GB, Cabrales A, Rissing SW, Binmore KG (2012) Suicidal punishment in the ant *Acromyrmex versicolor*. Evol Ecol Res 14:951–971
Queller DC (2003) Theory of genomic imprinting conflict in social insects. BMC Evol Biol 3:15
Raignier A, Van Boven JKA (1955) Etude taxonomoque, biologique et biométrique des Dorylus du sous-genre Anomma (Hymenoptera Formicidae). Ann Mus R Congo Belg 2:1–359
Ratnieks FLW, Foster KR, Wenseleers T (2006) Conflict resolution in insect societies. Annu Rev Entomol (Annual Reviews, Palo Alto)
Reichardt AK, Wheeler DE (1996) Multiple mating in the ant *Acromyrmex versicolor*: a case of female control. Behav Ecol Sociobiol 38:219–225
Robertson HG (1995) Sperm transfer in the ant *Carebara vidua* F. Smith (Hymenoptera: Formicidae). Insect Soc 42:411–418
Rubenstein DR (2012) Family feuds: social competition and sexual conflict in complex societies. Phil Trans R Soc B-Biol Sci 367:2304–2313
Ruttner F (1956) The mating of the honey bee. Bee World 3:2–15
Sauter A, Brown MJF, Baer B, Schmid-Hempel P (2001) Males of social insects can prevent queens from multiple mating. Proc R Soc B 268:1449–1454
Schmid-Hempel P (1998) Parasites in social insects. Princeton University Press, Princeton
Schmid-Hempel P, Crozier RH (1999) Polyandry versus polygyny versus parasites. Philos T Roy Soc B 354:507–515
Schoeters E, Billen J (2000) The importance of the spermathecal duct in bumblebees. J Insect Physiol 46:1303–1312
Snodgrass RE (1984) Anatomy of the honeybee. Comstock Publishing Associates, Ithaca
Sundström L (1994) Sex ratio bias, relatedness asymmetry and queen mating frequency in ants. Nature 367:266–268
Tarpy DR (2003) Genetic diversity within honeybee colonies prevents severe infections and promotes colony growth. Proc R Soc B 270:99–103
Trivers RL, Hare H (1976) Haplodiploidy and the evolution of the social insects. Science 191:249–263
Tschinkel WR (1987a) Fire ant queen longevity and age: estimation by sperm depletion. Ann Entomol Soc Am 80:263–266
Tschinkel WR (1987b) Relationship between ovariole number and spermathecal sperm count in ant queens: a new allometry. Ann Entomol Soc Am 80:208–211
Tsuji K (1996) Queen-male conflict over sperm use in social insects. Trends Ecol Evol 11:490–491
Weber NA (1972) Gardening ants: the attines. The American Philosophical Society, Philadelphia
West SA, Pen I, Griffin AS (2002) Cooperation and competition between relatives. Science 296:72–75
Wiernasz DC, Perroni CL, Cole BJ (2004) Polyandry and fitness in the western harvester ant, *Pogonomyrmex occidentalis*. Mol Ecol 13:1601–1606
Wilson EO (1971) The insect societies. Harvard University Press, Cambridge
Wilson EO (2000) Sociobiology: the new synthesis. Harvard University Press, Harvard
Winston ML (1991) The biology of the honey bee. Harvard University Press, Cambridge
Woyke J (1983) Dynamics of entry of spermatozoa into the spermatheca of instrumentally inseminated queen honey bees *Apis mellifera*. J Apic Res 22:150–154
Yue C, Schroder M, Gisder S, Genersch E (2007) Vertical-transmission routes for deformed wing virus of honeybees (*Apis mellifera*). J Gen Virol 88:2329–2336

Chapter 18
Mating Is a Give-and-Take of Influence and Communication Between the Sexes

Rafael L. Rodríguez

Abstract Reproductive interactions between males and females often involve exchanges of signals or stimulation between the sexes, as for example when male–female duetting is used in pair formation. Such exchanges may also be common during copulation. For instance, while males court the female during copulation, females often also touch the male or otherwise move in ways that seem easily perceived by the male. Such movements may offer feedback to males about how the female is reacting or going to react to his efforts. This may have important effects on the dynamics of mate choice and on its consequences for sexual selection and divergence; e.g., the feedback contained in female sexual response may allow males to adjust their behavior in adaptive ways. The presence of male displays and female preferences plus female feedback and male attentiveness may have important consequences for the patterns of assortative mating and fertilization that become established in a population. Here I highlight selected examples of various contexts in which male–female exchanges may occur, before and during copulation. My goal is to initiate a discussion about the sources of selection that may influence the evolution of female feedback to males and of male attentiveness to such feedback and to sketch out some potential consequences for the course of sexual selection.

18.1 Introduction

The logic of the theory of sexual selection is based on the recognition of two distinct evolutionarily stable strategies for sexual reproduction: the male strategy, which specializes in competing for access to the gametes of many females, and the female strategy, which specializes in securing from a limited number of

R.L. Rodríguez (✉)
Behavioral and Molecular Ecology Group, Department of Biological Sciences, University of Wisconsin–Milwaukee, Milwaukee, WI 53201, USA
e-mail: rafa@uwm.edu

A.V. Peretti and A. Aisenberg (eds.), *Cryptic Female Choice in Arthropods*,
DOI 10.1007/978-3-319-17894-3_18

males the genetic and non-genetic resources that increase progeny numbers and fitness (Andersson 1994; Kokko et al. 2006). From the nature of these strategies, it follows that males are selected to seek and attract females and induce in them favorable behavioral and physiological responses, whereas females are selected to respond differentially to potential mates, accepting and cooperating with some and rejecting others (Andersson 1994; Cordero and Eberhard 2003; Andersson and Simmons 2006). The resulting competition and discrimination within and between the sexes give rise to sexual selection, a powerful engine of evolution responsible for the most dramatic cases of rapid divergence and trait elaboration that are found in nature (Darwin 1871; West-Eberhard 1983, 2014; Andersson 1994; Safran et al. 2012; Rodríguez et al. 2013a; Seddon et al. 2013).

Although the logic of sexual selection is well understood, biologists are only beginning to fathom how the behavior of each sex exerts selection on the other sex. Recent work has started to reveal that the "traditional" view of the sex roles presented above may hide two-way flows of influence and communication between the sexes. As many chapters in this volume show, biologists have generally abandoned the view of females as passive observers of male displays for a better understanding of their active role in effecting their reproductive decisions. But we are still just beginning to realize how active females are during the reproductive process and also the extent to which males pick up on this for their own advantage. I think that a picture is emerging in which—in addition to selection on males to induce females to accept them and selection on females to respond selectively to males—there is also selection on females to influence male behavior and selection on males to attend to female responses and adjust their behavior accordingly (West-Eberhard 1983; Rodríguez and Barbosa 2014).

In this chapter, I present selected vignettes that illustrate a variety of forms that two-way avenues of influence and communication between males and females may take. These go beyond the realization that females as well as males may be under sexual and social selection (West-Eberhard 1983, 2014; Clutton-Brock 2007, 2009; Rubenstein and Lovette 2009) and may occur with or without sex role reversal (Gwynne 1991; Andersson 1994). The point is that even mating systems with "traditional" sex roles often feature reciprocal interactions that occur as males court and females make decisions about mating and fertilization. I begin with examples of interactions leading to copulation before examining interactions that take place during copulation. Some of the examples represent cases where males pick up on incidental cues provided by female responses, while others represent cases where males attend to specialized signals produced by females. I try to be clear in making this cue–signal distinction, but part of the message is that it may be too early to do so—many female signals may appear to us as by-products of female responses simply because we are only beginning to examine the possibility of male–female reciprocal interactions during the reproductive process.

This is not at all an exhaustive review. Instead, my goal is to highlight a sample of suggestive case studies that help refine our understanding of the behavioral, physiological, and evolutionary dynamics involved in traditional mating systems. I then initiate a discussion about the potential impact of these dynamics on sexual selection and its role in speciation.

18.2 Examples of Back-and-Forth Interactions Leading to Pair Formation

18.2.1 Females Advise Males on How to Court Them

Female behavior may often offer feedback to males about the effect that their courtship is having, allowing for real-time adjustments. A dramatic example has been documented in bowerbirds (Borgia and Presgraves 1995; Patricelli et al. 2002, 2006). Female bowerbirds prefer high-intensity male displays, but such displays may also startle the females and hinder pair formation. The best courtship display is therefore not necessarily the most intense, but the most intense that a given female can stand. Males would thus benefit from gaging the tolerance for high-intensity displays of the female they are courting and tailoring their behavior to her. In satin bowerbirds, as a female watches a male display, her behavior and posture indicate her reactions: She may crouch, and the more that she crouches, the less threatened and more receptive that she is (Fig. 18.1a). Experiments using robotic females to manipulate female behavior (Fig. 18.1b) show that males attend to female crouching and modulate their behavior accordingly: They increase display intensity when the female crouches more and reduce it when the female is startled, so that the more attentive males increase the success of their courtship efforts (Patricelli et al. 2002, 2006).

18.2.2 Male–Female Signal Exchanges Lead to Pair Formation

In some cases, female behavior does not only provide cues that males may attend to or not. Instead, female responses to male behavior are an integral part of the

(a) (b)

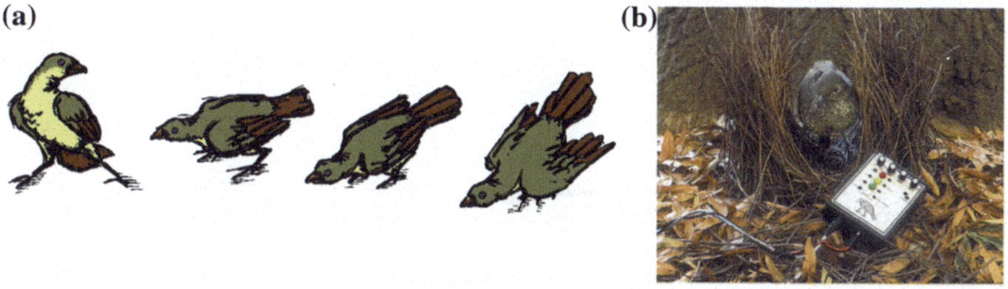

Fig. 18.1 Female behavior in satin bowerbirds provides feedback to males about their courtship efforts. **a** Variation in female posture indicates receptivity to high-intensity displays. The image on the *left* is of a female in the upright position upon arrival at the bower. The images to the *right* show stages of crouching, with the rightmost image showing the solicitation posture. Artwork by J. Albert Uy, reproduced with kind permission by J. Albert Uy and Gail Patricelli. **b** Experimental bower and robotic female (showing remote controller) used in tests of male attentiveness to female feedback. Photograph and permission by Gail Patricelli

pair formation process. In many insects and spiders, for instance, pair formation involves signal exchanges between males and females, or duets (Kraft 1982; Bailey 2003; Cocroft and Rodríguez 2005) (Fig. 18.2). In most insects that communicate with airborne sound, female signals are very short and mainly seem to convey their presence to the male (Bailey 2003). But in insects that communicate with substrate-borne vibrations, female response signals often resemble male signals in length and variability (Rodríguez and Barbosa 2014) (Fig. 18.2). This opens up the possibility that female duetting signals may contain feedback cues that males may attend to.

An example of information in female duetting signals occurs in *Enchenopa* treehoppers, phloem-feeding insects that communicate with plant-borne vibrational signals (Rodríguez and Cocroft 2006; Cocroft et al. 2008) (Fig. 18.2). *Enchenopa* females have strong mate preferences for male advertisement signals, and they express their preferences with their duetting behavior, being more likely to signal back to males that produce attractive signals (Rodríguez et al. 2004, 2006).

Fig. 18.2 Plant-borne vibrational communication in *Enchenopa binotata* treehoppers (Hemiptera: Membracidae). **a** Adult aggregation on the host plant *Ptelea trifoliata* (Rutaceae). **b** The basic unit of adult communication in *E. binotata*: a male-female signal duet. The male signal consists of a whine (a near-pure tone that slightly drops in frequency from beginning to end) followed by a series of pulses. The female signal consists of a single component

Females also produce more and longer signals when duetting with males that they find attractive, and males pick up on such variation in female behavior, being more likely to signal when they perceive longer female signals—i.e., those produced by females interacting with males they found attractive (Rodríguez et al. 2004, 2012; Rodríguez and Barbosa 2014).

18.2.3 Females Reassure Males About Their Receptivity

The above example shows that female duetting signals may convey information to the male not only about the female's presence but also about her receptivity. Such indications may be especially important in species in which the female's "intentions" toward the male may vary drastically. When females are larger than males, for instance, they may present a threat of injury or cannibalism, as in many spiders. In such cases, males approach females with extreme care, ready to flee at the slightest hint that she may attack (Kraft 1982; Uhl and Elias 2011). Male caution may reach such levels that males require encouragement from a particular female behavior before they will approach and attempt to copulate. A potential example of such reassurance by females occurs in *Schizocosa* wolf spiders (Stratton and Uetz 1981; Sullivan-Beckers and Hebets 2011, 2014). In these spiders, males court the female with vibrational–visual displays, and receptive females respond to male displays by turning in semicircles. These turns provide males with an indication about the females' receptivity: Courting males approach the female gradually, even when they have already located her, advancing slightly after each turn (Sullivan-Beckers and Hebets 2011).

Encouragement for the male to approach the female may also come from slight changes in her behavior, rather than from any one given behavior. In *Kukulcania hibernalis* crevice weaver spiders, for instance, females are initially aggressive toward courting males, moving swiftly within their web retreat or lunging out. But as males continue to court, females begin to move more slowly within their retreat, and this change seems to be what prompts the male to approach (Barrantes and Ramírez 2013).

18.2.4 When Males Succeeded in Attracting a Mate but Did not Realize It

Even in species with mating systems that lack clear male–female duetting, some form of communiqué from the female may be required to complete pair formation. I am only aware of anecdotal evidence for this, but the observations are intriguing. Consider a male frog signaling at a frog chorus. When a female finally arrives at the male's side, he often appears not to notice her, and she has to prod him before he realizes she is there and stops signaling to proceed to amplexus (Fig. 18.3).

Fig. 18.3 Two occasions in which a female gray tree frog *Hyla versicolor* (Anura: Hylidae) arrived at the site of a signaling male and the male did not realize it. **a** The female (*left*) is prodding the male in an apparent attempt to proceed to amplexus, while he continues to signal (note his expanded vocal sac). **b** The female (*left*) climbing on the male as he continues to signal

An informed guess places the frequency of such occurrences at about half of observed pair formation interactions in tree frogs (Höbel, personal communication). Similar "attention-getting" behavior by females has also been noted occasionally in the above *Enchenopa* treehoppers (Sullivan-Beckers, personal communication).

18.3 Examples of Back-and-Forth Interactions that Take Place During Copulation

The above cases illustrate varied contexts and ways in which female behavior can provide feedback to males during pair formation, and in some cases, males attend to such feedback and adjust their behavior. Such reciprocal interactions also occur during copulation; indeed, they are much more likely during copulation than at earlier stages and present a far greater diversity of contexts and opportunities. The reason is simple physical opportunity: During copulation, males are in direct contact with the female and are thus more likely to be able to detect movements and responses made by the females, perhaps even slight changes in behavior or physiology (Fig. 18.4). Moreover, female behavior during copulation sometimes fits the criteria that would class male behavior as copulatory courtship: Female movements are repeated and stereotyped; likely to be sensed by the male; mechanically unnecessary for the female to retain her position (or to dislodge the male from his position); and not performed in contexts other than copulation (Eberhard 1994, 2005). Female copulatory behavior fits these criteria in about 10 % of a sample of 131 species of insects and spiders that were scrutinized for behavior during copulation (Eberhard 1994), which probably represents an underestimate. Such female behavior may influence male–female interactions in a large variety of ways, which we

Fig. 18.4 Comparison of the opportunity for males to obtain feedback about female reproductive decisions between pair formation (**a**) and copulation (**b**). In these gray tree frogs, *Hyla versicolor* (Anura: Hylidae), the greater contact during copulation clearly affords greater opportunity for feedback. Of course, this depends on the number of reproductive decisions that remain to be made by the female once copulation begins. In the case of species with cryptic female choice, the list may be considerable. But even without cryptic female choice, female feedback, male attentiveness, and male–female coordination may yet be favored (Sect. 18.3.5)

are just beginning to explore. Thus, behavioral and physiological feedback from females to males seems much more likely during copulation than at earlier stages in the reproductive process.

18.3.1 Females Advise Males on How to Court Them: Copulatory Dialogues

A spectacular example of male–female give-and-take during copulation was documented in *Physocyclus globusus* pholcid spiders (Huber and Eberhard 1997; Peretti et al. 2006; Peretti and Eberhard 2010; Calbacho-Rosa and Peretti 2015). In this species, males court females during copulation by using their pedipalps to squeeze the female abdomen. Squeezes are probably quite hard, as they are performed with the thickest muscles on the male body. Females favor males that squeeze more, but they also appear to attempt to induce males to loosen their squeezes: Females stridulate during copulation, and they are more likely to stridulate while males are squeezing them, especially during long squeezes or if the male had recently not responded to stridulation. Thus, the best courtship squeezes are not necessarily the most intense, but the most intense that a female will tolerate. Indeed, the more attentive males (those that loosen squeezes more often when females stridulate) gain greater fertilization success (Peretti et al. 2006; Peretti and Eberhard 2010).

This example makes a striking parallel with the satin bowerbird case study mentioned above (Sect. 18.2.1), and I suspect that this kind of back-and-forth between the sexes may be widespread. Another potential example occurs in *Glossina pallidipes* tsetse flies (Briceño and Eberhard in preparation). In these flies, males use their cerci to squeeze the tip of the female abdomen during copulation. Females may vibrate their wings during squeezes, and males seem to respond to the females' wing vibration by shortening their squeezes (Briceño and Eberhard in preparation).

18.3.2 Females Warn Males of Likely Failure

Females sometimes appear to intimate when copulations are likely to fail. An example occurs in *Ozophora baranowskii* seed bugs (Rodríguez 1998, 1999). In these bugs, some copulations fail to result in sperm transfer, presumably because the female did not allow spermatophore transfer. During copulation, females tap males with their legs. The taps are not forceful. Instead, early in copulation, females tapped the male at higher rates in copulations in which no sperm transfer occurred and at lower rates in copulations that resulted in sperm transfer (Rodríguez 1998). These females thus appear to give males an early indication of their forthcoming decision. An analogous notice has been observed for female stridulation in some pholcid spiders, including the above *P. globosus*, in which females stridulate when an encounter is less likely to lead to copulation (Peretti et al. 2006; Dutto et al. 2011). I suspect that this type of feedback, before and during copulation, may turn out to be widespread (see e.g., Baena and Eberhard 2007).

18.3.3 Females Help the Males to Achieve Intromission

Females sometimes appear to help males in surprisingly active ways. Consider mating behavior in *Leiobunum vittatum* harvestmen (Fowler-Finn et al. 2014). Mating in this species begins when the male pounces on the female, wraps his legs around hers, and hooks his pedipalps on the base of her legs as she struggles and bobs. The female continues to struggle on and off during copulation. Nevertheless, she also lightly taps the male with her pedipalps in a way that would appear to be copulatory courtship (see above). Not only that, she also appears to help him achieve intromission, and he seems to be incapable of doing this by himself: When the male everts his penis to attempt intromission, the female places her pedipalps behind protrusions on the male's penis and appears to guide and pull him toward her. No intromission was observed without this behavior by the female (Fowler-Finn et al. 2014). Thus, it appears that in this species, males require the mechanical assistance from the females to achieve intromission.

There is a potential alternative interpretation of this female help. In the clade to which this harvestman belongs, males of some species provide nuptial gifts for females through their penis, and females acquire the nuptial gift by bringing the penis to their mouthparts (Macías-Ordóñez et al. 2010; Machado et al. 2015). In such cases, female tapping could be interpreted as solicitation of the nuptial gift and guiding the penis toward her as securing the nuptial gift. Also, because the females' mouthparts and genital opening are close by, it is difficult to distinguish oral from genital insertion (Fowler-Finn et al. 2014). But even if a nuptial gift is involved, no intromission (whether oral or genital) ever occurred without the females' apparent help (Fowler-Finn et al. 2014). Thus, even if females were soliciting a direct benefit from the males, they also helped them achieve intromission in every observed instance. This is all the more remarkable if we recall that before and during copulation, there occur male–female struggles that would make one expect little assistance from the female for the male.

18.3.4 Females Stimulate the Male's Genitalia

A major advance in sexual selection has been the discovery of stimulatory functions for male genitalia in the context of cryptic female choice (Eberhard 2009). It thus seems fitting that a new development involves females that deliver stimulation directly to the male genitalia during copulation. In *Cyanopterus sphinx* fruit bats, for example, females often lick the base of the male penis during intromission, and there is a positive relationship between this licking and the duration of copulation (Tan et al. 2009). Licking may have antibiotic effects or other advantages, but there may also be some benefit in extending the duration of copulation, such as increased sperm transfer (Tan et al. 2009). Such potential benefits are suggested by yet another twist in the tale: In *Pteropus giganteus* flying foxes, males lick the female genitalia before and after copulation, and there is a positive relationship between the duration of precopulatory licking and the duration of copulation (Maruthupandian and Marimuthu 2013).

18.3.5 Females Let the Male Know When They Are Ready to Lay Eggs

As the stages of the reproductive process advance from pair formation to fertilization, one could expect that the need for a jostle of persuasion and influence would diminish, because the higher the number of decisions that a female has already made in favor of a given male, the better aligned that their evolutionary interests should become (Alexander et al. 1997). Nevertheless, even when nothing remains but to proceed to fertilization, there still remain opportunities for male–female reciprocal interactions. An example of this has been noted anecdotally in *Hyla versicolor* tree frogs (Höbel, personal communication). In these frogs, females do not lay all of their eggs at once. Instead, they lay a few eggs at a time, in bouts separated by a few minutes, and it may take a few hours until a female has laid all of her eggs. Immediately before each bout of egg laying, females shiver their body and arch their back. It is not known if these movements are required for laying eggs or if they might constitute a specialized signal. Even in the former case, however, they could potentially alert the male (who all the while is in amplexus with the female; Fig. 18.4b) that egg laying is imminent, and this could help him time the release of his sperm appropriately. It is easy to imagine that the precision of this timing, and hence his fertilization success, could be severely hurt if such cues from the female were lacking.

18.4 Examples of Male–Female Interactions that I Have not Considered to Be Reciprocal

The examples in Sects. 18.2 and 18.3 go beyond documenting an active role for females in mate choice; they feature behavior that may convey feedback to the male, which he may use to his own advantage. To clarify the distinction between these two types of active participation by the female in the reproductive process, here I present examples of interactions that, although illuminating about the role of female behavior in mate choice in "traditional" mating systems, do not seem to involve potential female feedback and male attentiveness.

18.4.1 Females Signal to Attract More Potential Mates and Broaden Their Prospects for Choice

When males display to attract females, females do not simply watch the males, while they evaluate them and make their reproductive decisions. Instead, females move about from male to male or from site to site, seeking out and comparing the most attractive individuals and then making decisions (e.g., Uy et al. 2001; Murphy and Gerhardt 2002; Murphy 2012). Females may also seek to foster competition between males to facilitate their own comparisons and broaden the scope of their sampling. An excellent example of this occurs in *Gallinago media* great snipes (Sæther 2002). In these birds, males form leks where they gather to display and attract females. When females visit the leks, they produce loud signals, which attract males that then proceed to engage in fights. Males also lengthen their own signals in response to the females' signals, which makes them more attractive (Sæther 2002).

Another example of females seeking to increase the males they sample and to make them compete with each other occurs in the *Enchenopa* treehoppers mentioned in Sect. 18.2.2. Sometimes, when a male and a female are engaged in a duet and he is searching for her, he decreases the amplitude of his signals, but she increases the amplitude of hers (Rodríguez and Barbosa 2014). This would seem to increase the potential for detection by other males that may then approach the female and increase her prospects for choice. Similarly, female copulation calls—produced by females during copulation—seem to function to attract other males that may supplant the one currently copulating with the female (e.g., Løvlie et al. 2014).

18.4.2 Females Push or Kick the Male, Rather Than Gently Tap Him

Some of the female behaviors involved in the above interactions are subtle. But female tactics are not always restricted to gentle persuasion. In a close relative of the seed bug mentioned in Sect. 18.3.2, females push the male forcefully rather than tap him lightly and do so mostly toward the end of copulation (Rodríguez 1999). This behavior thus appears to be a forceful way to end copulation, rather than a signal seeking to induce a change in the male behavior. Such attempts by females to shake off or dislodge males during copulation are not uncommon (Eberhard 1994). A dramatic example occurs in *Callosobruchus maculatus* bean weevils (Crudgington and Siva-Jothy 2000; Edvardsson and Tregenza 2005; Hotzy and Arnqvist 2009). In these beetles, males have spiny genitalia that puncture the lining of the female genital tract (Crudgington and Siva-Jothy 2000), and spine length is positively correlated with fertilization success across populations (Hotzy and Arnqvist 2009). But puncturing damages the females' genitalia and lowers their longevity and fecundity, and females kick the males toward the end of copulation (Crudgington and Siva-Jothy 2000; Edvardsson and Tregenza 2005). Preventing the females from kicking results in longer and more damaging copulations (Crudgington and Siva-Jothy 2000), although there is evidence that female kicking itself may worsen the damage caused by male genitalia (Wilson and Tomkins 2014).

18.4.3 Females Cooperate with the Male to Form a Copulatory Plug

Female choice hinges on selective cooperation with some males and rejection of others, and such cooperation may involve many potential types of behavioral or physiological interactions with males (Eberhard 1996; Cordero and Eberhard 2003). But cooperative interactions may occur without exchanges of stimulation and influence or feedback from females that modifies male behavior. An example occurs in *Leucauge mariana* orb web spiders (Eberhard and Huber 1998; Aisenberg and Eberhard 2009; Aisenberg et al. 2015). In these spiders, males begin to form a plug on the female's genital plate by depositing a paste. But this alone is not sufficient to form an effective plug. The latter requires that the females add to the male's paste a liquid that she secretes, and females may be swayed to cooperate with the male in this manner by his copulatory courtship (Aisenberg and Eberhard 2009; Aisenberg et al. 2015). Similarly, in *Argiope keyserlingi* orb web spiders, all the female has to do is allow the male sufficient time to break off and lodge a fragment of his genitalia in her genital opening in a position that will make an effective plug; if she ends the copulation earlier, fragments of the male's genitalia are not lodged well and do not make an effective plug (Herbertstein et al. 2012; Schneider et al. 2015).

18.5 Discussion

Sexual selection research has achieved several major breakthroughs in the last few decades. These include the belated acceptance of Darwin's proposal that female mate choice is widespread in nature (Darwin 1871; West-Eberhard 1983; Andersson 1994; Andersson and Simmons 2006); the discovery that male–male competition, courtship, and mate choice continue after mating begins (Eberhard 1985, 1996, 2009; Birkhead and Møller 1998); and technical progress allowing the examination of genital behavior inside the body of the female (Briceño et al. 2010; Eberhard 2011; Briceño and Eberhard 2015). Another major breakthrough may lie in the realization that as females make mating and fertilization decisions, the changes in their behavior and physiology that effect those decisions provide males with indications about their motivation and attitude. Males may be able to use those cues to make inferences about likely female responses and adaptively modify their own behavior and tactics. Thus, male–female interactions that have historically been characterized as one-way avenues of communication between male signalers and female receivers may in fact involve two-way avenues of communication, stimulation, persuasion, and influence.

Here I have examined examples of such exchanges: four case studies dealing with precopulatory interactions and another six involving interactions during copulation. At both stages of the reproductive process, there seems to be ample opportunity for feedback from the female to the male. In some cases, such feedback has been shown to be used by males to modify their behavior and improve their mating or fertilization success. Of course, this short list of vignettes very likely underestimates the variety of contexts and circumstances in which reciprocal male–female interactions may be involved in pre- and postcopulatory mate choice.

18.5.1 The Evolution of Female Feedback and Male Attentiveness

Under what conditions does selection favor females that provide feedback to males? And under what conditions does selection favor males that attend to such feedback? There are varied potential benefits to this. It may increase the efficiency of sexual interactions and reduce associated costs. For instance, if a female intimates early during copulation that she will not allow sperm transfer, and if the male attends to her indication and gives the mating up, both may save fruitless expenditure of time and energy and decrease the risk of being caught by a predator. The male might also save sperm, e.g., if females reject males by ejecting freshly transferred sperm, as in the seed bug and pholcid spider case studies discussed above (Sects. 18.3.1 and 18.3.2) (Rodríguez 1998, 1999; Peretti et al. 2006; Peretti and Eberhard 2010). Alternatively, feedback may represent a way to resolve a trade-off that arises when intense stimulation is attractive but potentially harmful, as in the pholcid spiders above (Sect. 18.3.1). It may also help synchronize male and female activities once their evolutionary interests are more aligned (e.g., Sect. 18.3.5).

When addressing the above questions, two considerations suggest that we may not always need to identify benefits to females from providing feedback to males in order to explain the evolution of such feedback. The first consideration is that female responses that offer feedback may range from incidental cues to specialized signals. Indeed, feedback may originate as incidental cues that come under selection to influence male behavior to the females' advantage, thereby giving rise to specialized signals—which is one of the main hypotheses about the evolution of signals in communication systems (Greenfield 2002). We should thus expect to find in nature cases where males attend to incidental cues and other cases where males attend to signals and only in the latter cases would it make sense to ask about the benefits that may have selected for such feedback signals. Further, we do see in nature the full spectrum between females that provide feedback and females that use force, even among closely related species, as in the two seed bugs discussed above (Sects. 18.3.2 and 18.4.2) (Rodríguez 1998, 1999). We also find the full spectrum between attentive and forceful males. Research should seek to identify the causes of such variation in the nature of male–female interactions.

The second consideration involves the potential for a process of male–female coevolution analogous to Fisherian selection (Fisher 1958; West-Eberhard 1983; Mead and Arnold 2004; Fowler-Finn and Rodríguez 2015). In terms of female feedback and male attentiveness, this process would follow from the following conditions: (i) genetic variation in the female response (behavioral or physiological) that provides feedback; (ii) genetic variation in male attentiveness to such feedback; and (iii) assortative mating or fertilization, such that more attentive males tend to mate or fertilize females that provide feedback—e.g., males and females that are better able to coordinate their interactions may be more likely to mate and achieve fertilization (cf. Derlink et al. 2014). Under these conditions, there would arise a genetic correlation between female feedback and male attentiveness. This genetic correlation would, in turn, have the following consequences: Male attentiveness would be favored by the presence of feedback; the resulting increase in attentiveness would bring an increase in feedback as a correlated response; the increase in feedback in turn would continue to favor attentiveness and so on in self-reinforcing loops halted only by loss of genetic variation or costs to feedback or attentiveness (e.g., if providing feedback increased the risk of predation). Thus, the advantage to females may simply be the production of sons that are attentive to feedback and daughters that provide feedback.

18.5.2 Evolutionary Consequences of Female Feedback and Male Attentiveness

If "traditional" mating systems with competitive males and choosy females were to commonly feature feedback from females and attentiveness to that feedback by males, what would be the impact on our view of sexual selection and its consequences for divergence?

One potential consequence is for sexual conflict to be reduced. Consider males selected to deliver strong stimulation that improves their fertilization success but that may harm the female. In such cases, female feedback and male attentiveness may help resolve the stimulation–harm trade-off at an intermediate level of stimulation that proves effective but less damaging to the females (cf. West-Eberhard 2014).

Another consequence may be to increase the number of traits that may be involved in Fisherian selection. The standard theory for Fisherian selection features one trait in males (the display) and one trait in females (the preference) (Fisher 1958; Mead and Arnold 2004). Female feedback and male attentiveness may add traits to this mixture. For males, it seems straightforward to posit two traits: a display plus the ability to sense and react to feedback from the female. For females, it is similarly straightforward to posit two traits: a mate preference function (Ritchie 1996; Rodríguez et al. 2006, 2013b) and the behavior that expresses the mate preference and provides feedback to the male. Reality may be more complex than this, of course, but this scenario serves to highlight a series of potential outcomes. If a higher number of traits in each sex enter into the dynamics of Fisherian selection (if the required conditions are met; Sect. 18.5.1 above), there may be a greater likelihood of divergence in sexual traits and thus of reproductive isolation. For example, if feedback arises in one population and not in another, the pathways of male-female coevolution may diverge and lead to speciation.

It will be important to bear in mind that feedback and attentiveness may influence patterns of assortative mating and fertilization in a population. Male attentiveness to female feedback may reinforce the patterns of assortative mating that are established by female mate choice if males are more likely to pursue females that they know find them attractive (cf. Rodríguez et al. 2012; Rodríguez and Barbosa 2014). Alternatively, if males are able to tailor their courtship for different kinds of females, this may lead to patterns of mating that counter what would arise from female mate choice alone (cf. Kozak et al. 2009).

The above discussions have the implicit assumption that if selection were to favor male attentiveness and female feedback, it would favor high levels of attentiveness and feedback. But it is quite possible that intermediate levels would be optimal—say, enough male attentiveness to fit the specific female they are interacting with, but not so much that stimulation weakens too much, or enough feedback to increase the likelihood of success for an attractive male, but not so much that any male could reach high attractiveness. If so, there may be multiple combinations of levels in feedback/attentiveness that reach optimal outcomes, such as high feedback paired with low attentiveness, or low feedback paired with high attentiveness. The resulting evolutionary dynamics in such cases might resemble the scenario of coadaptation theory developed for parent–offspring conflict (Agrawal et al. 2001; Kölliker et al. 2012). This might mean that some components of a mating system exhibit Fisherian-like dynamics of self-reinforcing coevolution driven by positive genetic correlations, while other components are stabilized by negative genetic correlations.

In conclusion, there is much insight to be gained from exploring the nature of male–female interactions throughout the reproductive process. This is challenging, because it requires observational, experimental, and comparative studies to ask whether and how the behavioral and physiological responses of each sex modify the responses of the other sex. At our current state of knowledge, perhaps the first step should be to explore the variety of forms feedback and attentiveness take in nature.

Acknowledgments I am grateful to Anita Aisenberg and Alfredo Peretti for the invitation to contribute to this volume. Flavia Barbosa, Gerlinde Höbel, and one anonymous reviewer kindly provided constructive critiques of the manuscript. I also thank Gerlinde Höbel, Gail Patricelli, and J. Albert Uy for providing their photographs and illustrations. Financial support was provided in part by NSF grant IOS-1120790 to RLR and KD Fowler-Finn.

References

Agrawal AF, Brodie ED, Brown J (2001) Parent–offspring coadaptation and the dual genetic control of maternal care. Science 292:1710–1712

Aisenberg A, Barrantes G, Eberhard WG (2015) Post-copulatory sexual selection in two tropical orb-weaving *Leucauge* spiders. In: Peretti AV, Aisenberg A (eds) Cryptic female choice in arthropods. Springer, Cham

Aisenberg A, Eberhard WG (2009) Female cooperation in plug formation in a spider: effects of male copulatory courtship. Behav Ecol 20:1236–1241

Alexander RD, Marshall DC, Cooley JR (1997) Evolutionary perspectives on insect mating. In: Choe JC, Crespi BJ (eds) The evolution of mating systems in insects and arachnids. Cambridge University Press, Cambridge, pp 4–31

Andersson M (1994) Sexual selection. Princeton University Press, Princeton, NJ

Andersson M, Simmons LW (2006) Sexual selection and mate choice. Trends Ecol Evol 21:296–302

Baena ML, Eberhard WG (2007) Appearances deceive: female "resistance" behaviour in a sepsid fly is not a test of male ability to hold on. Ethol Ecol Evol 19:27–50

Bailey WJ (2003) Insect duets: underlying mechanisms and their evolution. Physiol Entomol 28:157–174

Barrantes G, Ramírez MJ (2013) Courtship, egg sac construction, and maternal care in *Kukulcania hibernalis*, with information on the courtship of *Misionella mendensis* (Araneae, Filistatidae). Arachnology 16:72–80

Birkhead TR, Møller AP (1998) Sperm competition and sexual selection. Academic Press, San Diego

Briceño RD, Eberhard WG Copulatory dialogue between male and female tsetse flies (Diptera: Muscidae: *Glossina pallidipes*) (in preparation)

Borgia G, Presgraves DC (1995) Coevolution of elaborated male display traits in the spotted bowerbird: an experimental test of the threat reduction hypothesis. Anim Behav 56:1121–1128

Briceño RD, Eberhard WG (2015) Species-specific behavioral differences in tsetse fly genital morphology and probable cryptic female choice. In: Peretti AV, Aisenberg A (eds) Cryptic female choice in arthropods. Springer, Cham

Briceño RD, Wegrzynek D, Chinea-Cano E, Eberhard WG, Dos Santos Rolo T (2010) Movements and morphology under sexual selection: tsetse fly genitalia. Ethol Ecol Evol 22:385–391

Calbacho-Rosa L, Peretti AV (2015) Copulatory and post-copulatory sexual selection in haplogyne spiders, with emphasis on Pholcidae and Oonopidae. In: Peretti AV, Aisenberg A (eds) Cryptic female choice in arthropods. Springer, Cham

Clutton-Brock TH (2007) Sexual selection in males and females. Science 318:1882–1885

Clutton-Brock TH (2009) Sexual selection in females. Anim Behav 77:3–11

Cocroft RB, Rodríguez RL (2005) The behavioral ecology of insect vibrational communication. Bioscience 55:323–334

Cocroft RB, Rodríguez RL, Hunt RE (2008) Host shifts, the evolution of communication and speciation in the *Enchenopa binotata* complex of treehoppers. In: Tilmon K (ed) Specialization, speciation, and radiation: the evolutionary biology of herbivorous insects. University of California Press, Berkeley, pp 88–100

Cordero C, Eberhard WG (2003) Female choice of sexually antagonistic male adaptations: a critical review of some current research. J Evol Biol 16:1–6

Crudgington HS, Siva-Jothy MT (2000) Genital damage, kicking and early death. Nature 407:855–856

Darwin C (1871) The descent of man, and selection in relation to sex. J. Murray, London

Derlink M, Pavlovcic P, Stewart AJA, Virant-Doberlet M (2014) Mate recognition in duetting species: the role of mate and female vibrational signals. Anim Behav 90:181–193

Dutto MS, Calbacho-Rosa L, Peretti AV (2011) Signalling and sexual conflict: female spiders use stridulation to inform males of sexual receptivity. Ethology 117:1040–1049

Eberhard WG (1985) Sexual selection and animal genitalia. Harvard University Press, Cambridge

Eberhard WG (1994) Evidence for widespread courtship during copulation in 131 species of insects and spiders, and implications for cryptic female choice. Evolution 48:711–733

Eberhard WG (1996) Female control: sexual selection by cryptic female choice. Princeton University Press, Princeton

Eberhard WG (2005) Sexually reversed copulatory courtship roles and possible nuptial feeding in the soldier beetle *Ditemnus acantholabus* (Champion 1915) (Coleoptera: Cantharidae). J Kansas Ent Soc 79:13–22

Eberhard WG (2009) Postcopulatory sexual selection: Darwin's omission and its consequences. Proc Natl Acad Sci USA 106:10025–10032

Eberhard WG (2011) Experiments with genitalia: a commentary. Trends Ecol Evol 26:17–21

Eberhard WG, Huber BA (1998) Courtship, copulation and sperm transfer in *Leucauge mariana* (Araneae, Tetragnathidae) with implications for higher classification. J Arachnol 26:342–368

Edvardsson M, Tregenza T (2005) Why do male *Callosobruchus maculatus* harm their mates? Behav Ecol 16:788–793

Fisher RA (1958) The genetical theory of natural selection. A complete variorum edition. Oxford University Press, New York

Fowler-Finn KD, Rodríguez RL (2015) The causes of variation in the presence of genetic covariance between sexual traits and preferences. Biol Revs (in press)

Fowler-Finn KD, Triana E, Miller O (2014) Mating in the harvestman *Leiobunum vittatum* (Arachnida: Opiliones): from premating struggles to solicitous tactile engagement. Behaviour 151:1663–1686

Gwynne DT (1991) Sexual competition among females: what causes courtship-role reversal? Trends Ecol Evol 6:118–121

Greenfield MD (2002) Signalers and receivers. Oxford University Press, New York

Herberstein ME, Wignall AE, Nessler SH, Harmer AMT, Schneider JM (2012) How effective and persistent are fragments of male genitalia as mating plugs? Behav Ecol 23:1140–1145

Hotzy C, Arnqvist G (2009) Sperm competition favors harmful males in seed beetles. Curr Biol 19:404–407

Huber BA, Eberhard WG (1997) Courtship, copulation, and genital mechanics in *Physocyclus globusus* (Araneae, Pholcidae). Can J Zool 75:905–918

Kokko H, Jennions MD, Brooks R (2006) Unifying and testing models of sexual selection. Annu Rev Ecol Evol Syst 37:43–66

Kölliker M, Royle NJ, Smiseth PT (2012) Parent–offspring co-adaptation. In: Royle NJ, Smiseth PT, Kölliker M (eds) The evolution of parental care. Oxford University Press, Oxford, pp 285–303

Kozak G, Reisland M, Boughman JW (2009) Sex differences in mate recognition and conspecific preference in species with mutual mate choice. Evolution 63:353–365

Kraft B (1982) The significance and complexity of communication in spiders. In: Witt PN, Rovner JS (eds) Spider communication. Mechanisms and ecological significance. Princeton University Press, Princeton, pp 15–66

Løvlie H, Zidar J, Berneheim C (2014) A cry for help: female distress calling during copulation is context dependent. Anim Behav 92:151–157

Macías-Ordóñez R, Machado G, Pérez-González A, Shultz AW (2010) Genitalic evolution in opiliones. In: Leonard JL, Córdoba-Aguilar A (eds) The evolution of primary sexual character in animals. Oxford University Press, New York, pp 285–306

Machado G, Requena GS, Toscano Gadea C, Stanley E, Macías-Ordoñez R (2015) Male and female mate choice in harvestmen: general patterns and inferences on the underlying processes. In: Peretti AV, Aisenberg A (eds) Cryptic female choice in arthropods. Springer, Cham

Maruthupandian J, Marimuthu G (2013) Cunnilingus apparently increases duration of copulation in the Indian flying fox, *Pteropus giganteus*. PLoS one 8(3):e59743

Mead LS, Arnold SJ (2004) Quantitative genetic models of sexual selection. Trends Ecol Evol 19:264–271

Murphy CG (2012) Simultaneous mate-sampling by female barking treefrogs (*Hyla gratiosa*). Behav Ecol 23:1162–1169

Murphy CG, Gerhardt HC (2002) Mate sampling by female barking treefrogs (*Hyla gratiosa*). Behav Ecol 13:472–480

Patricelli GL, Uy JAC, Walsh G, Borgia G (2002) Male displays adjusted to female's response. Nature 415:279–280

Patricelli GL, Coleman SW, Borgia G (2006) Male satin bowerbirds, *Ptilonorhynhcus violaceus*, adjust their display intensity in response to female startling: an experiment with robotic females. Anim Behav 71:49–59

Peretti A, Eberhard WG (2010) Cryptic female choice via sperm dumping favours male copulatory courtship in a spider. J Evol Biol 23:271–281

Peretti A, Eberhard WG, Briceño RD (2006) Copulatory dialogue: female spiders sing during copulation to influence male genitalic movements. Anim Behav 72:413–421

Ritchie MG (1996) The shape of female mating preferences. Proc Natl Acad Sci USA 93:14628–14631

Rodríguez RL (1998) Possible female choice during copulation in *Ozophora baranowskii* (Heteroptera: Lygaeidae): female behavior, multiple copulations, and sperm transfer. J Insect Behav 11:725–741

Rodríguez RL (1999) Male and female mating behavior in two *Ozophora* bugs (Heteroptera: Lygaeidae). J Kansas Ent Soc 72:137–148

Rodríguez RL, Barbosa F (2014) Mutual behavioral adjustment in vibrational duetting. In: Cocroft RB, Gogala M, Hill PSM, Wessel A (eds) Studying vibrational communication. Springer, Heidelberg, pp 147–169

Rodríguez RL, Cocroft RB (2006) Divergence in female duetting signals in the *Enchenopa binotata* species complex of treehoppers (Hemiptera: Membracidae). Ethology 112:1231–1238
Rodríguez RL, Haen C, Cocroft RB, Fowler-Finn KD (2012) Males adjust signaling effort based on female mate-preference cues. Behav Ecol 23:1218–1225
Rodríguez RL, Boughman JW, Gray DA, Hebets EA, Höbel G, Symes LB (2013a) Diversification under sexual selection: the relative roles of mate preference strength and the degree of divergence in mate preferences. Ecol Lett 16:964–974
Rodríguez RL, Hallet AC, Kilmer JT, Fowler-Finn KD (2013b) Curves as traits: genetic and environmental variation in mate preference functions. J Evol Biol 26:434–442
Rodríguez RL, Ramaswamy K, Cocroft RB (2006) Evidence that female preferences have shaped male signal evolution in a clade of specialized plant-feeding insects. Proc R Soc B 273:2585–2593
Rubenstein DR, Lovette IJ (2009) Reproductive skew and selection on female ornamentation in social species. Nature 462:786–790
Rodríguez RL, Sullivan LE, Cocroft RB (2004) Vibrational communication and reproductive isolation in the Enchenopa binotata species complex of treehoppers (Hemiptera: Membracidae). Evolution 58:571–578
Sæther SA (2002) Female calls in lek-mating birds: indirect mate choice, female competition for mates, or direct mate choice? Behav Ecol 13:344–352
Safran RJ, Flaxman SM, Kopp M, Irwin DE, Briggs D, Evans MR, Funk WC, Gray DA, Hebets EA, Seddon N, Scordato E, Symes LB, Tobias JA, Toews DPL, Uy JAC (2012) A robust new metric of phenotypic distance to estimate and compare multiple trait differences among populations. Curr Zool 58:426–439
Schneider J, Uhl G, Herberstein M (2015) Cryptic female choice within the genus *Argiope*: a comparative approach. In: Peretti AV, Aisenberg A (eds) Cryptic female choice in arthropods. Springer, Cham
Seddon N, Botero CA, Tobias JA, Dunn PO, MacGreggor HEA, Rubenstein DR, Uy JAC, Weir JT, Whittingham LA, Safran RJ (2013) Sexual selection accelerates signal evolution during speciation in birds. Proc R Soc B 280:20131065
Stratton GE, Uetz GW (1981) Acoustic communication and reproductive isolation in two species of wolf spiders. Science 214:575–577
Sullivan-Beckers L, Hebets EA (2011) Modality-specific experience with female feedback increases the efficacy of courtship signalling in male wolf spiders. Anim Behav 82:1051–1057
Sullivan-Beckers L, Hebets EA (2014) Tactical adjustment of signalling leads to increased mating success and survival. Anim Behav 93:111–117
Tan MT, Jones G, Zhu G, Ye J, Hong T, Zhou S, Zhang S, Zhang L (2009) Fellatio by fruit bats prolongs copulation time. PLoS one 4(10):e7595
Uhl G, Elias DO (2011) Communication. In: Herberstein ME (ed) Spider behaviour: flexibility and versatility. Cambridge University Press, Cambridge, pp 127–189
Uy JAC, Patricelli GL, Borgia G (2001) Complex mate searching in the satin bowerbird *Ptilonorhynchus violaceus*. Am Nat 158:530–542
West-Eberhard MJ (1983) Sexual selection, social competition, and speciation. Q Rev Biol 58:155–183
West-Eberhard MJ (2014) Darwin's forgotten idea: the social essence of sexual selection. Neurosci Biobehav Rev (in press)
Wilson CJ, Tomkins JL (2014) Countering counter adaptations: males hijack control of female kicking behavior. Behav Ecol 25:470–476

Index

A.V. Peretti and A. Aisenberg (eds.), *Cryptic Female Choice in Arthropods*,
DOI 10.1007/978-3-319-17894-3

D

N

Q

R

S

The manufacturer's authorised representative in the EU is Springer Nature Customer Service Centre GmbH, Europaplatz 3, 69115 Heidelberg, Germany. If you have any concerns regarding our products, please contact ProductSafety@springernature.com

Printed and bound by CPI Group (UK) Ltd, Croydon, CR0 4YY
25/09/2026
02222882-0007